PROCEEDINGS OF THE BRITISH ACADEMY • 198

EARLY FARMERS

The View from Archaeology and Science

Edited by

ALASDAIR WHITTLE AND PENNY BICKLE

Published for THE BRITISH ACADEMY
by OXFORD UNIVERSITY PRESS

OXFORD
UNIVERSITY PRESS

Great Clarendon Street, Oxford OX2 6DP
United Kingdom

Oxford University Press is a department of the University of Oxford. It furthers the University's objective of excellence in research, scholarship, and education by publishing worldwide. Oxford is a registered trade mark of Oxford University Press in the UK and in certain other countries

First published 2014

British Library Cataloguing in Publication Data
Data available

Library of Congress Cataloging in Publication Data
Data available

ISBN 978-0-19-726575-8

Contents

Illustrations

Colour plates

Figures

Tables

Contributors

Christina Adler is an Associate Lecturer in the Faculty of Dentistry at Sydney University, Australia.

Kurt W. Alt is a Professor at the Centre of Natural and Cultural History, Danube Private University at Krems-Stein, Austria, and Visiting Professor of the Institute of Anthropology, Johannes Gutenberg University of Mainz, Germany.

Adrian Bălăşescu is a Researcher at the National Museum of Romanian History, Bucharest.

Marie Balasse is a Research Scientist for the CNRS (Centre national de la recherche scientifique, France), affiliated to the MNHN (Muséum national d'Histoire naturelle) in Paris (UMR 7209, Archéozoologie, Archéobotanique: Sociétés, Pratiques et Environnements).

Eszter Bánffy is a Professor and Scientific Advisor at the Archaeological Institute, Research Centre for the Humanities, Hungarian Academy of Sciences, Budapest, Hungary. She is currently the Director of the German Archaeological Institute, Roman-Germanic Commission, Frankfurt am Main, Germany.

John C. Barrett is Professor of Archaeology at the University of Sheffield.

Maria Giovanna Belcastro is a Professor at the Dipartimento di Scienze Biologiche, Geologiche e Ambientali, Università di Bologna.

Robin Bendrey is a Postdoctoral Researcher at the University of Reading. He is a zooarchaeologist, currently working on the Central Zagros Archaeological Project.

Jean-François Berger is a Research Scientist for the CNRS, based in the department of Geography, History, Art History and Archaeology, Tourism, Urbanism at the University of Lyon 2, France.

Penny Bickle is a Lecturer in Archaeology at the University of York, specialising in the European Neolithic.

Victoria Keerl is a PhD student and Research Assistant in the Institute of Anthropology, Johannes Gutenberg University, Mainz, Germany.

Mechthild Klamm is a Consultant at the Department of Archaeological Conservation of the State Office from Heritage Management and Archaeology Saxony-Anhalt at Halle (Saale), Germany.

Corina Knipper is a Research Associate at the Curt-Engelhorn-Centre for Archaeometry at Mannheim, Germany.

Olaf Kürbis is an Archaeologist at the State Office for Heritage Management and Archaeology of Saxony-Anhalt, Germany.

Eva Lenneis is an *Extern* Scientist and Lecturer at the Institute for Prehistory and Historical Archaeology, University of Vienna, Austria.

Christian Lohr is a PhD student at the Römisch-Germanisches Zentralmuseum Mainz, Germany, investigating enclosures of the younger LBK.

Valentina Mariotti is a Lecturer at the Dipartimento di Scienze Biologiche, Geologiche e Ambientali, Università di Bologna. She is also affiliated to the CNRS laboratory, *Anthropologie bioculturelle Droit, Ethique et Santé* (AdéS, UMR 7268), Faculté de Médecine, University of Marseille, France.

Wendy Matthews is Associate Professor in Archaeology at the University of Reading.

Lynn Meskell is Professor of Anthropology and Director of the Archaeology Center at Stanford University.

Christian Meyer is a Bioarchaeologist at the State Office for Heritage Management and Archaeology of Saxony-Anhalt, Germany.

Richard Moussa is a Postdoctoral Researcher at the Research in Ecological Economics, Eco-Innovation and Tool Development for Sustainability laboratory (REEDS; EA 4456) at the University of Versailles Saint-Quentin-en-Yvelines and the Université de Paris Grand Ouest, France.

Nicole Nicklisch is a Research Associate at the State Office for Heritage Management and Archaeology, Saxony-Anhalt, and at the Danube Private University at Khems-Stein, Austria.

Lorette Noiret is a Research Fellow at the Centre for Mathematics and Physics in the Life Sciences and Experimental Biology (CoMPLEX), University College London, UK.

Tamsin O'Connell is a Lecturer in the Department of Archaeology and Anthropology, Division of Archaeology, University of Cambridge.

Martin O'Connor is a Professor, leading the Research in Ecological Economics, Eco-Innovation and Tool Development for Sustainability laboratory (REEDS; EA 4456) at the University of Versailles Saint-Quentin-en-Yvelines and the Université de Paris Grand Ouest, France.

Elena Ortu is a Researcher at the EPHE, Paris and CNRS, based at the University of Bordeaux 1, France.

Jessica Pearson is Senior Lecturer in Bioarchaeology at Liverpool University.

Pierre Pétrequin is an Emeritus Director of Research, CNRS, and the Director of Projet JADE and JADE II.

John Robb is a Reader at the Department of Archaeology and Anthropology, Cambridge University.

Aurélie Salavert is an Assistant Professor, based at the MNHN, Paris.

Hayley Saul is a Research Associate the Department of Archaeology, University of York.

Rick J. Schulting is a Lecturer in Scientific and Prehistoric Archaeology at the University of Oxford.

Dominique Schwartz is a Professor of Geography at the University of Strasbourg, France.

Alison Sheridan is the Principal Curator of Prehistory, National Museums Scotland, and President of the Prehistoric Society.

Lisa-Marie Shillito is a Research Fellow in the School of History, Classics and Archaeology, University of Edinburgh.

Jessica Smyth is a Research Associate at the University of Bristol. She is currently working for the project, NeoMilk: the milking revolution in temperate Neolithic Europe.

Clark Spencer Larsen is the Distinguished Professor of Social and Behavioral Sciences and Chair of the Department of Anthropology at the Ohio State University.

Anna Szécsényi-Nagy, is a Research Fellow at the Archaeological Institute, Research Centre for the Humanities, Hungarian Academy of Sciences, Budapest, Hungary.

Mary Anne Tafuri is a Lecturer in the Dipartimento di Biologia Ambientale at Sapienza University, Rome.

Carlos Tornero is a Postdoctoral Researcher at UMR 7209, CNRS/MNHN in Paris.

Anne Tresset is a Director of Research for the CNRS, based at the MNHN, Paris, France.

Joël Ughetto-Monfrin is a Research Engineer for the CNRS affiliated to the MNHN, Paris (UMR 7209).

Marc Vander Linden is an ERC Senior Research Associate at the Institute of Archaeology, University College London, UK.

Jean-Denis Vigne is a Director of Research for the CNRS, based at the MNHN, Paris, France, where he is unit director for the Archaeozoology and Archaeobotany laboratory (UMR 7209).

Valentina Voinea is a Researcher II, at the Museum for National History and Archaeology, Constanţa and Associate Lecturer at the University Ovidius, Constanţa.

Alasdair Whittle is a Distinguished Research Professor at Cardiff University and a Fellow of the British Academy.

James Williams is Analytical Research Laboratory Manager at the School of Chemistry, University of Bristol.

1

Introduction

Integrated and Multi-scalar Approaches to Early Farmers in Europe

PENNY BICKLE AND ALASDAIR WHITTLE

Early farmers in Europe: big questions

THE NEOLITHIC PERIOD WORLDWIDE can readily be identified as one of the great transformations in human history, with profound, long-term consequences still felt today. This was the time when agriculture began to be adopted as the principal form of subsistence, when more sedentary settlement was initiated, when larger communities started to be formed and wider social networks constructed, and when people created new views of themselves and their worlds. While the outlines of the big story are clear enough, such a thumbnail sketch conceals many pressing research questions. What was the order of changes? Did innovations all come at once, and what caused them? Who, in social terms, were the drivers of doing things differently, and of thinking new visions of existence into being? Once begun, how quickly did Neolithic societies develop, and what fostered their change? Was it agriculture and sedentism, or should we look among social relations, culture and worldviews, or interlocking sets of these factors?

In the specific case of Europe, there were no farmers at *c.* 7000 cal BC, but there were very few hunter-gatherers left after *c.* 4000 cal BC, and only those on the geographical periphery (from a modern perspective). If the Bronze Age is defined, on a necessarily sliding scale across the continent, as beginning in the fourth to third millennia cal BC, the three thousand years and more of Neolithic history saw radical change in agricultural techniques, the density of population and the size of settlements, the nature of sociality, and the character of belief systems. Many regions may also have seen considerable changes in the identity of populations, particularly perhaps but not exclusively in the early stages, when people from the outside were likely to have been the vectors of innovation.

Mapping out the lives of the earliest farmers continues to capture our imagination, in part because of the mixture of familiarity and unfamiliarity found in our archaeological datasets. Neolithic people appear often to have led settled

Proceedings of the British Academy **198**, 1–19.

Partly these issues arise due to the nature of research into the LBK, which has tended to prioritise either macro-scale changes or a focus on detailed site, or local, histories of change (understandably so for any research group, given the number of countries across which the LBK phenomenon as a whole is distributed), but rarely both together. In order to investigate issues of diversity and variability in practice, we set about analysing the isotopes at different scales, examining correlations on an individual and site level, as well as combining and contrasting data at a regional level and beyond. This meant attending to the overall patterns in the data, as well as trends identified at the cemetery or site level, and the regional scale. For example, at a small number of sites correlations between isotope ratios and burial with certain grave goods, such as *Spondylus* shells (imported from distant sources in the Adriatic or Aegean), could be identified that were not repeated at other sites. Individuals with unusual or outlying isotope ratios also posed a particular challenge. Outlying data points are often removed from analysis because they skew understanding of main trends, but in this case such values represent the practices and lives of past humans and, as such, form part of the total range of LBK lifeways, and we needed to take them seriously. The outliers identified in the LBK Lifeways project do not form a single coherent group, but rather each seems to present a myriad of different ways of being and living, whether choosing a more varied diet, or moving differently than the broader community. This has opened up dramatically the range of lifeways that we can discuss for the LBK, which have previously tended to be limited by a dichotomy between farmers and hunter-gatherers.

Overall, in the isotope data, we found a tension between broad similarity across multiple regions and more localised instances of variations – a central tendency among the data suggesting a shared way of doing things, within which there was the possibility of doing things differently. Carbon and nitrogen isotopes, which arise in the proportions of meat and plant proteins in the diet, showed broadly similar diets across the regions we studied, though with perhaps more meat consumption in the Upper Rhine Valley than the Danube catchment areas (Hedges *et al.* 2013). The strontium isotopes suggest that men and women had different mobility patterns, but that this sex-based variation was widely shared across the regions. As female strontium ratios showed greater variation in values than males, this suggests that LBK women were more likely to move during their lives than males, which we propose indicates the widespread practice of patrilocality (Bentley *et al.* 2012, Hedges *et al.* 2013; Figure 1.2). The isotope values which we analysed were thus created through widely shared practices of diet composition and lifetime mobility patterns – a balance of the individual lifeway, with a multitude of different relationships, and the collective action of the broader community. Recognising these multiple dimensions of practice behind the formation of isotope values opens up the possibility not only of populating our models of the past with mothers, fathers, daughters, sons, cousins and so on, but also of providing detail into how the LBK community was connected.

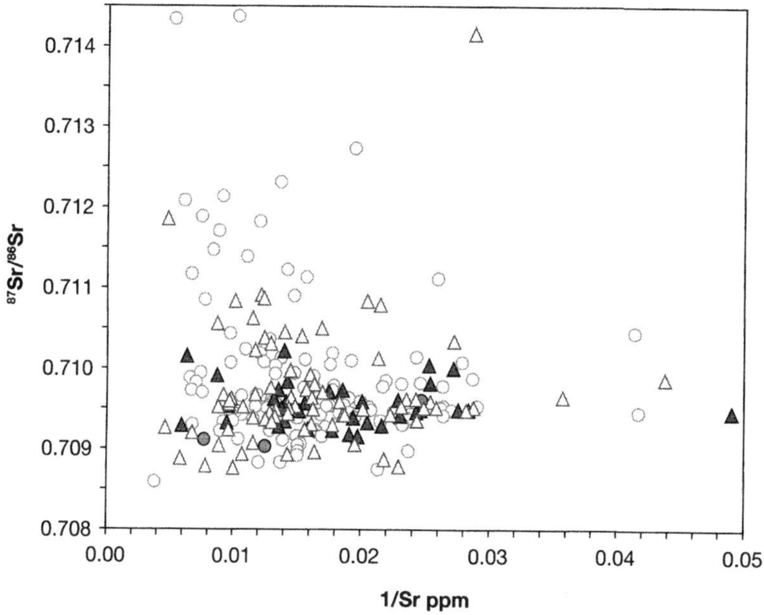

Figure 1.2 $^{87}Sr/^{86}Sr$ versus 1/Sr concentration ppm from adult individuals sampled as part of the LBK lifeways project. Circles show adult females, triangles show adult males. Filled symbols denote burial with an adze. From Bentley *et al.* (2012, 9329).

Such inferences are at their most powerful when integrated with the archaeological context, and considered alongside how LBK communities lived together, organised their subsistence activities and dealt with death. The post-built longhouse architecture of the LBK, with its complex arrangement of internal posts, is normally seen in settlements of multiple houses, replaced over successive generations. The organisation of the settlements has been much debated; were houses organised in loose clusters constituted by individual homesteads and their surrounding space or in coherent rows, and how does this impact on the kinds of social models we develop for settlements? Do LBK settlements represent loosely grouped families or more structured clan-like relationships? The organisation of cemeteries and funerary practices does not repeat these kinds of social organisation (Figure 1.3). Loose spatial groupings of graves may suggest that social relations such as kinship or lines of descent could have been important in determining treatment at death, but in terms of other aspects of burial rites (such as grave goods or body position) variability can be as great within each spatial group as between them. Rather, the age and sex of the deceased appear to have been stronger determining factors of variance in the funerary sphere, particularly in terms of the grave goods that accompanied them. Even here there is much inconsistency and we need not restrict ourselves to models of descent, but could bring in ideas borrowed from elsewhere

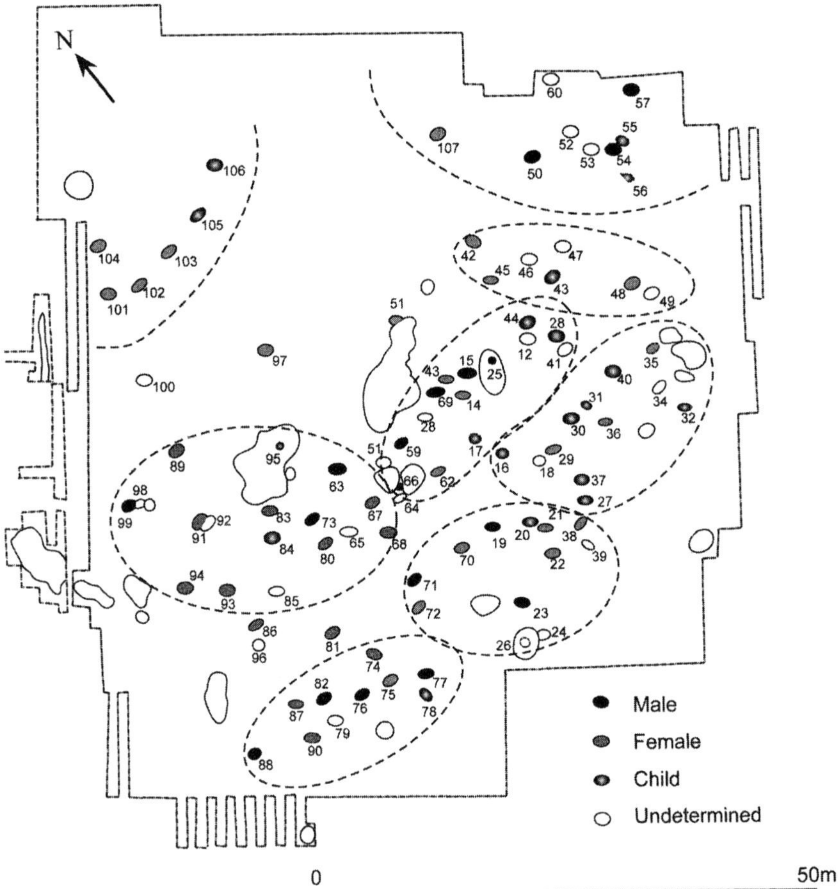

Figure 1.3 The Vedrovice LBK cemetery, showing possible clusters of burials. After Zvelebil and Pettitt (2008, 205).

such as the concept of sodalities, defined by Service (1962, 21) as a 'non-residential association that has some corporate functions or purposes'. In the case of the LBK and the Lifeways project, the strongly shared, narrow range of strontium isotope values which could be identified for adult males buried with a polished stone tool, principally adzes (Bentley *et al.* 2012), could perhaps suggest the presence of non-residential allegiances in the cemeteries, which connected individual households from different sites, rather than whole settlements or groups of communities being represented there.

In combination, the isotopic, osteological and archaeological data suggest that these varied scales of social cohesion could change over the life-course. To explore this further, individual lifeways can be constructed from the integration of different approaches and compared to other data at various scales. For example, at the cemetery at Rutzing in Upper Austria, grave 13 contained the inhumation of a young adult man, whose strontium isotope ratios suggested that he was born and then lived primarily on the lowland loess soils characteristically occupied by LBK populations (Figure 1.4). The stable isotope results from this individual included an average nitrogen value and a relatively high carbon value for this cemetery. So, while broadly sharing a diet with the rest of the community, there may have been some variation from the average in terms of carbon protein consumed. Possibly the variation in carbon values may be associated with woodland resource use and the high carbon value of this individual may indicate that actually he used fewer woodland resources than others buried at the Rutzing cemetery – and therefore sourced more of his diet from the open areas around an LBK settlement.

His burial context similarly blended practices shared across the local community with those that stood out as unique or exotic. The left-crouched position of his body in the grave and some of the grave goods that accompanied him speak to a classic or regular LBK burial rite. He was buried with *Spondylus* shell ornaments and a polished stone adze, but the west–east orientation of the burial (that is, with the head to the west) and the presence of 120 imitation deer teeth are both unique at the cemetery. In fact, all the objects in his grave make links beyond the immediate region, with raw materials from Bavaria, to the west of Upper Austria, and from the Bohemian Massif, to the north, while the *Spondylus* shell would have originated in the Aegean or Adriatic and the 120 imitation deer teeth are fairly rare in the LBK grave good assemblage, but much more frequent in contemporary Neolithic cultures to the east. In the years before his death, this individual had suffered from a pulmonary infection of some description, identifiable through pathologies on the rib bones, possibly tuberculosis, which must have reached an advanced stage to have appeared on the skeleton (identified by Linda Fibiger; Fibiger in Bickle *et al.* 2013). So while the isotopic and skeletal

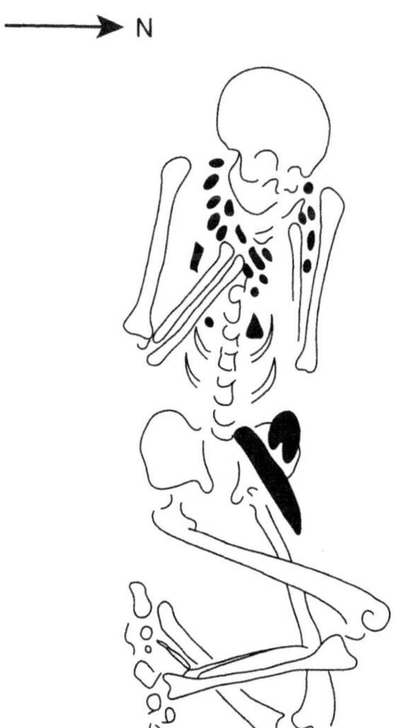

Figure 1.4 Burial 13 from Rutzing, Upper Austria. The grave goods shown in the picture are a *Spondylus* V-buckle and polished adze at the waist, and a *Spondylus* necklace on the upper torso. After Kloiber *et al.* (1971, 29).

Genetic studies and aDNA analysis, still relatively new techniques, map population dynamics at a larger chronological scale, but deal with the same themes of unravelling movement and relatedness. Analysis of organic residues encrusted on or absorbed in pots – lipid analysis – can indicate the source of fats and hence foodstuffs (and other materials) stored in ceramic vessels, giving important insights into consumption practices and the contexts for ceramic use. Bioarchaeological approaches which have a longer history in archaeological science are by no means static disciplines and recent developments in osteological studies, faunal analysis and archaeobotany have all expanded the remit of their insights into past lives. For example, faunal remains should inform not just on diet, but also on husbandry practices, human–animal relationships and belief systems, seasonality and environment, and the addition of aDNA analyses and isotope studies to this context has further increased the complexity of the questions we can ask.

Throughout this volume, different approaches are taken to combining archaeological data and analytical techniques. At one level, this can be seen as a presentation of the recurrent and dominant research questions for the early Neolithic, ranging from central matters of identity, subsistence, lifeways in general, and materiality. At another level, the issues discussed by the chapters range across how we can best combine varied forms of evidence that come from both quantitative and qualitative data (for example, Meskell and Pearson; Schulting and Fibiger), facilitate interpretation of results that stand as proxies for the archaeological phenomena under study (for example, Halstead, Barrett, Bogaard), choose the appropriate scales for analysis (Tafuri *et al.*, Brandt *et al.*, Bocquet-Appel *et al.*) and recast how we ask questions of human-material relationships (for example, Robb, Harris, Sheridan and Pétrequin).

The volume which has resulted thus provides a vivid snapshot of the possibilities – and limitations – of current archaeological investigation, and it also encourages timely reflection on the nature of the archaeological process in general, and of the place in this of archaeological science in particular. In arguing for this questioning archaeology, we present a series of chapters which involve or embrace archaeological science, but which are focused above all on research questions arising from the history of early farmers in Europe.

Organisation and scope of the volume

Six strands have been used to organise the chapters in this volume, though there are many other shared questions, interests and overlaps in methodology and approach which could also be found, and we make no claim, of course, to completeness. We begin with general reflections on the nature of the archaeological project, in particular in relation to the study of the Neolithic in Europe, and then move to origins and spread, farming itself, lifeways and materialities, and, finally, to comments on the use of proxies. It is worth pointing out, however, that several of the chapters in the sections below could have been placed in other strands. We hope that by placing chapters together which are asking similar questions, rather than just gathering related methodologies together, attention will fall on the shared objectives of different disciplinary streams in archaeology and on revealing what key questions have arisen over recent years.

First reflections

To provide guidance to what may lie ahead, Robb considers how our models of the Neolithic may develop over the coming years as past and recent divisions between different approaches to archaeology appear to decrease. He also raises central questions of scale. Barrett seeks more nuanced understanding of the dynamic relationship between humans, plants, animals and the land, in order to escape unhelpful dualisms and dichotomies in past debates about migration versus acculturation at the start of the Neolithic. In both these chapters, being Neolithic is cast as historical process, rather than reduced to adaptive capacity or mental state.

Origins and spread of the Neolithic

Two chapters arising from DNA studies start things off in more detail. Szécsényi-Nagy *et al.* investigate aDNA samples to map population dynamics for the established Neolithic of the Carpathian Basin, comparing the results from both western and eastern Hungary, and comparing these with the LBK of central Europe, and with later patterns. Brandt *et al.* focus on a tighter scale, combining aDNA with strontium and dietary isotopic analysis of human samples from the site of Karsdorf in eastern Germany, to generate not only a sense of genetic descent (cf. Brandt *et al.* 2013) but also of the perhaps surprising lack of relatedness among the people sampled, a possible fluidity of residence and some diversity of subsistence pathways at the scale of an established LBK settlement.

How human communities are formed, maintained and networked remains an essential interest in studies of the Neolithic. Population growth over the Neolithic (Bocquet-Appel 2011) and stronger evidence for collective activity, in the form of settlements for example, raise important questions about the forms of connections between and within different groups at various scales. Bocquet-Appel *et al.* draw on recent developments in multi-agent modelling to consider scales and forms of networks in the LBK, the first Neolithic across much of central Europe in the second half of the sixth millennium cal BC.

Farming

As one of the primary defining features of the Neolithic is the onset of a new suite of subsistence practices, it is not surprising that many of the recent advances in archaeological science applied to the Neolithic are aimed at finding out more about diet and agriculture. Recent developments have sought to place farming techniques and dietary patterns in broader social contexts and assess the role they played in bringing about cultural change. Chapters in this section deal with how subsistence and diet relate to each other, debating how cereal growing and stock-keeping, as well as continued use of wild resources, can be investigated through varied techniques.

The Neolithic not only altered human diets, but also those of animals; domestic livestock were managed, fed and cared for by human groups, who also controlled their breeding patterns and the landscape and environments they visited. Balasse *et al.* map the role of mobility and seasonality in stock-keeping in Romania through varied isotopic studies, demonstrating perhaps more limited ranges for cattle than had previously been modelled. How to connect different scales is considered by Tafuri *et al.*, who use isotopic analysis to assess the complexity of community membership and material expression of group identity in southern Italy. Hachem and Hamon investigate in detail one particularly well investigated LBK site, Cuiry-lès-Chaudardes in the Paris basin, combining analysis of house architecture, faunal remains and grinding tools to examine variation across the settlement. Bogaard investigates crop husbandry regimes with reference to a wide range of case studies from south-east, central and north-west Europe, through the application of stable isotope analysis to plant remains themselves, arguing for widespread manuring practices and that the role of crops has been systematically under-estimated in standard palaeodietary interpretation. Saul *et al.* argue, through organic residue analysis, that cuisine – how different foodstuffs were combined and mixed together, for social purposes beyond economic necessity – provides more nuanced views of the adoption of farming in northern Germany.

Lifeways

Studying human lifeways has been a growing theme in archaeology over recent years and discussing individual biographies, through exploring issues of health, violence, gender and changes over the life-course, is now becoming commonplace (for example, Zvelebil and Pettitt 2008). However, such issues influenced Neolithic communities not just on the scale of the individual or the local site, but also on the population level as well, and balancing out small-scale, personal biographies with changes on the population level are tackled by many of the chapters in this section.

Spencer Larsen presents the results of a large comparative study of the health of populations undergoing the tradition to farming, finding similar patterns across different global regions, but with very different outcomes. Pearson and Meskell present collaborative research from the site of Catalhöyük, Anatolia, exploring Neolithic notions of the body through both isotopic analysis and material representations in the form of figurines, buildings and funerary rites, with an emphasis on notions of maturity and ageing. Matthews *et al.* consider issues of change in the transition to agriculture in the Zagros and surrounding regions, through integrated microstratigraphic, phytolith and chemical analyses. They raise important issues of the use of wild and domestic plants, aspects of the early stages of animal management, including penning, foddering and use of dung for fuel, and change and variability within and between households and communities. The next two chapters focus on evidence for violence in the European Neolithic. Schulting and Fibiger assess the prevalence and demographic characteristics of evidence for violence on human skeletons, presenting regional and chronological trends (especially for southern Scandinavia and Britain) and exploring the different prevalence of violence amongst early farming groups. Meyer *et al.* present the evidence from mass graves from the early Neolithic in Germany, set against the range of other LBK mortuary practices, using a variety of biometrical and osteological studies to compare features from three sites in particular.

Materialities

This strand brings together researchers specialising in Neolithic artefacts and materials, approaching this topic both from the perspective of the role of tools and objects in social life, and engaging in scientific investigation, such as provenance studies and lipid analysis. The recent materialist turn in archaeology has brought about increased interest in understanding the properties of varied archaeological materials and objects as a way of engaging with social and human-material relationships (Ingold 2007). Harris considers how scientific approaches are changing interpretive agendas in studying materials, arguing that archaeologists should adopt an attitude to objects which sees them as animated. Smyth and Evershed survey the contribution which lipid analysis is making to understanding the extent of early milk consumption, and discuss the advantages of combining pottery analysis with scientific study of its contents. Sheridan and Pétrequin present the results of Projet JADE, which used innovative and non-invasive techniques as well as ethnographic research to source and trace the spread of jadeitite axes. They explore how such multi-disciplinary approaches could help to investigate the Mesolithic–Neolithic transition in Britain. Chapman, in contrast, considers Neolithic object makers as scientists themselves, arguing that they were concerned with principles such as geometry and precision.

Final comments

Halstead debates the inferences being made on the basis of different techniques, arguing that we need to pay closer attention to how our data act as proxies for different human behaviours, as well as to the assumptions that underpin how we extrapolate from the results of certain techniques.

is unlikely that one technique alone will be able to do justice to this complex and varied transmission (Milner *et al.* 2004). Furthermore, many archaeological resources, such as human remains, are finite, making it all the more important that we put our methodologies to best use, targeting well defined problems.

We have stressed the merits of integrating and combining approaches, and multi-scalar analysis, and we foresee such perspectives expanding even further in future research in archaeology, not only in the field of early farmers. We are not predicting, however, an end to controversy. Three topics catch our own attention, for example, and readers will think of others. The publication of new aDNA data from central Europe for early farmers and some hunter-gatherers (most recently by Brandt *et al.* 2013) raises rather fundamental questions about how we should correlate genetic signatures in the form of haplogroup or haplotype patterning with lived, cultural identities. Modern DNA evidence is still being used, alongside aDNA data, for example in creating genetic distance maps, which have been used so far mainly to underpin the inference that the early Neolithic saw the arrival of considerable numbers of new people stemming ultimately from the Near East – even though the modern dataset must include the whole of history since the Neolithic. Another topic to debate is high-level or big-scale modelling, as seen in this volume in the chapter by Bocquet-Appel *et al.*, and in related research on demography in Shennan *et al.* (2013). While high-level models or big-scale proxies may be useful to generate hypotheses and to stimulate debate, how long should they dominate before more locally based and direct evidence is examined? Thirdly, the new materialist or ontological turn presents very radical challenges to thinking about agency. We have put people – early farmers – first in our title, but others would prefer to raise a much more central role for things in the accounts which we write of the Neolithic, and to think through possible symmetries between people and objects; others again would add animals, to a triangular relationship between people, things and animals (see Harris, this volume, for discussion; with comment also in Robb, this volume). There is much still to debate.

Cross-cutting all the chapters are central questions of integration of techniques and scales of analysis. At what temporal, analytical and social scales can we best study people in the past in general, and given in this context that it is early farmers whom we most want to study – social practice rather than early farming as a decontextualised technique – what kind of narratives can we best construct? A full answer to these questions would take several volumes, and there is much about the agency, identity, sociality and worldview of Neolithic people which we cannot cover here. But our strands show how the expanding datasets now available to us can begin to be integrated into wider interpretation.

In the long-term perspective, agriculture has been around for far less time than hunting and gathering, yet the impact of farming has been global and, alongside the industrialisation of the last two centuries, it has helped to shape not only modern landscapes, environments and diets, but also our bodies, our genes, the diseases which we suffer and the structures of the societies in which we live (Barker 2011). In offering a particular temporal and geographical focus on early farmers in Europe, with one or two related studies of cognate interest, we hope the volume will speak to the broader question of how we can best practise an archaeology of the human past in general. Although techniques derived through archaeological science are the most frequent in this volume, we by no means wish to ignore or imply lesser importance for other dimensions and approaches drawn from interpretive archaeology or techniques seen as more traditionally archaeological. Rather, we call for better integration through a shared focus on debating what questions we should ask and on being more creative and expansive in how we answer them.

Acknowledgements

The LBK Lifeways project was supported by a major award from the Arts and Humanities Research Council. We are grateful to all the many colleagues who contributed to it, who are listed in full in Bickle and Whittle (2013). For the data referred to in this chapter, we are particularly indebted to our colleagues on the LBK Lifeways project Alex Bentley, Linda Fibiger, Julie Hamilton, Robert Hedges and Daniela Hofmann; to Joachim Wahl, Michael Francken and Claudia Gerling for information about, and allowing us to sample, the cemetery of Schwetzingen; and to the Landesmuseum Linz for information about, and allowing us to sample, the cemetery of Rutzing.

References

Barker, G. 2011. Archaeology: the cost of cultivation. *Nature* 473, 163–4.

Bentley, R.A., Bickle, P., Fibiger, L., Nowell, G., Dale, C., Hedges, R., Hamilton, J., Wahl, J., Francken, M., Grupe, G., Lenneis, E., Prefect, F. Teschler-Nicola, M., Arbogast, R.-M., Hofmann, D. and Whittle, A. 2012. Community differentiation and kinship among Europe's first farmers. *Proceedings of the National Academy of Sciences of the United States of America* 109, 9326–30.

Bickle, P. and Whittle, A. (eds) 2013. *The first farmers of central Europe: diversity in LBK lifeways*. Oxford: Oxbow Books.

Bickle, P., Bentley, R.A., Blesl, C., Fibiger, L., Hamilton, J., Hedges, R., Lenneis, E., Neugebauer-Maresch, C., Stadler, P., Teschler-Nicola, M., Tiefenböck, B. and Whittle, A. 2013. Austria. In P. Bickle and A. Whittle (eds), *The first farmers of central Europe: diversity in LBK lifeways*, 159–204. Oxford: Oxbow Books.

Bocquet-Appel, J.-P. 2011. When the world's population took off: the springboard of the Neolithic demographic transition. *Science* 333, 560–1.

Brandt, G., Haak, W., Adler, C.J., Szécsényi-Nagy, A., Karimnia, S., Möller-Rieker, S., Meller, H., Ganslmeier, R., Friederich, S., Dresely, V., Nicklisch, N., Pickrell, J.K., Sirocko, F., Reich, D., Cooper, A., Alt, K.W. and The Genotypic Consortium 2013.

Ancient DNA reveals key stages in the formation of central European mitochondrial genetic diversity. *Science* 342, 257–61.
Childe, V.G. 1936. *Man makes himself*. London: Watts and Co.
Childe, V.G. 1943. Archaeology as science. *Nature* 152, 22–3.
Childe, V.G. 1982 [1942]. *What happened in history*. London: Penguin.
Clarke, D. 1978. *Analytical archaeology* (second edition). London: Methuen and Co.
Gerling, C. 2012. Das linearbandkeramisches Gräberfeld von Schwetzingen, Rhein-Neckar-Kreis. *Fundberichte aus Baden-Württemberg* 32, 8–263.
Hedges, R., Bentley, R.A., Bickle, P., Cullen, P., Dale, C., Fibiger, L., Hamilton, J., Hofmann, D., Nowell, G. and Whittle, A. 2013. The supra-regional perspective. In P. Bickle and A. Whittle (eds), *The first farmers of central Europe: diversity in LBK lifeways*, 343–84. Oxford: Oxbow Books.
Henderson, J. 1989. Introduction. In J. Henderson (ed.), *Scientific analysis in archaeology*, 17–29. Oxford: Oxbow Books.
Ingold, T. 2007. Materials against materiality. *Archaeological Dialogues* 14, 1–16.
Jeunesse, C. 1997. *Pratiques funéraires au Néolithique ancien: sépultures et nécropoles danubiennes 5500–4900 av. J.-C.* Paris: Éditions Errance.
Johnson, M. 2010. *Archaeological theory: an introduction* (second edition). Oxford: Wiley-Blackwell.
Jones, A. 2002. *Archaeological theory and scientific practice*. Cambridge: Cambridge University Press.
Kloiber, Ä., Kneidinger, J. and Pertlwieser, M. 1971. Die neolithischen Siedlungen und die neolithischen Gräberfundplätze von Rutzing und Haid, Ortsgemeinde Hörsching, politischer Bezirk Linz-Land, Oberösterreich. Teil III. *Jahrbuch des Oberösterreichischen Musealvereins* 116, 23–50.
Milner, N., Craig, O.E., Bailey, G.N., Pederson, K. and Andersen, S.H. 2004. Something fishy in the Neolithic? A re-evaluation of stable isotope analysis of Mesolithic and Neolithic coastal populations. *Antiquity* 78, 9–22.
Pollard, A.M. 1995. Why teach Heisenberg to archaeologists? *Antiquity* 69, 242–7.
Pollard, A.M. 2011. Isotopes and impact: a cautionary tale. *Antiquity* 85, 631–8.
Pollard, A.M. and Bray, P. 2007. A bicycle made for two? The integration of scientific techniques into archaeological interpretation. *Annual Review of Anthropology* 36, 245–59.
Price, T.D., Bentley, R.A., Lüning, J., Gronenborn, D. and Wahl, J. 2001. Human migration in the Linearbandkeramik of central Europe. *Antiquity* 75, 593–603.
Renfrew, C. 1997. Setting the scene: Stonehenge in the round. In B. Cunliffe and C. Renfrew (eds), *Science and Stonehenge*, 3–14. Oxford: Oxford University Press.
Schiffer, M.B. 2013. *The archaeology of science: studying the creation of useful knowledge*. Heidelberg, New York, Dordrecht and London: Springer.
Service, E.R. 1962. *Primitive social organization: an evolutionary perspective*. New York: Random House.
Shennan, S., Downey, S.S., Timpson, A., Edinborough, K., Colledge, S., Kerig, T., Manning, K. and Thomas, M.G. 2013. Regional population collapse followed initial agriculture booms in mid-Holocene Europe. *Nature Communications* 1 October, doi 10.1038/ncomms3486.
Snow, C.P. 1959. *The two cultures and the scientific revolution*. Cambridge: Cambridge University Press.

Trigger, B.G. 1989. *A history of archaeological thought*. Cambridge: Cambridge University Press.

Wylie, A. 2002. *Thinking from things: essays in the philosophy of archaeology*. Berkeley and Los Angeles: University of California Press.

Zvelebil, M. and Pettitt, P. 2008. Human condition, life, and death at an early Neolithic settlement: bioarchaeological analyses of the Vedrovice cemetery and their biosocial implications for the spread of agriculture in central Europe. *Anthropologie* 46, 195–218.

The Neolithic: economy or culture?

The 'Neolithic' has traditionally been understood in about half a dozen ways, but in the last generation, the real debate has been between seeing the Neolithic as an economy and the Neolithic as a cultural orientation. Traditionally, the Neolithic was seen as the first, pre-metal farming communities in Europe. As Childe wrote:

> The escape from the impasse of savagery was an economic and scientific revolution that made the participants active partners with nature instead of parasites on nature . . . A society that acted thus was henceforth actively producing food, augmenting its own food supply. Potentially it could increase the supply to support a growing population.
>
> (Childe 1951, 48)

In Childe's influential view, when people began to farm, this new economy freed them from dependence upon nature, and it enabled all sorts of social and political developments which have shaped European society ever since.

In contrast to economic views, which remain important both within British archaeology and upon the Continent, many archaeologists contend that the Neolithic should really be understood as a set of cultural attitudes, propensities or material repertoires. For example:

> The British Neolithic consisted of a set of material resources . . . different communities could use this technology in order to create and reproduce economic regimes, social systems and interpretations of the world.
>
> (Thomas 1999, 222)

Similarly,

> If part of the process of becoming Neolithic was participation in new forms of social interaction, another part was a changing sense of belonging, of descent and of place in the scheme of things.
>
> (Whittle 1996, 8)

Proponents of the Neolithic-as-culture view often, though not always, tend to work in areas of Europe such as Britain and Scandinavia, in which the beginning of the Neolithic is marked by material culture and megalithic monuments more often than by abrupt economic change. They also reflect a generation of archaeology which emphasised humans' inhabitation of symbolic worlds, and which emphasised a local rather than generalising scale of analysis.

The battle lines for two decades, thus, have been: Is the Neolithic something big, continent-wide, and uniform, in other words an economy? Or do we really have to speak of many local Neolithics, local cultures, which vary immensely from each other?

At this point, it is worth stepping back and considering our general intellectual context. The 1980s and 1990s were marked in academia in general by the so-called 'Science Wars'. These were often bitter polemics between modernist advocates of science and post-modern critics. They culminated in the famous Sokal hoax, in which the well known physicist Alan Sokal wrote a parody article of slick-sounding gibberish arguing that we needed an 'emancipatory, socially constructed' mathematics; the parody article passed peer review and was published in a high-profile literary theory journal, allowing the scientist side to decry postmodernism as nothing more than slipshod, windy verbiage. Scientists and postmodernists differed epistemologically and philosophically, accusing each other of being (on one hand) oppressive capitalist proponents of scientism and (on the other hand) wishy-washy, 'anything-goes' relativists.

Archaeology's local version of the 'Science Wars' was the vitriolic argument between processualists and post-processualists which shaped archaeology in the 1980s and 1990s, and it adhered closely to the script. Processual archaeologists studied the 'nature' side of the dualism, while post-processualists viewed themselves as championing 'culture' (in the sense of human symbolic action). Many paradigmatic statements were written by Neolithic specialists, particularly on the post-processualist side; indeed, at points, in many British archaeology departments, it seemed as if 'archaeological theory' was synonymous with 'studying megaliths'. Early and important work such as by Hodder (1982, 1984), Shanks and Tilley (Shanks and Tilley 1982, 1987a, 1987b; Tilley 1984), and others (Richards and Thomas 1984, Bradley 1993, Barrett 1994) made the fundamental point that Neolithic people constituted their world symbolically through ritual practices. But both sides in the debate had taboos and blind spots. On the 'science' side, large-scale economic models for the Neolithic sometimes attempted to fit all Neolithic societies into a Procrustean bed. Even scholars who emphasised local economic variation had a surprisingly persistent blind spot about the fact that Neolithic people actually behaved culturally as well as eating. On the 'culture' side, arguments like Tilley's insistence that GIS (Geographical Information Systems) was incompatible with phenomenology and Thomas's claim that any attempt to define the Neolithic trans-regionally and economically was a totalising distortion really have to be read in this context. And fear of reductionist models meant that we have had a generation of studies with surprisingly little consideration of how people actually made a living; concepts of environment, ecology and demography tended to be dismissed out of hand because they were tainted by the smell of determinism. Similarly, the economy tended to be treated as an expression of identity, with very little idea of necessity – leaving one sometimes with an image of the people we study taking a 'gap year' before deciding to become Neolithic farmers.

Science and the humanities: a new archaeological vocabulary

In 2014, aside from a few old soldiers still fighting the last war, it is generally acknowledged that the processualist *versus* post-processualist debate has (thankfully) run its course (Hegmon 2003, Hodder 2007). Most archaeologists would acknowledge a broadly shared general theoretical platform which owes much to both schools, including a science-based methodology largely developed by processualism and a post-structuralist focus upon agency and meaning which was introduced by post-processualists but rapidly developed in both camps. Material culture plays a particularly important role in this new broad church. Several theorists have been influential, particularly since 2000: Tim Ingold's phenomenological approach (Ingold 2000), Daniel Miller's materiality approach (Miller 2005), and Bruno Latour's Actor Network Theory (Latour 2005). All three emphasise the complexity of the relationship between humans and material things. Many other archaeologists have come to the same realisation independently of such theorists, in a common-sense way. The central point here is that we cannot see human-material inter-actions in terms of a material world which acts upon passive humans, as in most environmental and economic views of the Neolithic. But neither can we see it in terms of humans who basically impose their cultural preferences upon an inert and constructed material world. By 'materiality' most theorists mean a consideration of the social nature of activities and projects which also acknowledges and works with (and sometimes against) their material characteristics.

This leads us to my main thesis in this section, which is that a new theoretical vocabulary for the Neolithic has been emerging in recent years, piecemeal and untrumpeted.

Consider animals. For many years, Neolithic faunal analysis was a way of determining what people ate, which in turn reflected simply an environmental setting or an adaptive food production strategy. The principal result was the vital statistic of 'percentage reliance on domesticates' which was the needle on the gauge indicating the degree of Neolithisation. On the other hand, in post-processual work, economy was seen basically as a matter of cultural preference, with little sense of necessity or real-world consequences. But gradually this dichotomy has been breaking down. Social zooarchaeologists such as Halstead (1981) had always taken a deeply social view of the animal economy. As Russell (1998, 2011) and Marciniak (2005) have argued convincingly, raising and eating domesticated animals are all about social relations of value, exchange and commensality as well as simply nourishing yourself. Similarly, it is clear that Neolithic farmers hunted wild animals not only for food, fur and feathers but also because of prestige, gender and the social valences of wild areas and resources. Isotopic studies (cf. Balasse *et al.*, this volume; Tafuri *et al.*, this volume) and residue studies (Smyth and Evershed, this volume) add detail on actual food consumption. The general point is the realisation that, even if an animal is ultimately eaten, raising it is a project which is done consciously and meaningfully, and which is structured by social needs, obligations and values as much as by the calories in the end project (see for example Hachem and Hamon, this volume). Basically, we have gone from Neolithic Diet to Neolithic Cuisine (cf. Saul *et al.*, this volume).

There has been a similar evolution in palaeobotany. Ancient plant use used to be the quintessential example of the (usually male) site director sending the flotation samples off to the (usually female) lab specialist, and getting back in return a shopping list of plants to stick in an appendix and never look at again. But, taking Bogaard's work on the Central European Neolithic as an indication of where the field is going (Bogaard 2004, Bogaard and Jones 2007), we now have detailed consideration of what gardening meant for the use of space, for social commitments, for scheduling and seasonality, and for sociality (cf. Bogaard, this volume). Gardening was Neolithic people's largest source of calories – but they did it as a fully embedded social project, in a distinctively Neolithic way.

Landscape studies tell a similar tale. For about a generation, there was a clear and often hostile dichotomy, which closely mimicked the split between physical and human geography. Specialists in Neolithic 'environments' did rather nitty-gritty studies of soils and landscape change using methods such as micromorphology, pollen and charcoal analysis, and GIS. Meanwhile, specialists in 'landscapes' talked about how humans created 'places' out of raw 'spaces' and experienced the landscape meaningfully by dwelling in it. The two groups generally talked past each other while studying the same topic. Over the last decade or two, the dichotomy has been breaking down. Environmental specialists increasingly see environments as produced by human action as well as by natural forces (for example, as in French and Lewis's work on Cranborne Chase; French and Lewis 2007). Conversely, GIS, traditionally regarded by phenomenologists as mechanistic and reductionist, is increasingly commonly used to reconstruct ritual landscapes, as in the Avebury project (Gillings *et al.* 2008) and Grima's work on the Neolithic temples of Malta (Grima 2008). The same is true for studies of space on a smaller scale, within settlements. The study of houses has evolved from simply treating house form as a cultural trait, to imposing rather blanket symbolic and structural interpretations based upon abstract house plans, to actually looking carefully at how houses were built, lived in, and destroyed. This 'dwelling perspective' has resulted in understanding how people actually inhabited houses, how house societies formed social units and how house biographies paralleled those of their inhabitants (Borić 2008, Hofmann and Smyth 2013).

Then there is trade and materials science, which has always been of special importance since Colin Renfrew – the original Neolithic research broker pioneering collaborations between science and social interpretations – worked with chemists to source obsidian in the 1960s (Cann and Renfrew 1964). Over the intervening years, studies of Neolithic 'trade' have come a long way from simply plotting how far objects moved from their sources; starting with Edmonds and Bradley's studies of the axe trade in Britain (Bradley and Edmonds 1993) and Pétrequin's work on Alpine axes (Pétrequin and Pétrequin 1993), and in Chapman's studies of shell rings in the Balkans (Chapman and Gaydarska 2006), the trend has increasingly been to contextualise exchange as a particular form of social action.

of the origins of farming. Even when one digs up an impeccably Middle Neolithic site, it tends to find its place in the larger narrative as establishing a 'typical' Neolithic, a good 500 or 1000 years into the Neolithic, the end point of the transition. This is a bit less so in areas such as Britain where features such as megaliths give us hooks to hang well theorised narratives about ritual landscapes and experience upon. But in many places, it is remarkable how archaeologists will spend huge amounts of time, effort and money to excavate Neolithic sites and then find so little to actually say about them. The reason lies in our disciplinary narratives. Much work in culture-historical traditions is essentially descriptive, documenting local particularities but not interpreting them. Post-processual work focuses interpretation upon a select, limited range of themes. In the social evolutionary tradition which remains implicit in much of the discipline, the big story lines are technology/economy and inequality. This leaves much of the Neolithic as a null category: early agricultural tribes are a baseline about which there is very little to say other than that they farmed and were egalitarian. Once you get past the Origins of Agriculture, you can safely fast-forward a couple of millennia until you get to Metals and Hierarchy.

This of course does a grave injustice to tribal social processes, which are full of fascinating ways of life easily more complex than anything in the world of hierarchical societies, which varied immensely and changed dramatically, and (most importantly) which created a world of human experience which people inhabited for millennia. Hence a major task for Neolithic studies is simply to provide ourselves with the lenses we need to see what is really important and interesting about the Neolithic. This is a task of vocabulary and narrative building. And it is a task about scales of analysis. The starting point is simply that there is no single inherent scale of analysis; if we analyse history at different scales we see different things, and scale is thus an analytical choice we impose on the material – this has been pointed out by anthropologists such as Bruce Knauft (1993) and feminist historians such as Joan Kelly (1977), who wrote an agenda-setting essay entitled *Did women have a Renaissance?* Moreover, to the extent that history involves processes unfolding at different scales, multiple, multi-scalar narratives are essential (Bailey 2007, Robb and Pauketat 2013). So far, so good; but now for the hard bit: in my experience, it is actually quite difficult to convince archaeologists of this in practice. Multi-scalar analysis gets lip-service, but most archaeologists turn out to have a strongly bounded comfort zone they are reluctant to stray out of. Usually this means either assuming that the scale of people is the real bottom line and looking at larger scale entities is reductionist or reifying, or feeling that the big-scale narrative is the real bottom line and all the quaint local variations and behaviours are just epiphenomenal froth. But as DeLanda (2006) notes, all scales of analysis are reductionist in some ways; the real question is simply whether they are usefully so.

What is missing, therefore, in Neolithic studies is the vocabulary to bridge the gap between our material on the ground, and various bigger interpretive pictures. We have had little to say about much of the Neolithic not because there is little to say about it, but because we have lacked useful words to say it with. And, importantly, this vocabulary has to be developed to suit the particular problems and theoretical affordances of Neolithic historical processes, not imposed from narrative frameworks with other agendas.

One moment on the ground: ethnographies of daily life

We might start with daily life, life on the ground. What was it like to live in the Neolithic, in a particular place and time? Here we are in good shape; writing daily life and what it meant for social reproduction has been one of our real successes, and the European Neolithic is probably better provided with more rich ethnographies of daily life than any other bit of world prehistory except possibly for the American Southwest (Barrett 1994; Tilley 1996, 1999; Thomas 1999; Robb 2007; Souvatzi 2008). Such syntheses typically pull together many different topics – foodways, deathways, making things, doing rituals, marking landscapes – to show how the practices of everyday life created absorbing, specific cultural worlds.

Genealogies of practice and bundles of institutions

An alternative approach, and one particularly apposite to scientific collaborations, is to trace a single practice over long time periods. How did foodways change throughout the Neolithic? How were they done in a particularly Neolithic way? Why did most Neolithic people eat little fish? When did Europeans start using dairy products? Why did they produce less than they were economically capable of producing? Did this really change in the late Neolithic? Evidence ranging from isotopes and residue analysis to settlement, architecture and iconography can be marshalled to investigate the genealogy of foodways. And so on, for the range of practices reviewed above; this is where the new socio-material vocabulary of the Neolithic really comes into its own. Moreover, while individual practices may have very long genealogies of stability and redefinition, there are obvious synergies between things such as form of settlement, food production, the definition of social units and landscapes, and investment in different forms of material culture. The most obvious example is the view of the Neolithic as village-dwelling, ritualistic farmers who worked at both making local styles of ornate pottery and exchanging objects such as axes and shell bracelets with complex biographies; while this is stereotypical, it does capture some of the interdependencies inherent in much of the Continental Neolithic as a set of traditions. Hence one way to implement the rather abstract definition of the Neolithic above as a set of human-material relationships is to see it as a recurrent bundle of interrelated practices and fields of action (cf. Whittle 2003). Such bundled practices were not rigidly fixed; they contained tensions which often gave Neolithic societies considerable creative energy, and they were continually redefined both from place to place and from generation to generation. But their basic synergies gave them historical stability; they offered a livable set of existential possibilities.

'Neolithic histories' and tipping points: developmental cycles of a few centuries

But what about Stonehenge? What about Varna? What about the Italian ditched villages? Within the Neolithic as a widely shared way of life, there are local cycles of development, many of which supply some of the most famous and florid Neolithic archaeology. Borrowing a phrase from Whittle (Bayliss and Whittle 2007), these are 'Neolithic histories' or local historical cycles which happened not because of an essentialist Neolithic *Zeitgeist* but from specific historical causes and actions. In them, people drew upon regionally shared ideas and idioms and reformulated them into an original local way of life, often intensifying or exaggerating them, sometimes creating the *ne plus ultra* of a more widely accepted way of doing things (as at Stonehenge, Avebury and the Golfe de Morbihan), sometimes inverting it to create local difference (as in the Maltese temples). For the European Neolithic, cycles of ritual intensification such as these, the Sardinian *domus de janas* tombs, and many others are the most obvious examples. But examples such as Varna hint at local experiments with new forms of politics and identity, and the ditched villages of the Tavoliere in Southern Italy give a clear example of how a specific kind of settlement and community could form a social world which lasted about a millennium and then collapsed abruptly. The common historical pattern is the development of a particular political or social system which takes locally available possibilities for action, recombines or exaggerates them in novel ways which last a few centuries, and then collapses, dissolves, or changes into something else. Attempts to stray too far from a rather centripetal generalised Neolithic way of life may have brought particular social tensions or points of fragility to the fore; the end is often quite abrupt, suggesting it hit a tipping point of some kind.

What kind of place was the Neolithic? Regional cultures

What kind of place was Neolithic Europe? This turns out to be a surprisingly interesting question. To start with, it is a question never raised in either processualism or post processualism; there was no conceptual space for it. Processualists summarised societies by cross-cultural generalities such as evolutionary 'stage' rather than by historical particularities; all 'early village societies' were fundamentally similar whether they were in Iran, Indiana, Indonesia, or Italy. Post-processualists, in contrast, were deeply suspicious of 'totalising' visions of imposed homogeneity, which in practice meant anything that smacked of system above the level of the local community. The result was to leave a void of ethnographic imagination; the temptation was to fill it by rewriting in the European Neolithic the last good ethnography we read. There was no sense of regional particularity.

But large-scale ethnographic regions exist. What makes the Pueblos the Pueblos, the New Guinea highlands the New Guinea highlands, and so on? This is the kind of stuff which goes in Chapter 1 of the ethnography, which post-modern ethnographers find distressingly normative and descriptive but find they cannot dispense with. It is all the widely shared, taken-for-granted things about how to inhabit a landscape, make a living, relate to people and understand the world; it is all the things your mind fills in when someone situates a site in 'medieval Europe' or 'Amazonia'. And, on important points, it cannot be generalised. Over 50 years ago a social anthropologist (Barnes 1962) wrote an article pointing out that the then-trendy models of segmentary lineages based on African societies did not work for New Guinea because they simply didn't have kinship structures like that there. Recently I heard a similar complaint by an Africanist who was criticising the tendency to apply Amazonian perspectivism models everywhere; animism, she said, is actually very rare in Africa. Neolithic Europe was its own place, not a pastiche of the ethnographically known tribal world. To take an obvious example, the presence of domesticated animals meant that Neolithic Europe differs from indigenous Melanesian and American societies not only economically but also in concepts of humanity, animality and landscape.

So what kind of place was Neolithic Europe, as an ethnographic landscape? If you walked across Europe from Russia to Britain at 4000 BC, would you find a familiar repertory of things, techniques, ideas and relations? Would some things change gradually (for instance, reliance upon agriculture, just as it did in indigenous North American societies as one went from the Mississippi to the Rockies)? Would a continuous and familiar Neolithic have an economic emphasis in some places, a cultural tone in others? What would change abruptly? What would someone from New Guinea, or the Amazon, or the Bronze Age find completely alien?

These are potentially answerable questions. And answering them is not only interesting, it is necessary. This is not reviving the old quest to define 'culture areas'; there is no suggestion of bounded areas, of trait listing and distribution maps. What thinking about ethnographic landscapes contributes is the essential backdrop to nuanced local studies. People did not make their lives from a *tabula rasa*; this broad regional cultural repertory or ethnographic landscape provided their potential resources and reflexes, their material and doxa. Moreover, ethnographic landscapes were the precondition for any social action happening beyond the borders of a settlement – they provided the mutual comprehensibility or incomprehensibility of neighbours, the matrix for the movement or isolation of people, things and ideas. Finally, history unfolds in specific places, not in the abstract. If we are interested in long-term change, because ethnographic landscapes set the limits to the possible and the unthinkable, they provide the parameters for path dependency, for specific historical trajectories rather than generalised evolutionary accounts.

the Neolithic, but by about 4000 BC they converged into a common lifestyle committed to a farming way of life. In a loose, haphazard way visible only at the largest scale, the bundle of related practices and institutions which was the Neolithic became a stable attractor: paradoxically, a more consolidated or homogeneous Neolithic lifestyle – an elastic envelope rather than a package – emerged as a long-term slowly assembled result of the transition, not as an initial condition of it.

Arguing this model in detail (Robb 2013) is really beside my point here. But what is relevant here is (first) that a model along these lines is really the logical extension of the emerging theoretical approach to the Neolithic noted above which focuses upon the materiality of social choices of action, and (secondly) that detailed scientific study of Neolithic practices – what exactly a regime of food production entailed, how a shift in mobility related to demography, how exactly exchanged materials and people travelled – is essential to such explanation.

The end of the Neolithic

The other big change archaeologists have focused upon, of course, is the end of the Neolithic. This presents us with paradoxes. On one hand, most archaeologists agree that there is a major change in how European society was constituted sometime between the fourth and third millennium. Europe at 3500 BC presents us with a distinctly Neolithic ethnographic landscape and Europe at 2000 BC presents us with a very different, discernibly Bronze Age world. On the other hand, there is no real agreement on how this change happened. There are at least half a dozen major ways of characterising it, including the traditional 'Indo-European' and 'Beaker folk' migrations (Mallory 1989), New Archaeological and Marxist attempts to construe it as the rise of inequality (Champion *et al.* 1984, Milisauskas 2002, Chapman 2008), Sherratt's *secondary products revolution* model of economic intensification (Sherratt 1981), and a range of models construing it as a qualitative change in social reproduction: a shift from ritualism to prestige competition (Shennan 1982, Braithwaite 1984, Thorpe and Richards 1984), from ancestry to genealogy (Barrett 1994, Thomas 1999), or a reorganisation of prestige, value and gender (Robb 2007).

Moreover, the exact nature and timing of the transition vary greatly across the continent and models tailored to one area tend not to work elsewhere. In Britain the change is from pre-existing megalithic worlds to Beakers and individual burial. But elsewhere – the Alps and Central Mediterranean, for instance – Beakers or analogous phenomena often accompany collective burials which arise at this time, marking a change from previous burial regimes. The fact that things which are diametrically opposed in one place support each other in other places gives a somewhat M.C. Escher feel to the prehistory. More importantly, it suggests that we are going at explaining this transition the wrong way. Very much as for models of the origins of farming, explanation has been polarised into regional and local approaches, with very much the same strengths and weaknesses. Regional approaches such as 'Indo-European' models and the 'secondary products revolution' insist that the transition has to happen the same everywhere, and they founder when it demonstrably does not. On the other hand, 'local' approaches generally focus narrowly upon only one local sequence – often the British sequence – usually outlined as an opposition of rather crystalline social logics. They do little to explain why regionality is so patently obvious in a transition marked by broad horizons of material culture, and why parallel but clearly independent developments happened in many places.

The point here is not to answer this question but to pose it. The fourth–third millennium transition happened gradually and piecemeal; almost everywhere it is marked by substantial local continuity. Yet it involved broadly shared repertories of material things and practices, many of which would have restructured landscapes of action around them. The result was not a single sequence, nor a collection of unrelated, purely local sequences, but a broad evolution from a Neolithic to a Bronze Age set of social possibilities. As with the early Neolithic, the result was not a single homogeneous way of life but an ethnographic landscape in which mutually known elements were reconfigured locally, often by people who knew what their neighbours were doing and were trying to construct regional similarity or local difference in particular contexts. As with the early Neolithic, the resulting ethnographic landscape contained structures for generating history, not as narrowly deterministic pathways ('inequality arose in the Bronze Age') but doorways to envelopes of possibilities and tendencies ('when inequality develops in the Neolithic, it tends to be ritual hierarchy; when inequality develops in the Bronze Age, it tends to be about display and exchange').

The last word

To conclude by beginning; every generation is destined to rewrite Shakespeare. If the same is true for the Neolithic, it has been about three decades since the main battle lines of Neolithic studies were drawn up, and they have remained more or less frozen since then: migration or acculturation, economic or cultural, continental or local, scientific or humanistic. But time has moved on. With the end of the Science Wars, the larger intellectual landscape underlying these fissures has shifted, and these positions are more than a little shop-worn; the polemics they generate are (in Hamlet's words) increasingly weary, stale, flat and unprofitable. There are more things in Heaven, Earth, and the Neolithic than we have yet been able to dream of; it is time for an overhaul. The research agenda outlined here starts from a widely accepted understanding of the materiality of social practices and draws out the implications for a series of problems particular to Neolithic studies. Whether this research agenda generates worthwhile work remains to be tested, but with this or another agenda, it is time to move Neolithic studies on. 'Tis a consummation devoutly to be wished.

References

Bailey, G. 2007. Time perspectives, palimpsests and the archaeology of time. *Journal of Anthropological Archaeology* 26, 198–223.

Barnes, J. 1962. African Models in the New Guinea Highlands. *Man* 62, 5–9.

Barrett, J. 1994. *Fragments from antiquity: an archaeology of social life in Britain, 2900-1200* BC. Oxford: Blackwell.

Bayliss, A. and Whittle, A. 2007. Histories of the dead: building chronologies for five southern British long barrows. *Cambridge Archaeological Journal* 17.1, *Supplement*.

Beckett, J.F. 2011. Interactions with the dead: a taphonomic analysis of burial practices in three megalithic tombs in County Clare, Ireland. *European Journal of Archaeology* 14, 394–418.

Bocquet-Appel, J.-P. 2011. The agricultural demographic transition during and after the agriculture inventions. *Current Anthropology* 52, S497–S510.

Bogaard, A. 2004. *Neolithic farming in Central Europe: an archaeobotanical study of crop husbandry practices*. London: Routledge.

Bogaard, A. and Jones, G. 2007. Neolithic farming in Britain and central Europe: contrast or continuity? In A. Whittle and V. Cummings (eds), *Going Over: The Mesolithic–Neolithic transition in Western Europe*, 357–75. Oxford: Oxford University Press for the British Academy.

Borić, D. 2008. First households and 'house societies' in European prehistory. In A. Jones (ed.), *Prehistoric Europe: theory and practice*, 109–42. Oxford: Wiley-Blackwell.

Bradley, R. 1993. *Altering the earth*. Edinburgh: Society of Antiquaries of Scotland.

Bradley, R. and Edmonds, M. 1993. *Interpreting the axe trade: production and exchange in Neolithic Britain*. Cambridge: Cambridge University Press.

Braithwaite, M. 1984. Ritual and prestige in the prehistory of Wessex, c. 2200–1400 BC: a new dimension to the archaeological evidence. In D. Miller D and C. Tilley (eds), *Ideology, power and prehistory*, 93–110. Cambridge: Cambridge University Press.

Cann, J.R. and Renfrew, C. 1964. The characterization of obsidian and its application to the Mediterranean region. *Proceedings of the Prehistoric Society* 30, 111–33.

Champion, T., Gamble, C., Shennan, S. and Whittle, A. 1984. *Prehistoric Europe*. London: Academic Press.

Chapman, J. and Gaydarska, B. 2006. *Parts and wholes: fragmentation in a prehistoric context*. Oxford: Oxbow Books.

Chapman, R. 2008. Producing inequalities: regional sequences in later prehistoric Southern Spain. *Journal of World Prehistory* 21, 195–260.

Childe, V.G. 1951. *Man makes himself*. New York: New American Library.

DeLanda, M. 2006. *A new philosophy of society: assemblage theory and social complexity*. London: Continuum.

French, C. and Lewis, H. 2007. *Prehistoric landscape development and human impact in the Upper Allen valley, Cranborne Chase, Dorset*. Cambridge: McDonald Institute for Archaeological Research.

Gillings, M., Pollard, J., Wheatley, D.W. and Peterson, R. 2008. *Landscape of the megaliths: excavation and fieldwork on the Avebury monuments, 1997–2003*. Oxford: Oxbow Books.
Grima, R. 2008. Landscape, territories, and the life-histories of monuments in Temple Period Malta. *Journal of Mediterranean Archaeology* 21, 35–56.
Halstead, P. 1981. Counting sheep in Neolithic and Bronze Age Greece. In I. Hodder, G. Isaac and N. Hammond (eds), *Pattern of the past: studies in honour of David Clarke*, 307–39. Cambridge: Cambridge University Press.
Hegmon, M. 2003. Setting theoretical egos aside: issues and theory in North American Archaeology. *American Antiquity* 68, 213–44.
Hodder, I. 1982. Sequences of structural change in the Dutch Neolithic. In I. Hodder (ed.), *Symbolic and structural archaeology*, 162–77. Cambridge: Cambridge University Press.
Hodder, I. 1984. Burials, houses, women and men in the European Neolithic. In D. Miller and C. Tilley (eds), *Ideology, power, and prehistory*, 51–68. Cambridge: Cambridge University Press.
Hodder, I. 2007. Can we move on now. *Cambridge Archaeological Journal* 17, 224–5.
Hofmann, D. and Smyth, J. (eds) 2013. *Tracking the Neolithic house in Europe: sedentism, architecture and practice*. New York: Springer.
Ingold, T. 2000. *The perception of the environment: essays on livelihood, dwelling, and skill*. London: Routledge.
Kelly, J. 1977. Did women have a Renaissance? In R. Bridenthal and C. Koonz (eds), *Becoming visible: women in European history*, 137–64. Boston: Houghton Mifflin.
Knauft, B. 1993. *South coast New Guinea cultures: history, comparison, dialectic*. New York: Cambridge University Press.
Latour, B. 2005. *Reassembling the social: an introduction to actor-network-theory*. Oxford: Clarendon.
Mallory, J. 1989. *In search of the Indo-Europeans*. London: Thames and Hudson.
Marciniak, A. 2005. *Placing animals in the Neolithic: social zooarchaeology of prehistoric farming communities*. London: UCL Press.
Milisauskas, S. (ed.) 2002. *European prehistory: a survey*. New York: Kluwer Academic/Plenum.
Miller, D. (ed.) 2005. *Materiality*. Durham: Duke University Press.
Nehlich, O., Montgomery, J., Evans, J., Schade-Lindig, S., Pichler, S.L., Richards, M.P. and Alt, K.W. 2009. Mobility or migration: a case study from the Neolithic settlement of Nieder-Morlen (Hessen, Germany). *Journal of Archaeological Science* 36, 1791–9.
Pétrequin, P. and Pétrequin, A.M. 1993. *Écologie d'un outil: la hache de pierre en Irian Jaya (Indonésie)*. Paris: CNRS.
Richards, C. and Thomas, J. 1984. Ritual activity and structured deposition in Later Neolithic Wessex. In R. Bradley and J. Gardiner (eds), *Neolithic Studies*, 189–218. Oxford: British Archaeological Reports.
Robb, J. and Pauketat, T. 2013. From moments to millennia: theorizing scale and change in human history. In J. Robb and T.R. Pauketat (eds), *Big histories, human lives: tackling problems of scale in archaeology*, 3–33. Santa Fe: SAR Press.
Robb, J.E. 2007. *The Early Mediterranean village: agency, material culture and social change in Neolithic Italy*. Cambridge: Cambridge University Press.
Robb, J.E. 2013. Material culture, landscapes of action, and emergent causation: a new model for the origins of the European Neolithic. *Current Anthropology* 54, 657–83.

of the Neolithic. These issues simply reiterate that the heart of the problem lies in the way we are inclined to view the relationship between the two populations that have traditionally described different stages of an evolutionary trajectory from hunter-gathering to farming, with all the value judgements that such a trajectory brings with it. These judgements include ideas of progress out of primitiveness, and the development of cultural and social complexity out of prior conditions of egalitarian simplicity. The 'Man the Hunter' symposium (Lee and DeVore 1968) changed the widely held perception that hunter-gatherer populations live in a state of ecological vulnerability. Much of the recent literature on the Neolithisation of the western Mediterranean, Atlantic and Baltic Europe has been characterised by the desire to rescue hunter-gatherers from the patronising judgement of primitiveness, and to allow them an agency in their own histories and in their transformation into the Neolithic.

It is important that we understand at the outset how the distinction between economic and cultural change has intersected with the issue of human agency in recent studies of Neolithisation. Although rarely stated so directly, reference to economic change has long tended to be freighted with a deterministic logic. Economies are too easily treated as if they simply linked human communities to the resources necessary for their survival, thus defining the economy as '[t]he primary human adaptation to the environment' (Higgs and Jarman 1975, 4). Consequently, economies are assumed to be designed for the efficient exploitation of those resources. Whilst this does not necessarily imply that economic processes will be designed to maximise production, the implication is none the less that economies, functioning efficiently within the bounds of an available technology to satisfy human requirements, will be of a kind that is determined by, and thus fully comprehensible with reference to, the available resource base and operating relative to a particular level of social organisation. Unsurprisingly this position is often accompanied by a Darwinian gloss that describes the 'adaptive fitness' of economic systems. Thus the theory of optimal foraging applied to hunter-gatherer behaviour provides for the modelling of 'foraging behaviours expected in different environmental circumstances' (Winterhalder 1981, 13). Lewis Binford developed a complex synthesis of hunter-gatherer subsistence strategies relative to available technologies and different ecological contexts, which drew upon a considerable body of ethnographic data. In that study he argued that subsistence strategies had to be understood before cultural variability could be satisfactorily explained and that social differentiation arose relative to the management of subsistence variability. Thus he concluded his analysis of hunter-gatherer organisational complexity with the comment that 'when I think about all of the factors that contribute to variability among hunter-gatherer groups, I am impressed with how much the character of the subsistence base has contributed to my explanations' (Binford 2001, 433).

In several different parts of the world during the Holocene, hunter-gatherer subsistence strategies eventually resulted in the domestication of plants and animals, and those domesticates then colonised out from the original home range that had supported their non-domesticated progenitors into new and quite different ecologies. If hunter-gatherer subsistence strategies had culminated in the domestication of plants and animals in one region, such as south-west Asia, then to what extent might pre-existing hunter-gatherer subsistence strategies have been responsible for the spread of those domesticated populations into the landscapes of temperate Europe? The emphasis upon subsistence economies tends to present the transformation from hunter-gatherers to farmers as if it were a response, if not to failure, then at least to the adaptive stress within the earlier system. Thus, for example, Binford introduced the concept of a 'packing threshold' for hunter-gatherer territories beyond which a shift in resource base will have been necessitated, with the potential to drive the economy towards the adoption or development of domesticates (Binford 2001, 435). Others have also suggested that the loss of a key hunter-gatherer subsistence resource might have prompted the adoption of domesticates, where those domesticates were available (Rowley-Conwy 1984). However, where subsistence mechanisms have been taken to carry their own internal logic of change, the overwhelming impression has been that the expansion of the Neolithic through south-east and central Europe was the consequence of colonisation by agriculturalists, with little or no input from any of the indigenous populations. The story told is that the history of Neolithic expansion was driven by population growth among farming populations with their subsequent colonisation into new lands (Ammerman and Cavalli-Sforza 1971).

Those archaeological narratives that treat the westward European expansion of domesticates as consequent upon the colonising settlement of farmers, combined perhaps with the failure of hunter-gatherer subsistence resources, have been countered more recently by attempts to provide hunter-gatherers with an active and strategic role in the history of their own transformation. Such attempts have moved the focus of the narrative away from economic modelling and towards the modelling of processes of cultural production, and the potential opportunities provided by exchanges between hunter-gatherers and agriculturalists to enable hunter-gatherers to design the material-cultural package of conditions necessary for their own transformation into the Neolithic (Zvelebil 1986; Thomas 1988, 2003). The teleological direction implicated in these narratives of exchange should be noticed; the proposed exchange relations that resulted ultimately in hunter-gatherers becoming agriculturalists never seemingly resulted in agriculturalists 'reverting' to hunting and gathering.

Given that both economic and cultural processes are socially embedded by virtue of both being lived out as relationships between people and material conditions, then characterising the turn to cultural processes as if it instigated a turn to a previously ignored 'social archaeology' is not really the issue; we do not need to recruit 'social' processes to explain the transformation (*contra* Bender 1978). More germane is the notion that economic processes were structured by the need for the available technology to be efficiently adapted to subsistence resources, whereas cultural processes were, by contrast, 'meaningfully constituted' (Hodder 1992, 12). Thus if hunter-gatherer communities were once treated as simply having had to adapt to the resources offered by nature for their subsistence, then hunter-gatherers could now be treated as having been actively involved in the maintaining, and indeed redirection of, their patterns of cultural behaviour. And if all materials, including the resources of nature, were workable according to people's commitment to particular cultural values, then a shift in those values could, as Ian Hodder argued (1990), bring about a change in subsistence practices; in effect the arrow of determinacy identified by Binford was reversed by Hodder.

The two positions therefore differ in their view of the practices of production and consumption, treating them as either having responded to the logic of material constraints necessary for survival, or as having responded to those constraints as if they were represented by, or were vitalised as, cultural systems of value. It would be easy enough to caricature this as the contrast between processual archaeology seeking economic determinants and post-processual archaeology embracing a cultural indeterminacy. In both cases, however, social structures are treated as the relationships between categories of people and things. For Binford such categories were given by the logic of needing to adapt to material conditions, whilst for Hodder they arose through the need to explore the order of things out of which a system of contrasting cultural values could be created. At base both these views simply concern the conditions that might determine the ways in which a human population orientated the practices necessary for its survival, practices that have been widely described in terms of a cycle that moved from the extraction to the consumption of resources. In an attempt to get beyond the economic/cultural dichotomy, I will argue that the cycles of consumption functioned to degrade an energy gradient and involved what might be termed a 'duty of care' towards selected reservoirs of that energy. By referring to a 'duty of care' I wish to highlight that the making of things ('economies of production') involves a moral commitment towards things ('cultural values') (Gudeman 2001). This perspective will, I suggest, enable us better to understand the changes that would have been necessary to enable domesticated grasses and livestock to colonise a wide range of soils and climatic regimes within Europe.

Living things

There has been much written of late concerning the agency of things. To refer to the agency of anything is to address its ability to effect change in existing material conditions; in other words, agency is the ability either to do work or to redirect the work of another agent. Clearly these are two rather different capabilities. Human agency does work that is rendered archaeologically visible by virtue of its surviving residues. The common assumption that by identifying a set of either the economic or the cultural consequences of that work we necessarily understand its motivation is, however, naïve. The agency of things on the other hand draws our attention to the fact that things might certainly do work themselves (machinery), or direct the work of other agents, such as humans, by providing, for example, a resource to be utilised (as an energy source or as a tool), or as a form of resistance (the mass of an object), or as encoded information to which the human agent might respond (cf. Gell 1998). What is odd about so much of the current literature, however, is not only the assumption that the agency of humans and the agency of things can be analysed as if they operated 'symmetrically' (cf. Witmore 2007), but also the fact that the agency of the rest of the biological world is either ignored or sidelined (Barrett 2014). Commenting upon Olsen's quest to understand how 'things and objects "mix" with human beings to form the configurations we call *society* and *history*' (Olsen 2010, 2, original emphasis), Ingold is forced to conclude that Olsen has managed to 'exclude animals and plants – which exist "alongside" (Olsen 2010, 9) but, by the same token, are not part of "material culture" – from the processes of social and historical life' (Ingold 2012, 431). But if our concern is with the mechanisms by which particular biological populations of plants, animals and humans colonised the diverse geographies of Europe from the ninth to the fifth millennia cal BC, then any focus purely on human agency and material culture is inadequate.

The point that Olsen and others appear to have missed is that living agencies differ from such inorganic things as machinery in one fundamental respect; living things are self-making through the process of metabolism whereby they seek out and degrade the energy sources that are necessary for their physical renewal and growth (Schneider and Sagan 2005). Life-forms are therefore sustained within various ecologies that can be characterised by the ways their energy gradients, powered ultimately by the sun, are degraded by forms of life (Schneider and Kay 1994). These energy gradients power the self-making of life-forms as nutrients are degraded in virtue of their flow through a food chain from plants, through herbivores to carnivores, and upon which, at every level, various other forms of biota and viruses are parasitic. Ecologies are therefore flows of energy, but they can also be regarded as 'texts' of encoded information that life utilises to sustain itself. In other words, life is a form of 'sense making' that seeks security and avoids the dangers that are afforded by the world that it enters (Thompson 2007). The work undertaken by living agencies is therefore a form of colonisation that creates the various ecological niches that sustain the meaningful process of self-making or *autopoiesis* (Maturana and Varela 1980, 1998; Odling-Smee *et al.* 2003), and conversely life is threatened when such niches suffer the erosion of their sources of energy and the diminishing of recognisable information. The 'dwelling perspective' that Ingold develops thus requires that we accept different forms of life to be active in the ways they 'read' themselves into their ecological roles (Ingold discovers in the world embody metaphysical qualities, and given also that consciousness consumes energy, then the human form of life tends to be locked into degrading the energy sources that both define, and are able to sustain, life's moral universe. This might explain something of the forward direction of history; the move from hunter-gathering to agriculture was a move up an energy gradient achieved by the commitment to newly recognised moral orders, and it consequently tended to block the possibility of following the reverse trajectory. Agriculturalists have at times certainly increased their foraging activities and thus moved into a lower energy gradient, but this has been enforced when domesticated resources have begun to fail, and it has normally fostered the loss of cultural cohesion among these communities.

The transformation from hunter-gatherer to agricultural ecologies therefore increased the energy gradient within which human populations were maintained by the concentration and the storage of nutrients in the greater body mass, and thus in the derived flesh, blood and milk, of domesticated animals, and in the grain harvested and threshed from dense stands of domesticated crops (Barrett 2011). This transformation was not likely to have been achieved by the direct imposition of a category distinction between the wild and the domesticated on to different populations of animals and plants; rather the category distinction arose as a consequence of a change in landscape perceptions. This was the change from the geographically extensive strategies of energy capture utilised by hunter-gatherers to the locally intensive nurturing and storage of energy derived from restricted areas of land that sustained those domesticates. This shift required two things. First, the new kind of geographical consciousness had to be derived from the practices of inhabitation resulting in a duty of care towards place and away from the care invested in a more dispersed network of seasonally available plants and migrating animals. Secondly, this geographical reorientation will have required an initial recognition of the free energy necessary to support the implementation of the changes in energy procurement that are defined by agriculture (Schneider and Sagan 2005, 68; Kauffman 1995).

These two requirements would certainly rule out any suggestion that hunter-gatherers could have 'adopted' farming as a consequence of a food crisis or some other condition of mal-adaptation (*contra* Cohen 1977). It also poses significant problems for models which claim that the colonisation of domesticates, necessarily supported by a system of agricultural management, took place into, and displaced hunter-gatherers from, what the agriculturalists might have perceived to have been virgin and, indeed, unknown territory. What would those pioneering agriculturalists have lived on before their domesticated resources had become established? Surely it is more likely that both requirements would have been satisfied in the consequences of earlier patterns of land-use.

Neolithisation

If the requirements for the successful colonisation of domesticates into the varied European ecologies can be established in this way, then the challenge to understand the detail of that process nonetheless remains. Consequently generalisations built upon limited examples must be treated with considerable caution (cf. Bickle and Hofmann 2007, Bickle and Whittle 2013). Agriculture, as a system of management, expressed the necessary duty of care towards the reproduction of domesticates simply because the land, whose fertility offered the free energy source that could be captured by the growth of crops and in the developing bodies of domesticates, before being degraded through storage and consumption, was in a sufficient state of preparedness. It was land that became the object of care.

We have long since abandoned models of *landnam* colonisation, with its images of a 'slash and burn' agricultural expansion into virgin forest. As the contributors to this volume make clear, the archaeology that attests to the developed stages of the central European *Linearbandkeramik* demonstrates that this particular form of agricultural life was lived as a moral commitment towards place. This appears, at least in some regions, to have been exercised through the long-term management of garden plots that were intensively worked, manured, and autumn-sown, and that were presumably accompanied by more dispersed patterns of grazing rights (Bogaard 2004, 96–114; this volume). If these garden plots were indeed tenured to households, as is implied, then the histories of those households' identities were objectified by the increasing elaboration of their domestic architecture and by the fixed locations of some of their cemeteries. It is indeed tempting to refer to these as 'house societies' where, as Lévi-Strauss proposed, the continuity of identities and alliances was expressed through the material and spiritual identity of the house itself (Lévi-Strauss 1983, Borić 2008, cf. Bloch 2008). But if the moral values of identity were thus expressed, then the existence of those same values had to be revealed by being in the world itself. It was through the practical duty of care towards the portions of land available for cultivation and the tracts of managed pasture for grazing that the lives of these households became possible. That reality was understood and sustained as a flow of energy from the fertility of the land through the stored plant and animal resources to the household and its traditions of household consumption.

The continuity of some households' land-tenure over sequences of settlement development has now been further demonstrated by Bogaard's mapping of the residues of ecologically specific weed flora resulting from the threshing, drying and storage of crops derived from specific soils by the individual household clusters or clan groupings that have been identified on the nucleated settlement of Vaihingen an der Enz in south-west Germany (Bogaard *et al.* 2011, Bogaard 2012). Bogaard's analysis, when set against the, presumably lengthy, life-span of the buildings, and the necessary reproduction of the seasonal labour demanded by hoe or spade agriculture, along with the equally intensive demands for harvesting and gleaning, are all reminders that a gender politics was at work in these histories. With labour roles likely to have been distributed by gender as well as by age, Bogaard's observation (2004, 163–4) that autumn sowing would have required a high labour investment at the same time of year as when wild fruits would have been available for gathering, implies either an increased burden being placed upon that portion of the population engaged in these tasks, or a contraction in the focus of its labour from more widely distributed resources to the household's lands.

The seemingly inevitable conclusion is that the clearance and management of forest margins and the clearance of scrub by hunter-gatherers established the patchwork of cleared, fertile land that was the necessary condition for its successful colonisation by domesticated plants and animals. The landscape therefore provided the discursive and practical medium by which human identities in relation to animal and plant populations were negotiated. Thus, whilst hunter-gatherer communities may have brought into view the geographically extensive patterns of information and energy embodied in the dispersed and seasonal cycles of animal migration and plant growth and thus revealed the moral and sacred order of those sustaining landscapes, they may also have created the more localised reserves of free energy that agricultural systems could subsequently inhabit. The fluidity by which people might have moved and by which cultural identities could have been reworked would also have enabled these changing landscapes to operate as the discursive media within which new alliances were forged and new populations were sustained (cf. Robb and Miracle 2007, 112–13; cf. Cummings and Harris 2011).

Conclusion

Hunter-gatherer and agricultural systems have been characterised here as different regions on a spectrum of energy gradients, where those regions were also capable of being read by their inhabitants to reveal the existence of diverse forces of moral order. I would argue that the analysis of the multivalent dynamic, by which the processes of degrading these different energy gradients might have resulted in common landscape readings by initially different populations of humans, plants and animals, should allow us to bypass current arguments that model the Mesolithic to Neolithic transformation as if it were the consequence of either population displacement or cultural exchange. However, the analysis of this more complex historical dynamic will only be possible if archaeology continues with the difficult task of building the intellectual and practical mechanisms that cross-cut the expectations and the techniques of what, in the past, have been distinguished as cultural and science-based archaeologies.

Acknowledgements

I am grateful to Mats Larsson and Julian Thomas for inviting me to participate in their session on the British and Scandinavian Neolithic at the Nordic TAG conference in Reykjavik 2013 where a very much earlier version of this chapter was presented. I have benefitted in developing the ideas presented here from the work of Irene Garcia-Rovira.

References

Ammerman, A.J. and Cavalli-Sforza, L.L. 1971. Measuring the rate and spread of early farming in Europe. *Man* 6, 674–88.

Barrett, J.C. 2011. The Neolithic revolution: an ecological perspective. In A. Hadjikoumis, E. Robinson and S. Viner (eds), *Dynamics of neolithisation in Europe: studies in honour of Andrew Sherratt*, 66–89. Oxford: Oxbow Books.

Barrett, J.C. 2014. The material constitution of humanness. *Archaeological Dialogues* 21, 65–74.

Bender, B. 1978. Gatherer-hunter to farmer: a social perspective. *World Archaeology* 10, 204–22.

Bickle, P. and Hofmann, D. 2007. Moving on: the contribution of isotope studies to the early Neolithic of Central Europe. *Antiquity* 81, 1029–41.

Bickle, P. and Whittle, A. 2013. LBK lifeways: a search for difference. In P. Bickle and A. Whittle (eds), *The first farmers of central Europe: diversity in LBK Lifeways*, 1–27. Oxford: Oxbow Books.

Binford, L.R. 2001. *Constructing frames of reference: an analytical method for archaeological theory building using ethnographic and environmental data sets*. London: University of California Press.

Bloch, M., 2008. Why religion is nothing special but is central. *Philosophical Transactions of the Royal Society* (B) 363, 2055–61.

Bogaard, A. 2004. *Neolithic farming in Central Europe: an archaeological study of crop husbandry practices*. London: Routledge.

Bogaard, A. 2012. *Plant use and crop husbandry in an early Neolithic village: Vaihingen an der Enz, Baden-Wurttemberg*. Bonn: Habelt.

Bogaard, A., Krause, R. and Strien, H-C. 2011. Towards a social geography of cultivation and plant use in an early farming community: Vaihingen an der Enz, south-west Germany. *Antiquity* 85, 395–416.

Borić, D. 2008. First households and 'house societies' in European prehistory. In A. Jones (ed.), *Prehistoric Europe: theory and practice*, 109–42. Chichester: Wiley-Blackwell.

Cohen, N. 1977. *The food crisis in prehistory: overpopulation and the origins of agriculture*. London: Yale University Press.

Cummings, V. and Harris, O. 2011. Animals, people and places: the continuity of hunting and gathering practices across the Mesolithic–Neolithic transition in Britain. *European Journal of Archaeology* 14, 361–82.

4

Multi-agent Modelling of the Trajectory of the LBK Neolithic

A Study in Progress

JEAN-PIERRE BOCQUET-APPEL, JÉRÔME DUBOULOZ, RICHARD MOUSSA, JEAN-FRANÇOIS BERGER, ANNE TRESSET, ELENA ORTU, JEAN-DENIS VIGNE, ROBIN BENDREY, STÉPHANIE BRÉHARD, DOMINIQUE SCHWARTZ, AURÉLIE SALAVERT, MARIA FERNANDA SANCHEZ-GOÑI, DAMIEN ERTLEN, YOANN GAUVRY, GOURGUEN DAVTIAN, MARC VANDER LINDEN, EVA LENNEIS, LORETTE NOIRET, AGNÈS GUILLAUMONT AND MARTIN O'CONNOR

Introduction

EXPLANATIONS OF THE FACTORS determining the trajectories of prehistoric human societies, from their emergence to their disappearance, rely above all on the chronological accuracy of the information available, regardless of the actual causes. When studies cover many millennia of evolutionary time, the degree of chronological accuracy is low and explanations usually involve the extra-human forces of environmental impacts, in particular the (non-linear) variation of insulation (see for example Potts 2012 for the evolution of early *Homo*); for short sub-millennium periods, explanations bring in the effects of human volition, at least since the well known example of Gibbon's *Decline and Fall of the Roman Empire* (1776–1789). The reverse approach is less common, in which explanations either bring in the effect, on the evolutionary time scale, of a type of behaviour such as fire control, whose biocultural impact on anatomy, culture and cognition has been noticeable since *Homo erectus* (Wrangham and Carmody 2010), or, conversely, the effect of climatic fluctuations on a short-term societal collapse, as in the well known examples of Moche and Maya societies (for a review see Carey 2012). When a researcher is dealing with an archaeological sequence representing the trajectory of a society, the explanatory level chosen will depend on the chronological accuracy of the sequence: macro-environmental when accuracy is low *vs.* human volition when it is high. But if the researcher is trying to understand the production of a

Proceedings of the British Academy **198**, 53–69.

society, given the macro-environmental forces at work, then choosing causes because of the lack of chronological definition of the data is obviously not a satisfactory approach. One way of considering both levels of explanation, and their possible interactions, is to construct computer models of artificial societies and run simulations according to reconstructed natural contexts. This approach was initiated more than ten years ago by Robert L. Axtell *et al.* (2002), with multi-agent modelling applied to the Kayenta Anasazi society. This approach laid the foundations of modelling archaeology in terms of societal trajectories, with the idea that 'agent models offer intriguing possibilities for overcoming the limitations of experimental archaeology through systematic analysis of alternative histories' (Axtell *et al.* 2002, 7275) (for a critical review of the application, see Janssen 2009). The approach has since been sustained (for example see Kohler and van der Leeuw 2007), and is the one we use here.

This chapter describes an ongoing research project in which the main purpose is to cross-analyse existing archaeological data with corresponding environmental data, in order to understand the reactiveness and degree of resilience of a prehistoric society to climate impacts by running simulations of multi-agent type. For reasons to do with the abundance and quality of available information, it was decided to focus on the agricultural society of the *Linearbandkeramik* culture or LBK (of the second half of the sixth millennium BC), which extended longitudinally across Europe from Hungary to Normandy and to Poland. The LBK was the main Neolithic society in temperate Europe. It formed in central Europe at around 5550 cal BC, on the pioneer front of the Balkan Neolithic cultures, which had become established earlier (Banffy and Cross 2010, Pavúk 2004). Although the oldest, it is probably the best documented of the European Neolithic cultures, and several thousand sites are known (Lüning 1988; Bogucki 2001). LBK houses, built from massive beams and up to 45m in length, are particularly recognisable, while the accompanying pits where household refuse was partially thrown allow accurate reconstructions of the LBK economy, technology and supply systems.

Agricultural societies are extremely vulnerable to climatic risks. Famines have regularly hit proto-historical and historical agricultural societies (see for example Bell 1970, Robbins 1975, Hassig 1981, Dirk 1980, Grolle 1997 and Alamgir 1981; for a summary of the twentieth century, see for example Devereux 2000, Castro 1952). We can confidently assume that famines have also repeatedly struck archaeological societies at different points across the globe since the beginning of agriculture. But for various reasons, their physical, cultural and biological traces are difficult and sometimes impossible to detect in the archaeological records. Controlled historical data, covering about five centuries in France since the thirteenth century, show 15 famines caused by climate alone (i.e. independently of limitations of the feudal system), each causing at least 300,000 deaths and even up to a million (Le Roy Ladurie 2004, Le Roy Ladurie *et al.* 2003), i.e. a major environmental social crisis every 40 years on average. That such crises occurred during the 650-year duration of the LBK culture and that one of them has been fatal to the LBK society is a hypothesis to be taken into consideration. Based on an ethno-historical inference and partial decontextualised ethno-archaeological models, part of an LBK model is described below, with an emphasis on the expansion mechanism. The model comprises sociological units corresponding to simple and to extended families living in individual farms and hamlets, an agricultural system where cereal crops and livestock are supplemented by hunting and gathering, a link between climate change and the efficiency of the agricultural system, a population expansion and densification mechanism starting from a pioneer front and a possibility of social dislocation through a short-term environmental crisis determining famine. The model also includes a reconstruction of the contemporary climate through a pollen series, as well as an ecological reconstruction including soil fertility.

To reconstruct and simulate the LBK society, the model incorporates: (1) the sociological units corresponding to individual farms, grouped into hamlets, (2) a system of economic production and agricultural performance in terms of carrying capacity, (3) a climatic and ecological reconstruction of the LBK period and of soil fertility in particular, (4) a link between climate variations and performance of the agricultural system, (5) a mechanism governing population expansion and densification from a pioneer front, and (6) an ethno-historical model of social dislocation caused by a short-term environmental crisis producing famine. To understand the development of the LBK cultural area through to its dissolution, the research is attempting to answer the following questions: What are the artificial archaeological remains that result from the socio-natural assumptions implicit in this LBK model, in terms of space-time distributions and densities of houses and hamlets? How does this LBK society, modelled with two environmental constraints, one representing climate variability during its actual historical trajectory and the other climate variability during a hypothetical straight and neutral trajectory, respond in terms of expansion rate, resilience and dissolution? The extent to which the model fits the data is measured in the usual way by testing the match between patterns of observed archaeological hamlets and simulated configurations. What are the assumptions that best fit the data? After the construction phase of the multi-agent LBK model described below, the current phase of intensive simulations began (not yet reported here).

The data

The geographical area in which the history of the LBK unfolded reached a maximum extent of 1.9 million square kilometres. It was bounded in the model by a rectangular frame in decimal coordinates (ymin: 42.8694, ymax: 55.1770, xmin: 6.7189, xmax: 33.3650), Lambert transformed, and divided into pixels of

Soil fertility estimates

In any agricultural system, soil fertility is of the first importance. Two parallel approaches were used to estimate soil fertility relatively to LBK agriculture: (1) collection of unpublished data and production of new palaeopedological data to fill observed gaps, and (2) a regressive approach using the reclassified existing soil maps as a starting point and applying one or more regression functions from specific studies to each fertility class defined. This approach rapidly produced a theoretical map of Neolithic soils, which can be translated into agrological potential (Figure 4.2). Other geo-natural variables were added to the estimated climatic and soil fertility variables (altitude, eight seasonal estimates for precipitation and temperature and their respective standard deviations, and exposure to sunlight), to which were added rivers (tributaries to the fourth degree), making 14 surface variables altogether.

Multi-agent modelling

This model works with the 'household' agent, represented archaeologically by an LBK house. Typically, this household corresponds, anthropologically speaking, to a simple or extended family consisting of one to three nuclear families, which are, relatively to the family of the two founding parents, their children/married stepchildren and their grandchildren, that is up to three generations coexisting in one to three nuclear families. Several LBK houses (households) in the same km^2 pixel, or in the eight nearest adjacent pixels, form a hamlet and its satellites.

A household owns three mini-herds of cattle, goats/sheep and pigs, which it cares/uses for economic purposes and eventually consumes. Together, the mini-herds of each species belonging to each household form the hamlet herd, that is to

Figure 4.2 Map of estimated soil fertility in the LBK cultural area.

say a reproductive unit shared by its inhabitants. Although they interact with the environment (pasture, manure, sensitivity to climate) and with the 'household' agents who possess them (kcal intake, working time, exchanges), these mini herds are not taken as agents in themselves.

This is because they are under the control of the farmers who take care of them and therefore have no autonomy in the model. One of the questions raised by LBK livestock is its importance within the farming system, which can be expressed by the ratio of the number of inhabitants in the hamlet to the number of heads of livestock. The initial ratio between humans and heads of livestock was set relatively to cattle, at $b = 0.5$ cattle/humans (Ebersbach and Schade 2004), followed by 0.3 goats and sheep and 0.1 pig. Sheep/goats and pigs vary in the same proportion as b. For a sensitivity analysis, the idea is to vary b by a 0.5–2.0 interval, as the zooarchaeological remains support the idea of relatively high consumption of cattle.

The economic and social system

Agriculture is organised into permanent small fields, which are created by slashing and burning forests. The system combines pulses, cereals and flax, requiring high labour input. Manure is important, cereals are sown in winter and/or spring and cereals and pulses are grown together. The agricultural operations conducted by farmers are simple sequences constrained by available manpower (producers). The whole forms an intensive farming system (Bogaard 2004, Bogaard *et al.* 2011).

Herds are managed by hamlets. Ruminants graze and browse in meadows and forest pastures, and are possibly given fodder during the winter; pigs are fed on household refuse and acorns. The animals are bred for their meat and their milk. The manure is collected. Meadows were rare so the animals are mainly fed on forest pruning, crop and domestic residues. Finally, hunting and gathering are seasonal activities that depend on the relative abundance of game, nuts (acorns and hazelnuts) and forest fruits according to season. Crops and livestock are sensitive to climatic variations. These positively or negatively affect the capacity of different soils to support vegetation. This 'fertility factor' oscillates, depending on the season on the climatic curve, between hydromorphy and drought according to the thresholds in the T/P ratio, which are determined by comparisons with hydrological estimations/simulations. Yields from cereal crops and herbaceous plants in general therefore vary as a proportion of these variations in 'fertility'. Beyond this routine situation, extreme events are considered: very wet and cold autumns, icy and very snowy winters, excessively wet or dry springs, very hot and dry or unusually wet summers. By causing serious damage at various stages of the vegetation cycle, these seasonal events can compromise harvests and even the natural vegetation available; excessive run-off or frost destroys seedlings or delays growth, soils and surface waters dry out, crops suffer from stem lodging or dehydration, seeds and fruits rot. Beyond routine variations, these meteorological events, which can seriously affect the survival of humans and animals, are simulated by setting the seasonal temperature and/or rainfall beyond the limits required by the biological characteristics of plants and defined by historical investigations (Le Roy Ladurie 2004, Le Roy Ladurie *et al.* 2003). The accumulation of successive unfavourable seasons can thus simulate the onset of historically known food shortages and famines. The parameter-values for seasonal crops (cereals and pulses), domesticated animals (cattle, sheep/goats, pigs), variations in yields, food needs, availability of herbaceous plants, winter fodder and acorns, game and wild plants, sensitivities to climate, working time for the various operations, returns from crops, meat and dairy products, all required a major ethno-historical documentation effort, which is beyond the scope of this chapter.

Regarding demographics, the number of nuclear family members is determined by the input-output, via the marriage system, of marriages, births and deaths, a set of age-specific fertility rates and a life table. The shape of the age-specific fertility table was derived from the average fertility curve for ten German villages with pre-1850 natural fertility (Knodel 1978, 491), but the average TFR was set to the LBK estimate of 6.6 children,[1] via a palaeodemographic estimator, using the ${}_{15}p_5$ indicator for LBK cemeteries (Bocquet-Appel 2002). The life table is the average of 15 European pre-industrial populations in northern and central Europe (1625–1780: Bocquet and Masset 1977).[2] This produces a life expectancy at birth of 26.7 years, and a life expectancy of 35.1 at 20 years of age. LBK marriages are not strictly patrilocal as commonly asserted (Bentley *et al.* 2012), but rather bilinear with a patrilocal trend, as shown by the strontium data. Furthermore, new nuclear families periodically split off from the founding family. These periodic splits, and the resulting house density in the hamlet and local density of hamlets, are determined by scalar stress (see below).

Chayanov's scalar stress as an expansion mechanism

Originally, the LBK geographical area was mainly covered in relatively open mixed oak primary forest (Kreuz 2008). Compared to pre-industrial European agricultural society, which totalled 140 million inhabitants in 1750 (Dupâquier 1998, 10), one can be sure that the demographic density of the agricultural society in the LBK area was very low. Assumption number one is that the LBK geographic expansion, given

[1] With the palaeodemographic indicator ${}_{15}p_5$ LBK: 298.41 / (298.41 + 886.61) = 0.25181, from 18 LBK cemeteries: Aisne series, Aiterhofen Ödmühle, Bad Württemberg, Bruchstedt, Nitra, Nordhausen, Octrois, Rutzing Haid, Sammelserie, Schönstedt, Sondershausen, Stuttgart, Trebur, Vedrovice, Vikletice, Vilnyanka, Wandersleben, Zengővárkony (see Bocquet-Appel 2002, 641). Total Fertility Rate = average of 6.6 children per female (Bocquet-Appel and Naji 2006, 356).

[2] Fifteen pre-industrial Life-tables: Susmilch 1765, Halley 1694, Norwich 1741–69, Northampton 1741–69, London 1759–68, Dupré de St-Maur 1774, Sweden 1755–57, 1761–63, France 1740–49, 1750–59, 1760–69, 1770–79, 1780–89, Geneva 1625–84 class II,III (Bocquet and Masset 1977).

the small population size, was not determined by a shortage of land limiting the carrying capacity of the production system. If ideological reasons for this expansion, if any, are set aside, reasons must be sought in another direction. Since the 1950s, several anthropologists have drawn attention to fissions in elementary population units when their size increases, even at a population density well below the carrying capacity of their production system (for a summary, see Bandy 2004). The proximate cause of these fissions was assigned to the flow of information exchanges between individuals, which increases exponentially with a linear increase in the group size, generating commensurate psychological tension among its members.

This tension is caused by an increase in disputes and a corresponding decrease in consensus, at each successive level of socio-geographical integration of an ethnic group. This psychological tension has been variously called Rappaport's irritation coefficient (Rappaport 1968) and scalar stress (Carneiro 1967, 1978; Johnson 1982), which is the term used in the remainder of this chapter. But the primary cause, as determined by the information flow, appears to be so general as to become an abstraction: beyond the sheer size of a group, there is its underlying population structure and the sensitivity of this structure to the economic constraints that condition its survival and reproduction, generating internal conflicts.

In a nuclear family in an agrarian community, a specific and long-identified source of tension is the pressure of number of consumers (*C*) on the workload of producers (*P*) (Chayanov 1923). The larger the number of consumers (*C*), the greater the amount of work that producers must provide, causing pressure on producers that generates the psychological strain of the so-called 'scalar stress' among members of a population unit. A simple demo-economic indicator expressing the economic and psychological pressure is given by the number of consumers *C* – all members of a nuclear family – relative to the number of producers *P* – individuals aged 15 +, i.e. *C/P* (Chayanov 1923). Over the 40–50 year duration of a nuclear family, this demo-economic indicator varies in the same direction as the number of consumers. The typical variation is as follows. When a nuclear family is started by the founding parents, it consists of the two parents, who are both economic producers and consumers; Chayanov's ratio R_c, is then $R_c = C/P = 2/2 = 1$. This indicates that there are as many consumers as producers and very low economic and psychological tension, given the reference value of 1. After a few years, with births and deaths of children and the presence of two to three simultaneously surviving children, who are also consumers, the value of R_c becomes 5/2 = 2.5. Compared to the initial value of $R_c = 1$, the new value indicates a workload 2.5 times higher for the producers, with an increase in the corresponding psychological pressure. As the surviving children become adolescents, R_c becomes (two parents plus two teens) = 4/4 =1, indicating that consumer pressure on producers has returned to its initial value. We see, therefore, that in the 40–50 years of the typical duration of a nuclear family, pressure on the producers does not remain constant but varies during its demographic history from low to high to low, which can be summarised by an *n* curve.

the attractiveness of a pixel from the point of view of the LBK was established, via an ordinary least square regression of the transformed variables of $f_x^{0.5}$ and $exp(F)$ on the dummy variables LBK settlement (presence/absence = 1/0) = $a + b_1 f_x^{0.5} + b_2\, exp(F)$, where a, and the b's are the regression coefficients. The entire set of scores for the 1.9 million pixels (standardised between 0–1), forms what is known as a surface of 'best patch' (Colour Plate 2).

Dislocation of the social system under environmental stress

A typical pattern of responses to famine has been formalised, mainly by Corbett (1988), Cutler (1984) and den Hartog (1981). To summarise, the pattern is expressed by: (1) high mobility of the population (especially males) seeking to collect, barter, earn, beg (and steal) food; and (2) the gradual disintegration of all forms of previously existing community support and family solidarity (Vaughan 1985). The sequence of responses to famines in the ethno-historical data has been incorporated as a partial model, decontextualised from any centralised supra-regional organisation and from the market economy. This partial model includes, in particular, a network of family support up to the second degree of kinship (uncle/aunt). It provides an approximation of the behaviour of early Neolithic farming households dealing with the scourge of famine (Bocquet-Appel 2012).

Modelling dynamics

The main rules and constraints of the model, which determine its operation, are as follows.

For each season in a crop year, all environmental information (climate, vegetation, soil fertility), sociological information (demographic input/output in the extended family households living in LBK houses) and information on the agricultural production system (agriculture, livestock, hunting and gathering) is updated in sequence (e.g. natural variables at the geographic coordinates x, y at time t: rainfall *(x, y, t)* and temperature *(x, y , t)*, determining the fertility *(x, y, t)* of geological soils, which, besides climate, affects agricultural production in the best patch; plus creation of random events in the landscape *(x, y, t)* such as storm or fire damage to forests). The sequential updating of all these variables directs the responses that agents must make to the actions they must perform.

A household agent must:

- ensure economic production (agricultural tasks, care for livestock, hunting and gathering) to meet the household's energy needs (Kcal), which are determined by its demographic structure;
- lead livestock to where it needs to be in the landscape to ensure its survival and reproduction and ensure an energy gain for the human agents (houses) in the

hamlet. All this has an impact on the demographic distribution of the herd, as determined by the probability of natural death, which depends on forage production and climate, and by probabilities of slaughter of the mini-herd for household consumption;

- ensure reproduction of the population (by finding spouses);
- cope with Chayanov's scalar stress C/P, possibly by splitting up the household. If this occurs through the departure of a nuclear family in the autumn season, then either this new nuclear family, which determines the construction of a new house, will densify the hamlet (under a density constraint expressing the effect of the scalar stress, formalised by a Poisson distribution), or it will be involved in colonising/densifying a region, within a circle of information of radius *c.* (say *c.* = 150km). Within this circle of information, the new nuclear family settles in the pixel most favourable to agricultural development ('best patch', see above), in the same colonisation direction as the founding parents;
- respond, if necessary, to a variety of typical climatic crises, either stochastically local or systematic on a larger geographical scale (regional to subcontinental), that have generated famines in experimental historical contexts in the Paris Basin and the northern half of France (Le Roy Ladurie *et al.* 2003, Le Roy Ladurie 2004).

The LBK population 'grows and multiplies', densifying existing hamlets or establishing new hamlets in the geographical directions and areas of the zones most favourable to the agricultural system determined by the 'best patch' areas, via Chayanov's scalar stress. The LBK population does not respond well to minor environmental changes in precipitation (x, y, t) and temperature *(x, y, t)* (and their chain of consequences for the economic system). The LBK society expanded geographically, driven by population growth and regulated according to the social rules already described. The colonisers moved from a pioneer front to settle territories, extending directionally their area of expansion over the best patch and densifying the area already occupied.

In the simulations/experiments stage now in progress, the long-term trends modelled are compared with LBK archaeological data and several scenarios for its expansion and eventual collapse are tested. These scenarios include and combine the long-term impacts of climate and agronomic constraints, together with assumptions about crisis contagion through the collapse of support and exchange networks or panic migration in the event of famine.

Concluding remarks

The main purpose of the Obresoc programme is to test the degree of resilience of the LBK society to major climatic changes. A cross-analysis of data is not sufficient, in itself, to understand the reasons for collapse of cultures and that of the archaeological societies underlying them. Because of the obscurity of ancient times, it has become usual to describe changes in the archaeological sequences by comparing batteries of items. But assumptions about the causes of changes in cultural phases that may be attributed to functions or dysfunctions in these ancient societies quickly come to a dead end, as there is no easy way to assess their impacts. Because tools are lacking to inject these hypotheses easily into the archaeological data, rather than attempting to understand the beginnings and endings of archaeological societies at the turning points of sequence changes, the main focus has been on the recognition sequence, so that the inter-sequences are left in the obscure areas of archaeology. The second issue addressed by this research is to understand how an archaeological society functioned, up to its end – considered as a sequence change – by injecting hypothetical or controlled and decontextualised ethno-historical scenarios by means of the multi-agent modelling.

Acknowledgements

Thanks to Mehdi Saqalli and Walter Finsinger for their participation in the Obresoc. The Obresoc project is funded by the French National Research Agency (ANR-09-CEP-004-01/OBRESOC).

References

Alamgir, M. 1981. An approach towards a theory of famine. In J.R.K. Robson (ed.), *Famine: its causes, effects, and management*, 19–40. New York: Gordon and Breach.

Arbogast, R., Jeunesse, C. and Schibler, J. (eds) 2001. *Rolle und Bedeutung der Jagd während des Frühneolithikums Mitteleuropas (Linearbandkeramik 5500–4900 v.Chr.)*. Rahden: Marie Leidorf.

Axtell, R.L., Epstein, J.M., Dean, J.S., Gumerman, G.J., Swedlund, A.C., Harburger, J., Chakravarty, S., Hammond, R., Parker, J. and Parker, M. 2002. Population growth and collapse in a multiagent model of the Kayenta Anasazi in Long House Valley. *Proceedings of the National Academy of Sciences of the United States of America* 99, 7275–9.

Bakels, C.C. 1978. *Four Linearbandkeramik settlements and their environment: a paleoecological Study of Sittard, Stein, Elsloo and Hienheim*. Analecta Praehistorica Leidensia 11.

Bandy, M.S. 2004. Fissioning, scalar stress, and social evolution in early village societies. *American Anthopologist* 106, 322–33.

Bánffy, E. and Oross, K. 2010. The earliest and the early phase of the LBK in Transdanubia. In D. Gronenborn and J. Petrasch (eds), *The spread of the Neolithic to Central Europe*, 255–72. Mainz: Römisch-Germanisches Zentralmuseum.

Bell, B. 1970. The dark ages in ancient history. I. The first dark age in Egypt. *American Journal of Archaeology* 75,1–26.

Bentley, R.A., Bickle, P., Fibiger, L., Nowell, G., Dale, C., Hedges, R., Hamilton, J., Wahl, J., Francken, M., Grupe, G., Lenneis, E., Teschler-Nicola, M., Arbogast, R.-M., Hofmann, D. and Whittle, A. 2012. Community differentiation and kinship among Europe's first farmers. *Proceedings of the National Academy of Sciences of the United States of America* 109, 9326–30.

Bocquet, J.P. and Masset, C.L. 1977. Estimateurs en paléodémographie. *L'Homme* 18, 65–90.

Bocquet-Appel, J.P. 2002. Paleoanthropological traces of a Neolithic demographic transition. *Current Anthropology* 43, 637–50.

Bocquet-Appel, J.P. 2012. Réponses aux famines des foyers domestiques agricoles des données ethno-historiques et modélisation archéologique. Manuscript Obresoc, unpublished.

Bocquet-Appel, J.P. and Naji, S. 2006. Testing the hypothesis of a worldwide Neolithic demographic transition. Corroboration from American cemeteries (with comments). *Current Anthropology* 47, 341–65.

Bogaard, A. 2004. *Neolithic farming in Central Europe: an archaeobotanical study of crop husbandry practices.* London: Routledge.

Bogaard, A., Krause, R. and Strien, H.-C. 2011. Towards a social geography of cultivation and plant use in an early farming community: Vaihingen an der Enz, south-west Germany. *Antiquity* 85, 395–416.

Bogucki, P. 2001. Recent research on early farming in central Europe. *Documenta Praehistorica* 28, 85–97.

Carey, M. 2012. Climate and history: a critical review of historical climatology and climate change historiography. *WIREs Climate Change* 3, 233–49.

Carneiro, R.L. 1967. On the relationship between size of population and complexity of social organization. *Southwestern Journal of Anthropology* 23, 234–43.

Carneiro, R.L. 1978. Political expansion as an expression of the principle of competitive exclusion. In R. Cohen and E.R. Service (eds), *Origins of the state: the anthropology of political evolution*, 205–23. Philadelphia: ISHI.

Chayanov, A.V. 1923. *The theory of peasant economy.* Berlin: P. Parey (and Madison: University of Wisconsin Press, 1986).

Childe, V.G. 1929. *The Danube in prehistory.* Oxford: Clarendon Press.

Cohen, M.N. 1977. *The food crisis in prehistory: overpopulation and the origins of agriculture.* New Haven: Yale University Press.

Corbett, J. 1988. Famine and household coping strategies. *World Development* 16, 1099–1112.

Cutler, P. 1984. Famine forecasting: prices and peasant behavior in Northern Ethiopia. *Disaster* 8, 48–56.

De Castro, J. 1952. *Geography of hunger.* Boston, MA: Little, Brown and Company.

den Hartog, A.P. 1981. Adjustment of food behavior during famine. In J.R.K. Robson (ed.), *Famine: its causes, effects and management*, 155–62. New York: Gordon and Breach.

Devereux, S. 2000. *Famine in the twentieth century.* IDS working paper 105.

Dirk, R. 1980. Social responses during severe food shortages and famine. *Current Anthropology* 21, 21–44.

Dubouloz, J. and Gauvry, Y. 2012. Typochronologie générale de la LBK. Manuscript Obresoc, unpublished.

Dupâquier, J. 1998. Introduction. In P. Bardet and J. Dupâquier (eds), *Histoire des populations de l'Europe, volume 2*, 8–17. Paris: Arthème Fayard.

Ebersbach, R. and Schade, C. 2004. Modeling the intensity of linear pottery land use: an example from the Mörlener Bucht in the Wetterau Basin, Hesse, Germany. In Stadtarchäologie Wien (ed.), *Enter the past: the e-way into the four dimensions of cultural heritage*, 337–48. Oxford: Archaeopress.

Gibbon, E. 1776–1789. *The history of the decline and fall of the Roman Empire* (vol. I, 1776; vols II, III, 1781; vols IV, V,VI, 1788–1789). London: Strahan & Cadell.

Grolle, J. 1997. Heavy rainfall, famine, and cultural response in the West African Sahel: the 'Muda' of 1953–54. *GeoJournal* 43, 205–14.

Gronenborn, D., Strien, H.C., Dietrich, S. and Sirocko, F. 2013. 'Adaptive cycles' and climate fluctuations: a case study from Linear Pottery Culture in western Central Europe. *Journal of Archaeological Science* Online first.

Hachem, L. 2011. *Le site néolithique de Cuiry-lès-Chaudardes, 1. De l'analyse de la faune à la structuration sociale*. Rahden: Marie Leidorf.

Hammel, E.A. 2005. Chayanov revisited: a model for the economics of complex kin units. *Proceedings of the National Academy of Sciences of the United States of America* 102, 7043–6.

Hassig, R. 1981. The famine of one rabbit: ecological causes and social consequences of a pre–Columbian calamity. *Journal of Anthropological Research* 77, 172–82.

Janssen, M.A. 2009. Understanding artificial Anasazi. *Journal of Artificial Societies and Social Simulation* 12, 13.

Johnson, G.A. 1982. Organizational structure and scalar stress. In C. Renfrew, M. Rowlands and B.A. Segraves-Whallon (eds), *Theory and explanation in archaeology*, 389–431. London: Academic Press.

Knodel, J. 1978. Natural fertility in pre-industrial Germany. *Population Studies* 32, 481–510.

Kohler, T.A. and van der Leeuw, S.E. 2007. *The model-based archaeology of socionatural systems*. Santa Fe: School for Advanced Research Press.

Kolle, H. 1995. *The Russian post-emancipation household: two villages in the Moscow area*. Bergen: Historisk institut.

Kovačiková, L., Bréhard, S., Šumberová, R., Balasse, M. and Tresset, A. 2012. New insights into the subsistence and early farming from Neolithic settlements in Central Europe: archaeozoological evidence from the Czech Republic. *Archaeofauna* 21, 71–97.

Kreuz, A. 2008. Closed forest or open woodland as natural vegetation in the surroundings of Linearbandkeramik settlements? *Vegetation History and Archaeobotany* 17, 51–64.

Kreuz, A., Marinova, E., Schäfer, E. and Wiethold, J. 2005. A comparison of early Neolithic crop and weed assemblages from the Linearbandkeramik and the Bulgarian Neolithic cultures: differences and similarities. *Vegetation History and Archaeobotany* 14, 237–58.

Le Roy Ladurie, E. 2004. *Histoire humaine et comparée du climat*. Paris: Librairie Arthème Fayard.

Le Roy Ladurie, E., Rousseau, D. and Vasak, A. 2003. *Les fluctuations du climat de l'an Mil à aujourd'hui*. Paris: Fayard.

Lüning, J. 1988. Frühe Bauern in Mitteleuropa im 6. und 5. Jahrtausend v. Chr. *Jahrbuch des Römisch-Germanisches Zentralmuseums Mainz* 35, 27–93.

Pavúk, J. 2004. Early Linear Pottery culture in Slovakia and the neolithisation of central Europe. In A. Lukes and M. Zvelebil (eds), *LBK dialogues: studies in the formation of the Linear Pottery culture*, 71–82. Oxford: Archaeopress.

Pelzer, K.J. 1945. *Pioneer settlement in the Asiatic tropics*. New York: Institute of Pacific Relations.

Polla, M. 2003. Peasant and hunter households in Oulanka, northern Russia, 1710–1910. *The History of the Family* 8, 163–81.

Potts, R. 2012. Environmental and behavioral evidence pertaining to the evolution of Early Homo. *Current Anthropology* 53, S299–S317.

Rappaport, R.A. 1968. *Pigs for the ancestors: ritual in the ecology of a New Guinea People*. New Haven, CT: Yale University Press.

Robbins, R.G. 1975. *Famine in Russia 1891–1892*. New York: Columbia University Press.

Salavert, A. 2010. *Apport de l'archéobotanique à la compréhension des sociétés néolithiques. Analyses anthracologiques et carpologiques de neuf sites rubanés de Moyenne-Belgique (5200–5000 av. J.-C.)*. Unpublished thèse de doctorat, Université de Paris 1 Panthéon-Sorbonne, Institut Royal de Sciences Naturelles de Belgique.

Sanders, W.T., Parsons, J.R. and Santley, R.S. 1979. *The basin of Mexico: ecological processes in the evolution of a civilization*. New York: Academic Press.

Schmitzberger, M. 2009. *Haus- und Jagdtiere im Neolithikum des österreichischen Donauraumes*. Unpublished PhD thesis, Universität Wien.

Szoltysek, M. 2008. Three kinds of preindustrial household formation system in historical Eastern Europe: a challenge to spatial patterns of the European family. *History of the Family* 13, 223–57.

Vaughan, M. 1985. Famine analysis and family relations: 1949 in Nyasaland. *Past and Present* 108, 177–205.

Wrangham, R. and Carmody, R. 2010. Human adaptation to the control of fire. *Evolutionary Anthropology* 19, 187–99.

5

Ancient DNA Evidence for a Homogeneous Maternal Gene Pool in Sixth Millennium cal BC Hungary and the Central European LBK

ANNA SZÉCSÉNYI-NAGY, VICTORIA KEERL, JÁNOS JAKUCS, GUIDO BRANDT, ESZTER BÁNFFY AND KURT W. ALT

With contributions by Kitti Köhler, Balázs Mende, Viktória Kiss, Krisztián Oross and Tibor Marton (Institute of Archaeology, Research Centre for the Humanities, Hungarian Academy of Sciences); György Pálfi, Erika Molnár and Antónia Marcsik (Institute of Anthropology, University of Szeged); Katalin Sebők (Institute of Archaeology, University of ELTE, Budapest); András Czene, Ágnes Kovács and Róbert Patay (Directorate of Pest County Museums); Rozália Kustár (Viski Károly Museum, Kalocsa); Krisztina Somogyi and Tibor Paluch (Ásatárs Ltd); László Domboróczki (Dobó István Castle Museum, Eger); and Zsuzsanna K. Zoffmann

Abbreviations

aDNA	Ancient DNA
rCRS	revised Cambridge reference sequence
HVS	Hyper variable segment of the mitochondrial genome
LBKS	*Linearbandkeramik* culture in Moravia and Lower Austria
LBKSW	*Linearbandkeramik* culture in south and south-west Germany
LBKC	*Linearbandkeramik* culture in central Germany
LBKT	*Linearbandkeramik* in western Hungary/Transdanubia
MDS	Multidimensional scaling
mt	Mitochondrial
PCA	Principal component analysis
SHA	Shared haplotype analysis
SZA	Szakálhát culture

Proceedings of the British Academy **198**, 71–93.

Introduction

For large parts of Europe, the *Linearbandkeramik* culture (LBK) was the first to introduce the Neolithic technologies of agriculture, animal husbandry and a sedentary lifestyle. The LBK originated around 5600/5500 cal BC in Transdanubia, western Hungary, presumably from the preceding Starčevo culture (Oross and Bánffy 2009). Shortly after its formative phase, the LBK expanded northwards and westwards, rapidly reaching as far as central Germany (Bánffy 2004; Gronenborn 1999, 2007). Contemporary to the LBK in Transdanubia (LBKT), a similar culture formed on the Great Hungarian Plain (Alföld): the Alföld LBK. That was rooted in the Körös-Criş culture, which in turn was mostly contemporaneous with the early Neolithic Starčevo culture in Transdanubia (Whittle *et al.* 2002, Oross and Siklósi 2012). Both Körös-Criş and Starčevo cultures have their origin in the Balkans; they may have differentiated out of the same south-east European Early Neolithic substrate. In contrast to its Transdanubian relative, the Alföld LBK never expanded beyond the Great Hungarian Plain. However, in its younger period it began to differentiate into several regional cultural units. As a result of this process, the Szakálhát culture (SZA) emerged in the southern and central part of the Alföld, parallel to the Notenkopf and Zseliz phases of the LBKT in Transdanubia (M. Virág 2009). The Szakálhát culture was in many senses a transitory unit between Alföld LBK and the Late Neolithic of the Balkans, spreading from the southern Alföld along the Tisza towards the north. Typical southeastern elements like the tell settlement structure started to flourish along with some characteristic north Balkan (Vinča culture) elements (Horváth 2006). Archaeological evidence indicates some contacts between the people of the Szakálhát culture and those of the late LBK in Transdanubia (Hegedüs 1985, M. Virág 2009), yet up till now there has been hardly any investigation of their possible genetic links.

There have been some ancient DNA (aDNA) studies on central European LBK remains. Haak *et al.* (2005) analysed a total of 24 LBK individuals from present-day Germany, Austria and Hungary, and a further 21 individuals from Derenburg in Saxony-Anhalt, central Germany (Haak *et al.* 2010). The N1a haplogroup, which was found in high frequencies at the sites studied, is very rare in modern Europe, and the first conclusion drawn from these findings was that early Neolithic farmers did not have much impact on the maternal gene pool of Europe (Haak *et al.* 2005). N1a as a 'signature haplogroup' of the LBK was later supplemented by other mitochondrial haplogroups, representing a genetic 'Neolithic package', including the haplogroups T2, J, K, V and HV (Haak *et al.* 2010, Brandt *et al.* 2013). This stands in marked contrast to the maternal gene pool of pre-Neolithic central European hunter-gatherer populations, which are dominated by the U4, U5 and U8 subhaplogroups (Bramanti *et al.* 2009, Fu *et al.* 2013)

For a more accurate assessment of genetic continuity or discontinuity at the beginning of the LBK, and in order to detect possible genetic differentiation between geographic regions of the LBK dissemination area, the Transdanubian LBK, which was influential in the further LBK development and expansion in western Europe, has to be compared with the central European LBK aDNA dataset (Haak *et al.* 2005, Bramanti 2008, Haak *et al.* 2010, Brandt *et al.* 2013). Furthermore, the genetic composition of the Alföld LBK is largely unknown, since only one authentic result from the Alföld LBK has been published to date (Haak *et al.* 2005). Therefore the aim of this study was to examine the central European Neolithic using new data from these two Hungarian regions in order to investigate the formation of LBK mitochondrial variability and to gain a more complex insight into possible genetic regionality within the LBK.

Samples were taken from multiple Middle Neolithic sites of the Transdanubian and Alföld regions of Hungary, and analysed for their mitochondrial DNA (mtDNA) variability. Results from the Szakálhát culture and from the LBK in Transdanubia were compared with previously published ancient DNA datasets, using population genetic methods.

Materials and methods

We selected and sampled 41 skeletons from six sites of the Szakálhát culture on the Great Hungarian Plain and 61 skeletons from eight LBK sites in Transdanubia (Figure 5.1; Table 5.1).

Sample preparation, DNA extraction, amplification of the mitochondrial DNA, cloning and sequencing were carried out in the aDNA laboratories of the Institute of Anthropology at the Johannes Gutenberg University of Mainz, following our standard protocols (Brandt *et al.* 2013).

We genotyped individuals by sequencing the hyper-variable segment I and II (HVS-I/II) of the control region, using our previously published protocols (Szécsényi-Nagy *et al.* in prep., Brandt *et al.* 2013). Twenty-two single-nucleotide coding region polymorphisms of the mitochondrial genome were additionally analysed, following the protocol of Haak *et al.* (2010).

For comparative population genetic analyses, we used ancient mtDNA data from the archaeogenetic literature. The published LBK sites were subdivided into groups based on their geographic location, in order to detect possible regional differences in the LBK maternal gene pool: a southern LBK group (LBKS, n = 7) containing Lower Austrian and Czech (Moravian) sites (Haak *et al.* 2005, Bramanti 2008); a south-western LBK group (LBKSW, n = 12), containing sites from south-west Germany (Haak *et al.* 2005); and a LBK group in central Germany (LBKC, n = 89), containing sites of the Middle-Elbe-Saale region (Haak *et al.* 2010, Brandt *et al.* 2013). The analysis also involved mtDNA data from a succession of four cultures (Rössen: n = 11; Schöningen: n = 33; Baalberge: n = 19; and Salzmünde: n = 29) from the fifth–fourth millennia cal BC in central Germany (Brandt *et al.*

Table 5.1 Continued

	Culture/Phase	*Site name*	*Sum of LBK/SZA skeletons*	*Sampled for aDNA*	*Year of excavation*	*Grave characteristics*	*Archaeological ref.*
5.	LBK/Zseliz	Harta-Gátőrház	6	5	2002–2003	Four of the six uncovered crouched burials were found in a separate grave, while two of them were buried into abandoned clay pits. Three of the deceased were provided with grave goods.	Kustár 2003, Somogyvári 2008
6.	LBK/Zseliz	M85 Enese elkerülő 02. Kóny, Proletár-dűlőII	5	4	2008	Four of the five burials were crouched, laid on the left side (in feature 612/647 a double child burial was found), another adult deceased was in an extended position on their back. Two burials were provided with grave goods.	Varga 2010, Oross–Marton 2012, for 14C dates see Szécsényi-Nagy *et al.* in prep.
7.	LBK/Keszthely	Szemely-Hegyes	3	2	2006–2007	Two burials could be securely assigned to the LBK based on radiocarbon dates and a third (not sampled) on the evidence of the grave goods.	unpublished, T. Paluch personal comm., for 14C dates see Szécsényi-Nagy *et al.* in prep.
8.	LBK/Keszthely	Tolna-Mözs	2	2	2008–2009	In view of the grave goods and radiocarbon measurement two graves could be securely assigned to the LBK period. A left crouched burial was	Marton–Oross 2012, for 14C dates see Szécsényi-Nagy *et al.* in prep.

geographic regions of the LBK dissemination area, the Transdanubian LBK, which was influential in the further LBK development and expansion in western Europe, has to be compared with the central European LBK aDNA dataset (Haak *et al.* 2005, Bramanti 2008, Haak *et al.* 2010, Brandt *et al.* 2013). Furthermore, the genetic composition of the Alföld LBK is largely unknown, since only one authentic result from the Alföld LBK has been published to date (Haak *et al.* 2005). Therefore the aim of this study was to examine the central European Neolithic using new data from these two Hungarian regions in order to investigate the formation of LBK mitochondrial variability and to gain a more complex insight into possible genetic regionality within the LBK.

Samples were taken from multiple Middle Neolithic sites of the Transdanubian and Alföld regions of Hungary, and analysed for their mitochondrial DNA (mtDNA) variability. Results from the Szakálhát culture and from the LBK in Transdanubia were compared with previously published ancient DNA datasets, using population genetic methods.

Materials and methods

We selected and sampled 41 skeletons from six sites of the Szakálhát culture on the Great Hungarian Plain and 61 skeletons from eight LBK sites in Transdanubia (Figure 5.1; Table 5.1).

Sample preparation, DNA extraction, amplification of the mitochondrial DNA, cloning and sequencing were carried out in the aDNA laboratories of the Institute of Anthropology at the Johannes Gutenberg University of Mainz, following our standard protocols (Brandt *et al.* 2013).

We genotyped individuals by sequencing the hyper-variable segment I and II (HVS-I/II) of the control region, using our previously published protocols (Szécsényi-Nagy *et al.* in prep., Brandt *et al.* 2013). Twenty-two single-nucleotide coding region polymorphisms of the mitochondrial genome were additionally analysed, following the protocol of Haak *et al.* (2010).

For comparative population genetic analyses, we used ancient mtDNA data from the archaeogenetic literature. The published LBK sites were subdivided into groups based on their geographic location, in order to detect possible regional differences in the LBK maternal gene pool: a southern LBK group (LBKS, n = 7) containing Lower Austrian and Czech (Moravian) sites (Haak *et al.* 2005, Bramanti 2008); a south-western LBK group (LBKSW, n = 12), containing sites from south-west Germany (Haak *et al.* 2005); and a LBK group in central Germany (LBKC, n = 89), containing sites of the Middle-Elbe-Saale region (Haak *et al.* 2010, Brandt *et al.* 2013). The analysis also involved mtDNA data from a succession of four cultures (Rössen: n = 11; Schöningen: n = 33; Baalberge: n = 19; and Salzmünde: n = 29) from the fifth–fourth millennia cal BC in central Germany (Brandt *et al.*

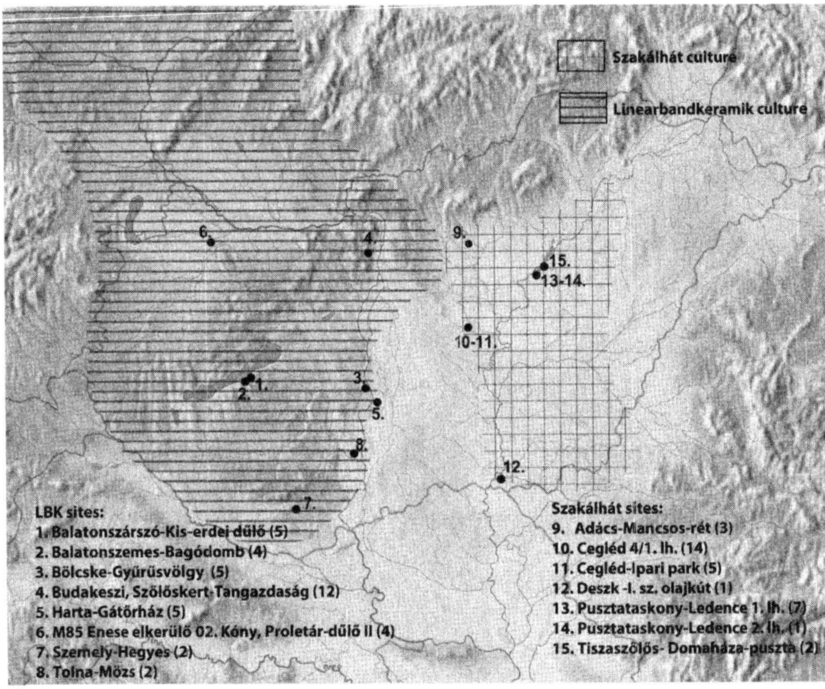

Figure 5.1 Map of the Carpathian Basin. The map shows the studied LBK and Szakálhát culture dissemination areas (based on Raczky and Anders 2003, and Oross and Bánffy 2009) and the sampled archaeological sites. The numbers of successfully typed skeletons are in brackets after the site names.

2013), and four third–second millennia cal BC cultures in central Germany (Bernburg: n = 17; Corded Ware: n = 44; Bell Beaker: n = 35; and Únětice: n = 94), which represent the later Neolithic and the Early Bronze Age of central Europe (Haak *et al.* 2008, Lee *et al.* 2012, Brandt *et al.* 2013). Three south-west European populations from the sixth–fifth millennia cal BC of Portugal (n = 18; Chandler *et al.* 2005), the Basque Country and Navarra (n = 43; Hervella *et al.* 2012), and Catalonia (n = 20; Lacan *et al.* 2011; Gamba *et al.* 2012) represent the Early Neolithic of the Iberian peninsula. Two Upper Palaeolithic/Mesolithic hunter-gatherer metapopulations, one from central and northern Europe (n = 14; Bramanti *et al.* 2009, Szécsényi-Nagy *et al.* in prep., Fu *et al.* 2013), and one from the Iberian peninsula (n = 14; Chandler *et al.* 2005, Hervella *et al.* 2012, Sánchez-Quinto *et al.* 2012), were also included in the analyses.

The single published and authentic Alföld LBK mtDNA result from Ecsegfalva 23 (Haak *et al.* 2005) was left out from the population genetic analyses, because in this study we were concentrating on a specific descendent culture of the Alföld LBK. The aDNA results of Guba *et al.* (2011) from the Great Hungarian Plain were

Table 5.1 Summary of the sampled archaeological sites from the Carpathian Basin

	Culture/Phase	*Site name*	*Sum of LBK/SZA skeletons*	*Sampled for aDNA*	*Year of excavation*	*Grave characteristics*	*Archaeological ref.*
1.	LBK/Bicske-Biňa, Keszthely, Zseliz	Balatonszárszó-Kis-erdei-dűlő	43	23	2001–2006	24 of the 43 excavated burials in an LBK settlement were left-crouched, seven bodies were crouched on the right side, two skeletons lain in extended position; seven burials were provided with grave goods.	Oross *et al*. 2004, Oross–Marton 2012
2.	LBK/Bicske-Biňa, Keszthely	Balatonszemes-Bagódomb	6	6	1999–2001	Six burials were excavated. The bodies were buried in crouched position. One deceased was provided with grave good (small stone blade), while an adult man was covered with ochre.	Bondár–Honti–Kiss 2000, Kiss 2002, Kiss–Sebők 2007
3.	LBK/Notenkopf, Zseliz	Bölcske-Gyűrűsvölgy	5	5	2008	Three skeletons were found on the edge of clay pit nr. 36, while two truncated skeletons came to light in a double grave; The body of a child was covered with vessel fragments, typical for the Notenkopf style.	Sebők 2009
4.	LBK/Notenkopf, Zseliz	Budakeszi, Szőlőskert-Tangazdaság	15	14	2006	15 burials were uncovered, most of them lain in a crouched position, without any grave goods. One deceased was provided with six different vessels, typical for the Notenkopf style.	Czene–Ottományi 2007, Czene 2008, for 14C dates see Szécsényi-Nagy *et al*. in prep.

Table 5.1 Continued

	Culture/Phase	*Site name*	*Sum of LBK/SZA skeletons*	*Sampled for aDNA*	*Year of excavation*	*Grave characteristics*	*Archaeological ref.*
5.	LBK/Zseliz	Harta-Gátőrház	6	5	2002–2003	Four of the six uncovered crouched burials were found in a separate grave, while two of them were buried into abandoned clay pits. Three of the deceased were provided with grave goods.	Kustár 2003, Somogyvári 2008
6.	LBK/Zseliz	M85 Enese elkerülő 02. Kóny, Proletár-dűlőII	5	4	2008	Four of the five burials were crouched, laid on the left side (in feature 612/647 a double child burial was found), another adult deceased was in an extended position on their back. Two burials were provided with grave goods.	Varga 2010, Oross–Marton 2012, for 14C dates see Szécsényi-Nagy *et al.* in prep.
7.	LBK/Keszthely	Szemely-Hegyes	3	2	2006–2007	Two burials could be securely assigned to the LBK based on radiocarbon dates and a third (not sampled) on the evidence of the grave goods.	unpublished, T. Paluch personal comm., for 14C dates see Szécsényi-Nagy *et al.* in prep.
8.	LBK/Keszthely	Tolna-Mözs	2	2	2008–2009	In view of the grave goods and radiocarbon measurement two graves could be securely assigned to the LBK period. A left crouched burial was	Marton–Oross 2012, for 14C dates see Szécsényi-Nagy *et al.* in prep.

							found, accompanied by a fragment of a vessel pedestal, *Spondylus* pendant and *Spondylus* bead, while the other left crouched skeleton came to light, without any grave goods.	
9.	Szakálhát	Adács-Mancsos-rét	4	4	2005		Four burials were uncovered within the Neolithic settlement. The deceased were buried among the settlement features, some of them had different type of *Spondylus* pendants and beads as grave goods.	Domboróczki 2005
10.	Szakálhát	Cegléd 4/1. lh	15	15	2003–2004		15 burials were uncovered within the Neolithic settlement. All of the deceased were crouched on the left side. Six of them had grave goods: *Spondylus* beads were found in four graves *Spondylus* discs in two graves.	Kovács 2011
11.	Szakálhát	Cegléd-Ipari park	10	10	2000		Ten burials of the late Szakálhát culture were uncovered. The bodies were crouched on left side in seven burials, on the right side in one, while the placement of two bodies was uncertain because only fragmented bones were found. Two of the ten burials were provided with vessels, and two of them had ochre.	Tari 2001

Table 5.1 Continued

	Culture/Phase	*Site name*	*Sum of LBK/SZA skeletons*	*Sampled for aDNA*	*Year of excavation*	*Grave characteristics*	*Archaeological ref.*
12.	Szakálhát	Deszk -I. sz. olajkút	1	1	1966	The burial in grave 5 could be assigned to the Szakálhát – Bucovăţ culture based on the radiocarbon data (6225±55 BP). The skeleton was crouched on the left side, laid on its back.	Trogmayer 1969, Whittle *et al.* 2002
13–14.	Szakálhát	Pusztataskony-Ledence 1-2. lh.	10	9	2010	Ten of the several uncovered burials dated to the Szakálhát culture. The graves were found within the Neolithic settlement.	unpublished, P. Raczky–A. Anders–K. Sebők personal comm.
15.	Szakálhát	Tiszaszőlős-Domaháza-puszta	2	2	2003	Two of the six uncovered burials are from the latest phase of the Szakálhát culture (late Szakálhát – early Tisza phase) based on the radiocarbon data. They laid on their right side, in crouched position.	Domboróczki 2010

also excluded from the analysis because of the misdating of some skeletons (Bánffy *et al.* 2012).

Population genetic comparisons were made based on mitochondrial haplogroup frequencies and HVS-I sequence data. We used a suite of population genetic analyses, including principal component analysis (PCA), multidimensional scaling (MDS) and shared haplotype analysis (SHA), in order to identify affinities of our Carpathian Basin samples within the mtDNA variability of European prehistoric populations. The details and settings of these analyses are presented in our previous papers (Brandt *et al.* 2013, Szécsényi-Nagy *et al.* in prep.).

The PCA was carried out based on the relative frequencies of 22 (sub)haplogroups (H, H5, HV, HV0, V, I, J, K, N, N1a, R, T1, T2, U, U2, U3, U4, U5a, U5b, U8, W and X), which were observed from the ancient samples, while the HVS-I sequences (nucleotide position 16056–16400) were used for the computation of pairwise genetic distances under consideration of the Tamura and Nei substitution model (Tamura and Nei 1993) and an associated gamma value of 0.177. The distances were visualised by MDS. In both analyses the SZA and LBKT datasets were compared with the 16 published cultural formations or groups detailed above. In order to exclude biases induced by potential maternal kinship within the datasets, which may have led to an overestimation of haplogroup frequencies and genetic affinities, we included reduced datasets – of two or more individual at a given sites sharing HVS-I sequences (putative maternal relatives), only one was counted (indicated with the * symbol) – in both PCA and MDS analyses. Redundant haplotypes with identical HVS-I and, if provided, HVS-II sequences that were found in several individuals at the same site, were counted only once in these datasets.

The proportion of matching haplotypes between the different LBK groups (LBKT, LBKC, LBKS and LBKSW) and the SZA was evaluated by SHA. In order to identify potentially Mesolithic lineages within the LBK groups, the northern and central European hunter-gatherer dataset was also included in the analyses. Haplotype matches were established by direct counting. In order to minimise effects of maternal kinship within the datasets, the proportion of shared haplotypes was reported instead of the proportion of individuals sharing haplotypes. Since archaeological evidence indicates that the Carpathian Basin and more specifically, the LBKT, was the starting point for the westwards expansion of the LBK, the central European LBK groups were examined for lineages which they shared with the LBKT, SZA, and a combination of the two as an approximation of a 'Carpathian Basin genetic substrate'.

Results

We successfully reproduced the mitochondrial HVS-I sequences of 39 LBKT specimens (63.9% success rate) and 33 SZA individuals (80.48% success rate).

The mitochondrial variability of the two sample sets is quite similar. The major characteristic haplogroups of the mtDNA 'Neolithic package' (Brandt *et al.* 2013) are present in both cultures: T2, K, J, V, HV and N1a (Table 5.2). Further common haplogroups of the two cultures from Hungary are T1 and H. T1 in central Germany was first detected in the Baalberge culture, but its consistent presence is described from the Late Neolithic Corded Ware culture onwards (Brandt *et al.* 2013). However, it might have been a signal of the first farmers in the Carpathian Basin, in accordance with modern population genetic studies (Richards *et al.* 2000, Pala *et al.* 2012). The T1 occurrence in the aDNA dataset of the early Neolithic Starčevo culture also supports its connection to the mtDNA 'Neolithic package' (Szécsényi-Nagy *et al.* in prep.).

The major part of H lineages in the LBKT (58.3%) and SZA (33.3%) are haplotypes concordant with the revised Cambridge Reference Sequence (rCRS), which could not be further resolved by HVS-I and not even by HVS-II sequences (Andrews *et al.* 1999, Van Oven and Kayser 2009). Although there are no Mesolithic hints of the haplogroup H from central or south-east Europe (as opposed to south-west and north-east Europe: Chandler *et al.* 2005, Hervella *et al.* 2012, Der Sarkissian *et al.* 2013), it might be considered as a hunter-gatherer component of the Neolithic gene pool in the Carpathian Basin. The extended use of new sequencing technology (next-generation sequencing), which can yield whole mitochondrial genomes, may further increase the resolution of haplogroup H in the future and help address this issue (Brotherton *et al.* 2013). Besides next-generation sequencing, the use of multiplex assays, targeting the most common haplogroup H clades (Martinez-Cruz *et al.* 2012) would also be a future goal for our project.

Disparities between the LBKT and SZA are the inconsistent presence of haplogroups U4, U2, U5a and W. The three U subhaplogroups might reflect genetic traces of a hunter-gatherer legacy, which only rarely resurface in the Early and Middle Neolithic Carpathian Basin. Population genetic studies based on present-day mitochondrial variability supposed an origin of haplogroup U in the Near East and a spread over Europe and western Asia by anatomically modern humans during the early Upper Palaeolithic (Richards *et al.* 2000, Fu *et al.* 2013). Haplogroups U2, U4 and U5 were also reported with varying frequencies from distant hunter-gather communities of south-west, central, northern and eastern Europe (Chandler *et al.* 2005, Bramanti *et al.* 2009, Krause *et al.* 2010, Hervella *et al.* 2012, Sánchez-Quinto *et al.* 2012, Fu *et al.* 2013, Der Sarkissian *et al.* 2013). Derivatives of the LBKT U5a and SZA U4 haplotypes were found in several hunter-gatherer communities of central and north-east Europe (Bramanti *et al.* 2009, Der Sarkissian *et al.* 2012). U2 has been present in eastern Europe since the Upper Palaeolithic (Krause *et al.* 2010, Der Sarkissian *et al.* 2012) and (re)emerged in central Germany in the course of population dynamic events towards the Early Bronze Age in the third and second millennia cal BC (Brandt *et al.* 2013). However, no exact match of the LBKT U2 haplotype has been found. We should not over-interpret the absence

Table 5.2 Haplogroup frequencies of the Szakálhát culture and the *Linearbandkeramik* culture in Transdanubia

Country and region	*Culture*	*Ref.*	*Abbr.*	*n*	*Mitochondrial haplogroup frequencies (%)*																
					H	*H5*	*HV*	*HV0*	*V*	*J*	*K*	*N1a*	*T1*	*T2*	*U2*	*U3*	*U4*	*U5a*	*U5b*	*W*	*X*
Hungary, Transdanubia	LBK culture	1	LBKT	39	25.6	5.1	2.6	0	2.6	7.7	12.8	10.3	2.6	25.6	2.6	0	0	2.6	0	0	0
Hungary, Alföld	Szakálhát culture	this study	SZA	33	12.1	6.1	0	3	6.1	15.2	15.2	3	6.1	27.3	0	0	3	0	0	3	0
Lower-Austria, Moravia	LBK culture	2, 5	LBKS	7	14.3	14.3	0	0	0	14.3	28.6	0	0	28.6	0	0	0	0	0	0	0
Southwest Germany	LBK culture	3	LBKSW	12	8.3	8.3	8.3	0	0	8.3	16.7	8.3	0	33.3	0	8.3	0	0	0	0	0
Central Germany	LBK culture	3, 4, 6	LBKC	89	15.7	0	4.5	0	5.6	12.4	20.2	13.5	0	20.2	0	0	0	2.2	1.1	3.4	1.1

References: 1: Szécsényi-Nagy *et al.* in prep; 2: Haak *et al.* 2005; 3: Haak *et al.* 2005; 4: Haak *et al.* 2010; 5: Bramanti 2008; 6: Brandt *et al.* 2013.

of the haplogroup W in the LBKT, since it appeared before in Transdanubia, in the Early Neolithic Starčevo culture (Szécsényi-Nagy *et al.* in prep.), and it occurs also in the other LBK groups (Brandt *et al.* 2013). Additionally, its frequency is low in the SZA and LBKC dataset, indicating an overall scattered Neolithic occurrence.

The haplogroup compositions were evaluated with PCA, in order to define haplogroup frequency based similarities between the studied cultural formations (Figure 5.2). The first component (PC1) on the plot separates the south-west and central and north European hunter-gatherer metapopulations and the Iberian Neolithic cultures from the central European and Hungarian sixth–fourth millennia cal BC datasets. The second component (PC2) divides the south-west Neolithic and hunter-gatherer datasets from the Late Neolithic cultures of central Europe (Corded Ware, Bell Beaker) and the Early Bronze Age Únětice culture. SZA and LBKT

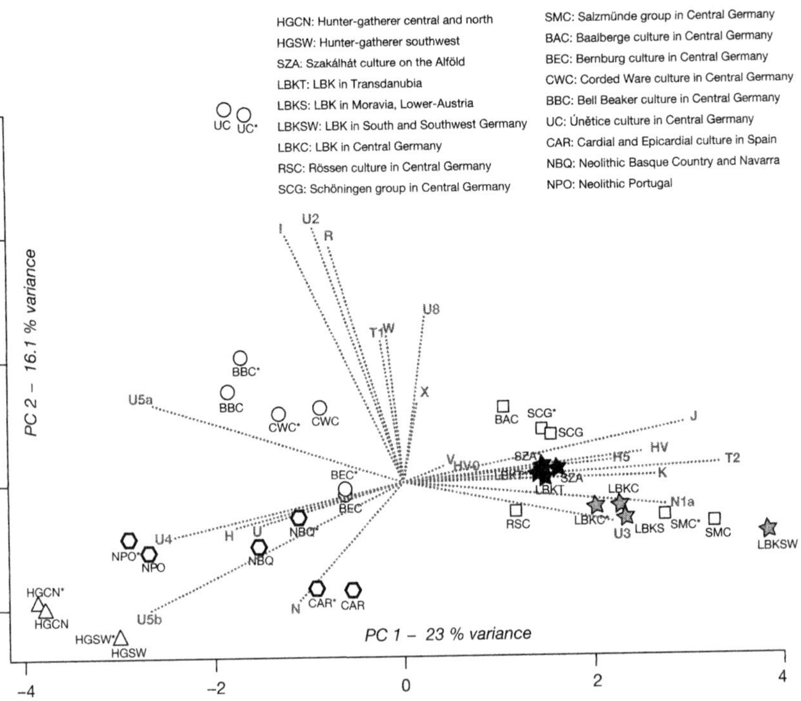

Figure 5.2 Principal component analysis of 18 prehistoric mtDNA datasets. Shadings and symbols denote populations of different periods or European regions: hunter-gatherer (triangles), Middle Neolithic Carpathian Basin cultures (black stars), central European LBK (grey stars) and fifth–fourth millennia cal BC cultures (rectangle), central German third–second millennia cal BC cultures (circles), sixth–fifth millennia cal BC cultures of the Iberian peninsula (hexagons). Each population is represented by a dataset containing all individuals and by a reduced set excluding potential relatives (*). The contribution of each haplogroup is superimposed as a grey component loading vector. The first two principal components together display 39.1% of the total genetic variation.

have an almost identical position on the PCA plot, resulting from the almost similar haplogroup composition. The LBK groups and the SZA form one larger group with the subsequent fifth–fourth millennia cal BC cultures of central Germany, indicating common ancestry and the continued LBK tradition of the latter ones (Brandt *et al.* 2013).

The MDS shows a similar picture to the PCA plot (Figure 5.3). The closeness of the LBK groups to each other, which is due to the small genetic distances (F_{st} values) among them, is striking on the MDS plot. The SZA is located in the cluster of the LBK groups, which is associated with the fifth–fourth millennia cal BC central German Neolithic cultures, supporting a maternal genetic affinity on HVS-I that extends to the sequence level. The two hunter-gatherer metapopulations, the Iberian Neolithic sample sets and the Late Neolithic/Early Bronze Age central German datasets are noticeably set apart from the LBK-SZA group along the first and second coordinates.

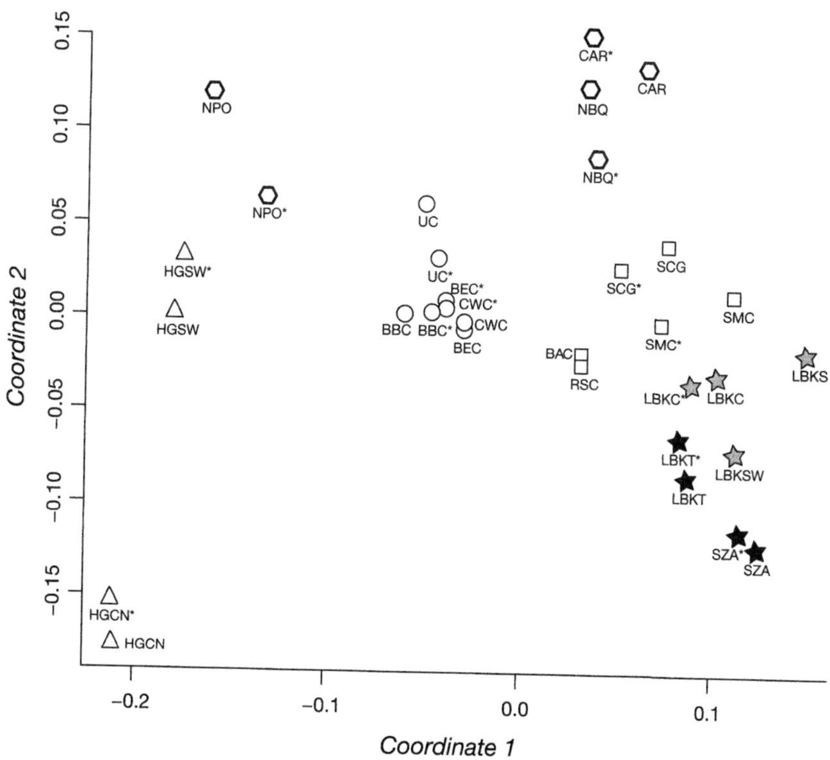

Figure 5.3 Multidimensional scaling of 18 prehistoric cultural formations. Genetic distances (F_{st}) between the STA and LBKT samples and 16 published prehistoric datasets were computed based on HVS-I sequences and visualised by multidimensional scaling (MDS) with a stress value of 0.1167. Shadings and symbols denote populations of different periods or European regions according to Figure 5.2. Each population is represented by a dataset containing all individuals and another, which did not contain potential maternal relatives (*). Culture abbreviations are the same as in Figure 5.2.

The SHA shows that SZA and LBKT share 43.5% of their haplotypes. In the more westerly LBK groups, the percentage of lineages shared with the LBKT ranges from 57.14% in LBKS to 38.24% in LBKC. The percentage of lineages shared with SZA lies between 41.18% in LBKC and 25% in LBKSW. When combining SZA and LBKT into one 'Carpathian Basin' group, the proportion of shared haplotype remains unchanged for LBKS and LBKSW, but increases noticeably to 55.88% in the LBKC (Figure 5.4). Most of the lineages shared between different LBK groups belong to haplogroups H, J, K and T2, which are found in high frequencies throughout all LBK culture groups. N1a, although present in most LBK geographic regions, constitutes almost none of the shared lineages, because the N1a haplotypes in the Neolithic dataset are highly variable. Only LBKC contains lineages (5.88%) that directly match the central/north European hunter-gatherer metapopulation. The shared haplotype analysis suggests a strong affinity of SZA to LBKT, which may be either due to a common substrate, i.e. the Starčevo-Körös-Çris complex, or intensive admixture. The homogeneity of these two groups may explain why lineages from the SZA group that are not observed in LBKT nevertheless appear in the central European LBK. The consistently high percentage of shared lineages across all LBK groups indicates a common ancestry.

In summary, both frequency- and sequence-based analytic approaches reveal a strong similarity between the LBK groups studied. This affinity is not noticeably influenced by the geographic distances between groups. Our findings may thus reveal a maternal 'genetic signature' that mirrors the cultural character defining the LBK. The maternal contribution of Mesolithic populations, on the other hand, is negligible. While this may in part be due to the fact that there is only a limited number of pre-Neolithic haplotypes currently available, they show a remarkable homogeneity throughout Europe.

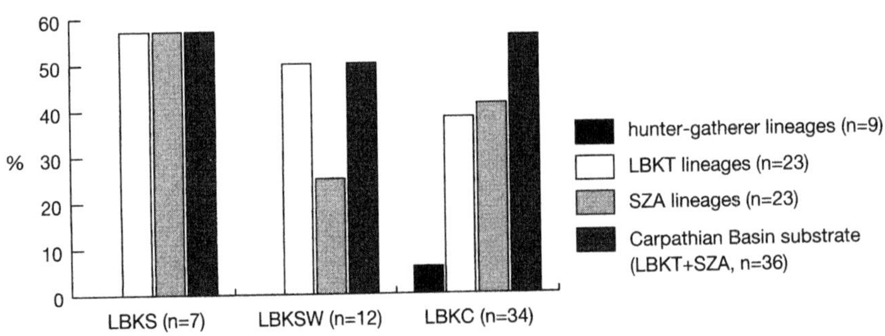

Figure 5.4 Shared haplotype analysis. Proportion of shared haplotypes between hunter-gatherers, Carpathian Basin SZA and LBK cultures, and central European LBK groups. Numbers in brackets denote the total number of different mitochondrial haplotypes observed in each group. Abbreviations: LBKT: LBK in Transdanubia; LBKS: LBK in Lower Austria and Moravia; LBKSW: LBK in south and south-west Germany; LBKC: LBK in central Germany; SZA: Szakálhát culture.

Discussion

Given the fast expansion time of LBK material culture, there has always been a discussion about the dispersal of farming in Europe. Various theories for Neolithisation have been proposed, which vary along a spectrum between a massive invasion/demic diffusion of the first farmers (Ammerman and Cavalli-Sforza 1984) and the theory of cultural diffusion, without a significant physical expansion (Whittle 1996, Budja 1999). Integrationist approaches consider further processes such as 'leapfrog' colonisation, contact or frontier mobility, which allow for admixture with local populations from its earliest phase onwards (Zvelebil 2001). One could also expect a clinal pattern of admixture in space and time, with increasing rates of admixture from the source region (in our case from Transdanubia as the earliest region where the LBK became established), to the parts of the LBK territory that are geographically distant and/or were settled at a later time. Similar patterns have been observed on a larger scale in modern European populations using both autosomal and uniparental markers (Ammerman and Cavalli-Sforza 1984; Chikhi *et al.* 1998, 2002; Semino *et al.* 2000). Our results indicate that at least on the east–west axis of LBK territory, a clinal pattern of (maternal) admixture cannot be presumed, as direct hints of Mesolithic introgression are generally rather sparse. However, a more comprehensive hunter-gatherer dataset would be needed to address this question in depth.

A massive migration/colonisation of LBK farmers without significant genetic hunter-gatherer contribution would require a fast-growing LBK population (Petrasch 2001), but this radical population increase is not evident from the Starčevo and earliest LBK archaeological records in western Hungary (Kalicz 2010). Instead, another model such as given by the principles of ideal despotic distribution (Shennan 2009) is more congruent with our results. Moreover, recent large-scale archaeological excavations have supplied new insight into larger early LBK settlements in Hungary, which probably indicate larger source communities for colonisation than previously supposed (Oross and Bánffy 2009, Marton and Oross 2012, Oross and Marton 2012).

We agree, on the other hand, that colonisation or migration and acculturation capture only a fraction of the total possible transition forms, and that it is important to study Neolithisation on regional level in order to prevent over-generalisations (Zvelebil 2001, Robb and Miracle 2007). Furthermore, our study only deals with the maternal genetic variability. Due to the often extensive degradation of ancient specimens, autosomal and Y-chromosomal information on LBK populations are still lacking, but would complete the picture.

From a maternal perspective, our results show strong genetic ties between the LBK in Transdanubia and the Szakálhát culture, possibly due to common ancestry in the Starčevo-Körös-Criş cultural complex. This theory is supported by a Starčevo mitochondrial dataset composed of 44 individuals, which is proven to represent

61.54% and 55.56% of the LBKT and the central European LBK mitochondrial substratum respectively (see Szécsényi-Nagy *et al.* in prep. for SHA details). Alternatively, the Transdanubian LBK and the Szakálhát culture could have been genetically influenced by the Middle Neolithic of the Balkans (the Vinča culture), an also conceivable process, which arises out of the archaeology (Kalicz 1994, Horváth 2006, Paluch 2011). In this case, the similarity between the Transdanubian LBK and the Szakálhát culture may have developed during the Middle Neolithic of the Carpathian Basin. In order to further elucidate this question, data from the Körös, the early Alföld LBK, the formative and/or Bicske-Biňa phase of the Transdanubian LBK, and the Vinča culture are all needed.

Our working hypothesis, that the LBK dataset can be divided into smaller geographically based groups such as the Transdanubian, Moravian/Lower Austrian, south-west and central German, has not been supported by the population genetic analyses. They cluster together, both by the haplogroup- and haplotype-based analyses (PCA-MDS), and show no significant genetic distance (F_{st} value) from each other. Furthermore, the LBK groups share a high proportion of maternal lineages (38–57%). However, influences by the stochastic effect of smaller sample sizes in the Moravian/Lower Austrian and south-west German LBK groups cannot be excluded entirely.

The genetic analysis of 72 Neolithic Szakálhát and LBK samples reveals that potential maternal kinship within the Middle Neolithic Carpathian Basin had no major role in the burial system. Out of the seven pairs of potentially maternally related individuals (three in SZA and four in LBKT), who shared HVS-I and in some cases HVS-II sequences, only one (two children at Kóny) was buried in a double grave. These findings indicate an overall scattered trend of kinship-based mortuary customs, which is not unexpected in settlement burials of the Neolithic.

Additionally, the high diversity of different maternal lineages rather supports than rules out the assumption of patrilocality in the LBK as a general social system, which is based on stable isotopic results (Bentley 2012, Bentley *et al.* 2012). In the Middle Neolithic Carpathian Basin, such a post-marital residence rule could have existed as well, which would result in the observed high mtDNA diversity in the Szakálhát and LBKT cultures. The LBK migration could not be directly identified by stable isotopes (Sr, O) to date. The first immigrant LBK farmers in central Europe have not yet been included or identified in geochemical researches; even at the intensively studied site of Vedrovice (Zvelebil and Pettitt 2008, Bentley *et al.* 2012), only 15 samples were analysed from the earliest settlement burials (Smrčka *et al.* 2006). Our LBKT dataset is mainly composed of individuals who lived after the supposed migration waves towards north and west. The early phase of the LBK culture (Bicske-Biňa phase) is detectable at only three sampled sites, out of the seven sampled in Transdanubia, but the successfully typed specimens are either not from this early phase (Balatonszárszó; Tolna-Mözs) (Oross and

Marton 2012, Marton and Oross 2012) or could not be dated so precisely (Balatonszemes). These circumstances might let us assume that the starting point for the central European LBK dispersal in western Hungary is better explained by the Starčevo than the LBKT dataset. This issue is mirrored in the possible late Mesolithic genetic traditions in the LBKT, mentioned above, whereas the Starčevo data show an almost pure early 'Neolithic package' mtDNA composition, even more congruent with the central European LBK (Szécsényi-Nagy *et al.* in prep.).

A minor issue arises in the Alföld region as well, when considering the Mesolithic comparison data. While the U4 and U5 haplogroups are associated with north and central European hunter-gatherer-populations, the assumption that the Mesolithic populations on the Alföld had the same genetic substrate is as yet unproven. The only authentic Mesolithic settlements from the Jászság area are without skeletal remains (Kertész 2002). There are presumed local Mesolithic traditions in the highlands north of the Alföld (Domboróczki 2003), and anthropogenic signals in the northeastern fringe of the Alföld region (Bánffy 2006). However, the local forager contribution to the Neolithisation of the Alföld is still hypothetical from an archaeological point of view, and human remains for further analyses are exceedingly rare.

Our results support a maternal genetic affiliation between the Szakálhát culture and the LBK in Transdanubia, and a genetic continuity between the LBK in the Carpathian Basin and in central Europe. We have shown that demonstrable hunter-gatherer impacts on the Neolithic datasets studied are negligible to date. Based on the haplogroup similarities and lineage matches between Starčevo, LBKT, and LBK in central Europe, we argue for an initial migration wave of the LBK from Transdanubia (Szécsényi-Nagy *et al.* in prep.). This new early Neolithic genetic input in central Europe remained as a predominantly unchanged substrate for the following two millennia (Brandt *et al.* 2013).

Acknowledgements

This study is part of a three-year research project funded by the German Research Foundation (DFG) aimed at investigating the population dynamics of the Neolithic Carpathian Basin (AL 287-10-1). The project is a cooperation between the Institute of Anthropology at the Johannes Gutenberg University of Mainz and the Institute of Archaeology, Research Centre for the Humanities, at the Hungarian Academy of Sciences in Budapest.

References

Ammerman, A.J. and Cavalli-Sforza, L.L. 1984. *The Neolithic transition and the genetics of populations in Europe*. Princeton: Princeton University Press.

Andrews, R.M., Kubacka, I., Chinnery, P.F., Lightowlers, R.N., Turnbull, D.M. and Howell, N. 1999. Reanalysis and revision of the Cambridge reference sequence for human mitochondrial DNA. *Nature Genetics* 23(2), 147.

Bánffy, E. 2004. *The 6th millennium* BC *boundary in western Transdanubia and its role in the Central European Neolithic transition (the Szentgyörgyvölgy-Pityerdomb settlement)* Budapest: Archaeological Institute of the Hungarian Academy of Science.

Bánffy, E. 2006. Eastern, Central and Western Hungary – variations of Neolithisation models. *Documenta Praehistorica* 33, 125–42.

Bánffy, E., Brandt, G. and Alt, K.W. 2012. 'Early Neolithic' graves of the Carpathian Basin are in fact 6000 years younger – appeal for real interdisciplinarity between archaeology and ancient DNA research. *Journal of Human Genetics* 57, 467–9.

Bentley, R.A. 2013. Mobility and the diversity of early Neolithic lives: isotopic evidence from skeletons. *Journal of Anthropological Archaeology* 32, 303–12.

Bentley, R.A., Bickle, P., Fibiger, L., Nowell, G.M., Dale, C.W., Hedges, R.E.M., Hamilton, J., Wahl, J., Francken, M., Grupe, G., Lenneis, E., Teschler-Nicola, M., Arbogast, R.-M., Hofmann, D. and Whittle, A. 2012. Community differentiation and kinship among Europe's first farmers. *Proceedings of the National Academy of Sciences of the United States of America* 109, 9326–30.

Bondár, V., Honti, Sz. and Kiss, V. 2000. A tervezett M7-es autópálya Somogy megyei szakaszának megelőző régészeti feltárása (1992–1999). *Előzetes jelentés I. A Somogy Megyei Múzeumok Közleményei* 14, 99–100.

Bramanti, B. 2008. Ancient DNA: genetic analysis of aDNA from sixteen skeletons of the Vedrovice. *Anthropologie* (Brno) 46, 153–60.

Bramanti, B., Thomas, M.G., Haak, W., Unterlaender, M., Jores, P., Tambets, K., Antanaitis-Jacobs, I., Haidle, M.N., Jankauskas, R., Kind, C.-J., Lueth, F., Terberger, T., Hiller, J., Matsumura, S., Forster, P. and Burger, J. 2009. Genetic discontinuity between local hunter-gatherers and central Europe's first farmers. *Science* 326, 137–40.

Brandt, G., Haak, W., Adler, C.J., Roth, C., Szécsényi-Nagy, A., Karimnia, S., Möller-Rieker, S., Meller, H., Ganslmeier, R., Friederich, S., Dresely, V., Nicklisch, N., Pickrell, J.K., Sirocko, F., Reich, D., Cooper, A., Alt, K.W. and Genographic Consortium 2013. Ancient DNA reveals key stages in the formation of Central European mitochondrial genetic diversity. *Science* 342, 257–61.

Brotherton, P., Haak, W., Templeton, J., Brandt, G., Soubrier, J., Jane Adler, C., Richards, S.M., Sarkissian, C. Der, Ganslmeier, R., Friederich, S., Dresely, V., Van Oven, M., Kenyon, R., Van der Hoek, M.B., Korlach, J., Luong, K., Ho, S.Y.W., Quintana-Murci, L., Behar, D.M., Meller, H., Alt, K.W. and Cooper, A. 2013. Neolithic mitochondrial haplogroup H genomes and the genetic origins of Europeans. *Nature Communication* 4, 1764.

Budja, M. 1999. The transition to framing in Mediterranean Europe – an indigenous response. *Documenta Praehistorica* 26, 119–41.

Chandler, H., Bryan S. and Zilhão, J. 2005. Using ancient DNA to examine genetic continuity at the Mesolithic–Neolithic transition in Portugal. In P. Arias, R. Ontañón and C. García

Moncó (eds), *Actas dell III Congreso del Neolítico en la Península Ibérica*, 781–6. Santander: Monografías del Instituto internacional de Investigaciones Prehistóricas de Cantabria 1, Santander.

Chikhi, L, Destro-Bisol, G., Bertorelle, G., Pascali, V. and Barbujani, G. 1998. Clines of nuclear DNA markers suggest a largely Neolithic ancestry of the European gene pool. *Proceedings of the National Academy of Sciences of the United States of America* 95, 9053–8.

Chikhi, L., Nichols, R.A., Barbujani, G. and Beaumont, M.A. 2002. Y genetic data support the Neolithic demic diffusion model. *Proceedings of the National Academy of Sciences of the United States of America* 99, 11008–13.

Czene, A. 2008. Újkőkori falu Budakeszin. In M. Gyöngyössy (ed.), *Képek a múltból – Az elmúlt évek ásatásaiból Pest megyében*, 14–15. Szentendre: Pest Megyei Múzeumok Igazgatósága.

Czene, A. and Ottományi, K. 2007. Budakeszi, Szőlőskert-Tangazdaság (MRT 7. k. 4/8 lh.). *Régészeti Kutatások Magyarországon 2006* (2007), 162.

Der Sarkissian, C., Balanovsky, O., Brandt, G., Khartanovich, V., Buzhilova, A., Koshel, S., Zaporozhchenko, V., Gronenborn, D., Moiseyev, V., Kolpakov, E., Shumkin, V., Alt, K.W., Balanovska, E., Cooper, A. and Haak, W. 2013. Ancient DNA reveals prehistoric gene-flow from Siberia in the complex human population history of north east Europe. *PLoS Genetics* 9, e1003296.

Domboróczki, L. 2003. The radiocarbon data from Neolithic archaeological sites in Heves County (north-eastern Hungary). *Agria* 39, 5–76.

Domboróczki, L. 2005. Adács-Mancsos-rét. *Régészeti Kutatások Magyarországon 2005*, 175.

Domboróczki, L. 2010. Report on the excavation at Tiszaszőlős-Domaháza-puszta and a new model for the spread of the Körös culture. In J.K. Kozłowski and P. Raczky (eds), *Neolithization of the Carpathian basin: northernmost distribution of the Starčevo/Körös culture,* 137–76. Kraków and Budapest: Polish Academy of Sciences and Institute of Archaeological Sciences of the Eötvös Loránd University.

Fu, Q., Mittnik, A., Johnson, P.L.F., Bos, K., Lari, M., Bollongino, R., Sun, C., Giemsch, L., Schmitz, R., Burger, J., Ronchitelli, A.M., Martini, F., Cremonesi, R.G., Svoboda, J., Bauer, P., Caramelli, D., Castellano, S., Reich, D., Pääbo, S. and Krause, J. 2013. A revised timescale for human evolution based on ancient mitochondrial genomes. *Current Biology* 23, 1–7.

Gamba, C., Fernández, E., Tirado, M., Deguilloux, M.F., Pemonge, M.H., Utrilla, P., Edo, M., Molist, M., Rasteiro, R., Chikhi, L. and Arroyo-Pardo, E. 2012. Ancient DNA from an Early Neolithic Iberian population supports a pioneer colonization by first farmers. *Molecular Ecology* 21, 45–56.

Gronenborn, D. 1999. A variation on a basic theme: the transition to farming in southern central Europe. *Journal of World Prehistory* 13, 123–210.

Gronenborn, D. 2007. Beyond the models: 'Neolithisation' in Central Europe. In A. Whittle and V. Cummings (eds), *Going over: the Mesolithic–Neolithic transition in North-West Europe*, 73–98. London: Oxford University Press for the British Academy.

Guba, Z., Hadadi, É., Major, Á., Furka, T., Juhász, E., Koós, J., Nagy, K. and Zeke, T. 2011. HVS-I polymorphism screening of ancient human mitochondrial DNA provides evidence for N9a discontinuity and East Asian haplogroups in Neolithic Hungary. *Journal of Human Genetics* 56, 784–96.

Haak, W., Forster, Peter, Bramanti, B., Matsumura, S., Brandt, G., Tänzer, M., Villems, R., Renfrew, C., Gronenborn, D., Alt, K.W. and Burger, J. 2005. Ancient DNA from the first European farmers in 7500-year-old Neolithic sites. *Science* 310(5750), 1016–18.

Haak, W., Brandt, G., De Jong, H.N., Meyer, C., Ganslmeier, R., Heyd, V., Hawkesworth, C., Pike, A.W.G., Meller, H. and Alt, K.W. 2008. Ancient DNA, strontium isotopes, and osteological analyses shed light on social and kinship organization of the Later Stone Age. *Proceedings of the National Academy of Sciences of the United States of America* 105, 18226–31.

Haak, W., Balanovsky, O., Sanchez, J.J., Koshel, S., Zaporozhchenko, V., Adler, C.J., Der Sarkissian, C.S.I., Brandt, G., Schwarz, C., Nicklisch, N., Dresely, V., Fritsch, B., Balanovska, E., Villems, R., Meller, H., Alt, K. W. and Cooper, A. 2010. Ancient DNA from European early neolithic farmers reveals their near eastern affinities. *PLoS Biology* 8(11), e1000536.

Hegedűs , K. 1985. The settlement of the Neolithic Szakálhát group at Csanytelek-Újhalastó. *Móra Ferenc Múzeum Évkönyve* 1982-83 (1985), 7–54.

Hervella, M., Izagirre, N., Alonso, S., Fregel, R., Alonso, A., Cabrera, V.M. and De la Rúa, C. 2012. Ancient DNA from hunter-gatherer and farmer groups from Northern Spain supports a random dispersion model for the Neolithic expansion into Europe. *PloS One* 7(4), e34417.

Horváth, F. 2006. Comments on the connections between the Vinča complex and the Cerpathian Basin. In N. Tasić and C. Grozdanov (eds), *Homage to Milutin Garašanin*, 309–24. Belgrade: Serbian Academy of Sciences and Arts, Macedonian Academy of Sciences and Arts.

Kalicz, N. 1994. A Dunántúli (Közép-európai) vonaldíszes kerámia legidősebb leletei és a korai Vinca kultúra. In G. Lőrinczy (ed.), *A kőkortól a középkorig (Tanulmányok Trogmayer Ottó 60. születésnapjára)*, 67–84. Szeged: Csongrád Megyei Múzeumok Igazgatósága

Kalicz, N. 2010. An Grenze „zweier Welten" –Transdanubien (Ungarn) im Frühneolithikum. In D. Gronenborn and J. Petrasch (eds), *Die Neolithisierung Mitteleuropas (The spread of the Neolithic to central Europe)*, 235–54. Mainz: Römisch-Germanisches Zentralmuseum.

Kertész, R. 2002. Mesolithic hunter-gatherers in the northwestern part of the Great Hungarian Plain. *Praehistoria* 3, 281–304.

Kiss, V. 2002. Előzetes jelentés a Balatonszemes, Bagó-dombon végzett megelőző feltárásról (Somogy megye). *Régészeti kutatások Magyarországon 2000*, 39–46.

Kiss, V. and Sebők, K. 2007. Balatonszemes-Bagódomb. In K. Belényesy, Sz. Honti and V. Kiss (eds), *Gördülő idő. Régészeti feltárások az M7 autópálya Somogy megyei szakaszán Zamárdi és Ordacsehi között*, 113–21. Kaposvár-Budapest: Somogy Megyei Múzeumok Igazgatósága and MTA Régészeti Intézete.

Kovács, Á. 2011. Középső neolit temetkezések Cegléd-Abony térségében. *Studia Comitatensia* 31, 49–81.

Krause, J., Briggs, A.W., Kircher, M., Maricic, T., Zwyns, N., Derevianko, A. and Pääbo, S. 2010. A complete mtDNA genome of an early modern human from Kostenki, Russia. *Current Biology* 20, 231–6.

Kustár, R. 2003. Harta-Gátőrház. *Régészeti Kutatások Magyarországon 2003*, 227.

Lacan, M., Keyser, C., Ricaut, F.-X., Brucato, N., Tarrús, J., Bosch, A., Guilaine, J., Crubézy, E. and Ludes, B. 2011. Ancient DNA suggests the leading role played by men in the Neolithic dissemination. *Proceedings of the National Academy of Sciences of the United States of America* 108, 18255–9.

Lee, E.J., Makarewicz, C., Renneberg, R., Harder, M., Krause-Kyora, B., Müller, S., Ostritz, S., Fehren-Schmitz, L., Schreiber, S., Müller, J., Von Wurmb-Schwark, N. and Nebel, A. 2012. Emerging genetic patterns of the European Neolithic: perspectives from a late Neolithic Bell Beaker burial site in Germany. *American Journal of Physical Anthropology* 148, 571–9.

Martínez-Cruz, B., Harmant, C., Platt, D.E., Haak, W., Manry, J., Ramos-Luis, E., Soria-Hernanz, D.F., Bauduer, F., Salaberria, J., Oyharçabal, B., Quintana-Murci, L. and Comas, D. 2012. Evidence of pre-Roman tribal genetic structure in Basques from uniparentally inherited markers. *Molecular Biology and Evolution* 29, 2211–22.

Marton, T. and Oross, K. 2012. Siedlungsforschung in linienbandkeramischen Fundorten in Zentral- und Südtransdanubien – Wiege, Peripherie oder beides? In F. Kreienbrink, M. Cladders, H. Stäuble, T. Tischendorf and S. Wolfram (eds), *Siedlungsstruktur und Kulturwandel in der Bandkeramik. Beiträge der internationalen Tagung "Neue Fragen zur Bandkeramik oder alles beim Alten?!" Leipzig, 23. bis 24. September 2010*, 220–39. Dresden: Druckhaus Dresden.

Oross, K. and Bánffy, E. 2009. Three successive waves of Neolithisation: LBK development in Transdanubia. *Documenta Praehistorica* 36, 175–89.

Oross, K. and Marton, T. 2012. Neolithic burials of the Linearbandkeramik settlement at Baltonszárszó and their European context. *Acta Archaeologica Academiae Scientiarum Hungaricae* 63, 257–99.

Oross, K. and Siklósi, Zs. 2012. Relative and absolute chronology of the Early Neolithic in the Great Hungarian Plain. In A. Anders and Z. Siklósi (eds), *The first Neolithic sites in central/south-east European transect, Volume III. The Körös Culture in eastern Hungary*, 129–59. Oxford: Archaeopress.

Oross, K., Marton T. and Fábián, S. 2004. Balatonszárszó-Kis-erdei-dűlő középső neolit településének temetkezései (Előzetes jelentés). In G. Ilon (ed.), *ΜΩΜΟΣ 3. Őskoros Kutatók III. Összejövetelének konferenciakötete*, 283–92. Szombathely: Vas Megyei Múzeumok Igazgatósága.

Pala, M., Olivieri, A., Achilli, A., Accetturo, M., Metspalu, E., Reidla, M., Tamm, E., Karmin, M., Reisberg, T., Hooshiar Kashani, B., Perego, U. a, Carossa, V., Gandini, F., Pereira, J.B., Soares, P., Angerhofer, N., Rychkov, S., Al-Zahery, N., Carelli, V., Sanati, M.H., Houshmand, M., Hatina, J., Macaulay, V., Pereira, L., Woodward, S.R., Davies, W., Gamble, C., Baird, D., Semino, Ornella, Villems, R., Torroni, A. and Richards, M.B. 2012. Mitochondrial DNA signals of late glacial recolonization of Europe from Near Eastern refugia. *American Journal of Human Genetics* 90, 915–24.

Paluch, T. 2011. *Maroslele-Pana: egy középső neolitikus lelőhely a kultúrák határvidékén*. Szeged: Móra Ferenc Múzeum.

Petrasch, J. 2001. Seid fruchtbar und mehret euch und füllet die Erde und machet sie euch untertan: Überlegungen zur demographischen Situation der bandkeramischen Landnahme. *Archäologisches Korrespondenzblatt* 31, 13–26.

Raczky, P. and Anders, A. 2003. The internal relations of the Alföld Linear Pottery culture in Hungary and the characteristics of human representation. In E. Jerem and P. Raczky (eds), *Morgenrot der Kulturen: frühe Etappen der Menschheitsgeschichte in Mittel- und Südosteuropa*, 155–82. Budapest: Archaeolingua.

Richards, M., Macaulay, V., Hickey, E., Vega, E., Sykes, B., Guida, V., Rengo, C., Sellitto, D., Cruciani, F., Kivisild, T., Villems, R, Thomas, M., Rychkov, S., Rychkov, O., Rychkov, Y., Gölge, M., Dimitrov, D., Hill, E., Bradley, D., Romano, V., Calì, F., Vona, G., Demaine, A., Papiha, S., Triantaphyllidis, C., Stefanescu, G., Hatina, J., Belledi, M., Di Rienzo, A., Novelletto, A., Oppenheim, A., Nørby, S., Al-Zaheri, N., Santachiara-Benerecetti, S., Scozari, R., Torroni, A. and Bandelt, H.J. 2000. Tracing European founder lineages in the Near Eastern mtDNA pool. *American Journal of Human Genetics* 67, 1251–76.

Robb, J. and Miracle, P. 2007. Beyond 'migration' versus 'acculturation': new models for the spread of agriculture. In A. Whittle and V. Cummings (eds), *Going over: the Mesolithic–Neolithic transition in North-West Europe*, 99–115. London: Oxford University Press for the British Academy.

Sánchez-Quinto, F., Schroeder, H., Ramirez, O., Avila-Arcos, M.C., Pybus, M., Olalde, I., Velazquez, A.M. V, Marcos, M.E.P., Encinas, J.M.V., Bertranpetit, J., Orlando, L., Gilbert, M.T.P. and Lalueza-Fox, C. 2012. Genomic affinities of two 7,000-year-old Iberian hunter-gatherers. *Current Biology* 22, 1494–9.

Sebők, K. 2009. Bölcske, Gyűrűsvölgy. *Régészeti Kutatások Magyarországon 2008*, 154.

Semino, O., Passarino, G., Oefner, P.J., Lin, A.A., Arbuzova, S., Beckman, L.E., Benedictis, G. De, Francalacci, P., Kouvatsi, A., Limborska, S., Marcikiæ, M., Mika, A., Mika, B., Primorac, D., Santachiara-Benerecetti, A.S., Cavalli-Sforza, L.L. and Underhill, P.A. 2000. The genetic legacy of Paleolithic Homo sapiens sapiens in extant Europeans: a Y chromosome perspective. *Science* 290, 1155–9.

Shennan, S. 2009. Evolutionary demography and the population history of the European Early Neolithic. *Human Biology* 81, 339–55.

Smrčka, V., Mihaljevic M., Zobcova J., Humpolova A. and Berkovec T. 2006. Multielementarní chemicka analýza z kosterních pozůstatků lidí a zvířat neolitickeho sídliště ve Vedrovicich (Česka Republika). *Ve službach archeologie* 7, 329–40.

Somogyvári, Á. 2008. Régészeti kutatások újabb eredményei Bács-Kiskun megyében, különös tekintettel a késő rézkor időszakára. In P. Ekler (ed.), *Közgyűjteményi tudományos napok II-III. Eger-Kecskemét*, 70–82. Budapest: Országos Széchényi Könyvtár and Gondolat Kiadó.

Szécsényi-Nagy, A., Brandt, G., Keerl, V., Jakucs, J., Haak, W., Moeller-Rieker, S., Köhler, K., Mende, B.G., Fecher, M., Oross, K., Marton, T., Osztás, A., Kiss, V., Pálfi, G., Molnár, E., Sebők, K., Czene, A., Paluch, T., Šlaus, M., Novak, M., Pećina-Šlaus, N., Ősz, B., Voicsek, V., Somogyi, K., Tóth, G., Kromer, K., Bánffy, E., Alt, K.W. In prep. Tracing the genetic origin of Europe's first farmers reveals insights into their social organization.

Tamura, K. and Nei, M. 1993. Estimation of the number of nucleotide substitutions in the control region of mitochondrial DNA in humans and chimpanzees. *Molecular Biology and Evolution* 10, 512–26.

Tari, E. 2001. Cegléd, Ipari Park. *Régészeti Kutatások Magyarországon 2000* (2001), 121.

Trogmayer, O. 1969. Die Bestattungen der Körös-Gruppe. *Móra Ferenc Múzeum Évkönyve* 1969, 2, 5–15.

Van Oven, M. and Kayser, M. 2009. Updated comprehensive phylogenetic tree of global human mitochondrial DNA variation. *Human Mutation* 30, E386–94.

Varga, K. 2010. Kóny – Proletár dűlő (Győr-Moson-Sopron megye). In J. Kvassay (ed.), *Évkönyv és jelentés a KÖSZ 2008. évi feltárásairól/Field service for cultural heritage 2008 yearbook and review of archaeological investigations*, 38. Budapest: Field Service for Cultural Heritage.

M. Virág, Zs. 2009. The eastern periphery of the central European LBC in the region of Budapest. In J.K. Kozłowski (ed.), *Interactions between different models of neolithization north of the central European agro-ecological barrier*, 9–31. Kraków: Polska Akademia Umiejętności.

Whittle, A. 1996. *Europe in the Neolithic: the creation of new worlds*. Cambridge: Cambridge University Press.

Whittle, A., Bartosiewicz, L., Borić, D., Pettitt, P. and Richards, M. 2002. In the beginning: new radiocarbon dates for the Early Neolithic in northern Serbia and south-east Hungary. *Antaeus* 25, 63–117.

Zvelebil, M. 2001. The agricultural transition and the origins of Neolithic society in Europe. *Documenta Praehistorica* 28, 1–26.

Zvelebil, M. and Pettitt, P. 2008. Human condition, life, and death at an Early Neolithic settlement: bioarchaeological analyses of the Vedrovice cemetery and their biosocial implications for the spread of agriculture in Central Europe. *Anthropologie* (Brno) 46, 195–218.

6

Settlement Burials at the Karsdorf LBK Site, Saxony-Anhalt, Germany

Biological Ties and Residential Mobility

GUIDO BRANDT, CORINA KNIPPER, NICOLE NICKLISCH, ROBERT GANSLMEIER, MECHTHILD KLAMM AND KURT W. ALT

Introduction

THE LBK REPRESENTS THE EARLY FARMERS OF CENTRAL EUROPE and is one of the most intensively studied archaeological entities world-wide. Its characteristic longhouses and elaborately decorated pottery have attracted the interest of archaeological research since the late nineteenth century (Buttler and Haberey 1936, Klopfleisch 1883). Being a major socio-economic turning point, the establishment of agriculture and domestic animal husbandry in central Europe and their long-term maintenance have been the focus of intensive research (Scharl 2004). Building on numerous excavations that have been conducted over decades, profound investigations have shed light on settlement practices, community sizes and social organisation, demographic and economic aspects, and burial practices (Alt *et al.* 2004, Bickle and Whittle 2013, Bocquet-Appel and Bar-Yosef 2008, Lüning *et al.* 2005, Zimmermann 2009, Zimmermann *et al.* 2009, Strien 2000, Trautmann 2006). Beginning in the late twentieth century, newly developed scientific techniques that investigate human and faunal skeletal remains have contributed substantially to our understanding of lifestyle and mobility. Recent ancient DNA (aDNA) studies have provided direct insights into the gene pool of contemporary Mesolithic and Neolithic communities, revealing genetic discontinuities between indigenous hunter-gatherers and early farmers (Bramanti *et al.* 2009, Haak *et al.* 2010) alongside a possible origin of the genetic 'Neolithic package' in the Near Eastern Fertile Crescent (Haak *et al.* 2010, Brandt *et al.* 2013). Strontium isotope analyses have identified non-local individuals among burial communities (Price *et al.* 2001, Bentley *et al.* 2003, Turck *et al.* 2012, Price *et al.* 2006, Bickle *et al.* 2011, Nehlich *et al.* 2009) and explored domestic animal husbandry strategies (Bentley and Knipper 2005, Knipper 2011), while light stable isotope ratios have informed about

Proceedings of the British Academy **198**, 95–114.

dietary practices (Dürrwächter *et al.* 2006, Asam *et al.* 2006, Oelze *et al.* 2011, Bickle *et al.* 2011). Recently, LBK research has culminated in large-scale interdisciplinary and inter-regional projects, such as the 'Lifeways project' organised from Cardiff University (Bickle and Whittle 2013).

The research presented here is part of an interdisciplinary study that investigates Neolithic culture and population changes in the central German Middle Elbe-Saale (MES) region. It focuses on the LBK site of Karsdorf, where settlement burials were found in close association with the typical longhouses (Behnke 2012). The arrangement of features with human skeletons raised questions about possible biological relations, age and sex distribution as well as mobility and dietary patterns of the burial groups. These facets of LBK lifestyle were addressed by aDNA, osteological, light stable isotope (Oelze *et al.* 2011), and strontium isotope analyses.

The LBK settlement and burials at Karsdorf

Karsdorf is situated in the MES region of central Germany, in the south of Saxony-Anhalt. The location in the Unstrut valley is characteristic for LBK settlements in this area. Despite a lack of systematic research, a number of sites are known from the river banks, the confluence with smaller streams, or in the vicinity of springs that originate from the geological border of the upper Buntsandstein with the lower Muschelkalk (Colour Plate 3). Fertile loess soils just above the zones of probable flooding at elevations of 110–130 m a.s.l. provided optimal agricultural conditions. Therefore, the Unstrut valley was very probably intensively managed with cultivated plots near numerous settlements. In contrast, isolated stone axes do not indicate an equally dense settlement of nearby higher elevation areas such as the Querfurt Plateau to the east with less frequent running water courses or the Buntsandstein hills to the west with elevations of 220–300 m a.s.l. Usage of these zones for pasture or other off-site activities should, however, be taken into consideration.

Excavations were conducted at Karsdorf between 1996–2010 and discovered settlements and graves from the Neolithic to the Iron Age (Behnke 2007, 2011, 2012). The LBK is represented by 24 longhouses oriented north-west–south-east (named A–X). Most of them were arranged in four stacked rows, including one in the south (1), two in the middle (2 and 3), and one in the north of the site (4). The remaining houses were built between these rows (Colour Plate 4). The length of the buildings varied between 10 and 25 m. While the southern and northern rows contained one of the two longest buildings each, mid-sized houses 10–20 m long formed the central part of the settlement. According to the current state of research the site dates from the earlier (Flomborn phase) to the middle LBK (terminology after Meier-Arendt 1966; Phases 2 to 4 of the central German chronology; Schwarz in prep.). The decorated pottery from the house-associated pits in the central part of the settlement suggests a chronological succession from the west to the east, in which the older houses H, J, O, and S were replaced by the houses P and Q. The remaining houses could not be dated, because the lack of sufficient archaeological finds prevented classification into the LBK chronology. Five of six AMS dates on bone collagen fall between about 5140–4960 cal BC, while one yielded an older date of 5355–5300 cal BC (Table 6.1). Earlier findings of discrepancies between radiocarbon dates of charcoal and bone (Denaire 2009) make us aware of a possibly older absolute chronological range which is likely to match the estimates based on the published pottery of about 5200–5070 cal BC (Hans-Christoph Strien, pers. comm., September 2013).

Special features of the LBK site of Karsdorf are graves in association with particular houses, which can therefore be regarded as settlement burials (Veit 1996). Overall, 34 inhumations were discovered, of which 28 individuals were entombed in settlement and oblong pits, mainly on the west side of the houses H, J, K, O, P, Q, S, and T, which are located in the centre of the settlement (house rows 2 and 3). The pits near houses H, P, and Q contained multiple burials. The remaining six individuals (14, 88, 220, 430, 509, 529) were interred separately beside or between the houses (Table 6.1; Colour Plate 4).

Most of the individuals were buried in a flexed position on their left or right sides, oriented to the north-east or north-west (Table 6.1). Six individuals were inhumed in supine and four in prone position, of which only three showed a fully stretched body. The furnishing of the graves at Karsdorf can be regarded as rather sparse. Only nine of the 34 burials contained grave goods, such as one or more ceramic vessels (Ind. 115, 122, 131, 158, 324, 509), silex artefacts (Ind. 303), a shell pendant (Ind. 302), an antler pendant (Ind. 192) and an axe (Ind. 131) (Table 6.1).

Physical anthropology

Osteological investigations were conducted on 31 individuals and focused on the distribution of age and sex. Age at death of the adults was determined by cranial sutures, dental wear and features of the innominate (Ferembach *et al.* 1980, Lovejoy 1985, Lovejoy *et al.* 1985, Brooks and Suchey 1990), while the age of subadults was ascertained by dentition, and ossification patterns, as well as by long bone length (Stloukal and Hanáková 1978, Ferembach *et al.* 1980, Ubelaker 1989). Morphological and metrical criteria were used for sex estimation of the adults (Phenice 1969, Ferembach *et al.* 1980, Murail *et al.* 2005).

The assemblage is composed of 20 adults (64.5%), one juvenile (15–18 years; 3.2%) and ten subadults (32.3%), of which six can be classified as infans I (0–6 years; 19.4%) and four as infans II (7–14 years; 12.9%). The majority of the infans I children died between 1.5–4 years and of the infans II between 6–9 years. The proportion of subadults, the non-existent neonates in particular, is lower than

Table 6.1 Archaeological, anthropological, palaeogenetic, and strontium isotope data of the LBK burials at the settlement of Karsdorf

Archaeology								Anthropology		Palaeogenetics		
Feature	HK-No.	house	position	grave goods	¹⁴C–ID	BP	cal BC 1σ 68.3%)	age	sex	HVS–I (16046–16401)	HVS–II (34–397)	Haplogroup
88	2004:26057a	K/L	n.d	–				n.a	n.a			
95	2004:26062a	K	↯	–				15–18	I	rCRS	263G 309.1C 309.2C 315.1C	H
115	2004:26074a	J	↯	2×C				2–3	n.d.			
324	2004:26218a	H	n.d	1×C				1½–2	n.d.			
400	2004:26218a	H	↯	–	KIA-40356	6116±30	5068–4992	35–45	♀	16224C 16311C	73G 263G 309.1C 315.1C	K
430	2004:26267a	H/M	↯	–				20–22	♀	rCRS	152C 263G 315.1C	H
482	2004:26297a	M	n.d	–				n.a	n.a			
509	2004:26318a	P/M	↯ v	1×C				30–40	♂	16126C 16153A 16294T 16296T		T2e
529	2004:26333a	P/M	↯	–				ca. 8	I	16096T 16126C	73G 185A 228A 263G 295T 315.1C	J1c
537	2004:26340a	O	↘ d	–	KIA-40357	6116±32	5056–4959	25–30	♂	16069T 16126C	73G 185A 228A 263G 295T 315.1C	J1c
605	2004:26370a	H	↯	–	KIA-40358	6127±31	5075–4997	40–50	♂	16126C 16294T 16296T 16304C 16321T		T2b
606	2004:26371a	H	↯	–				6–7	n.d.			
611	2004:26374a	H	↖ d	–				24–26	♂	rCRS	263G 315.1C	H
14	2005:L3736a	.	↯ v	–				45–55	♂	16069T 16126C 16274A		J
122	2005:37516a	P	↯	1×C	KIA-40359	6168±32	5140–5089	50–65	♂	16069T 16126C	185A 188G 228A 263G 295T 315.1C	J1c2
131	2005:37518a	O	↯	1×A, 3×C				21–24	♂?	16192T 16256T 16270T		U5a
155	2005:37539a	O	↯	–				65–75	♀	16311C		HV
220	2005:37585a	T/X	↯	–	MAMS–1348	6333±23	5355–5300	n.a	n.a.	16147a 16154C 16172C 16223T 16248T 16320T 16355T		N1a1a3
297	2005:37667a	Q	↯	–				1½–2	n.d.			
298	2005:37668a	Q	↯	–				2–3	n.d.	rCRS	263G 315.1C	H
299	2005:3767a	Q	↯ v	–	KIA-40360	6142±31	5079–5020	22–24	♀?	16093C		H
300	2005:37673a	Q	NO-SW	–				4-–45	♂	16192T 16270T		U5b
301	2005:37674a	Q	↯	–				1½–2	n.d.	16224C 16311C	73G 146C 152C 263G 309.1C 315.1C 32	K2a5
302	2005:37676a	Q	↯ d	1×SP				45–50	♀	rCRS	263G 309.1C 315.1C (B)	H
303	2005:37679a	Q	↯ d	1×S				50–65	♂	16224C 16311C 16319A		K1b1a
304	2005:37683a	P	↯ d	–				35–45	♂	16126C 16189C 16294T 16296T		T2f
305	2005:37684a	P	↯	–				2–4	n.d.			
306	2005:37686a	O	↯	–				8–9	n.d.			
310	2005:37688a	O	↗ v	–				>20	n.d.			
157	2006:14418a	S	↯	–				ca. 8	I	16093C 16224C 16311C		K1a
158	2006:14419a	S	↯	3×C				>30	♀?			
169	2006:14422a	S	↯	–				50–65	♀			
170	2006:14423a	S	↯	–				45–60	♂	rCRS	263G 315.1C	H
192	2006:14430a	S	↯ d	1×AP				27–35	I	16093C 16209C 16224C 16311C		K1a

Strontium isotopes			
Tooth	Sr ppm	²⁷Sr/²⁶Sr	±2σ
16	365	0.70826	0.00003
48	56	0.71005	0.00001
18	270	0.70832	0.00017
26	65	0.70940	0.00003
27	118	0.70929	0.00009
16	134	0.70863	0.00009
17	269	0.70852	0.00013
Right femur	234	0.70857	0.00005
85	109	0.70981	0.00006
16	81	0.70957	0.00006
17	162	0.70876	0.00024
18	277	0.70861	0.00012
37	422	0.70819	0.00017
38	522	0.70828	0.00013
Left femur	218	0.70867	0.00003
17	192	0.70890	0.00010
18	233	0.70866	0.00017
27	425	0.70866	0.00005
17	136	0.70981	0.00001
Left fibula	151	0.70950	0.00001
36	288	0.70820	0.00026
Rigt femur	164	0.70881	0.00002
47	120	0.70988	0.00014
Right femur	100	0.70941	0.00001
84	251	0.70832	0.00003
85	253	0.70825	0.00001
47	103	0.70980	0.00012
48	196	0.70811	0.00024
Right femur	195	0.70878	0.00006
48	183	0.70943	0.00010
17	256	0.70885	0.00022
17	199	0.70845	0.00022
18	265	0.70837	0.00007
38	86	0.71011	0.00011
55	70	0.70945	0.00004
16	77	0.70947	0.00001
27	209	0.70835	0.00005
37	109	0.70981	0.00003

flexed position, left side ↯
flexed position, right side ↯
stretched position ←
dorsal **d**
ventral **v**

C = ceramic
A = axe
S = silex
SP = shell pendant
AP = antlers pendant

♂ = Male
♀ = Female
I = Indifferent
n.d. = not determined
n.a. = not analyzed

The nomenclature of the teeth follows the FDI notation

expected from demographic models (United Nations 1955, Weiss 1973), but similar to further LBK sites in the MES region (Fritsch *et al.* 2008, Nicklisch in prep.) and other areas in Germany (Baum 1990, Siemoneit 1997, Gerling and Francken 2007). Among the adults were more males (n = 11; 55%) than females (n = 7; 35%), while the sex of two adults could not be determined (10%). The maximum age at death of males ranged between 40 and 59 years and of females from 40–49 years respectively. The oldest individual is a woman (Ind. 155, 65–75 years).

According to palaeodemographic estimations (Modderman 1970, Smolla 1974), the total number of burials is too small in relation to the settlement size and the period of occupation suggested by the radiocarbon dates and ceramic decorations. This implies that the excavated individuals do not represent the total living population of the settlement, but are rather a selection, and more graves, perhaps in a so far unknown cemetery, are to be expected near the site.

Due to the sparse grave furnishings, only little correlation between grave goods and the sex of the deceased is possible. However, grave goods, such as an axe in a man's grave and a shell buried with a woman, imply sex-specific grave furnishings. In association with the houses S, P, Q, and H, women, men, and children were buried together, in some cases even in the same pit. This raised questions about potential biological relationships among these individuals.

Palaeogenetics

We selected samples from 23 skeletons, including five subadults, one juvenile, eleven males, five females and one indeterminate adult of the Karsdorf community (Table 6.1) and genotyped the mtDNA by sequencing the hyper-variable segment I and II (HVS-I/II) of the control region and 22 haplogroup-defining single-nucleotide polymorphisms of the coding region (Haak *et al.* 2010). Sample preparation, DNA extraction, amplification, cloning and sequencing were carried out in the aDNA laboratories of the Institute of Anthropology at the Mainz University, following previously described standard protocols (Brandt *et al.* 2013). We were able to successfully reproduce endogenous HVS-I profiles from all sampled individuals (Table 6.1). In addition, we generated 11 HVS-II sequences from individuals with identical HVS-I sequences in order to increase the phylogenetic resolution for the evaluation of potential maternal kinship.

The palaeogenetic investigations yielded 20 different haplotypes that can be distinguished by their HVS-I and HVS-II sequences (Table 6.1; Brandt *et al.* 2013), indicating a high diversity of maternal lineages of the Karsdorf community (87%). These haplotypes belong to eight different haplogroups, including H (30.4%), K (21.7%), J (17.4%), T2 (13%), N1a (4.3%), HV (4.3%), U5a (4.3%), and U5b (4.3%; Table 6.1).

Maternal kinship

We used the mtDNA profiles to investigate maternal kinship among the Karsdorf individuals. As mentioned above, the palaeogenetic analyses revealed highly diverse mtDNA lineages that resulted in a small number of possibly related individuals. Overall, we identified only two haplotypes with identical HVS-I and HVS-II sequences that are shared by two or more individuals and may indicate maternal kinship. The first lineage belongs to the haplogroup H and was found in the individuals 170, 298 and 611, while the second haplotype, which is characteristic for the haplogroup J1c, relates the individuals 529 and 537 (Table 6.1). Both lineages are common in other datasets of the LBK (Haak *et al.* 2005, Haak *et al.* 2010, Brandt *et al.* 2013, Bramanti 2008), subsequent Neolithic cultures (Brandt *et al.* 2013) and present-day populations of central Europe. Consequently, even if the mtDNA data suggest potential maternal kinship among these individuals, a closer relationship seems to be rather unlikely considering the overall high haplotype frequency in ancient and modern-day populations.

This is also reflected by the archaeological context, because the possibly related individuals were predominantly buried in pits that were associated with different houses that were very probably not contemporary (Colour Plate 4). The individuals 170, 298 and 611 were buried close to houses S, Q, and H respectively, while the grave of individual 537 is associated with house O and individual 529 was buried at some distance from house P, between the house rows 2 and 3. This separation precludes the assumption of a closer biological relationship spanning one or two generations. Nevertheless, the fact that a few identical haplotypes were found among different phases may indicate a marginal amount of continuing mtDNA lineages, and therefore maternal gene flow over time and generations. More importantly, the palaeogenetic analyses reveal no evidence for maternal kinship among adjacent individuals of each house. This suggests that the individuals most probably do not represent families and that kinship was not the primary reason for being buried in association to a house, albeit paternal kinship cannot be excluded and the small number of analysed children allows only limited statements on the parent–child level.

Considering the timespan of the Karsdorf site of more than 300 years (including the earliest radiocarbon date), one would expect more identical maternal lineages if the settlement was occupied by a succession of several generations of the same community members. Thus, the sparse indications for maternal kinship suggest a dynamic and mobile group of which several members were buried elsewhere and/or which integrated individuals who originated from other communities.

Population comparisons

In order to investigate whether the Karsdorf population differs genetically from previous Mesolithic or contemporaneous LBK groups in other European regions, we compared our samples with respective data from the palaeogenetic literature. In detail, we used a hunter-gatherer meta-population of central and northern Europe (HGcn, n = 17) that includes data from Croatia (Szécsényi-Nagy *et al.* in prep.), the Czech Republic (Fu *et al.* 2013), Germany (Bramanti *et al.* 2009, Fu *et al.* 2013), Lithuania (Bramanti *et al.* 2009), Luxembourg (Fu *et al.* 2013) and Poland (Bramanti *et al.* 2009). Furthermore, we assembled the currently available LBK data into six sample sets that either represent particular LBK sites in central Germany or distant regions of the LBK distribution. These datasets comprise the sites of Derenburg (DER, n = 22: Haak *et al.* 2005, 2010), Halberstadt (HAL, n = 31: Haak *et al.* 2005, Brandt *et al.* 2013), Oberwiederstedt I (OBW, n = 8: Haak *et al.* 2005, Brandt *et al.* 2013), and Naumburg and Eisleben (NAU, n = 5: Haak *et al.* 2005, Brandt *et al.* 2013) in the MES region, as well as a south-western LBK group (LBKsw, n = 12) encompassing the sites of Flomborn, Schwetzingen, Vaihingen, and Seehausen (Haak *et al.* 2005), and a southern LBK group (LBKs, n = 7) including the sites of Asparn Schletz 2 (Haak *et al.* 2005) and Vedrovice (Bramanti 2008).

We applied a suite of population genetic methods, including shared haplotype analyses, haplotype-based genetic distance (F_{st}) computations and a haplogroup-based Fisher test to identify significant differences in the mtDNA gene pool between the Karsdorf population and the comparative data. We used the HVS-I sequences (nucleotide position 16,056–16,400) to conduct the shared haplotype and the genetic distance analyses in Arlequin 3.5.1 (Excoffier and Lischer 2010) using the Tamura and Nei substitution model (Tamura and Nei 1993) and an associated gamma value of 0.186. Significant variations in F_{st}-values were tested by performing 10,000 permutations. Furthermore we used the frequencies of 15 haplogroups that were observed among the ancient data (H, HV, J, K, N1a, T2, U, U3, U4, U5a, U5b, U8, V, W, X) to identify significant differences at the haplogroup level using Fisher's exact test for frequency data (Fisher 1922). This test was carried out in R 2.13.1 (The R Foundation for Statistical Computing 2011, http://www.r-project.org/), performing 10,000 permutations.

The genetic distances and the Fisher test reveal that the different LBK groups have highly similar mtDNA compositions, which is indicated by the absence of significant differences on the haplotype as well as on the haplogroup level (Figure 6.1). Conversely, all LBK groups show significantly different mtDNA profiles compared to the hunter-gatherers, which is in accordance to the genetic discontinuity at the time of the Neolithic transition in central Europe as described previously (Bramanti *et al.* 2009, Haak *et al.* 2010). Nevertheless, about 8% of the Karsdorf haplotypes show direct matches with the HGcn. These lineages belong to the haplogroups U5a and U5b that are characteristic for pre-Neolithic communities in central Europe (Bramanti *et al.* 2009, Malmström *et al.* 2009) and presumably indicate a Mesolithic legacy for the respective Karsdorf individuals (131 and 300).

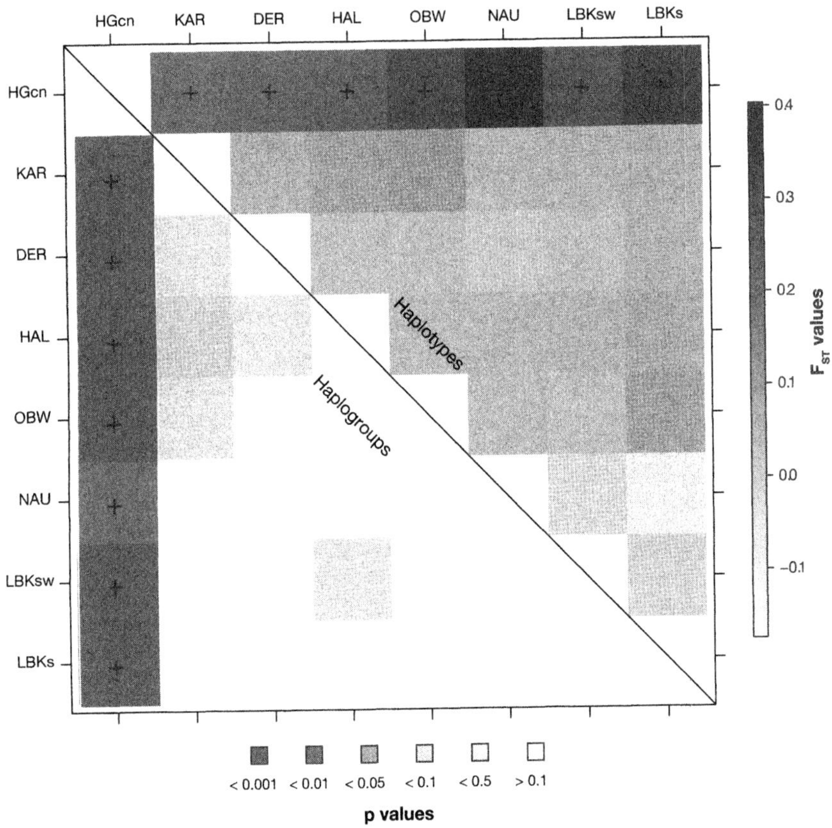

Figure 6.1 Comparative analyses of the haplogroup and haplotype composition among hunter-gatherers and distant LBK groups from central Europe. The lower triangle summarizes the results of the Fisher test based on discrete haplogroup frequencies. The shading indicates significant variation in haplogroup composition between pairs of datasets (white and light grey areas = non-significant, dark grey areas = significant). The upper triangle shows the genetic distances (F_{st}) between pairs of datasets based on the haplotype data of the HVS-I sequences (nucleotidposition 16,056–16,400). The shading indicates the degree of genetic distance between the groups ranging from white and light grey (small distances/high similarities) to dark grey (large distance/dissimilarities). Significant differences are indicated by +. Abbreviations are mentioned in the text. (Graphic: G. Brandt).

The non-significant differences among the LBK groups are founded on a high amount of shared lineages. About 30–60% of the Karsdorf haplotypes show direct matches with the remaining LBK groups (Figure 6.2). These shared lineages predominantly belong to the haplogroups N1a, K, J, T2, and HV that have – complemented by V, W, and X – previously been described as the mitochondrial 'Neolithic package', which originated in the Near East Fertile Crescent and reached central Europe in the sixth millennium cal BC with the first farmers of the LBK (Haak *et al.* 2010, Brandt *et al.* 2013). In the Karsdorf community, this package

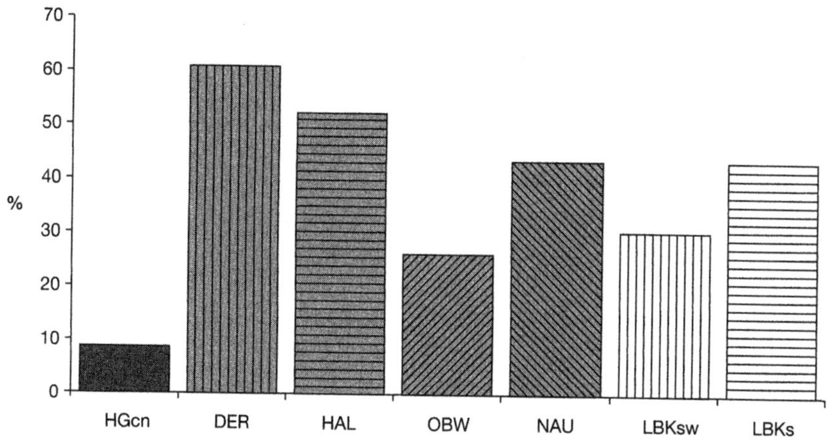

Figure 6.2 Results of the shared haplotype analysis. The bar plot indicate the amount of shared mtDNA-lineages (%) between the Karsdorf community and hunter-gatherers from central and north Europe (dark grey) as well as distant LBK groups from the Mittelelbe-Saale region (light grey) and the southern/southwestern LBK dissemination area (white). Abbreviations are mentioned in the text. (Graphic: G. Brandt).

comprises about 60% and can be found in even larger amounts (70–100%) in the remaining LBK sample sets (not shown) and also in the early Neolithic of the Carpathian Basin (Szécsényi-Nagy *et al.*, this volume). Together these results support no regional differences between distant LBK groups and indicate that the Neolithic transition in central Europe was accompanied by the introduction of a very homogenous mtDNA signature, which affected vast areas of the LBK distribution.

Strontium isotope analysis

Strontium isotope analysis was conducted on enamel from 21 individuals including 3 children and the juvenile, all 11 males, 5 females and 1 adult of indeterminate sex. Sample preparation and analysis followed the procedure described in Knipper *et al.* (2012). The tooth enamel was separated during aDNA sample preparation, mechanically and chemically cleaned, and ashed. Strontium separation and analyses were performed at the Curt-Engelhorn-Center for Archaeometry at Mannheim. The Sr concentrations of the eluants of the Sr separation were determined with a Quadrupole ICP-MS. The strontium concentrations given in Table 6.1 and plotted in Colour Plate 5 represent about 80% of the original strontium content of the samples. They are rather an indicator of samples that are highly enriched in strontium than a precise measure of the strontium concentrations. Strontium isotope analysis was performed using a magnetic sector field MC-ICP-MS (Multi-Collector

Inductively Coupled-Plasma Mass Spectrometer) VG Axiom and corrected according to the exponential mass fractionation law to $^{88}Sr/^{86}Sr = 8.375209$.[1]

If possible, we prepared samples of two anatomically different teeth. In contrast to the aDNA analysis, where duplicates are required to confirm the reliability of the results, strontium isotope samples from different teeth of the same individual do not necessarily yield similar data. Enamel only incorporates strontium during its mineralisation period and remains afterwards unchanged (Bentley 2006). Because the crowns of different teeth form at different intervals, deviating strontium isotope ratios of early- and late-forming teeth indicate an alteration of the origin of a person's diet and imply movement during childhood (Schweissing and Grupe 2003, Knipper 2011). The enamel of the deciduous teeth analysed in this study represents the last weeks *in utero* until about the eighth month after birth. The crowns of the first permanent molars formed in the first three years of life, those of the second molars between about the third and the seventh year, and those of the third molars between about the seventh or eighth and the 14th–18th year (AlQahtani *et al.* 2010). In addition, six samples of cortical bone were prepared in the same way as the teeth.

The $^{87}Sr/^{86}Sr$ ratios of the enamel samples vary between 0.70811 (Ind. 299; ♀) and 0.71011 (Ind. 304; ♂). Their Sr concentrations range from 56 ppm (Ind. 95; juvenile, sex indet.) to 522 ppm (Ind. 605; ♂). The bones yielded Sr isotope ratios between 0.70857–0.70950 and Sr concentrations between 100–234 ppm. The bivariate plot revealed two groups (Colour Plate 5). The first (A) is characterised by low $^{87}Sr/^{86}Sr$ values between 0.70811–0.70867 and variable Sr concentrations between 134–522 ppm. The second (B) has more radiogenic $^{87}Sr/^{86}Sr$ ratios between 0.70929–0.71011 and Sr concentrations of 56–183 ppm. Both analysed teeth of most individuals plot into the same group. Eleven burials (52.4%) have Sr isotope ratios in group A, eight (38.1%) in group B and two individuals (9.5%) revealed an isotopic shift with data in both clusters. Eight males have Sr isotope ratios in group A, and three in group B. The isotope ratios of the females are more balanced with two representatives in each cluster, and one individual with an isotopic shift. Children also occur in both clusters. Representatives of both isotope ranges appear in all burial groups and are associated with different houses (Colour Plate 6). There are no spatial or possibly chronological tendencies. The bones plot in the same clusters as the enamel samples of the same individuals and are not remarkably enriched in Sr (Colour Plate 5). This suggests rather effective cleaning procedures and very limited diagenetic overprint.

[1] During the course of the measurement standard materials yielded the following $^{87}Sr/^{86}Sr$ ratios: NBS 987: 0.71024 ± 0.00004 (n = 13); Eimer & Amend (E&A): 0.70800 ± 0.00003 (n = 1); 0.70801 ± 0.00006 (n = 3); 0.70802 ± 0.00004 (n = 4).

Linking strontium isotope ratios to geological conditions

The two strontium isotope data ranges can be explained by the geological conditions in the near environs of the Karsdorf settlement. The MES region of Central Germany is shaped by widely spread loess deposits, on which very fertile soils developed (Bachmann *et al.* 2008). The site in the Unstrut river valley is situated on a large alluvial cone which accumulated gravel and weathering debris of Triassic shelly limestone *(Muschelkalk)* which originated from an escarpment less than 1 km east of the site (Colour Plate 3). The river valley itself is formed by Holocene alluvial sediments and Pleistocene gravel and sands. Loess dominates in the wider parts of the valley to the north and on the Querfurter Platte, the plateau above the about 80 m high escarpment. To the west, to the south, and beyond the Unstrut River, loess and red sandstone *(Buntsandstein)* prevail.

A recent study sampled modern vegetation, water, soils, and rocks from up to some 20 km south of Karsdorf and reported $^{87}Sr/^{86}Sr$ ratios of between about 0.7080–0.7090 from the alluvial plains of the Unstrut and Saale Rivers, gypsum-dominated layers of the upper red sandstone *(Buntsandstein)* as well as shelly limestone *(Muschelkalk)* quarries and forested escarpments (Maurer *et al.* 2012, tab. 1–2). This range corresponds to range A amongst the Sr isotope data from Karsdorf. Its closest sources are the Unstrut floodplain and the limestone (*Muschelkalk* and *Upper Buntsandstein*) escarpment including its weathering products in the immediate vicinity of the site. The biologically available strontium from Pleistocene loess at locations without influence of limestone weathering is more radiogenic with $^{87}Sr/^{86}Sr$ ratios of modern and archaeological samples ranging from about 0.7090–0.7103 (Maurer *et al.* 2012, Knipper *et al.* 2012, Haak *et al.* 2008). This corresponds to range B of the Karsdorf data. 'Exotic', remarkably different, isotope ratios are absent among the investigated burials. There is no indication of individuals who grew up in geologically distinct uplands (Knipper 2011, Oelze *et al.* 2012) or further north in central Germany where glacial moraines yield more radiogenic strontium (Knipper *et al.*, unpublished data).

Implications of the strontium isotope data for land-use and mobility

The formation of two data groups instead of one discrete linear array and the placement of both enamel datasets from most individuals in the same group imply the reflection of two distinct economic areas that provided strontium with different isotope ratios and concentrations (Montgomery *et al.* 2007). The proximity of the calcareous weathering products and loess deposits to the site offers a number of alternative socio-economic interpretations. In a first scenario, the different households at the Karsdorf settlement would have husbanded plots on either one or the other geological unit near the settlement. Based on archaeological plant macro-remains, Bogaard *et al.* (2011) have recently proposed a model of arable land distribution amongst the households at the very densely populated LBK site of Vaihingen/Enz in southwest Germany. According to this model, different house groups maintained different cultivation areas at different distances from the site, and some households had an advantage over others (Bogaard *et al.* 2011, 405–8). An application of this model to Karsdorf would suggest that individuals with Sr isotope ratios in range A grew up in households that cultivated plots on calcareous soils, very probably in the Unstrut valley in the near vicinity of the settlement, and that those with Sr isotope ratios in range B grew up in households that predominantly cultivated plots on loess, possibly beyond the landmarks of the Unstrut River or about 80 m above the site on the Querfurt plateau 1–2 km away. Such considerations about land management by the Karsdorf community have to respect the existence of other LBK settlements in the near vicinity, of which three have been located so far, but not intensively studied to confirm their contemporaneous existence (Colour Plate 3).

The model developed at Vaihingen builds on 'clan' groups that have been differentiated by a combination of aspects of material culture and architecture and comprise multiple households each (Bogaard *et al.* 2011, tab. 1). At Karsdorf, representatives of both isotope groups have been found in association with all the houses (Colour Plate 6). If long-term cultivation of certain land plots was organised at the household level and the burials represent the members of these households, one would expect to see tendencies of predominance of either one or the other range among the burial groups. Therefore, the Karsdorf data do not seem to reflect the Vaihingen model, unless the household groupings are blurred by post-childhood movement among households with different management regimes and subsequent burial in the group to which an individual was affiliated in adulthood rather than at birth.

A second plausible interpretation is that land plots managed by the Karsdorf community concentrated between the *Muschelkalk/Upper Buntsandstein* escarpment and the Unstrut, and that individuals with Sr isotope data in range A characterise people who were indigenous to the very site. This, of course, cannot exclude the presence of undistinguishable people from other limestone locations. Sr isotope ratios in range B may then represent individuals who grew up in (nearby) settlements who subsisted on resources from the loess. This interpretation of the data spectrum implies strong ties and frequent residential changes among LBK communities.

Finally, because of the paucity of grave goods and well datable finds in clear association with the houses, assurance that the burials were inhumed during or shortly after the use of each house is beyond chronological resolution. In case of house S, there even is some indication that at least one of the burials was constructed several generations after the building was abandoned (Behnke 2012). Even though this seems much less likely than a direct association between the houses and their former inhabitants, the still visible ruins may have served as a cemetery that received dead from nearby settlements with varying catchment areas that are reflected in the Sr isotope ratios.

Karsdorf as part of a regional network of LBK settlements

Central Germany is characterised by dry climate conditions and wide distributions of loess on which *Schwarzerden* (chernozems) with black humus horizons of up to 60 cm developed. The area is among the most fertile worldwide and a preferred settlement landscape during the LBK (Behrens 1973; Preuß 1996, Typentafeln 1–113/Strichzeichnungen, Karten 1–11, Beilagen 1–17; Link 2011). Nevertheless, the current state of research is largely shaped by rescue excavations. This also affects the distribution of known LBK sites in the vicinity of Karsdorf and makes future discoveries very likely. Yet, the current knowledge of LBK sites indicates multiple options for interaction at the local (Unstrut valley and confluences; Colour Plate 3), regional (MES) and inter-regional (distribution area of the LBK as a whole) level, as well as for integration into settlement networks, as they have for instance been reconstructed in the Rhineland (Claßen 2005, Zimmermann *et al.* 2004).

The total number of burials at Karsdorf is too small to reflect the supposed complete living population of all excavated houses and from the whole duration of the settlement. This indicates a selection of people for burial near the houses, while others have likely been entombed at a different, yet unknown location. The deficiency of people is also supported by the mtDNA data. The house-associated inhumations are no former maternally related family groups, and the few individuals with identical mtDNA haplotypes were buried near different houses. A succession of multiple generations of the same community should have led to more shared maternal lineages. The heterogeneous haplotype composition implies that some individuals have left the community and others have joined it. The potential newcomers probably originated from other LBK groups, since the haplotype and haplogroup composition of the Karsdorf community is not significantly different from other LBK sites (Figure 6.1). Frequent dynamics and changing affiliations are plausible prerequisites of the prevalence of the mtDNA 'Neolithic package' and the LBK as an archaeological culture over several hundred years.

Interaction among LBK groups is also a possible reason for the formation of the two Sr isotope data ranges. They correspond to the baseline data for calcareous soils in the Unstrut river valley and on loess and are in agreement with differentiated land use around the Karsdorf site. However, the consideration of all data – including the low number of shared mtDNA lineages and the lacking dominance of either one or the other data range within the same house-associated burial group – let it appear more likely that the Sr isotope ratios reflect personal interaction. Residential changes may have taken place at the local level, among the communities in the valleys of the Unstrut River and its tributaries (Colour Plate 3). Although the current state of research cannot identify archaeologically contemporaneous sites, Karsdorf was probably part of an affiliating settlement system. The closely comparable geological properties of the sites have probably produced $^{87}Sr/^{86}Sr$ ratios similar to those of the Karsdorf community with dominance of either one or the other data range depending on the very local conditions. Thus, individuals of both data groups may have originated from any other of the LBK sites mapped in Colour Plate 3. Moreover, Karsdorf is situated in the central German settlement region (Link 2011, fig. 2) with numerous sites, of which many were located in more typical loess locations than the sites in the Unstrut valley. Therefore, especially some of the individuals with $^{87}Sr/^{86}Sr$ ratios in range B may be persons who joined the community from locations with the majority of cultivated plots on the loess.

Earlier Sr isotope and material culture research indicated the predominance of patrilocal residential rules during the LBK, in which descent followed the male line, and women moved into the communities of the men (Bentley *et al.* 2012, Bentley 2007, Whittle and Bickle 2013, Krahn 2003, Eisenhauer 2003). At Karsdorf, the majority of the males have Sr isotope ratios in range A, while the data of the females are more variable with relatively more individuals in range B. Furthermore, the combination of the Sr isotope data with the results of previous carbon and nitrogen isotope analyses (Oelze *et al.* 2011) discloses a rather tight data cluster of males with $\delta^{15}N$ values of around 9‰ and $^{87}Sr/^{86}Sr$ ratios in range A (Colour Plate 7). These sex-specific tendencies and the similarity of the Sr isotope data of the youngest child (Ind. 298; 2–3 years) with the majority of the males may be evaluated as being in agreement with the predominance of patrilocal residential rules. In this case, individual 299 (22–24 y.; ♀?) with an M2 in range B and an M3 in range A may be an example of a woman who has moved during adolescence.

However, two 8-year-old children (Ind. 157 and 529) and three men (Ind. 122, 304, 300) among the individuals in the loess range and indication for childhood mobility of the only juvenile (Ind. 95) suggest that entering personal relationships and doing this following patrilocal residential rules were not the only reason for residential relocations. Moreover, matrilocality cannot be ruled out completely if the similarity of the Sr isotope ratios of the two 8-year-old children and three women is considered. Furthermore, both Sr isotope ranges occur in association with all houses, which argues against a possibly non-local founder group with limited contacts afterwards, but for frequent exchange within the dense network of LBK sites.

Overall, the aDNA and isotope investigations at the settlement burial community of Karsdorf indicated that mobility and exchange among LBK groups were the rule rather than an exception. Within a consolidated settlement network, fluctuation among communities may have been a crucial component of LBK life which ensured the spread of this archaeological culture over thousands of kilometres and its endurance over several hundred years. There is no indication for the burial of genetically closely related individuals near their houses. Even though the dataset is small, the study is one step towards understanding the multiple facets of lifestyle and mobility of the early farmers in central Europe.

Acknowledgements

We are indebted to Michael Brauns, Sigrid Klaus and Bernd Höppner for their help with Sr isotope analysis at the Curt-Engelhorn Centre for Archaeometry at Mannheim and Sabine Moeller-Rieker for cloning and sequencing at the Institute of Anthropology in Mainz. We very much appreciate the very fruitful discussions with Hans-Christoph Strien about the archaeological and chronological context and the interpretation of the data presented here. This research was supported by the German Research Foundation (Al 287/7-1 and Al 287-7-2; Me 3245/1-1 and Me 3245/1-2) and the Geocycles Earth System Research Centre at the University of Mainz.

Bibliography

AlQahtani, S.J., Hector, M.P. and Liversidge, H.M. 2010. Brief communication: the London atlas of human tooth development and eruption. *American Journal of Physical Anthropology* 142, 481–90.

Alt, K.W., Arbogast, R.-M., Jeunesse, C. and van Willigen, S. (eds) 2004. *Archéologie funéraire du Néolithique danubien*. Zimmersheim: Association pour la Promotion de la Recherche Archéologique en Alsace.

Asam, T., Grupe, G. and Peters, J. 2006. Menschliche Subsistenzstrategien im Neolithikum: Eine Isotopenanalyse bayerischer Skelettfunde. *Anthropologischer Anzeiger* 64, 1–23.

Bachmann, G.H., Ehling, B.-C., Echner, R. and Schwab, M. (eds) 2008. *Geologie von Sachsen-Anhalt.* Stuttgart: E. Schweizerbart'sche Verlagsbuchhandlung.

Baum, N. 1990. Aiterhofen-Ödmühle. Paläodontologie eines bandkeramischen Gräberfeldes in Niederbayern. *Praehistorische Zeitschrift* 65, 157–203.

Behnke, H.J. 2007. Süßes aus der Bronzezeit – Siedlungsgunst im Unstruttal bei Karsdorf, Burgenlandkreis. Die Ausgrabungen 2004. *Archäologie in Sachsen-Anhalt* N.F. 4, 2006, I, 63–86.

Behnke, H.J. 2011. Erste Siedler der Linienbandkeramik in der Karsdorfer Feldflur. Ergebnisse der Ausgrabungen im Jahr 2005. *Archäologie in Sachsen-Anhalt* N.F. 5, 184–99.

Behnke, H.J. 2012. Siedlungsgunst im Unstruttal bei Karsdorf, Burgenlandkreis: Ergebnisse der Grabungen 2006 und 2007. *Archäologie in Sachsen-Anhalt* N.F.6, 35–70.

Behrens, H. 1973. *Die Jungsteinzeit im Mittelelbe-Saale-Gebiet.* Berlin: VEB Deutscher Verlag der Wissenschaften.

Bentley, R.A. 2006. Strontium isotopes from the earth to the archaeological skeleton: a review. *Journal of Archaeological Method and Theory* 13, 135–87.

Bentley, R.A. 2007. Mobility, specialisation and community diversity in the Linearbandkeramik: isotopic evidence from skeletons. In A. Whittle and V. Cummings (eds), *Going over: the Mesolithic–Neolithic transition in north-west* Europe, 117–40. Oxford: Oxford University Press for the British Academy.

Bentley, R.A. and Knipper, C. 2005. Transhumance at the early Neolithic settlement at Vaihingen (Germany). *Antiquity,* 79, online: http://www.antiquity.ac.uk/projgall/bentley/index.html.

Bentley, R.A., Chikhi, L. and Price, T.D. 2003. The Neolithic transition in Europe: comparing broad scale genetic and local scale isotopic evidence. *Antiquity* 77, 63–6.

Bentley, R.A., Bickle, P., Fibiger, L., Nowell, G.M., Dale, C.W., Hedges, R.E.M., Hamilton, J., Wahl, J., Francken, M., Grupe, G., Lenneis, E., Teschler-Nicola, M., Arbogast, R.-M., Hofmann, D. and Whittle, A. 2012. Community differentiation and kinship among Europe's first farmers. *Proceedings of the National Academy of Sciences of the United Sates of America* 109, 9326–30.

Bickle, P. and Whittle, A. (eds) 2013. *The first farmers of central Europe: diversity in LBK lifeways.* Oxford: Oxbow Books.

Bickle, P., Hofmann, D., Bentley, R.A., Hedges, R.E.M., Hamilton, J., Laiginhas, F., Nowell, G., Pearson, D.G., Grupe, G. and Whittle, A. 2011. Roots of diversity in a Linearbandkeramik community: isotope evidence at Aiterhofen (Bavaria, Germany). *Antiquity* 85, 1243–53.

Bocquet-Appel, J.P. and Bar-Yosef, O. (eds) 2008. *The Neolithic demographic transition and its consequences.* Dordrecht, London: Springer.

Bogaard, A., Krause, R. and Strien, H.-C. 2011. Towards a social geography of cultivation and plant use in an early farming community: Vaihingen an der Enz, south-west Germany. *Antiquity* 85, 395–416.

Bramanti, B. 2008. Ancient DNA: Genetic analysis of aDNA from sixteen skeletons of the Vedrovice collection. *Anthropologie* 46, 153–60.

Bramanti, B., Thomas, M.G., Haak, W., Unterländer, M., Jores, P., Tambets, K., Antanaitis-Jacobs, I., Haidle, M.N., Jankauskaus, R., Kind, C.-J., Lüth, F., Terberger, T., Hiller, J., Matsumura, S., Forster, P. and Burger, J. 2009. Genetic discontinuity between local hunter-gatherers and central Europe's first farmers. *Science* 326, 137–40.

Brandt, G., Haak, W., Adler, C.J., Roth, C., Szécsényi-Nagy, A., Karimia, S., Möller-Rieker, S., Meller, H., Ganslmeier, R., Friederich, S., Dresely, V., Nicklisch, N., Pickrell, J.K., Sirocko, F., Reich, D., Cooper, A., Alt, K.W. and The Genographic Consortium. 2013. Ancient DNA reveals key stages in the formation of central European mitochondrial genetic diversity. *Science* 342, 257–61.

Brooks, S.T. and Suchey, J.M. 1990. Skeletal age determination based on the os pubis: a comparision of the Acsádi-Nemeskéri and Suchey-Brooks methods. *Journal of Human Evolution* 5, 227–38.

Buttler, W. and Haberey, W. 1936. *Die bandkeramische Ansiedlung bei Köln-Lindenthal.* Berlin: de Gruyter.

Claßen, E. 2005. Siedlungsstrukturen der Bandkeramik im Rheinland. In J. Lüning, C. Frirdich and A. Zimmermann (eds), *Die Bandkeramik im 21. Jahrhundert. Symposium in der Abtei Brauweiler bei Köln vom 16.9.–19.9.2002*, 113–24. Rahden: Marie Leidorf.

Denaire, A. 2009. Radiocarbon dating of the western European Neolithic: comparison of the dates on bones and dates on charcoals. *Radiocarbon* 51, 657–74.

Dürrwächter, C., Craig, O.E., Collins, M.J., Burger, J. and Alt, K.W. 2006. Beyond the grave: variability in Neolithic diets in Southern Germany? *Journal of Archaeological Science* 33, 39–48.

Eisenhauer, U. 2003. Jüngerbandkeramische Residenzregeln: Patrilokalität in Talheim. In J. Eckert, U. Eisenhauer and A. Zimmermann (eds), *Archäologische Perspektiven: Analysen und Interpretationen im Wandel. Festschrift für Jens Lüning zum 65. Geburtstag*, 561–73. Rahden: Marie Leidorf.

Excoffier, L. and Lischer, H.E. 2010. Arlequin suite ver 3.5: a new series of programs to perform population genetics analyses under Linux and Windows. *Molecular Ecology Resources* 10, 564–7.

Ferembach, D., Schwidetzki, I. and Stloukal, M. 1980. Recommendations for age and sex diagnoses of skeletons. *Journal of Human Evolution* 9, 517–49.

Fisher, R.A. 1922. On the interpretation of $\chi 2$ from contingency tables, and the calculation of P. *Journal of the Royal Statistical Society* 85, 87–94.

Fritsch, B., Claßen, E., Müller, U. and Dresely, V. 2008. Die linienbandkeramischen Gräberfelder von Derenburg 'Meerenstieg II' und Halberstadt 'Sonntagsfeld', Lkr. Harz. *Jahresschrift für Mitteldeutsche Vorgeschichte* 92, 25–229.

Fu, Q., Mittnik, A., Johnson, P.L.F., Bos, K., Lari, M., Bollongino, R., Sun, C., Giemsch, L., Schmitz, R., Burger, J., Ronchitelli, Anna M., Martini, F., Cremonesi, Renata G., Svoboda, J., Bauer, P., Caramelli, D., Castellano, S., Reich, D., Pääbo, S. and Krause, J. 2013. A revised timescale for human evolution based on ancient mitochondrial genomes. *Current Biology* 23, 553–9.

Gerling, C. and Francken, M. 2007. Das linienbandkeramische Gräberfeld von Schwetzingen. *Bulletin der Schweizerischen Gesellschaft für Anthropologie* 13, 43–50.

Haak, W., Forster, P., Bramanti, B., Matsumura, S., Brandt, G., Tänzer, M., Villems, R., Renfrew, C., Gronenborn, D., Alt, K.W. and Burger, J. 2005. Ancient DNA from the first European farmers in 7500-year-old Neolithic sites. *Science* 310, 1016–18.

Haak, W., Brandt, G., de Jong, H.N., Meyer, C., Ganslmeier, R., Heyd, V., Hawkesworth, C., Pike, A.W.G., Meller, H. and Alt, K.W. 2008. Ancient DNA, Strontium isotopes, and osteological analyses shed light on social and kinship organization of the Later Stone Age. *Proceedings of the National Academy of Sciences of the United States of America* 105, 18226–31.

Haak, W., Balanovsky, O., Sanchez, J.J., Koshel, S., Zaporozhchenko, V., Adler, C.J., Der Sarkissian, C.S.I., Brandt, G., Schwarz, C., Nicklisch, N., Dresely, V., Fritsch, B., Balanovska, E., Villems, R., Meller, H., Alt, K.W., Cooper, A. and the Genographic Consortium 2010. Ancient DNA from European early Neolithic farmers reveals their Near Eastern affinities. *PLoS Biology* 8, e1000536.

Klopfleisch, F. 1883. *Vorgeschichtliche Altertümer der Provinz Sachsen I.* Halle: Hendel.

Knipper, C. 2011. *Die räumliche Organisation der linearbandkeramischen Rinderhaltung: naturwissenschaftliche und archäologische Untersuchungen.* Oxford: Archaeopress.

Knipper, C., Maurer, A.-F., Peters, D., Meyer, C., Brauns, M., Galer, S.G., von Freeden, U., Schöne, B. and Alt, K.W. 2012. Mobility in Thuringia or mobile Thuringians: a strontium isotope study from early Medieval central Germany. In E. Kaiser, J. Burger and W. Schier (eds), *Migrations in prehistory and early history: stable isotopes and population genetics*, 293–317. Berlin: De Gruyter.

Krahn, C. 2003. Überlegungen zum Interaktionssystem der bandkeramischen Siedlungen auf der Aldenhovener Platte. In J. Eckert, U. Eisenhauer and A. Zimmermann (eds), *Archäologische Perspektiven. Analysen und Interpretationen im Wandel. Festschrift für Jens Lüning zum 65. Geburtstag*, 515–44. Rahden: Marie Leidorf.

Link, T. 2011. Böhmische Dörfer? Zur Stellung der Dresdener Elbtalweitung zwischen sächsischer und böhmischer Bandkeramik. In T. Doppler, B. Ramminger and D. Schimmelpfennig (eds), *Grenzen und Grenzräume? Beispiele aus Neolithikum und Bronzezeit*, 11–24. Kerpen-Loogh: Welt und Erde.

Lovejoy, C.O. 1985. Dental wear in the Libben population: its functional pattern and role in the determination of adult skeletal age at death. *American Journal of Physical Anthropology* 68, 47–56.

Lovejoy, C.O., Meindl, R.S., Pryzbeck, T.R. and Mensforth, R.P. 1985. Chronological metamorphosis of the auricular surface of the ilium: a new method for the determination of adult skeletal age at death. *American Journal of Physical Anthropology* 68, 15–28.

Lüning, J., Frirdich, C. and Zimmermann, A. (eds) 2005. *Die Bandkeramik im 21. Jahrhundert. Symposium in der Abtei Brauweiler bei Köln vom 16.9.–19.9.2002*. Rahden: Marie Leidorf.

Malmström, H., Gilbert, T.M., Thomas, M.G., Brandström, M., Storå, J., Molnar, P., Andersen, P.K., Bendixen, C., Holmlund, G., Götherström, A. and Willerslev, E. 2009. Ancient DNA reveals lack of continuity between neolithic hunter-gatherers and contemporary Scandinavians. *Current Biology* 19, 1758–62.

Maurer, A.-F., Galer, S.J.G., Knipper, C., Beierlein, L., Nunn, E.V., Peters, D., Tütken, T., Alt, K.W. and Schöne, B. 2012. Bioavailable $^{87}Sr/^{86}Sr$ in different environmental samples – effects of anthropogenic contamination and implications for isoscapes in past migration studies. *Science of the Total Environment* 433, 216–29.

Meier-Arendt, W. 1966. *Die bandkeramische Kultur im Untermaingebiet*. Bonn: Habelt.

Modderman, P.J.R. 1970. *Linearbandkeramik aus Elsloo und Stein*. Gravenhage: Staatsuitgeverij.

Montgomery, J., Evans, J.A. and Cooper, R.E. 2007. Resolving archaeological populations with Sr-isotope mixing models. *Applied Geochemistry* 22, 1502–14.

Murail, P., Bruzek, J., Houët, F. and Cunha, E. 2005. DSP: a tool for probabilistic sex diagnosis using worldwide variability in hip-bone measurements. *Bulletins et Mémoires de la Société d'Anthropologie de Paris* 17, 167–76.

Nehlich, O., Montgomery, J., Evans, J., Schade-Lindig, S., Pichler, S.L., Richards, M.P. and Alt, K.W. 2009. Mobility or migration – a case study from the Neolithic settlement of Nieder-Mörlen (Hessen, Germany). *Journal of Archaeological Science* 36, 1791–9.

Nicklisch, N. in preparation. *Paläodemographische und epidemiologische Untersuchungen an neolithischen und frühbronzezeitlichen Bestattungen aus dem Mittelelbe-Saale-Gebiet im Kontext populationsdynamischer Prozesse*. PhD thesis, University of Mainz.

Oelze, V.M., Siebert, A., Nicklisch, N., Meller, H., Dresely, V. and Alt, K.W. 2011. Early Neolithic diet and animal husbandry: stable isotope evidence from three Linearbandkeramik (LBK) sites in central Germany. *Journal of Archaeological Science* 38, 270–9.

Oelze, V.M., Nehlich, O. and Richards, M.P. 2012. 'There's no place like home' – no isotopic evidence for mobility at the early Bronze age cemetery of Singen, Germany. *Archaeometry* 54, 752–78.

Phenice, T.W. 1969. A newly developed visual method of sexing in the Os pubis. *American Journal of Physical Anthropology* 30, 297–301.

Preuß, J. (ed.) 1996. *Das Neolithikum in Mitteleuropa. Kulturen-Wirtschaft-Umwelt vom 6. bis 3. Jahrtausend v.u.Z. Bd. 3*. Wilkau-Hasslau: Beier and Beran.

Price, T.D., Wahl, J. and Bentley, R.A. 2006. Isotopic evidence for mobility and group organization among Neolithic farmers at Talheim, Germany, 5000 BC. *European Journal of Archaeology* 9, 259–84.

Price, T.D., Bentley, R.A., Lüning, J., Gronenborn, D. and Wahl, J. 2001. Prehistoric human migration in the Linearbandkeramik of central Europe. *Antiquity* 75, 593–603.

Scharl, S. 2004. *Die Neolithisierung Europas: ausgewählte Modelle und Hypothesen*. Rahden: Marie Leidorf.

Schwarz, R. In preparation. Chronologie und Verbreitung der neolithischen Kulturen in Sachsen-Anhalt. In H. Meller (ed.), *Kataloge zur Dauerausstellung im Landesmuseum für Vorgeschichte Halle 3: Neolithikum II*. Halle: Landesmuseum für Denkmalpflege und Archäologie Sachsen-Anhalt, Landesamt für Vorgeschichte.

Schweissing, M.M. and Grupe, G. 2003. Stable strontium isotopes in human teeth and bone: a key to migration events of the late Roman period in Bavaria. *Journal of Archaeological Science* 30, 1373–83.

Siemoneit, B. 1997. *Das Kind in der Linienbandkeramik. Befunde aus Gräberfeldern und Siedlungen in Mitteleuropa*. Rahden: Marie Leidorf.

Smolla, G. 1974. Paläodemographische und paläopathologische Grundlagen. In W. Bernhard and A. Kandler (eds), *Bevölkerungsbiologie*, 333–343. Stuttgart: Fischer.

Stloukal, M. and Hanáková, H. 1978. Die Länge der Längsknochen altslavischer Bevölkerungen unter besonderer Berücksichtigung von Wachstumsfragen. *Homo* 29, 53–69.

Strien, H.-C. 2000. *Untersuchungen zur Bandkeramik in Württemberg*. Bonn: Habelt.

Szécsényi-Nagy, A., Brandt, G., Keerl, V., Jakucs, J., Fecher, M., Moeller-Rieker, S., Köhler, K., Mende, B.G., Oross, K., Marton, T., Osztás, A., Kiss, V., Pálfy, G., Molnár, E., Raczky, P., Anders, A., Sebők, K., Czene, A., Kustár, R., Paluch, T., Somogyi, K., Šlaus, M., Novak, M., Zoffmann, Z., Tóth, G., Ősz, B., Voicsek, V., Bánffy, E. and Alt, K.W. In preparation. Ancient mitochondrial and Y chromosomal DNA reveals the western Carpathian Basin as a corridor between the Near East and Central Europe in the 6th millennium BC.

Tamura, K. and Nei, N. 1993. Estimation of the number of nucleotide substitutions in the control region of mitochondrial DNA in humans and chimpanzees. *Molecular Biology and Evolution* 10, 512–26.

Trautmann, I. 2006. *The significance of cremations in Early Neolithic communities in central Europe*. PhD Dissertation, Universität Tübingen. http://tobias-lib.ub.uni-tuebingen.de/volltexte/2007/3005/ [last access: September 2013].

Turck, R., Kober, B., Kontny, J., Haack, F. and Zeeb-Lanz, A. 2012. 'Widely travelled people' at Herxheim? Sr isotopes as indicators of mobility. In E. Kaiser, J. Burger and W. Schier (eds), *Population dynamics in prehistory and early history: new approaches using stable isotopes and genetics*, 149–63. Berlin: De Gruyter.

Ubelaker, D.H. 1989. *Human skeletal remains: excavation, analysis, interpretation*. Washington, D.C.: Taraxacum.

United Nations Department of Social Affairs. 1955. *Age and sex patterns of mortality: model life-tables for underdeveloped countries*. New York: United Nations.

Veit, U. 1996. *Studien zum Problem der Siedlungsbestattungen im europäischen Neolithikum*, Münster, New York: Waxmann.

Weiss, K.M. 1973. *Demographic models for anthropology*. Washington: Society for American Archaeology.

Whittle, A. and Bickle, P. 2013. Performing LBK lifeways. In P. Bickle and A. Whittle (eds), *The first farmers of central Europe: diversity in LBK lifeways*, 385–401. Oxford: Oxbow Books.

Zimmermann, A. (ed.) 2009. *Studien zum Alt- und Mittelneolithikum im Rheinischen Braunkohlenrevier. Beiträge zur neolithischen Besiedlung der Aldenhovener Platte VII.* Rahden: Marie Leidorf.

Zimmermann, A., Richter, J., Frank, T. and Wendt, K.P. 2004. Landschaftsarchäologie II – Überlegungen zu Prinzipien einer Landschaftsarchäologie. *Bericht der Römisch-Germanischen Kommission* 85, 37–95.

Zimmermann, A., Wendt, K.P., Frank, T. and Hilpert, J. 2009. Landscape archaeology in central Europe. *Proceedings of The Prehistoric Society* 75, 1–53.

7

Cattle and Sheep Herding at Cheia, Romania, at the Turn of the Fifth Millennium cal BC

A View from Stable Isotope Analysis

MARIE BALASSE, CARLOS TORNERO, STÉPHANIE BRÉHARD, JOËL UGHETTO-MONFRIN, VALENTINA VOINEA AND ADRIAN BĂLĂŞESCU

Introduction

THE PRESENT STUDY WAS CONDUCTED within the framework of the SIANHE project involving stable isotope analysis of well documented zooarchaeological assemblages from sites located in south-east Romania and dated from the early sixth millennium to the second half of the fifth millennium cal BC (Figure 7.1). This time period encompasses the earliest occurrence of food producing communities in southern Romania with the Starčevo-Criş I culture, attested at Măgura-*Boldul lui Moş Ivănuş* (Andreescu and Mirea 2008, Mirea 2011); the settling of Chalcolithic communities west of the Black Sea from the turn of the fifth millennium cal BC, illustrated by the Hamangia settlement at Cheia (Voinea and Neagu 2008); and the advent of the tell sites of the Chalcolithic culture of Gumelniţa A2 in the second half of the fifth millennium cal BC, present at Borduşani-Popină (Popovici *et al.* 2008), Hârşova (Popovici *et al.* 2001) and Vităneşti (Andreescu *et al.* 2003). Stable isotope analyses are conducted on these assemblages in order to investigate cattle, caprine and pig husbandry practices in connection with the orientation of animal production, as evidenced by the relative importance of domestic stock in the faunal spectra and from its demographic management.

Specifically, the stable isotope studies conducted within the framework of the SIANHE project address (1) the environment of animal husbandry, or the place of domestic animals in the ecosystem (the surrounding landscape and trophic web), from which the scale of husbandry may also be approached; (2) the management of animal diet, including seasonal provisioning with fodder; and (3) birth seasonality and birth season, from which the seasonal availability of

Proceedings of the British Academy **198**, 115–42. © The British Academy 2014.

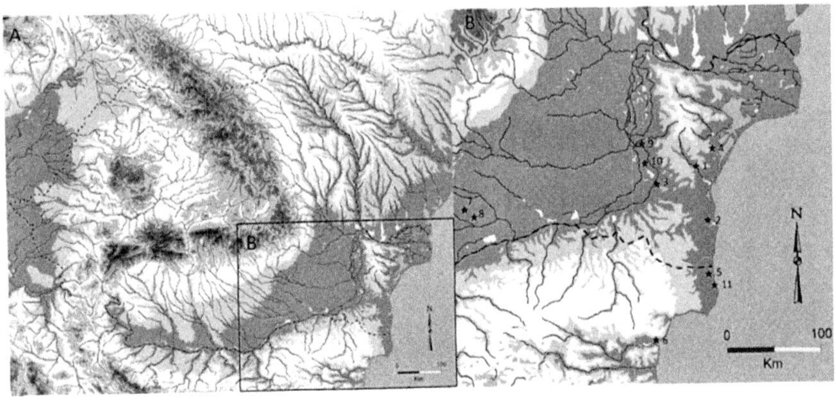

Figure 7.1 Map of Romania (A) and the study area (B), with location of the sites mentioned in the text. 1: Cheia; 2: Techirghiol; 3: Cernavoda; 4: Baia-Hamangia, Ceamurlia de Jos, Golovița; 5: Durankulak; 6: Varna; 7: Măgura; 8: Vităneşti; 9: Hârşova; 10: Borduşani-Popină. Sites 1–5 belong to the Hamangia culture.

animal resources such as tender meat and milk may be inferred. These points are addressed in a diachronic perspective. These parameters are also strongly underlain by the degree of interweaving of animal and plant husbandry (Bogaard 2004, 2005; Halstead 2006). Results from the stable isotope analysis conducted on the faunal assemblage of Măgura-*Boldul lui Moş Ivănuş* brought new elements to these considerations. At Măgura, the analysis of bone collagen carbon and nitrogen isotope ratios ($\delta^{13}C$ and $\delta^{15}N$) suggested that cattle and caprines fed in places with similar stable isotope ecology to those of the hunted wild large herbivores, whereas domestic pigs had similar dietary behaviour to their wild counterparts. These findings suggest extensive management strategies. Nevertheless, investigation of cattle and sheep diet at the seasonal scale, from sequential analysis ($\delta^{18}O$ and $\delta^{13}C$) in tooth enamel, provided evidence for winter leaf-foddering in one bovine, involving possibly the keeping of the animal by the settlement, and late winter supplementary feeding of sheep with plant material collected in the summer. In this context, late winter also corresponded to the sheep birth season, and therefore lactation time, as inferred from tooth enamel $\delta^{18}O$ analysis. The reduced size of the early Neolithic assemblage of Măgura-*Boldul lui Moş Ivănuş* requires great caution when interpreting these results, which may still be considered as the evidence for the existence of such husbandry practices in this time period, although the scale and the frequency of such husbandry practices remain unknown (Balasse *et al.* 2013). In the present study, similar questions and analytical strategies were addressed and applied to the faunal assemblage of the Hamangia III site of Cheia.

Cheia and its faunal assemblage

Defined by Berciu (1966), when archaeological excavations were conducted at the eponymous site, the Hamangia culture in Romania and Bulgaria has seen limited archaeological research over the last half century. The particularities of the Hamangia culture, including its genesis, its periodisation, the degree of mobility of these communities and the intercultural exchange dynamics all remain to be further deepened. In this effort, a particularly important step was the work conducted at the cemetery of Durankulak, where the layout of space and the funeral inventory showed the existence of a complex social organisation (Todorova 2002). For a long time, Hamangia communities were characterised by conservatism and isolation. The intercultural relationships were considered unidirectional, from neighbouring communities to the Hamangia area, as exchanges or even displacement of communities (Boian, Precucuteni, Marica) (Berciu 1966, Haşotti 1997, Slavchev 2004/2005). However, the archaeological finds of the last two decades have revealed the presence of specific Hamangia items – clay figurines and fine pieces of *Spondylus* – beyond the line of the Danube in the Romanian plain, in north-east Muntenia and southern Moldova (Marinescu-Bîlcu *et al.* 1995).

The archaeological discoveries at the settlement of Cheia (2001–2011) allow for a re-evaluation of the paleoeconomy of Hamangia communities. The site is located in the karst area of the Central Dobruja plateau, between the Danube (28 km west as the crow flies) and the Black Sea (14 km east; Figure 7.1). The stratigraphic sequences distinguished in sections of the central zone of the settlement demonstrate a periodic habitation, with reoccupations after a sufficiently long period of abandonment for the old deserted dwellings to be totally destroyed (Voinea and Neagu 2008). Radiocarbon dating was conducted on bones among the corpus selected for stable isotope analyses. Results confirm a chronological attribution to the turn of the fifth millennium cal BC (Table 7.1), corresponding to the beginning of the Chalcolithic period in Dobruja.

Researched over an area of 600 square metres, representing approximately a fifth of the settlement, the site has provided an impressive amount of osteological material, over 11,000 fragments (NISP = 11,163), which is the largest faunal assemblage for the Hamangia culture so far. Faunal remains belong to molluscs, fish, reptiles, birds and mammals (Bălăşescu and Radu 2003, 2004; Bălăşescu 2008 and unpublished results; Radu 2008). Mammals are prevalent (83% of remains, but only 3920 of those were identified to the specific level), followed by fish (10%) and birds (4%). Mammalian fauna are heavily dominated by domestic animals (89%), illustrating the importance of animal husbandry for the Cheia community. Domestic stock includes cattle (49%) and caprines (sheep and goats, 39%). Domestic pig could not be positively identified within the suid remains, but its presence in the unidentified *Sus* remains (1% of the whole assemblage) cannot be ruled out. Wild animal remains belong to a large spectrum (14 species). Mainly

Table 7.1 Results from radiocarbon dating of bone collagen from Cheia

Context	*Specimen #*	*Identification*	*Code*	*BP*	*1 sigma*	*2 sigma*
SU 3074	CHE Bos 3074-1	*Bos primigenius*	UBA-18099	6141±29	5206–5016 cal BC	5208–5003 cal BC
SU 3094	CHE Cervus 5	*Cervus elaphus*	UBA-18100	6064±26	5005–4937 cal BC	5047–4855 cal BC
SU 3094	CHE Bos 80	*Bos taurus*	UBA-21109	6093±39	5188–4944 cal BC	5207–4856 cal BC
SU 3118	CHE Bos 66	*Bos taurus*	UBA-21110	6138±44	5207–5003 cal BC	5214–4961 cal BC

big and medium-sized animals were hunted: deer, a small wild horse, perhaps hydruntine or hemione, aurochs and wild boar.

Zooarchaeological studies carried out on the rare other representative faunal assemblages in Romania and Bulgaria for the Hamangia culture, at Techirghiol and Durankulak, show great similarities with the finds at Cheia: a great predominance of husbandry over hunting, exploitation of cattle and caprines, and only a rare presence of domestic pig (Manhart 1998, Spassov 2002, Bălăşescu and Radu 2004). To our knowledge, no palaeobotanical studies to date have been conducted on Hamangia contexts, and the crop husbandry practices of these communities remain to be defined. Botanical remains including wood charcoal and charred seeds were retrieved during the excavation at Cheia. Their study is in progress and might reveal valuable information on this question.

Material and methods

Stable isotope analyses were conducted on bone collagen ($\delta^{15}N$ and $\delta^{13}C$) and tooth enamel bioapatite ($\delta^{18}O$ and $\delta^{13}C$). Collagen was extracted from 67 bone specimens from red deer, roe deer, wild horse (hydruntine/hemione), aurochs, wild boars, cattle and sheep (Table 7.2). Among bovines and suids, domestic and wild taxa were distinguished using morphometrical data (Degerbøl and Fredskild 1970, El Susi 1996, Manhart 1998). All bones belong to different individuals and were from adult animals. Tooth enamel bioapatite was extracted from 12 domestic cattle lower third molars (M3), and the second (M2) and third molars (M3) from 13 sheep (26 teeth). Bone collagen from the cattle mandibles was also extracted, adding 12 more collagen samples, for comparison of $\delta^{13}C$ values from bone collagen and enamel bioapatite.

Table 7.2 Results from bone collagen analysis: collagen extraction yield, collagen carbon (%C) and nitrogen contents (%N), atomic C/N ratio, stable carbon ($\delta^{13}C$) and nitrogen isotope ratios ($\delta^{15}N$)

Sample	*Identification*	*Skeleton part*	*Bone wt (mg)*	*Yield (mg/g)*	*%C*	*%N*	*C/N*	*$\delta^{13}C$ (‰)*	*$\delta^{15}N$ (‰)*
CHE Cervus 1	*Cervus elaphus*	metatarsal R	354	144	41.1	15.1	3.2	–20.1	5.5
CHE Cervus 2	*Cervus elaphus*	metatarsal R	301	91	39.9	14.7	3.2	–20.2	5.4
CHE Cervus 3	*Cervus elaphus*	metatarsal R	299	67	35.0	12.6	3.2	–20.0	5.5
CHE Cervus 4	*Cervus elaphus*	metatarsal R	304	97	40.4	14.8	3.2	–21.6	5.1
CHE Cervus 5	*Cervus elaphus*	metatarsal R	345	141	41.0	15.0	3.2	–20.6	5.5
CHE Cervus 6	*Cervus elaphus*	metatarsal R	328	117	41.0	15.0	3.2	–20.7	5.6
CHE Equus 1	*Equus* sp.	tibia R	303	144	41.6	15.3	3.2	–21.1	3.6
CHE Equus 2	*Equus* sp.	tibia R	306	161	41.2	15.1	3.2	–20.7	3.6
CHE Equus 3	*Equus* sp.	tibia R	331	147	40.6	14.8	3.2	–21.1	4.1
CHE Equus 4	*Equus* sp.	tibia R	302	40	37.2	13.6	3.2	–20.7	3.4
CHE Equus 5	*Equus* sp.	tibia R	325	166	41.4	15.2	3.2	–21.2	3.9
CHE Equus 6	*Equus* sp.	tibia R	318	47	38.0	14.0	3.2	–21.0	3.6
CHE Equus 7	*Equus* sp.	tibia R	310	25	38.1	14.1	3.2	–21.2	3.3
CHE Sus 1	*Sus scrofa*	metacarpal R	295	172	41.3	15.2	3.2	–20.9	5.6
CHE Sus 2	*Sus scrofa*	metacarpal R	300	131	40.9	15.0	3.2	–21.0	6.2
CHE Sus 3	*Sus scrofa*	metacarpal R	306	134	41.3	15.2	3.2	–21.7	6.6
CHE Sus 4	*Sus scrofa*	metacarpal R	306	161	41.2	15.2	3.2	–20.3	4.6
CHE Sus 5	*Sus scrofa*	metacarpal R	303	146	42.0	15.4	3.2	–19.6	4.0
CHE Sus 6	*Sus scrofa*	metacarpal R	318	157	41.4	15.2	3.2	–20.9	5.9
CHE Capreolus 1	*Capreolus capreolus*	humerus L	315	129	41.5	15.3	3.2	–19.6	6.2

Table 7.2 Continued

Sample	*Identification*	*Skeleton part*	*Bone wt (mg)*	*Yield (mg/g)*	*%C*	*%N*	*C/N*	*δ^{13}C (‰)*	*δ^{15}N (‰)*
CHE Capreolus 2	*Capreolus capreolus*	humerus L	347	133	37.9	13.7	3.2	–21.7	5.8
CHE Capreolus 3	*Capreolus capreolus*	humerus L	305	132	36.2	13.2	3.2	–18.3	5.5
CHE Capreolus 4	*Capreolus capreolus*	humerus L	307	142	37.8	13.8	3.2	–19.2	4.9
CHE Capreolus 5	*Capreolus capreolus*	humerus L	302	123	36.9	13.4	3.2	–20.4	5.8
CHE Capreolus 6	*Capreolus capreolus*	humerus L	292	111	40.0	14.6	3.2	–20.4	4.3
CHE Capreolus 7	*Capreolus capreolus*	humerus L	294	141	40.0	14.6	3.2	–21.3	5.8
CHE Capreolus 8	*Capreolus capreolus*	humerus L	296	113	40.3	14.7	3.2	–21.6	5.4
CHE Capreolus 9	*Capreolus capreolus*	humerus L	308	130	37.7	13.8	3.2	–20.3	6.2
CHE Capreolus 10	*Capreolus capreolus*	humerus L	307	123	36.7	13.2	3.2	–21.4	7.0
CHE Bos 3074–1	*Bos primigenius*	humerus L	303	43	41.9	15.4	3.2	–21.8	5.5
CHE Bos 3067–4	*Bos taurus*	humerus R	300	129	44.3	16.2	3.2	–20.3	6.1
CHE Bos 3092–5	*Bos taurus*	humerus R	304	68	42.6	15.7	3.2	–19.9	6.3
CHE Bos 3118–6	*Bos taurus*	humerus R	308	16	38.2	14.2	3.1	–20.7	5.4
CHE Bos 3106–7	*Bos taurus*	humerus R	316	86	42.5	15.6	3.2	–20.8	6.4
CHE Bos 3107–8	*Bos taurus*	humerus R	319	89	40.9	15.1	3.2	–19.6	6.3
CHE Bos 3079–9	*Bos taurus*	humerus R	365	122	44.0	16.2	3.2	–19.8	6.3
CHE Bos 3108–10	*Bos taurus*	humerus R	310	99	43.5	16.0	3.2	–19.3	6.4
CHE Bos 3079–12	*Bos taurus*	humerus R	323	49	40.6	14.8	3.2	–20.2	5.4
CHE Bos 3099–13	*Bos taurus*	humerus R	301	15	37.7	14.1	3.1	–20.3	5.0

Table 7.2 Continued

Sample	*Identification*	*Skeleton part*	*Bone wt (mg)*	*Yield (mg/g)*	*%C*	*%N*	*C/N*	*δ¹³C (‰)*	*δ¹⁵N (‰)*
CHE Bos 3106–14	*Bos taurus*	humerus R	306	110	43.6	16.0	3.2	–20.0	5.3
CHE Bos 3092–15	*Bos taurus*	humerus R	304	129	43.7	16.1	3.2	–20.0	5.6
CHE Bos 3106–16	*Bos taurus*	humerus R	330	88	43.3	16.0	3.2	–20.2	5.5
CHE Bos L1–18	*Bos taurus*	humerus R	298	132	40.9	15.0	3.2	–19.9	5.3
CHE Bos 3090–19	*Bos taurus*	humerus R	298	141	40.5	14.9	3.2	–20.3	5.2
CHE Bos 3105–20	*Bos taurus*	humerus R	303	89	40.6	14.8	3.2	–20.2	5.4
CHE Bos 3099–21	*Bos taurus*	humerus R	303	34	41.2	15.2	3.2	–20.1	5.2
CHE Bos 3106–22	*Bos taurus*	humerus R	321	46	42.7	15.7	3.2	–19.8	5.0
CHE Bos 3118–23	*Bos taurus*	humerus R	316	41	39.2	14.3	3.2	–20.4	5.0
CHE Bos 3106–24	*Bos taurus*	humerus R	297	89	40.2	14.9	3.2	–19.6	6.1
CHE Bos 3103–26	*Bos taurus*	humerus R	315	132	44.7	16.5	3.2	–19.7	5.1
CHE Bos 3106–27	*Bos taurus*	humerus R	296	63	36.9	13.5	3.2	–18.9	5.5
CHE Bos 3092–28	*Bos taurus*	humerus R	306	109	43.1	15.7	3.2	–20.3	5.5
CHE Bos 3099–29	*Bos taurus*	humerus R	334	73	42.8	15.7	3.2	–19.4	5.6
CHE Ovis 1	*Ovis aries*	mandible L	360	80	42.3	15.5	3.2	–20.0	6.1
CHE Ovis 2	*Ovis aries*	mandible L	332	83	42.7	15.7	3.2	–19.9	6.6
CHE Ovis 3	*Ovis aries*	mandible L	353	108	42.4	15.5	3.2	–20.4	6.1
CHE Ovis 4	*Ovis aries*	mandible L	344	114	42.5	15.6	3.2	–19.8	6.1
CHE Ovis 5	*Ovis aries*	mandible L	366	120	42.6	15.8	3.1	–19.2	6.3
CHE Ovis 6	*Ovis aries*	mandible L	356	108	42.8	15.7	3.2	–19.7	5.0
CHE Ovis 7	*Ovis aries*	mandible L	351	93	42.6	15.4	3.2	–19.9	5.3
CHE Ovis 9	*Ovis aries*	mandible L	352	122	43.1	15.8	3.2	–20.1	5.9

Table 7.2 Continued

Sample	*Identification*	*Skeleton part*	*Bone wt (mg)*	*Yield (mg/g)*	*%C*	*%N*	*C/N*	*δ¹³C (‰)*	*δ¹⁵N (‰)*
CHE Ovis 10	*Ovis aries*	mandible L	354	97	42.9	15.8	3.2	–20.2	5.8
CHE Ovis 11	*Ovis aries*	mandible L	372	128	43.2	16.1	3.1	–18.5	6.9
CHE Ovis 12	*Ovis aries*	mandible L	352	111	43.2	16.0	3.1	–19.0	5.7
CHE Ovis 13	*Ovis aries*	mandible L	355	101	42.0	15.4	3.2	–18.9	5.8
CHE Ovis 14	*Ovis aries*	mandible L	347	128	43.3	15.9	3.2	–19.8	5.1
CHE Ovis 15	*Ovis aries*	mandible L	350	105	43.2	15.9	3.2	–18.8	5.8
CHE Bos 54	*Bos taurus*	mandible L	400	121	43.1	15.7	3.2	–19.1	5.4
CHE Bos 55	*Bos taurus*	mandible L	356	105	41.1	15.0	3.2	–20.2	4.7
CHE Bos 57	*Bos taurus*	mandible R	420	124	42.5	16.2	3.1	–19.2	5.0
CHE Bos 61	*Bos taurus*	mandible R	339	73	39.0	14.3	3.2	–19.6	5.4
CHE Bos 62	*Bos taurus*	mandible R	309	77	41.6	15.6	3.1	–19.9	4.9
CHE Bos 66	*Bos taurus*	mandible R	332	122	41.2	15.2	3.2	–20.3	5.7
CHE Bos 68	*Bos taurus*	mandible R	405	137	42.9	15.7	3.2	–18.8	5.0
CHE Bos 76	*Bos taurus*	mandible R	345	45	39.8	14.7	3.2	–19.5	4.6
CHE Bos 79	*Bos taurus*	mandible R	368	66	37.0	13.3	3.2	–20.6	6.2
CHE Bos 80	*Bos taurus*	mandible R	384	112	42.1	15.5	3.2	–19.2	5.7
CHE Bos 85	*Bos taurus*	mandible R	361	159	42.6	15.8	3.2	–20.2	5.4
CHE Bos 107	*Bos taurus*	mandible R	342	135	42.7	16.1	3.1	–19.7	5.5

Collagen was extracted from 290 to 420 mg of bone powder, following steps described in Bocherens *et al.* (1991). The analysis was conducted on 400 to 500 μg of collagen on an elemental analyser Thermo Flash 112 interfaced to a Thermo DeltaVAdvantage IRMS. The analytical precision, determined from nine to 13 analyses of an alanine standard within each run, varies from 0.11 to 0.20‰ for $\delta^{13}C$ and from 0.04 to 0.13‰ for $\delta^{15}N$, and from 0.1 to 0.4% for C content and from 0.1 to 0.2% for N content.

Tooth enamel was sampled sequentially on the first lobe of 12 cattle third molars (M3), and on the second lobe of 26 sheep M2 and M3. Each sample is positioned in the tooth crown by its distance from the enamel-root junction (ERJ). After pre-treatment (Balasse *et al.* 2003), enamel samples weighing approximately 600 μg were reacted with orthophosphoric acid at 70°C for 240 seconds in individual vessels in an automated cryogenic distillation system (Kiel IV device), interfaced with a DeltaVAdvantage IRMS. Over the period of analysis, 85 runs of our laboratory internal carbonate standard (Marbre LM) gave an analytical precision of 0.02‰ (1σ) for $\delta^{13}C$ and 0.07‰ for $\delta^{18}O$ values. The analytical precision within each run, calculated from seven to eight measurements of the standard Marbre LM, varies from 0.01 to 0.04‰ for $\delta^{13}C$ and from 0.04 to 0.14‰ for $\delta^{18}O$.

Seasonality of birth may be inferred from inter-individual variability in the sequence of the seasonal cycle tracked in a given tooth (Fricke and O'Neil 1996, Balasse *et al.* 2003). Comparing the position in the tooth crown of the minimum or maximum $\delta^{18}O$ value in this cycle can help to assess inter-individual variability. The sequences measured in cattle third molars were modelled using the method proposed in Balasse *et al.* (2012b) in order to calculate objectively the position of the maximum value (x_0). Moreover, the model allows the estimation of the length of tooth crown formed over a year, corresponding to the period of the sinusoidal cycle. Inter-individual variability in tooth size (as reflected by the period) may interfere with the distance of the tooth crown where the minimum and maximum $\delta^{18}O$ values are measured. The model may be used to normalise the distance to the period, in order to allow a direct comparison between individuals (Balasse *et al.* 2012b)

Results

Bone collagen stable isotope values

Bone collagen extraction yields vary from 15 to 172 mg/g (mean 105 ± 38 mg/g). The carbon contents of these extracts vary from 35 to 44.7% (mean of 41 ± 2.2%); the nitrogen contents vary from 12.6 to 16.5% (mean 15.1 ± 0.8%). Collagen atomic C/N ratios range from 3.1 to 3.2 (Table 7.2). All extracts satisfy the criteria defined for collagen yielding reliable stable carbon and nitrogen isotope ratios (DeNiro 1985, Ambrose 1990).

Results from bone collagen stable isotope measurements are reported in Table 7.2 and on Figure 7.2. In wild fauna, $\delta^{13}C$ values vary from –21.8‰ to –18.3‰ and $\delta^{15}N$ values from 3.3 to 7.0‰. The $\delta^{13}C$ and $\delta^{15}N$ values measured in domestic cattle and sheep are all included within these ranges of variation, although they tend towards the highest $\delta^{13}C$ values; indeed $\delta^{13}C$ values lower than –21‰ all belong to wild animals, and the mean $\delta^{13}C$ values for wild (–20.7 ± 0.82‰) and domestic species (sheep and cattle: –19.8 ± 0.52‰) are statistically significantly different (Student t-test, $p < 0.001$). Considering mean $\delta^{13}C$ values of –27‰ and –12.5‰ for modern C_3 and C_4 plants respectively (Vogel *et al.* 1978, Cerling *et al.* 2003, Kohn 2010), to which a correction of –1.5‰ must be applied to compensate for the fossil fuel effect (Freyer and Belacy 1983), and a 5‰ ^{13}C-enrichment between the protein fraction of diet and collagen (Ambrose and Norr 1993), the $\delta^{13}C$ values measured in the faunal assemblage from Cheia globally reflect a C_3 plant-dominated environment. In our effort to estimate the composition of pastures grazed by the fauna from Cheia, we follow Kohn (2010), who highlighted that the estimation of the mean $\delta^{13}C$ value for C_3 plants was biased against data from dry environments (with mean annual precipitation lower than 500 mm/year) to which the majority of plants with $\delta^{13}C$ values above –25.5‰ belong, in modern ecosystems. Once corrected for the fossil fuel effect, we consider the values of –24‰ for diet, or –19‰ in collagen, or –10.2‰ in enamel bioapatite (applying a 14.1‰ isotope enrichment between diet and enamel bioapatite: Cerling and Harris 1999), as reflecting the feeding on plants from dry ecosystems. Some of the bone collagen $\delta^{13}C$ values measured in domestic cattle, sheep and roe deer (up to –18.3‰ in CHE Capreolus 3) suggest feeding on an annual basis on a community of plants with a mean $\delta^{13}C$ value as high as –23.3‰, potentially reflecting either the contribution of C_3 plants from dry ecosystems or even a locally increased availability of C_4 plants.

The widest amplitude of variation of $\delta^{13}C$ values is recorded in roe deer (3.4‰), reflecting diversified habitat or feeding behaviour. The amplitude of variation of the $\delta^{15}N$ values measured in wild boars (2.5‰) is comparable to that measured in roe deer (2.7‰), and their range of variation is comprised within the range of variation defined for strict herbivores at the site (Figure 7.2). This suggests a mostly herbivorous diet for the analysed specimens.

The smallest amplitude of variation in $\delta^{13}C$ values is observed in the hydruntine/hemione (0.5‰), possibly reflecting a very selective feeding behaviour. The hydruntine/hemione also delivered the lowest $\delta^{15}N$ values (3.3 to 4.1‰), contrasting with a mean value of 5.4‰ for the remaining herbivores. A similar distinction was observed at Vităneşti, where five *Equus ferus* specimens yielded $\delta^{15}N$ values of 3.4 to 4.2‰ whereas the remaining fauna had values ranging between 4.9 and 9.1‰ (Balasse *et al.* 2013), and at the Bulgarian sites of Varna and Durankulak where four *Equus hydruntinus* specimens yielded depleted $\delta^{15}N$ values (3.5 to 3.8‰) in comparison to the remaining herbivores (5.3 to 10.2‰; Honch *et al.* 2006). In both cases, a specific digestive physiology and/or ecological niche were advanced. From the present datasets it is not possible to favour one explanation but this observation would certainly deserve more attention.

Enamel bioapatite stable carbon and oxygen isotope values

For cattle, results from tooth enamel $\delta^{18}O$ and $\delta^{13}C$ analysis are reported in the Appendix and in Figures 7.3–7.4. For sheep, only the ranges of variation of $\delta^{13}C$ values for each tooth are shown (Table 7.3) and typical sequences of $\delta^{18}O$ and $\delta^{13}C$ values in the second and third molars are shown for two individuals in Figure 7.5. Amplitudes of intra-tooth variations in $\delta^{13}C$ values are compared on Figure 7.6. In cattle, the $\delta^{18}O$ values range from –7.8 ‰ to –1.6‰ and intra-tooth amplitude of variation varies from 2.3 to 4.3‰ except in CHE Bos80 M3, where it is higher (5.4‰). In all third molars, variations in $\delta^{18}O$ values over tooth crown formation

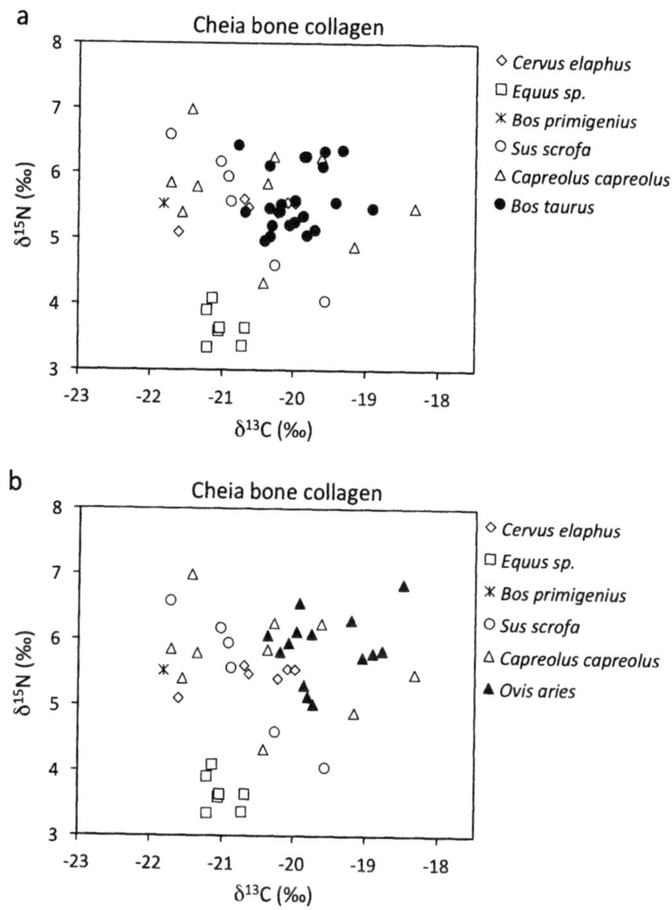

Figure 7.2 Results from bone collagen stable nitrogen ($\delta^{15}N$) and carbon isotope analysis ($\delta^{13}C$). Position of domestic cattle (a) and sheep (b) compared to wild fauna.

most likely reflect the seasonal cycle (Fricke and O'Neil 1996). Considering all cattle third molars except one, the $\delta^{13}C$ values vary between –11.6 and –9.6‰, and the amplitude of intra-tooth variation is limited (0.7 to 1.4‰). In CHE Bos80 M3, the amplitude of variation reaches 2.4‰, due to seasonally higher values reaching –8.1 ‰.

The $\delta^{13}C$ values measured in sheep molars vary between –13.8 and –8.6 ‰ (diet $\delta^{13}C$ values = –27.5 to –22.4‰) and the amplitude of intra-individual variation is in most cases higher than in cattle teeth (0.8 to 3.5‰). As observed earlier from bone collagen $\delta^{13}C$ values, the highest $\delta^{13}C$ values recorded in cattle and sheep tooth enamel (–8.1 and –8.6‰), corresponding to values of –21.9 ‰ and –22.4‰ for diet, reflect a contribution of C_3 plant resources from arid ecosystems or, to a lesser extent, a contribution of C_4 plants.

Modelling of $\delta^{18}O$ sequences measured in cattle third molars

Nine of the 12 cattle $\delta^{18}O$ sequences could be adequately modelled (because of a high asymmetry in the measured sequence, CHE Bos54 M3 and CHE Bos61 M3 could only be modelled over the last 20 and 22 mm respectively, resulting in very short sequences; the sequence from CHE Bos107 M3 was equally too short). Results for the best fit for combined variation of the equation parameters are shown in Table 7.4. The Pearson correlation coefficient was used to assess the fitting of the modelled to the measured datasets ($r > 0.92$ in all cases).

The location of the maximum $\delta^{18}O$ value in tooth crown (x_0) varies from 8.6 to 25.7 mm from ERJ. When normalised to the period of the cycle, this figure varies

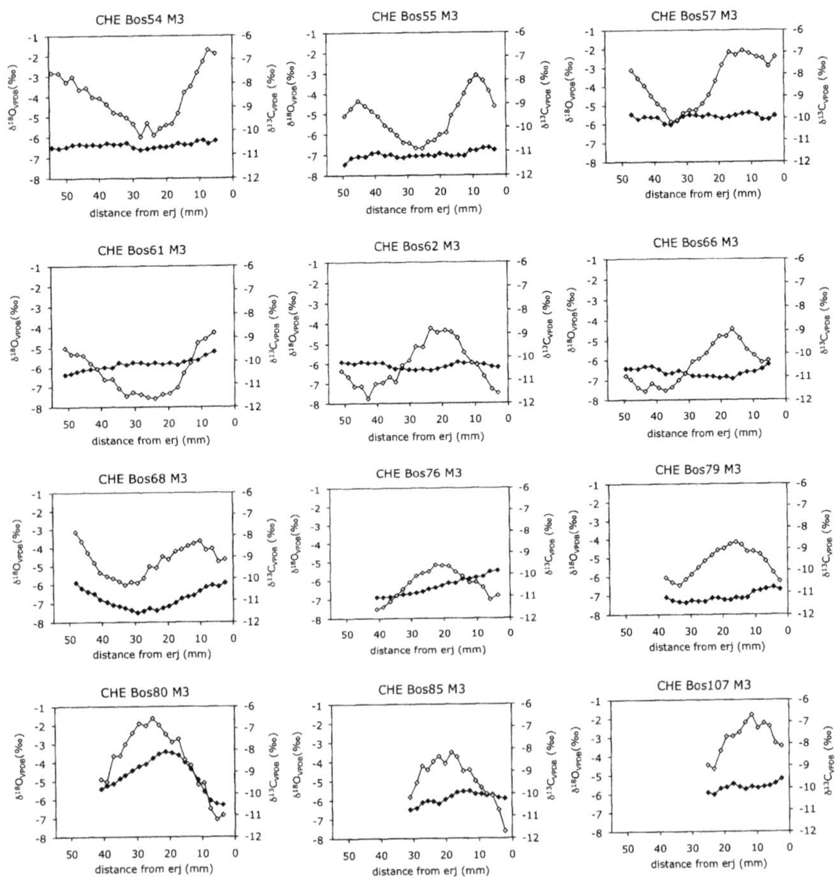

Figure 7.3 Results from sequential stable carbon ($\delta^{13}C$, black symbols) and oxygen ($\delta^{18}O$, open symbols) isotope analysis in cattle third molar (M3) enamel bioapatite. Each sample is located in the tooth crown relative to its distance from the enamel-root junction (erj).

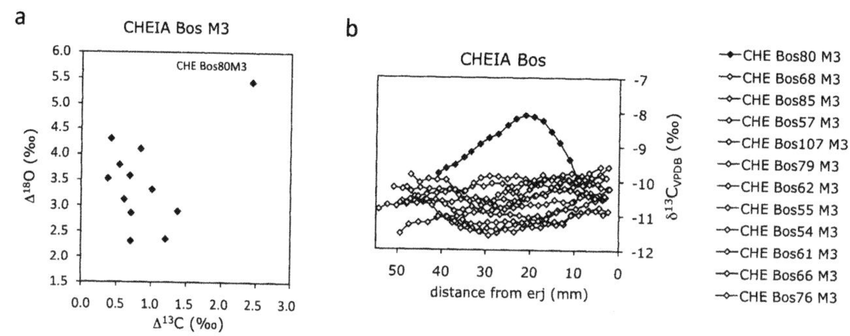

Figure 7.4 (a) Amplitude of intra-tooth variation in ($\delta^{13}C$) and ($\delta^{18}O$) values measured in cattle tooth enamel bioapatite. (b) Sequences of $\delta^{13}C$ values measured in cattle third molar enamel. CHE Bos80 M3 (in black) recorded higher $\delta^{13}C$ values than remaining individuals.

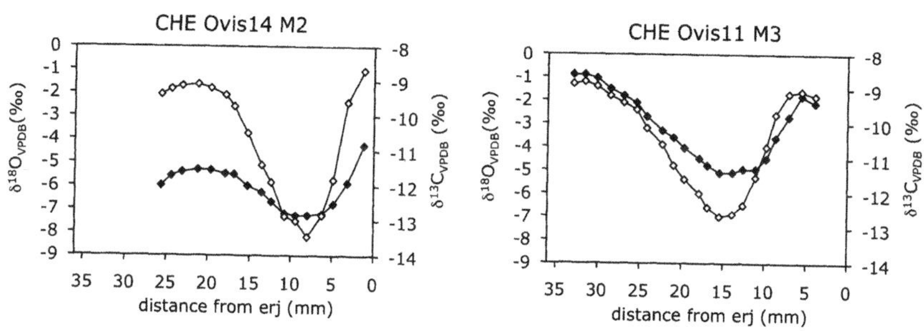

Figure 7.5 Results from sequential stable carbon ($\delta^{13}C$, black symbols) and oxygen ($\delta^{18}O$, open symbols) isotope analysis in enamel bioapatite from two sheep second (M2) and third (M3) molars.

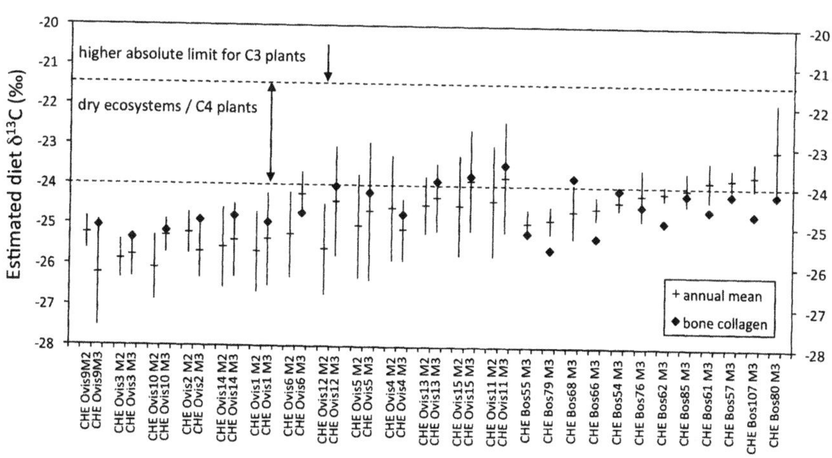

Figure 7.6 Stable carbon isotope composition ($\delta^{13}C$) of sheep and cattle diet, estimated from the $\delta^{13}C$ values measured in tooth enamel bioapatite (range of variation and annual mean) and in mandible bone collagen. Dry ecosystem and extreme limit for C_3 plants from Kohn (2010), corrected for fossil fuel effect.

Table 7.3 Results from stable carbon analysis ($\delta^{13}C$) of enamel bioapatite carbonate. Maximum (max) and minimum value (min) in each tooth, amplitude of variation (Δ). Diet $\delta^{13}C$ value is estimated from enamel bioapatite $\delta^{13}C$ value using a 14.1‰ isotope enrichment (Cerling and Harris 1999) and from bone collagen $\delta^{13}C$ value using a 5‰ spacing (Ambrose and Norr 1993).

Tooth specimen	*Enamel $\delta^{13}C$ (‰)*			*Diet $\delta^{13}C$ (enamel bioapatite)*			*Bone collagen*	*Diet $\delta^{13}C$ (‰)*
	max	*min*	*Δ*	*max*	*min*	*(max−min)/2*	*$\delta^{13}C$ (‰)*	*(bone collagen)*
CHE Ovis1 M2	−11.0	−12.9	2.0	−24.7	−26.7	−25.7		
CHE Ovis1 M3	−10.5	−12.8	2.3	−24.2	−26.5	−25.4	−20.0	−25.0
CHE Ovis2 M2	−11.0	−12.0	1.1	−24.7	−25.7	−25.2		
CHE Ovis2 M3	−11.3	−12.6	1.3	−25.0	−26.3	−25.7	−19.9	−24.9
CHE Ovis3 M2	−11.7	−12.6	1.0	−25.4	−26.4	−25.9		
CHE Ovis3 M3	−11.5	−12.6	1.1	−25.2	−26.3	−25.8	−20.4	−25.4
CHE Ovis4 M2	−9.5	−12.2	2.7	−23.3	−25.9	−24.6		
CHE Ovis4 M3	−10.6	−12.2	1.6	−24.4	−25.9	−25.1	−19.8	−24.8
CHE Ovis5 M2	−10.0	−12.6	2.6	−23.7	−26.3	−25.0		
CHE Ovis5 M3	−9.2	−12.6	3.5	−22.9	−26.4	−24.7	−19.2	−24.2
CHE Ovis6 M2	−10.4	−12.6	2.2	−24.2	−26.3	−25.3		
CHE Ovis6 M3	−9.9	−11.1	1.1	−23.7	−24.8	−24.3	−19.7	−24.7
CHE Ovis9 M2	−11.1	−11.9	0.8	−24.8	−25.6	−25.2		
CHE Ovis9 M3	−11.2	−13.8	2.6	−24.9	−27.5	−26.2	−20.1	−25.1
CHE Ovis10 M2	−11.6	−13.2	1.6	−25.3	−26.9	−26.1		
CHE Ovis10 M3	−11.1	−12.0	0.8	−24.9	−25.7	−25.3	−20.2	−25.2
CHE Ovis11 M2	−9.2	−12.0	2.8	−23.0	−25.8	−24.4		
CHE Ovis11 M3	−8.6	−11.4	2.8	−22.4	−25.2	−23.8	−18.5	−23.5
CHE Ovis12 M2	−10.7	−13.0	2.3	−24.5	−26.7	−25.6		
CHE Ovis12 M3	−9.3	−12.0	2.8	−23.0	−25.8	−24.4	−19.0	−24.0
CHE Ovis13 M2	−10.0	−11.5	1.5	−23.8	−25.2	−24.5		
CHE Ovis13 M3	−9.7	−11.4	1.7	−23.4	−25.2	−24.3	−18.9	−23.9
CHE Ovis14 M2	−10.8	−12.8	2.0	−24.6	−26.6	−25.6		
CHE Ovis14 M3	−10.7	−12.6	1.9	−24.5	−26.3	−25.4	−19.8	−24.8
CHE Ovis15 M2	−9.5	−12.0	2.5	−23.3	−25.8	−24.5		
CHE Ovis15 M3	−8.8	−11.4	2.6	−22.6	−25.2	−23.9	−18.8	−23.8
CHE Bos55 M3	−10.9	−11.5	0.7	−24.6	−25.3	−24.9	−20.2	−25.2
CHE Bos79 M3	−10.8	−11.5	0.7	−24.5	−25.2	−24.9	−20.6	−25.6
CHE Bos68 M3	−10.2	−11.6	1.4	−24.0	−25.3	−24.6	−18.8	−23.8
CHE Bos66 M3	−10.5	−11.1	0.6	−24.3	−24.8	−24.6	−20.3	−25.3
CHE Bos54 M3	−10.4	−10.8	0.4	−24.2	−24.6	−24.4	−19.1	−24.1
CHE Bos76 M3	−9.8	−11.1	1.3	−23.6	−24.8	−24.2	−19.5	−24.5
CHE Bos62 M3	−10.2	−10.6	0.4	−24.0	−24.3	−24.1	−19.9	−24.9
CHE Bos85 M3	−9.9	−10.7	0.8	−23.6	−24.5	−24.1	−20.2	−25.2
CHE Bos61 M3	−9.6	−10.6	1.0	−23.4	−24.4	−23.9	−19.6	−24.6
CHE Bos57 M3	−9.8	−10.3	0.5	−23.5	−24.1	−23.8	−19.2	−24.2
CHE Bos107 M3	−9.6	−10.3	0.7	−23.4	−24.1	−23.7	−19.7	−24.7
CHE Bos80 M3	−8.1	−10.5	2.4	−21.9	−24.3	−23.1	−19.2	−24.2

from 0.23 to 0.51, meaning that the maximum $\delta^{18}O$ value occurs at 23 to 51% of the end of the periodic cycle recorded in tooth crowns. Inter-individual variability in this parameter is 28% of an annual cycle or approximately 3.5 months, illustrating the length of the calving period at Cheia (Figure 7.7).

Table 7.4 Results from the calculation of the best fit (method of least squares) for combined variation of X (period), A (amplitude), x_0 (= position of $\delta^{18}O_{max}$: delay) and M (mean) when the mode 1 is applied to the cattle third molar $\delta^{18}O$ data sets. The model uses the equation: $\delta^{18}O_m = A. \cos(2\Pi - (x\text{-}x_0)/X) + M$.

Specimen	*X*	*A*	x_0	*M*	*Pearson r*	x_0/X
CHE Bos80M3	54.3	2.8	25.7	−4.7	0.99	0.47
CHE Bos68M3	38.7	1.2	12.5	−4.8	0.96	0.32
CHE Bos85M3	43.0	2.0	20.0	−5.5	0.96	0.47
CHE Bos57M3	42.3	1.8	11.7	−3.9	0.99	0.28
CHE Bos79M3	37.0	1.1	16.3	−5.3	0.99	0.44
CHE Bos62M3	40.7	1.5	20.9	−5.9	0.98	0.51
CHE Bos55M3	37.4	1.4	8.6	−5.4	0.92	0.23
CHE Bos66M3	41.5	1.3	17.4	−6.1	0.98	0.42
CHE Bos76M3	47.7	1.2	20.6	−6.4	0.98	0.43

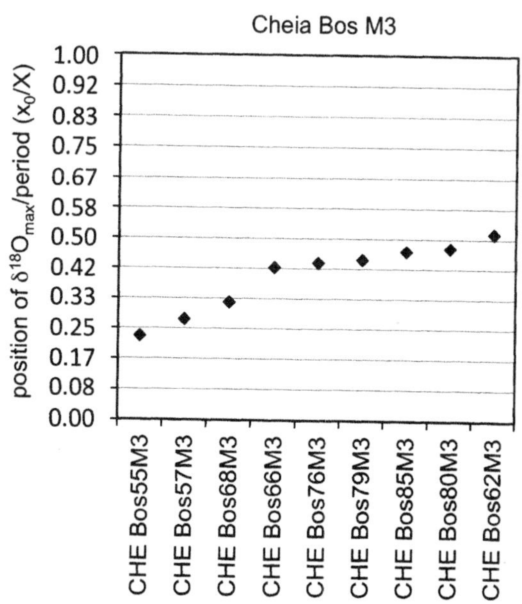

Figure 7.7 Position of the maximum values ($\delta^{18}O_{max}$) in the $\delta^{18}O$ sequences measured in the cattle third molars, normalised to the period (X) of the cycle defined in each tooth, using the modelling procedure described in Balasse *et al.* (2012b).

Discussion

Location of pastures for domestic herds

At Cheia, the $\delta^{13}C$ values measured in bone collagen from wild and domestic species globally refer to an open C_3 plant-dominated environment. Some bone specimens from Cheia yielded bone collagen $\delta^{13}C$ values very close to or higher than –19‰. Such high $\delta^{13}C$ values could reflect either access to C_3 plant communities from dry environments, or a contribution of C_4 plants. The present day Dobruja is among the driest regions in Romania, with mean annual precipitation levels of 350–510 mm/year (Panaitescu *et al.* 2011), falling within ecosystems where C_3 plants could be characterised by high $\delta^{13}C$ values (above –25.5‰; Kohn 2010). Palynological records from Shabla-Ezerets, Bulgaria (approximately 100 km south of Cheia; Figure 7.1), suggest that mid-Holocene vegetation along the Black Sea coast would have been wooded steppes similar to today's (Connor *et al.* 2013). It is therefore likely that the surroundings of Cheia, at the time of the Chalcolithic occupation, included dry ecosystems.

Most $\delta^{13}C$ values measured in wild fauna hunted at Cheia, however, ranging from –22 to –20‰ (except for three roe deer and one wild boar), do not especially refer to arid conditions. The significant difference observed between $\delta^{13}C$ values measured in wild and domestic animals suggests that hunting and herding may have been conducted in places with different ecologies. C_4 plants contribute to a very small part of wild biomass in Romania today, representing 1–2% of plant species (Collins and Jones 1985, Pyankov *et al.* 2010), though the representation in species number is not always an accurate prediction of local biomass availability of C_4 plants. *Chenopodiaceae* plants, among which C_4 plants dominate, are locally widespread on the western coast of the Black Sea, where they are favoured on soils with high salinity (Ivanova *et al.* 2000). Therefore an explanation could be the exploitation of salt marshes on the Black Sea coast as pasture areas for domestic herds. Cheia is approximately 14 km distant from the coast. However, at the time of its occupation, the Tasaul Lake, 7–8 km from the site, was a littoral lagoon connected to the Black Sea (Popescu and Caraivan 2002/2003, Radu 2008). Similar salt marsh ecosystems could have occurred at this lake. Seven to eight kilometres is within daily walking distance for cattle and sheep raised in semi-arid lands (Squires *et al.* 1972, Western and Dunne 1979, Homewood and Rodgers 1991), although close to the maximum. This resource could therefore have been accessible on a daily or alternate day basis.

It is interesting to note that similarly high $\delta^{13}C$ values (from –19.4 to –17.7‰) were measured in caprines from the Chalcolithic sites of Varna and Durankulak, in the first half of the fifth millennium cal BC (Honch *et al.* 2006). They could also reflect exploitation of the Black Sea littoral salt marshes as pasture areas.

The wild fauna would have been hunted in the hinterlands, except for a few roe deer specimens. Roe deer, which delivered two of the highest $\delta^{13}C$ values measured in bone collagen (−19.2 and −18.3‰) have a wide range of habitats, including also (salt) marshes, where they may have acquired this unexpected isotope signature. Interestingly, the bovine humerus attributed from morphometrical criteria to the aurochs (CHE Bos30 74-1) yielded the lowest $\delta^{13}C$ value of the analysed assemblage, contrasting with the high $\delta^{13}C$ values measured in the domestic counterparts with expected similar feeding behaviour. Although the sole datum for aurochs requires cautious interpretation, this would be also in favour of geographical separation of domestic flocks and wild herds. Because domestic pig was not identified among the suid bones, it is not possible to say whether domestic pigs and wild boars followed a similar scheme.

Sheep and cattle diet $\delta^{13}C$ at the seasonal scale

In sheep, the $\delta^{13}C$ values measured in tooth enamel confirm the data from bone collagen, and suggest that the contribution of plants with higher $\delta^{13}C$ values is seasonal. From the coevolution of $\delta^{18}O$ and $\delta^{13}C$ values in tooth crowns, this contribution was at its highest in the summer (Figure 7.5). This contribution is especially high in CHE Ovis 6, 12, 5, 4, 13, 15 and 11 (Figure 7.6) and may still be underestimated. Indeed, a six-month delay in enamel mineralisation for sheep second and third molars (Zazzo *et al.* 2010, Blaise and Balasse 2011, Balasse *et al.* 2012b) may attenuate the annual amplitude of variation of seasonally driven changes in diet, when recovered using the present analytical protocol (Balasse 2003). For these individuals, it is possible to envisage a pattern of seasonal mobility, with summer pastures including the littoral ecosystems with a higher occurrence of C_4 plants. Because the mortality profile suggests the presence of caprines at the site the whole year round (Bălăşescu 2008), this mobility should not be seen as a transhumance but rather as a change in the treks for grazing or watering.

In cattle, a pattern of seasonal variation is not clearly observed in the third molars (Figures 7.3 and 7.5; except in CHE Bos80 M3), possibly reflecting different feeding behaviours or feeding areas for cattle and sheep, with greater seasonal mobility for sheep. However, because the duration of enamel mineralisation in cattle third molars may differ from that in sheep molars, a direct comparison of the intra-tooth amplitude of variation of $\delta^{13}C$ values between both species may not be relevant. A longer duration for enamel mineralisation could lead in cattle third molars to a reduction of the amplitude of variation of the measured signal. Because this parameter is still unknown, it is difficult to conclude on this point. When comparing tooth enamel annual average $\delta^{13}C$ values ((max+min)/2), it comes out that one third of cattle third molars yielded average values reflecting $\delta^{13}C$ values ≥ −24‰ for diet (Figure 7.6). These individuals grazed on C_3 plants from dry ecosystems, although we cannot rule out a possible contribution of C_4 plants. The timing of incorporation of this contribution at the annual scale could also be a less significant factor than in sheep, but on a more regular basis.

Exogenous origin for one domestic bovine?

The third molar from one domestic bovine (CHE Bos80 M3, directly dated to Hamangia III; Table 7.1) yielded the highest $\delta^{13}C$ value in enamel (–8.1‰) referring to diet with a $\delta^{13}C$ value of –21.9‰. In modern ecosystems, the $\delta^{13}C$ values of C_3 plants very rarely exceed –23‰ (Kohn 2010). This limit value would be close to –21.5‰ in pre-industrial ecosystems. Therefore CHE Bos80 must have included C_4 plants in its diet over the summer recorded in its third molar, to a relative amount unattained in any other sampled sheep and cattle teeth. CHE Bos80 M3 also delivered the highest amplitude of variation in $\delta^{18}O$ (Figure 7.4a), although the range of variation of $\delta^{18}O$ measured in this tooth is similar to the range defined by remaining cattle. In comparison, the $\delta^{13}C$ value measured in bone collagen from the mandible of CHE Bos80, reflecting integration of a diet signal over several years (from the tooth wear pattern, the age at death was estimated at 6–7 years; Ducos 1968), falls within the lowest values measured in the M3, and is similar to those measured in mandibular bone from other individuals (Figure 7.6). This suggests that the great contribution of C_4 plants to the diet of CHE Bos80 over the summer recorded in its M3 may have been exceptional over its lifetime. Different scenarios may be envisaged. A climatic event such as a very dry summer could have constrained the animal to feed on a higher proportion of C_4 plants because of the scarcity of C_3 pastures under such circumstances. Alternatively, this bovine could have been acquired by exchange from a community whose cattle herds spent summertime on pastures with a higher availability of C_4 plants. This place could be the salt marshes of the Black Sea littoral or the riparian and marsh ecosystems of the Danube floodplain (approximately 28 km away), where the presence of C_4 plants is suggested in the second half of the fifth millennium cal BC by seasonally high $\delta^{13}C$ values measured in cattle and caprine tooth enamel from the Borduşani-Popină site (Gillis *et al.* 2013; Figure 7.1). The occupation of both ecosystems by Hamangia communities is attested in the archaeological record (Figure 7.1).

Inter-annual mobility between pasture areas?

Among sheep and cattle, the difference between the mean $\delta^{13}C$ values for diet over lifespan (from bone collagen) and the mean value over a year (from the range of variation of $\delta^{13}C$ values in enamel) varies from –1.1 to 1.6‰, and from 0.1 to 1.6‰ in absolute values (Table 7.3; Figure 7.6). This difference is partly explained by the precision around the average isotope enrichment values used to estimate diet $\delta^{13}C$ from collagen and bioapatite $\delta^{13}C$; moreover, different plant biochemical components have different $\delta^{13}C$ values (Heaton 1999) and bioapatite $\delta^{13}C$ value reflects that of the whole diet, whereas collagen $\delta^{13}C$ value is biased towards that of protein (Ambrose and Norr 1993). It is difficult to estimate to what extent the observed difference is a consequence of these factors. Because in adults the isotope signature of teeth and bone refers to different moments of life, the difference may also refer to inter-annual variability in environmental factors, as well as inter-annual variability in pasture areas. Although the perennial character of habitat structures recovered at Cheia suggests a year-round occupation, a pattern of periodic abandonment of the settlement is also suggested, for sufficiently long periods of time to lead to complete destruction of the dwelling houses (Voinea and Neagu 2008). A pattern of mobility running over several years could possibly explain the variability in $\delta^{13}C$ value between bone and teeth.

Cattle birth seasonality

The cattle mortality profile at Cheia suggests a mixed exploitation for meat and milk (Bălăşescu 2008). Distribution of cattle births is a key parameter for milk production. Indeed lactation is initiated by calving; consequently the availability of milk throughout the year is partly related to the length of the calving season.

Within most Neolithic faunal assemblages, the number of tooth specimens available for tooth enamel sequential analysis is certainly a major limiting factor in establishing the seasonality of birth. The dataset presented here for Cheia on 12 cattle molars, among which only nine could be adequately modelled, may appear limited (although already totalling 256 analyses). The site delivered a substantial number of tooth rows and this preliminary dataset will be enlarged in the future.

Besides, it must be kept in mind that most zooarchaeological assemblages are palimpsests of herds. The significance of the length of the birth period as defined from the presently applied procedure depends on the period of time represented by the faunal assemblage and the stability of environmental parameters and herding practices in time and in space. In cattle, birth seasonality is mainly driven by food availability. The cattle mating season is therefore strongly related to the annual cycle of vegetation development, unless herders provide significant supplementation when food is scarce. At Cheia, it is likely that the mating season would remain the same under unchanged environmental conditions. The possible presence of an exogenous specimen (CHE Bos80 M3) must be considered, but it is not thought that this individual could have been acquired by long-distance exchange, from a location with a radically different annual cycle of vegetation development. Finally, such considerations would be especially relevant when results suggest an extended birth period. In fact, the length of the calving period at Cheia was estimated at three to four months (CHE Bos80 falls within this range), which is rather restricted and quite similar to the three-month calving period observed in free-ranging Highland cattle in northern France (Lecomte and Le Neveu 1986). For comparison, the same protocol, applied to cattle teeth from the site of Bercy (northern France, dating to the beginning of the fourth millennium cal BC) led to the estimation of a period of birth spanning approximately six months (Balasse *et al.* 2012a). For now, it is difficult to disentangle the environmental constraints and cultural choices influencing cattle birth distribution at these Neolithic sites. Acquisition of additional datasets from other sites from different environmental and chronological contexts may help in the future to evaluate the diversity of this parameter.

Similar work performed on Cheia sheep molars has shown a lambing period extended over less than four months. Comparison with results from modern reference datasets suggests that sheep births occurred from late winter to early summer (Tornero *et al.* 2013). Determining the season of birth in past cattle herds would require $\delta^{18}O$ reference sequences in cattle molars. None is available yet, although some work is being conducted to achieve this aim. Once the season of cattle birth is known, comparison with the season of birth for caprines should help us to envision the complementarity of cattle and caprine husbandry throughout the annual round.

Conclusion

The diversity of possible schemes for herding strategies and the equifinality of the biological systems behind the stable isotope signals eventually retrieved by zooarchaeologists in bone and tooth remains lead to a great complexity for interpretation. Uncertainties remain in the interpretation of high $\delta^{13}C$ values measured in domestic animals from Cheia. Although different possibilities are not mutually exclusive, do they reflect a locally dry environment, or the occurrence of C_4 plants in the littoral salt marshlands? The answer to this question would be a key to understanding the pattern of land use among these Hamangia communities. The wide range of variation in $\delta^{13}C$ values measured in roe deer, a highly adaptable species with a wide habitat spectrum, may refer to the diversity of accessible ecological systems. Diverse scenarios may be envisaged for herding. Cattle and sheep may have been kept in the vicinity of a site surrounded by dry ecosystems, while the absence of a 'dry environment' signal in the $\delta^{13}C$ of wild fauna (except in roe deer) could signify that hunting, contributing only to a small part of animal resources, was mostly practised further away from the coast in the hinterlands. This first scenario would suggest small-scale herding. The pattern of recurrent yearly occupation of the site, with prolonged phases of abandonment, could reflect the cyclical need for a regeneration of local pastures damaged by overgrazing. Alternatively, the $\delta^{13}C$ values measured in wild fauna may reflect the local environment, while higher $\delta^{13}C$ values measured in domestic animals would reflect the exploitation of different ecosystems for grazing areas, including a possible contribution of C_4 plants which would have to be found on the Black Sea littoral. This second scenario would rather suggest extensive herding. The use of littoral salted marshlands cannot actually be ruled out from the first scenario either. New light on this specific question could be shed by a complementary study including measurements of strontium and or sulphur isotope ratios, which in certain circumstances may be used to distinguish between marine littoral and terrestrial ecosystems (Ericson 1985, Richards *et al.* 2001).

Table 7.5 Results from stable carbon ($\delta^{13}C$) and oxygen isotope compositions ($\delta^{18}O$) measured in tooth enamel bioapatite. Samples are positioned in the tooth crown using the distance from the enamel-root junction (erj).

Sample	*mm from erj*	$\delta^{13}C$	$\delta^{18}O$	*Sample*	*mm from erj*	$\delta^{13}C$	$\delta^{18}O$	*Sample*	*mm from erj*	$\delta^{13}C$	$\delta^{18}O$
CHE Bos80 M3−1	40.8	−9.8	−4.9	CHE Bos 76 M3−1	40.8	−11.0	−7.5	CHE Bos62 M3−1	51.0	−10.2	−6.4
CHE Bos80 M3−2	39.1	−9.6	−5.0	CHE Bos 76 M3−2	38.8	−11.1	−7.4	CHE Bos62 M3−2	48.9	−10.2	−6.7
CHE Bos80 M3−3	37.1	−9.5	−3.7	CHE Bos 76 M3−3	36.6	−11.0	−7.1	CHE Bos62 M3−3	47.0	−10.3	−7.2
CHE Bos80 M3−4	35.1	−9.3	−3.6	CHE Bos 76 M3−4	34.7	−11.0	−6.8	CHE Bos62 M3−4	44.9	−10.2	−7.1
CHE Bos80 M3−5	33.5	−9.2	−3.0	CHE Bos 76 M3−5	32.6	−10.9	−6.5	CHE Bos62 M3−5	42.8	−10.3	−7.8
CHE Bos80 M3−6	31.3	−8.9	−2.4	CHE Bos 76 M3−6	30.6	−10.9	−6.1	CHE Bos62 M3−6	40.6	−10.2	−7.0
CHE Bos80 M3−7	29.3	−8.8	−1.9	CHE Bos 76 M3−7	28.6	−10.9	−5.8	CHE Bos62 M3−7	38.3	−10.3	−7.0
CHE Bos80 M3−8	27.3	−8.7	−2.0	CHE Bos 76 M3−8	26.6	−10.8	−5.6	CHE Bos62 M3−8	36.2	−10.4	−6.7
CHE Bos80 M3−9	25.1	−8.4	−1.6	CHE Bos 76 M3−9	24.7	−10.7	−5.5	CHE Bos62 M3−9	34.3	−10.5	−6.9
CHE Bos80 M3−10	23.3	−8.2	−2.0	CHE Bos 76 M3−10	22.6	−10.6	−5.2	CHE Bos62 M3−10	32.3	−10.5	−6.1
CHE Bos80 M3−11	21.2	−8.1	−2.5	CHE Bos 76 M3−11	20.6	−10.5	−5.2	CHE Bos62 M3−11	30.1	−10.6	−5.8
CHE Bos80 M3−12	19.2	−8.1	−2.9	CHE Bos 76 M3−12	18.6	−10.4	−5.2	CHE Bos62 M3−12	28.1	−10.6	−5.1
CHE Bos80 M3−13	17.2	−8.2	−2.7	CHE Bos 76 M3−13	16.5	−10.4	−5.5	CHE Bos62 M3−13	25.9	−10.5	−5.2
CHE Bos80 M3−14	15.3	−8.6	−3.9	CHE Bos 76 M3−14	14.4	−10.2	−5.8	CHE Bos62 M3−14	23.5	−10.6	−4.2
CHE Bos80 M3−15	13.4	−8.9	−4.2	CHE Bos 76 M3−15	12.3	−10.2	−6.1	CHE Bos62 M3−15	21.4	−10.5	−4.4
CHE Bos80 M3−16	11.2	−9.4	−5.2	CHE Bos 76 M3−16	10.3	−10.2	−6.1	CHE Bos62 M3−16	19.0	−10.4	−4.3
CHE Bos80 M3−17	9.3	−9.9	−5.1	CHE Bos 76 M3−17	8.2	−10.1	−6.4	CHE Bos62 M3−17	17.1	−10.4	−4.4
CHE Bos80 M3−18	7.5	−10.3	−6.5	CHE Bos 76 M3−18	6.1	−9.9	−7.0	CHE Bos62 M3−18	15.2	−10.2	−4.7
CHE Bos80 M3−19	5.6	−10.5	−7.1	CHE Bos 76 M3−19	3.8	−9.8	−6.8	CHE Bos62 M3−19	13.3	−10.3	−5.4
CHE Bos80 M3−20	3.6	−10.5	−6.8					CHE Bos62 M3−20	11.0	−10.3	−6.0
								CHE Bos62 M3−21	9.4	−10.3	−6.1
								CHE Bos62 M3−22	7.3	−10.3	−6.6
								CHE Bos62 M3−23	4.9	−10.4	−7.3
								CHE Bos62 M3−24	3.0	−10.5	−7.5

Table 7.5 Continued

Sample	*mm from erj*	$\delta^{13}C$	$\delta^{18}O$	*Sample*	*mm from erj*	$\delta^{13}C$	$\delta^{18}O$	*Sample*	*mm from erj*	$\delta^{13}C$	$\delta^{18}O$
CHE Bos54 M3−1	56.4	−10.8	−2.6	CHE Bos68 M3−1	47.9	−10.2	−3.1	CHE Bos55 M3−1	49.5	−11.5	−5.1
CHE Bos54 M3−2	54.4	−10.7	−2.8	CHE Bos68 M3−2	46.1	−10.5	−3.6	CHE Bos55 M3−2	47.5	−11.3	−4.7
CHE Bos54 M3−3	52.3	−10.7	−2.8	CHE Bos68 M3−3	44.2	−10.6	−4.3	CHE Bos55 M3−3	45.3	−11.2	−4.4
CHE Bos54 M3−4	50,0	−10.7	−3.3	CHE Bos68 M3−4	42.2	−10.7	−4.8	CHE Bos55 M3−4	43.1	−11.2	−4.6
CHE Bos54 M3−5	48.2	−10.6	−3.0	CHE Bos68 M3−5	40.6	−11.0	−5.4	CHE Bos55 M3−5	41.1	−11.1	−4.8
CHE Bos54 M3−6	46,0	−10.6	−3.7	CHE Bos68 M3−6	38.5	−11.1	−5.6	CHE Bos55 M3−6	39.1	−11.0	−5.1
CHE Bos54 M3−7	43.8	−10.6	−3.6	CHE Bos68 M3−7	36.7	−11.2	−5.7	CHE Bos55 M3−7	37.1	−11.2	−5.6
CHE Bos54 M3−8	41.9	−10.6	−4.0	CHE Bos68 M3−8	34.8	−11.3	−5.8	CHE Bos55 M3−8	35.2	−11.1	−5.8
CHE Bos54 M3−9	39.7	−10.6	−4.1	CHE Bos68 M3−9	33.0	−11.4	−6.0	CHE Bos55 M3−9	33.3	−11.3	−6.0
CHE Bos54 M3−10	37.7	−10.5	−4.4	CHE Bos68 M3−10	31.1	−11.5	−5.9	CHE Bos55 M3−10	31.3	−11.3	−6.4
CHE Bos54 M3−11	35.7	−10.6	−4.8	CHE Bos68 M3−11	29.3	−11.6	−5.9	CHE Bos55 M3−11	29.5	−11.2	−6.4
CHE Bos54 M3−12	33.7	−10.6	−4.9	CHE Bos68 M3−12	27.4	−11.5	−5.7	CHE Bos55 M3−12	27.5	−11.2	−6.7
CHE Bos54 M3−13	31.8	−10.5	−5.0	CHE Bos68 M3−13	25.7	−11.4	−5.0	CHE Bos55 M3−13	25.5	−11.2	−6.7
CHE Bos54 M3−14	29.8	−10.7	−5.3	CHE Bos68 M3−14	23.6	−11.5	−5.1	CHE Bos55 M3−14	23.7	−11.2	−6.4
CHE Bos54 M3−15	27.7	−10.8	−6.0	CHE Bos68 M3−15	21.6	−11.4	−4.5	CHE Bos55 M3−15	21.8	−11.2	−6.3
CHE Bos54 M3−16	25.7	−10.8	−5.3	CHE Bos68 M3−16	19.8	−11.3	−4.6	CHE Bos55 M3−16	20.1	−11.1	−6.0
CHE Bos54 M3−17	23.8	−10.7	−5.9	CHE Bos68 M3−17	17.6	−11.1	−4.2	CHE Bos55 M3−17	18.1	−11.1	−5.9
CHE Bos54 M3−18	21.9	−10.7	−5.6	CHE Bos68 M3−18	15.8	−10.9	−4.1	CHE Bos55 M3−18	16.6	−11.2	−5.1
CHE Bos54 M3−19	20.1	−10.7	−5.4	CHE Bos68 M3−19	13.8	−10.9	−3.9	CHE Bos55 M3−19	14.4	−11.2	−4.6
CHE Bos54 M3−20	18.3	−10.7	−5.4	CHE Bos68 M3−20	12.0	−10.8	−3.8	CHE Bos55 M3−20	12.7	−11.2	−4.0
CHE Bos54 M3−21	16.5	−10.6	−4.8	CHE Bos68 M3−21	10.1	−10.6	−3.6	CHE Bos55 M3−21	10.8	−11.0	−3.4
CHE Bos54 M3−22	14.5	−10.6	−3.8	CHE Bos68 M3−22	8.2	−10.4	−4.1	CHE Bos55 M3−22	9.0	−11.0	−3.1
CHE Bos54 M3−23	12.5	−10.6	−3.5	CHE Bos68 M3−23	6.4	−10.3	−4.1	CHE Bos55 M3−23	6.9	−10.9	−3.4
CHE Bos54 M3−24	10.7	−10.5	−2.8	CHE Bos68 M3−24	4.4	−10.4	−4.8	CHE Bos55 M3−24	5.2	−10.9	−3.9
CHE Bos54 M3−25	8.8	−10.4	−2.3	CHE Bos68 M3−25	2.5	−10.2	−4.6	CHE Bos55 M3−25	3.5	−11.0	−4.6
CHE Bos54 M3−26	7.3	−10.6	−1.7								
CHE Bos54 M3−27	5.2	−10.4	−1.9								

CHE Bos61 M3−1	51.0	−10.6	−5.0	CHE Bos66 M3−1	49.5	−10.7	−6.8	CHE Bos57 M3−1	47.1	−9.8	−3.1
CHE Bos61 M3−2	49.2	−10.5	−5.3	CHE Bos66 M3−2	47.9	−10.7	−7.0	CHE Bos57 M3−2	45.1	−10.1	−3.6
CHE Bos61 M3−3	47.3	−10.5	−5.3	CHE Bos66 M3−3	45.7	−10.7	−7.4	CHE Bos57 M3−3	43.1	−10.0	−4.0
CHE Bos61 M3−4	45.4	−10.4	−5.4	CHE Bos66 M3−4	43.5	−10.6	−7.5	CHE Bos57 M3−4	41.1	−10.0	−4.5
CHE Bos61 M3−5	43.2	−10.4	−5.8	CHE Bos66 M3−5	41.4	−10.6	−7.2	CHE Bos57 M3−5	39.1	−10.0	−4.9
CHE Bos61 M3−6	41.2	−10.3	−6.1	CHE Bos66 M3−6	39.3	−10.7	−7.4	CHE Bos57 M3−6	37.0	−10.3	−5.2
CHE Bos61 M3−7	39,0	−10.3	−6.6	CHE Bos66 M3−7	37.3	−10.9	−7.5	CHE Bos57 M3−7	35.0	−10.3	−5.9
CHE Bos61 M3−8	36.7	−10.3	−6.6	CHE Bos66 M3−8	35.2	−10.8	−7.3	CHE Bos57 M3−8	33.1	−10.2	−5.8
CHE Bos61 M3−9	34.8	−10.1	−7.1	CHE Bos66 M3−9	33.2	−10.7	−7.0	CHE Bos57 M3−9	31.1	−9.9	−5.4
CHE Bos61 M3−10	32.5	−10.2	−7.5	CHE Bos 66 M3−10	31.1	−10.9	−6.7	CHE Bos57 M3−10	29.2	−9.9	−5.3
CHE Bos61 M3−11	30.6	−10.1	−7.3	CHE Bos66 M3−11	28.8	−11.0	−6.1	CHE Bos57 M3−11	27.2	−9.9	−5.2
CHE Bos61 M3−12	28.3	−10.1	−7.4	CHE Bos66 M3−12	26.8	−11.0	−5.9	CHE Bos57 M3−12	25.1	−9.9	−4.9
CHE Bos61 M3−13	26.2	−10.2	−7.5	CHE Bos66 M3−13	24.4	−11.0	−5.6	CHE Bos57 M3−13	23.2	−9.9	−4.5
CHE Bos61 M3−14	24.1	−10.1	−7.6	CHE Bos66 M3−14	22.4	−11.0	−5.2	CHE Bos57 M3−14	21.1	−9.9	−3.7
CHE Bos61 M3−15	21.8	−10.1	−7.4	CHE Bos66 M3−15	20.4	−11.1	−4.8	CHE Bos57 M3−15	19.1	−10.0	−2.9
CHE Bos61 M3−16	19.6	−10.1	−7.3	CHE Bos66 M3−16	18.3	−11.0	−4.8	CHE Bos57 M3−16	16.8	−9.9	−2.2
CHE Bos 61 M3−17	17.3	−10.2	−7.0	CHE Bos66 M3−17	16.3	−11.1	−4.4	CHE Bos57 M3−17	14.8	−9.9	−2.3
CHE Bos61 M3−18	15.3	−10.0	−6.3	CHE Bos66 M3−18	14.1	−10.9	−4.9	CHE Bos57 M3−18	12.7	−9.8	−2.1
CHE Bos61 M3−19	13.2	−10.0	−5.8	CHE Bos66 M3−19	11.9	−10.8	−5.5	CHE Bos57 M3−19	10.7	−9.8	−2.2
CHE Bos61 M3−20	11.0	−10.0	−4.8	CHE Bos66 M3−20	9.7	−10.8	−5.8	CHE Bos57 M3−20	8.4	−9.8	−2.4
CHE Bos61 M3−21	8.7	−9.8	−4.6	CHE Bos66 M3−21	7.4	−10.7	−6.1	CHE Bos57 M3−21	6.7	−10.1	−2.5
CHE Bos61 M3−22	6.1	−9.6	−4.3	CHE Bos66 M3−22	5.4	−10.5	−6.0	CHE Bos57 M3−22	4.8	−10.1	−2.9
								CHE Bos57 M3−23	2.8	−9.9	−2.4
CHE Bos79 M3−1	37.6	−11.2	−6.1	CHE Bos85 M3−1	31.0	−10.7	−5.8	CHE Bos107 M3−1	25.0	−10.2	−4.5
CHE Bos79 M3−2	35.6	−11.4	−6.3	CHE Bos85 M3−2	29.0	−10.7	−5.1	CHE Bos107 M3−2	23.1	−10.3	−4.7
CHE Bos79 M3−3	33.5	−11.4	−6.5	CHE Bos85 M3−3	27.2	−10.4	−4.2	CHE Bos107 M3−3	21.0	−10.0	−3.7
CHE Bos79 M3−4	31.5	−11.5	−6.2	CHE Bos85 M3−4	25.5	−10.3	−4.4	CHE Bos107 M3−4	19.1	−10.0	−3.0
CHE Bos79 M3−5	29.7	−11.4	−5.9	CHE Bos85 M3−5	23.8	−10.3	−4.0	CHE Bos107 M3−5	17.2	−9.8	−2.9
CHE Bos79 M3−6	27.6	−11.4	−5.5	CHE Bos85 M3−6	22.1	−10.4	−3.7	CHE Bos107 M3−6	15.2	−10.0	−2.7
CHE Bos79 M3−7	25.5	−11.4	−5.2	CHE Bos85 M3−7	20.1	−10.2	−4.1	CHE Bos107 M3−7	13.4	−10.1	−2.2
CHE Bos79 M3−8	23.3	−11.3	−4.9	CHE Bos85 M3−8	18.3	−10.1	−3.5	CHE Bos107 M3−8	11.5	−10.0	−1.8

Table 7.5 Continued

Sample	*mm from erj*	$\delta^{13}C$	$\delta^{18}O$	*Sample*	*mm from erj*	$\delta^{13}C$	$\delta^{18}O$	*Sample*	*mm from erj*	$\delta^{13}C$	$\delta^{18}O$
CHE Bos79 M3−9	21.4	−11.3	−4.6	CHE Bos85 M3−9	16.6	−9.9	−3.8	CHE Bos107 M3−9	9.6	−10.0	−2.5
CHE Bos79 M3−10	19.5	−11.3	−4.5	CHE Bos85 M3−10	14.7	−9.9	−4.5	CHE Bos107 M3−10	7.7	−9.9	−2.2
CHE Bos79 M3−11	17.5	−11.3	−4.3	CHE Bos85 M3−11	12.8	−9.9	−4.4	CHE Bos107 M3−11	6.0	−9.9	−2.4
CHE Bos79 M3−12	15.8	−11.2	−4.2	CHE Bos85 M3−12	11.1	−10.0	−5.0	CHE Bos107 M3−12	4.2	−9.8	−3.3
CHE Bos79 M3−13	14.1	−11.3	−4.3	CHE Bos85 M3−13	9.3	−10.0	−5.4	CHE Bos107 M3−13	2.3	−9.6	−3.5
CHE Bos79 M3−14	12.4	−11.2	−4.7	CHE Bos85 M3−14	7.7	−10.1	−5.7				
CHE Bos79 M3−15	10.6	−11.0	−4.6	CHE Bos85 M3−15	5.9	−10.0	−5.8				
CHE Bos79 M3−16	8.5	−10.9	−4.8	CHE Bos85 M3−16	4.0	−10.2	−6.5				
CHE Bos79 M3−17	6.8	−10.8	−5.2	CHE Bos85 M3−17	2.1	−10.2	−7.6				
CHE Bos79 M3−18	4.4	−10.8	−5.8								
CHE Bos79 M3−19	2.4	−10.9	−6.2								

Acknowledgements

This work is funded by the ERC Starting Grant SIANHE *Stable Isotope Investigation of Neolithic Husbandry in the diverse climatic and environmental settings of Europe* (GA 202881). Zooarchaeological analyses conducted by A. Bălăşescu were supported by a grant from the Romanian National Authority for Scientific Research, CNCS – UEFISCDI (Project number: PN-II-ID-PCE-2011-3-1015).

References

Ambrose, S.H. 1990. Preparation and characterization of bone and tooth collagen for isotopic analysis. *Journal of Archaeological Science* 17, 431–51.

Ambrose, S.H. and Norr, L. 1993. Experimental evidence for the relationship of the carbon isotope ratios of whole diet and dietary protein to those of bone collagen and carbonate. In J.B. Lambert and G. Grupe (eds), *Prehistoric human bone: archaeology at the molecular level*, 1–37. Berlin: Springer.

Andreescu, R. and Mirea, P. 2008. Teleorman Valley. The beginning of Neolithic in southern Romania. *Acta Terrae Septemcastrensis* 7, 57–75.

Andreescu, R., Mirea, P. and Apope, S. 2003. Cultura Gumelniţa în Vestul Munteniei. Aşezarea de la Vităneşti. jud. Teleorman. *Cercetări Arheologice* 12, 71–88.

Bălăşescu, A. 2008. Consideraţii cu privire la exploatarea mamiferelor în aşezarea Hamangia III de la Cheia. *Pontica* 41, 49–56.

Bălăşescu, A. and Radu, V. 2003. Studiul arheozoologic preliminar al materialului faunistic de la Cheia (jud. Constanţa). Campania 2001. *Pontica* 33/34, 25–30.

Bălăşescu, A. and Radu, V. 2004. *Oameni şi animale. Strategii şi resurse la comunităţile preistorice Hamangia şi Boian*. Seria Cercetări Pluridisciplinare 9. Bucureşti: Biblioteca Muzeului Naţional, Editura Cetatea de Scaun.

Balasse, M. 2003. Potential biases in sampling design and interpretation of intra-tooth isotope analysis. *International Journal of Osteoarchaeology* 13, 3–10.

Balasse, M., Smith, A.B., Ambrose, S.H. and Leigh, S.R. 2003. Determining sheep birth seasonality by analysis of tooth enamel oxygen isotope ratios: the Late Stone Age site of Kasteelberg (South Africa). *Journal of Archaeological Science* 3, 205–15.

Balasse, M., Boury, L., Ughetto-Monfrin, J. and Tresset, A. 2012a. Stable isotope insights ($\delta^{18}O$, $\delta^{13}C$) into cattle and sheep husbandry at Bercy (Paris, France, IVth millennium BC): birth seasonality and winter leaf foddering. *Environmental Archaeology* 17, 29–44.

Balasse, M., Obein, G., Ughetto-Monfrin, J. and Mainland, I. 2012b. Investigating seasonality and season of birth in past herds: a reference set of sheep enamel stable oxygen isotope ratios. *Archaeometry* 54, 349–68.

Balasse, M., Bălăşescu, A., Janzen, A., Ughetto-Monfrin, J., Mirea, P. and Andreescu, R. 2013. Early herding at Magura-Boldul lui Mos Ivanus (early sixth millennium BC, Romania): environments and seasonality from stable isotope analysis. *European Journal of Archaeology* 16, 221–46.

Berciu, D. 1966. *Cultura Hamangia*. Bucureşti: Editura Academiei Republicii Sociáliste România.

Blaise, E. and Balasse, M. 2011. Seasonality and season of birth of modern and late Neolithic sheep from south-eastern France using tooth enamel $\delta^{18}O$ analysis. *Journal of Archaeological Science* 38, 3085–93.

Bocherens, H., Fizet, M., Mariotti, A., Lange-Badré, B., Vandermeersch, B., Borel, J.P. and Bellon, G. 1991. Isotopic biogeochemistry (^{13}C, ^{15}N) of fossil vertebrate collagen: implications for the study of a fossil food web including Neandertal Man. *Journal of Human Evolution* 20, 481–92.

Bogaard, A. 2004. The nature of early farming in central and south-east Europe. *Documenta Praehistorica* 3, 49–58.

Bogaard, A. 2005. 'Garden agriculture' and the nature of early farming in Europe and the Near East. *World Archaeology* 37, 177–96.

Cerling, T.E. and Harris, J.M. 1999. Carbon isotope fractionation between diet and bioapatite in ungulate mammals and implications for ecological and paleoecological studies. *Oecologia* 120, 347–63.

Cerling, T.E., Harris, J.M. and Passey, B. 2003. Diets of East African bovidae based on stable isotope analysis. *Journal of Mammalogy* 84, 456–70.

Collins, R.P. and Jones, M.B. 1985. The influence of climatic factors on the distribution of C_4 species in Europe. *Vegetatio* 64, 121–9.

Connor, S.E., Ross, S.A., Sobotkova, A., Herries, A.I.R., Mooney, S.D., Longford, C. and Iliev, I. 2013. Environmental conditions in the SE Balkans since the last Glacial Maximum and their influence on the spread of agriculture into Europe. *Quaternary Science Reviews* 68, 200–15.

Degerbøl, M. and Fredskild, B. 1970. *The urus (*Bos primigenius *Bojanus) and the Neolithic domesticated cattle (*Bos taurus domesticus *Linné) in Denmark. Zoological and palynological investigations*. Det Kongelige Danske Videnskabernes Selskab, Biologiske skrifter 17(1). København: Ed. Munksgaards Forlag.

Ducos, P. 1968. *L'origine des animaux domestiques en Palestine*. Publications de l'Institut de Préhistoire de l'Université de Bordeaux 6. Paris: Éditions Masson.

DeNiro, M.J. 1985. Post-mortem preservation and alteration of in vivo bone collagen isotope ratios in relation to paleodietary reconstructions. *Nature* 317, 806–9.

El Susi, G. 1996. *Vânatori, pescari si crescatori de animale în Banatul mileniilor VI*. Timisoara: Editura Mirton.

Ericson, J.E. 1985. Strontium isotope characterization in the study of prehistoric human ecology. *Journal of Human Evolution* 14, 503–14.

Freyer, H. D. and Belacy, N.1983. $^{13}C/^{12}C$ records in northern hemisphere trees during the past 500 years. Anthropogenic impact and climatic superposition. *Journal of Geophysical Research* 88, 6844–52.

Fricke, H.C. and O'Neil, J.R. 1996. Inter- and intra-tooth variation in the oxygen isotope composition of mammalian tooth enamel: some implications for palaeoclimatological and palaeobiological research. *Palaeogeography, Palaeoclimatology, Palaeoecology* 126, 91–100.

Gillis, R., Bréhard, S., Balasescu, A., Ughetto-Monfrin, J., Popovici, D., Vigne, J.-D. and Balasse, M. 2013. Sophisticated cattle dairy husbandry at Bordusani-Popină (Romania, 5th mill BC): the evidence from complementary analysis of mortality profiles and stable isotopes. *World Archaeology* 45, 447–72.

Halstead, P. 2006. Sheep in the garden: the integration of crop and livestock husbandry in early farming regimes of Greece and southern Europe. In D. Serjeantson and D. Field (eds), *Animals in the Neolithic of Britain and Europe*, 42–55. Oxford: Oxbow Books.

Haşotti, P. 1997. *Haşotti, Epoca neolitică în Dobrogea*. Constanţa: Bibliotheca Tomitana I.

Heaton, T.H.E. 1999. Spatial, species and temporal variations in the 13C/12C ratios of C_3 plants: implications for palaeodiet studies. *Journal of Archaeological Science* 26, 637–49.

Homewood K.M. and Rodgers W.A. 1991. *Maasailand ecology. Pastoralist development and wildlife conservation in Ngorongoro, Tanzania*. Cambridge: Cambridge University Press.

Honch, N.V., Higham, T.F.G., Chapman, J., Gaydarska, B. and Hedges, R.E.M. 2006. A palaeodietary investigation of carbon ($^{13}C/^{12}C$) and nitrogen ($^{15}N/^{14}N$) in human and faunal bones from the Copper Age cemeteries of Varna I and Durankulak, Bulgaria. *Journal of Archaeological Science* 33, 1493–1504.

Ivanova, A., Nechev, J. and Stefanov, K. 2000. Lipid composition of some halophyte plants from the Black Sea coast in Bulgaria. *Comptes rendus de l'Académie bulgare des Sciences* 53(3), 83–6.

Kohn, M.J. 2010. Carbon isotope compositions of terrestrial C_3 plants as indicators of (paleo)ecology and (paleo)climate. *Proceedings of the National Academy of Sciences of the United States of America* 107, 19691–5.

Lecomte, T. and Le Neveu, C. 1986. *Le Marais Vernier. Contribution à l'étude et à la gestion d'une zone humide*. Doctoral thesis. Université de Rouen Haute-Normandie.

Manhart, H. 1998. *Die vorgeschichtliche Tierwelt von Koprivec und Durankulak und anderen prähistorischen Fundplatzen in Bulgarien aufgrund von Knochenfunden aus archäologischen Ausgrabungen*. Documenta Naturae 116. München: Paläobotanisch-Botanistratigraphische Arbeitsgruppe.

Marinescu-Bîlcu, S., Matei, G. and Renţa, E. 1995. *Ciulniţa. 1994, 1995*. Cronica cercetărilor din România. http://www.cimec.ro/scripts/arh/cronica.

Mirea, P. 2011. Between everyday and ritual use – 'small altars' or 'cult tables' from Măgura 'Buduiasca',Teleorman County (I): the Early Neolithic finds. *Buletinul Muzeului Judetean Teleorman. Seria Arheologie* 3, 41–58.

Panaitescu, L., Lungu, M. and Nita, S. 2011. Agro-meteorological conditions for the cultivation of autumn wheat in South Dobrudja. *Present Environment and Sustainable Development* 5, 169–78.

Popescu, D. and Caraivan, G. 2002/2003. Evoluţia zonei lacului Taşaul în cuaternarul târziu. *Pontica* 35/36, 53–9.

Popovici, D., Randoin, B. and Rialland, Y. 2001. Le tell néolithique et chalcolithique d'Hârsova (Roumanie). In J. Guilaine (ed.), *Communautés villageoises du Proche-Orient à l'Atlantique (8000–2000 avant notre ère)*, 119–52. Paris: Éditions Errance.

Popovici, D., Haită, C., Bălăşescu, A., Radu, V., Vlad, F. and Tomescu, J. 2008. *Archaeological pluridisciplinary researches at Borduşani-Popină. Muzeul National de Istorie a României*. Seria Cercetări Pluridisciplinare VI. Târgovişte: Editura Cetatea de Scaun.

Pyankov, V.I., Ziegler, H., Akhani, H., Deigele, C. and Lüttge, U. 2010. European plants with C_4 photosynthesis: geographical and taxonomic distribution and relations to climate parameters. *Botanical Journal of the Linnean Society* 163, 283–304.

Radu, V. 2008. Studiul materialului faunistic (moluşte, peşti, ţestoase şi păsări) prelevat din nivelul Hamangia III de la Cheia (jud. Constanţa). *Pontica* 41, 57–64.

Richards, M.P., Fuller, B.T. and Hedges, R.E.M., 2001. Sulphur isotopic variation in ancient bone collagen from Europe: implications for human palaeodiet, residence mobility, and modern pollutant studies. *Earth and Planetary Science Letters* 191, 185–90.

Slavchev, V. 2004/2005. Monuments of the Final Phase of Cultures Hamangia and Sava on the territory of Bulgaria. *Pontica* 37/38, 9–20.

Squires, V.R., Wilson, A.D. and Daws, G.T. 1972. Comparison of the walking activity of some Australian sheep. *Proceedings of the Australian Society for Animal Production* 9, 376–80.

Spassov, N.I., 2002. The animal bones from the prehistoric necropolis near Durankulak (NE Bulgaria) and the latest record of *Equus hydruntinus* Regalia. In H. Todorova (ed.), *Durankulak, Band 2: die prähistorischen Gräberfelder von Durankulak,* 313–24. Berlin and Sofia: Deutsches Archäologisches Institut.

Todorova, H. (ed.) 2002. *Durankulak, Band II: die prähistorischen Gräberfelder von Durankulak.* Berlin and Sofia: Deutsches Archäologisches Institut.

Tornero, C., Balasescu, A., Ughetto-Monfrin, J., Voinea, V. and Balasse, M. 2013. Seasonality and season of birth in early Eneolithic sheep from Cheia (Romania): methodological advances and implications for animal economy. *Journal of Archaeological Science* 40, 4039–55.

Vogel, J.C., Fuls, A. and Ellis, R.P. 1978. The geographical distribution of Kranz grasses in South Africa. *South African Journal of Science* 74, 209–15.

Voinea, V. and Neagu, G. 2008. Archaeological research at Hamangia III settlement from Cheia (2004–2008). *Pontica* 41, 9–26.

Western, D. and Dunne, T. 1979. Environmental aspects of settlements site decisions among pastoral Maasai. *Human Ecology* 7, 75–98.

Zazzo, A., Balasse, M., Passey, B.H., Moloney, A.P., Monahan, F.J. and Schmidt, O. 2010. The isotope record of short- and long-term dietary changes in sheep tooth enamel: implications for quantitative reconstruction of paleodiets. *Geochimica et Cosmochimica Acta* 74, 3571–86.

8

Herding Practices in the Ditched Villages of the Neolithic Tavoliere (Apulia, South-east Italy)

A Vicious Circle? The Isotopic Evidence

MARY ANNE TAFURI, JOHN ROBB, MARIA GIOVANNA BELCASTRO, VALENTINA MARIOTTI, PAOLA IACUMIN, ANTONIETTA DI MATTEO AND TAMSIN O'CONNELL

Introduction

THE UNDERSTANDING OF NEOLITHISATION IN EUROPE entails models and perspectives that are far from finding a consensus. Most archaeologists, however, agree upon the idea that the very notion of the Neolithic has shifted from being a material culture-related concept to a way of life (cf. Halstead 2011). Despite the large-scale models proposed for exploring the introduction of farming to Europe, most of the time the Neolithic is approached at a small-scale level.

In the Mediterranean, the understanding of Neolithisation is no different, with an extremely fragmented picture, which often relies on a bottom-up approach. In Italy, the spread of farming is traditionally associated with key areas such as the south-east coast of the Peninsula, where a great number of studies have concentrated. In the Tavoliere plain of northern Apulia (Figure 8.1), between the late 1940s and 1950s, a series of aerial photographs revealed the traces of buried contexts, which later were revealed to be the archaeological evidence of an intense occupation of the area already at very early phases of the Neolithic (Bradford 1949, 1950, 1957). A large number of the so-called *villaggi trincerati*, medium- to large-sized ditched villages were identified and later excavated (Tinè 1975, 1983; Cassano and Manfredini 1983, 2005; Tunzi Sisto *et al.* 2006). On the basis of plant and animal remains, they were traditionally associated with food-producing economies, characterised by intensive farming and increasingly specialised herding (Pessina and Tinè 2008). The strong reliance on agriculture of these first Neolithic inhabitants and the importance of early crops as well as livestock and other foodstuffs constructed the diet and social practices of these human populations. The existence

Proceedings of the British Academy **198**, 143–57. © The British Academy 2014.

of different econiches in a rather small area, which includes the alluvial plain (the Tavoliere itself), the seacoast, and highlands such as the Gargano and the Preappennines, favours an approach based on isotopic studies. Stable carbon and nitrogen isotope analysis are among the best ways to identify and define the different isoscapes in Neolithic south-east Italy; they can also contribute to the reconstruction of economic models for early and middle Neolithic populations, by directly assessing food consumption.

Despite the large number of sites registered in the Tavoliere, only a few have been excavated systematically and can provide human and animal skeletal remains for an isotopic investigation. For this study we selected three sites, all dated to the last centuries of the sixth millennium cal BC (approximately 5500–5400 cal BC); two of them (Passo di Corvo and Masseria Candelaro) are ditched villages located in the proximity of the Candelaro river, while the third (Grotta Scaloria) is a cave used for ritual and funerary purposes placed at the northernmost limit of the Tavoliere, at the foot of the Gargano mountains. For comparative analysis, our dataset includes isotopic values from early to middle Neolithic sites in the regions surrounding the Tavoliere (Figure 8.1).

Site descriptions

The site of Passo di Corvo lies on a terrace of the Amendola plain, about 10 km north-east of the city of Foggia and 25 km south-west of Manfredonia. The area of the village is about 40 ha, inside of which a number of C-shaped ditches were excavated between 1966–1982 (Tinè 1983). An outer ditch encloses an area of at least 90–130 ha (Figure 8.2). Two areas contained the human remains of at least 12 individuals, with one further individual (a female; t.11) found inside one of the ditches. The site should be considered as exceptional within the Neolithic evidence of the Tavoliere. The overall size of the area, and the over 80 C-compounds are unique. According to Tinè (1983, 183–9), at its peak of occupancy the site would have housed 30–35 families, structurally organised so as to allow the planning and construction of the massive ditch that limits the occupied area. This interpretation is today debated and the very function of the C-compounds is questioned (Monaco, pers. comm.). In any case we are probably looking at several groups of sedentary families, thriving on an agricultural economy with the contribution of the herding of domestic species such as cattle, ovicaprids and pigs. The very role of the ditches was believed by the excavators (Tinè 1983) to be functional to the maintenance of the village, either for draining water or for acquiring soil for farming purposes; again, later approaches have questioned this interpretation and the function of these structures was seen in a socio-cultural perspective, where ditches might have favoured groups' self-identity and other forms of social cohesion (Cassano and Manfredini 1983). A ritual function has also been put forward

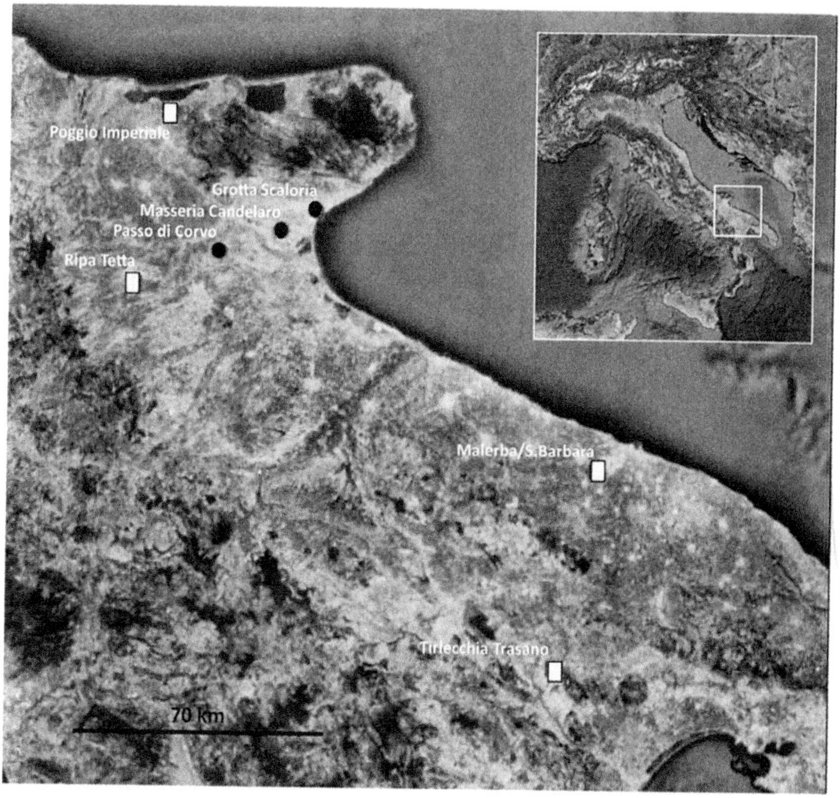

Figure 8.1 Map of the Apulian Tavoliere with sites investigated (circles = sites fully investigated; squares = comparative sites).

(Antoniazzi *et al.* 1990); most perspectives are plausible and, often, compatible (for a review see Skeates 2000).

Located about 14 km south-west of Manfredonia, Masseria Candelaro is placed on a small elevation (*Coppa* in Apulian) on the terrace of the river Candelaro, today about 1.5 km away. The site was excavated between the 1970s–1990s (Cassano and Manfredini 1983, 2005). Various ditches were brought to light, the largest measuring about 300 m in diameter. The earliest evidences are dated to the Early Neolithic. The site revealed several dwelling units, but the excavators suggest that during the middle Neolithic parts of the ditches were used for funerary purposes. Several burials, together with a cache with eight skulls, were excavated at the site (Salvadei and Santandrea 2003).

The Scaloria Cave is well known within Italian prehistory as a Neolithic ritual site. It is located about 2 km inland, close to the modern town of Manfredonia, at the foot of the Gargano mountains. The site was excavated in 1978–1979 (Tinè and Isetti 1980a, 1980b; Gimbutas 1981), but results were never fully published. While

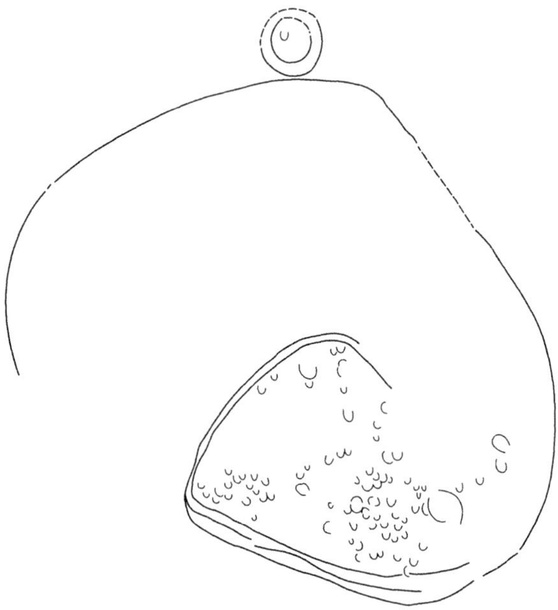

Figure 8.2 The plan of the ditched village of Passo di Corvo; the inner ditch (with C-compounds) and the outer ditch are visible (after Tinè 1983, modified).

the lower part of the cave (Scaloria bassa) was associated with a ritual function, excavations of the upper part of the cave (Scaloria alta) yielded an assemblage of commingled, highly fragmented, human remains. Since 2007, the Scaloria archives and collections are being re-analysed. The human skeletal assemblage counts at least 31 individuals (Knüsel *et al.* in press), with a systematic isotopic study carried out (Tafuri *et al.* in press).

Stable carbon and nitrogen analysis

Isotope content in bone collagen can help to determine the relative protein contribution to human and animal diet over a period of about ten years preceding death (Ambrose and Norr 1993, Hedges *et al.* 2007). Carbon isotope ratio ($\delta^{13}C$) is able to distinguish between marine ($\delta^{13}C$ enriched) and terrestrial ($\delta^{13}C$ depleted) diet. It can also reveal the type of plant consumed, with particular reference to the kind of photosynthetic pathway followed (C_4 plants are normally ^{13}C enriched while C_3 species are ^{13}C depleted). In Europe, where C_4 plants are infrequent, carbon isotopic ratios are also used to assess the intake of marine foods, since marine environments are enriched in $\delta^{13}C$ relative to temperate terrestrial ecosystems. Nitrogen isotopic values reflect the trophic level of an organism, considering that there is an approximate 3–5‰ increase in $\delta^{15}N$ along the food chain (Minagawa and Wada 1984, Hedges and Reynard 2007). An individual's nitrogen isotopic value indicates its position in the terrestrial food chain (herbivore, omnivore, carnivore), which for humans can be used as an indication of the relative importance of plant or animal protein in the diet (O'Connell and Hedges 1999). The type of animal protein consumed cannot be distinguished, that is, the difference between meat and secondary products, nor its quality (Privat *et al.* 2005, Katzenberg and Krouse 1989). Nitrogen isotopic values can also distinguish marine/freshwater versus terrestrial food intake, since aquatic species have much higher nitrogen isotopic values (Schoeninger *et al.* 1983).

For this study, we collected 82 samples of human bone and 27 of terrestrial animal bone for analysis (Table 8.1). Collagen extraction followed a modified Longin (1971) method (Brown *et al.* 1988). In brief, cortical bone (0.5 g) was cleaned by sand abrasion and demineralised in 0.5M aq. HCl at 4°C for several days. The samples were then rinsed to neutral pH and gelatinised in pH3 HCl at 70°C for 48 hours. The collagen solution was filtered off with 5–8 μm Ezee filters, frozen and then freeze-dried. Each of the collagen extracts was weighed (*c.* 1 mg) in triplicate into tin capsules, and stable carbon and nitrogen isotope ratios were measured using an automated elemental analyser coupled in continuous-flow mode to an isotope-ratio-monitoring mass-spectrometer (Costech elemental analyser coupled to a Thermo Finnigan MAT253 mass spectrometer). Analysis was carried out at the Godwin Labatory, University of Cambridge. Based on replicate analyses of international and laboratory standards, measurement errors are less than ±0.2‰ for $\delta^{13}C$ and $\delta^{15}N$. The collagen yield, the percentage of carbon and nitrogen, and the atomic C:N ratio of each sample were also recorded to check collagen quality (DeNiro 1985, Ambrose 1990, van Klinken 1999).

Results

Results of stable carbon and nitrogen isotopes ratio are reported in Table 8.1. At Masseria Candelaro mean carbon values are −19.2‰ for the humans and −21.1‰ for the animals, while mean nitrogen ratios are 9.3‰ and 6.3‰ for humans and animals respectively. At Passo di Corvo, mean carbon values are −19.3‰ for humans and −19.7‰ for fauna samples; mean nitrogen values are unexpectedly high with 13.3‰ and 10.2‰ for the humans and animals respectively. Grotta Scaloria mean ratios appear to be in line with those of Masseria Candelaro; carbon values are −19.3‰ and −19.9‰, while mean nitrogen ratios are 8.4‰ and 6.0‰ for human and faunal specimens respectively. At all sites, there is a >2‰ nitrogen enrichment between animals and humans (mostly herbivores).

The $\delta^{13}C$ values suggest a predominantly terrestrial diet, based on the consumption of C_3 plants (Figure 8.3); faunal specimens show greater variability,

Table 8.1 Stable carbon ($\delta^{13}C$) and nitrogen ($\delta^{15}N$) isotope data of human and animal bone collagen. C:N ratio is reported as a collagen quality indicator.

Sample	*Species*	$\delta^{13}C_{(PDB)}$	$\delta^{15}N_{(AIR)}$	*C:N*
Passo di Corvo				
PC T3	human	−19.2	13.0	3.2
PC T8	human	−19.1	12.9	3.2
PC T4	human	−19.2	12.9	3.2
PC T5B	human	−19.2	12.6	3.2
PC T7	human	−19.0	13.9	3.2
PC T9	human	−19.1	14.6	3.2
PC T6	human	−19.0	13.5	3.2
PC T11	human	−19.3	12.5	3.2
PC T10	human	−19.0	15.4	3.2
PC T10	human	−19.3	14.1	3.2
PC T11	human	−19.1	11.9	3.2
PC T12	human	−19.2	13.2	3.2
PC T13	human	−19.1	13.8	3.3
PC no code	human	−20.9	11.2	4.7
PC Bos_1	cattle	−18.3	11.1	3.2
PC Bos _2	cattle	−20.0	9.3	3.2
PC Ovis _3	sheep/goat	−20.5	9.8	3.2
PC Ovis _4	sheep/goat	−20.3	9.3	3.3
PC Ovis _5	sheep/goat	−19.3	11.7	3.3
Grotta Scaloria				
GS-2	human	−19.6	7.5	3.2
GS-3	human	−19.5	7.9	3.3
GS-4	human	−19.8	6.9	3.2
GS-5	human	−19.6	9	3.2
GS-6	human	−19.5	9.4	3.3
GS-7	human	−19.3	6.8	3.2
GS-8	human	−19.2	8.7	3.2
GS-9	human	−19.4	8.3	3.2
GS-11	human	−19.0	7.7	3.2
GS-12	human	−19.3	7.9	3.2
GS-13	human	−19.1	8.6	3.2
GS-14	human	−19.1	8.9	3.2
GS-16	human	−19.2	8.1	3.2
GS-17	human	−19.4	7.9	3.2
GS-18	human	−19.1	8.9	3.2
GS-19	human	−19.2	7.3	3.2
GS-20	human	−19.5	7.3	3.2
GS-21	human	−19.5	8.1	3.2
GS-22	human	−19.1	7.3	3.2
GS-23	human	−19.5	8.2	3.2
GS-24	human	−19.9	8.2	3.2
GS-25	human	−19.2	8.8	3.2
GS-26	human	−19.0	8.2	3.2
GS-27	human	−19.0	8.8	3.2
GS-28	human	−19.3	8.7	3.2

Table 8.1 Continued

Sample	*Species*	$\delta^{13}C_{(PDB)}$	$\delta^{15}N_{(AIR)}$	*C:N*
GS-29	human	−19.1	8.2	3.2
GS-30	human	−19.2	8.1	3.2
GS-31	human	−19.4	8.6	3.2
GS-32	human	−19.2	8.6	3.2
GS-33	human	−19.0	9.2	3.2
GS-34	human	−19.5	8.9	3.2
GS-35	human	−19.8	8.1	3.2
GS-37	human	−19.0	9	3.2
GS-38	human	−19.5	8	3.2
GS-39	human	−19.8	8.1	3.3
GS-40	human	−19.1	8.5	3.1
GS-41	human	−19.3	8.3	3.2
GS-A	human	−19.6	9.2	3.4
GS-B	human	−19.3	9.0	3.3
GS-C	human	−19.8	8.6	3.3
GS-D	human	−19.8	10.6	3.3
GS-E	human	−19.4	8.8	3.4
GS-F	human	−18.9	9.5	3.3
GS-188	human	−19.2	8.9	3.2
GS-42	sheep/goat	−17.7	7.2	3.2
GS-43	sheep/goat	−19.4	5.6	3.2
GS-44	sheep/goat	−19.7	5.4	3.2
GS-45	sheep/goat	−20.4	7.7	3.3
GS-46	sheep/goat	−20.4	5.1	3.2
GS-47	cattle	−20.8	6.3	3.2
GS-48	cattle	−20.0	6.4	3.2
GS-51	pig	−20.7	6.3	3.2
GS-52	red deer	−20.3	5	3.1
GS-54	sheep/goat	−18.7	6.5	3.2
GS-55	sheep/goat	−20.3	6.9	3.2
GS-56	cattle	−17.6	5.8	3.2
GS-57	sheep/goat	−17.7	6.2	3.2
GS-58	roe deer	−21.2	4.7	3.2
GS-59	red deer	−20.5	4.8	3.3
GS-60	red deer	−21.1	5.3	3.2
GS-61	sheep/goat	−20.4	6.4	3.2
GS-62	sheep/goat	−20.2	7.9	3.2
GS-63	sheep/goat	−20.7	6	3.2
GS-65	red deer	−21.1	4.1	3.3
Masseria Candelaro				
MC 1	human	−19.4	9.0	3.3
MC 2	human	−19.4	10.3	3.3
MC 3	human	−19.4	9.4	3.3
MC 4	human	−18.9	9.3	3.0
MC 5	human	−19.3	10.1	3.4
MC 6	human	−19.2	9.2	3.5
MC 7	human	−19.2	9.2	3.4

Table 8.1 Continued

Sample	*Species*	$\delta^{13}C_{(PDB)}$	$\delta^{15}N_{(AIR)}$	*C:N*
MC 8	human	−19.1	8.8	3.0
MC 9	human	−19.2	9.8	3.4
MC 10	human	−19.2	10.3	3.4
MC 11	human	−18.9	11.4	2.9
MC 12	human	−18.7	9.2	3.2
MC 13	human	−18.9	9.0	3.4
MC 14	human	−19.2	8.5	3.0
MC 15	human	−19.3	9.5	3.0
MC 16	human	−19.5	8.6	3.0
MC 17	human	−19.0	8.9	3.0
MC 18	human	−19.3	7.9	3.2
MC 19	human	−18.8	8.8	3.1
MC 20	human	−19.4	9.4	3.0
MC 21	human	−19.4	8.8	2.9
MC 22	human	−19.3	9.2	3.0
MC 23	human	−18.2	8.2	3.1
MC 24	human	−19.9	9.2	3.0
MCXVII	sheep/goat	−21.5	6.4	3.7
MCXVII	pig	−20.8	6.3	3.2

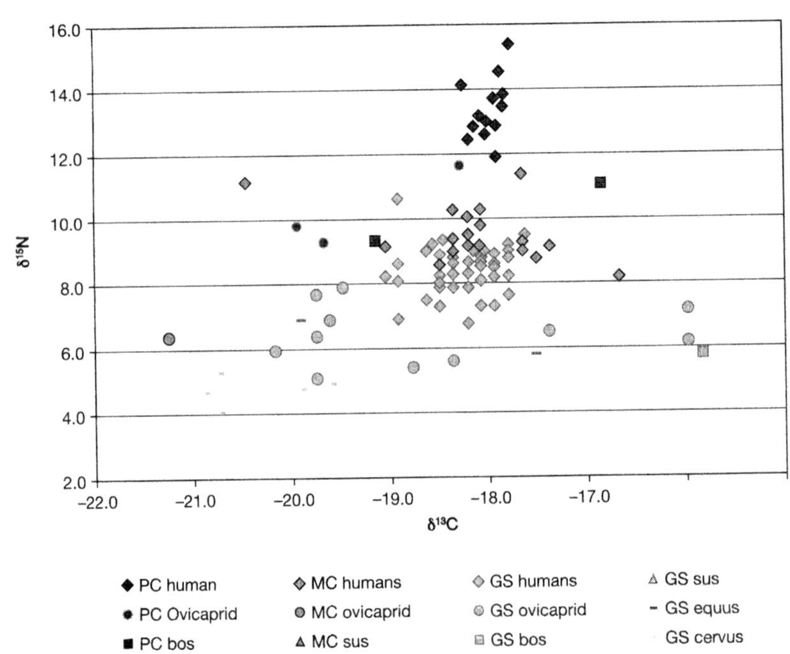

Figure 8.3 Stable carbon ($\delta^{13}C$) and nitrogen ($\delta^{15}N$) isotope data of humans and animals from Passo di Corvo (PC), Masseria Candelaro (MC) and Grotta Scaloria (GS).

which is unsurprising, especially considering the presence of wild fauna in the dataset (namely, the deer, though the cattle and ovicaprids at Grotta Scaloria show the broadest range). At Masseria Candelaro and Grotta Scaloria $\delta^{15}N$ is again indicative of a terrestrial diet, with the consumption of animal proteins by the humans. At Masseria Candelaro higher $\delta^{15}N$ and $\delta^{13}C$ in the humans as opposed to the fauna might indicate the contribution of marine resources to the diet, but faunal data come from only two specimens (an ovicaprid and a swine), which makes our interpretation only tentative.

Passo di Corvo shows $\delta^{15}N$ (for both human and animal specimens) that is exceptionally high. Such nitrogen values are normally associated with a significant consumption of marine resources, but two aspects challenge this interpretation: (1) $\delta^{15}N$ is high for the humans *and* the herbivores analysed (two cattle and three ovicaprids), which would imply that both humans and animals at the site consumed aquatic species; and (2) high nitrogen isotopic ratios do not correlate with $\delta^{13}C$-enriched values – a condition that normally occurs when marine resources are consumed – while, on the contrary, carbon data at Passo di Corvo are consistent with those at the other sites, showing the normal consumers-consumed offset (*c.* 1‰).

We can include in the dataset mean nitrogen isotopic values of human and animal samples from early to middle Neolithic ditched villages surrounding those of the Tavoliere, namely Ripa Tetta, in the northernmost portion of the plain; Tirlecchia and Trasano in the province of Matera; Malerba and S. Barbara in the area of Bari, and Poggio Imperiale north of the Gargano mountains (Figure 8.4). At all sites $\delta^{15}N$ is consistent with data from Masseria Candelaro and Scaloria. High $\delta^{15}N$ values –

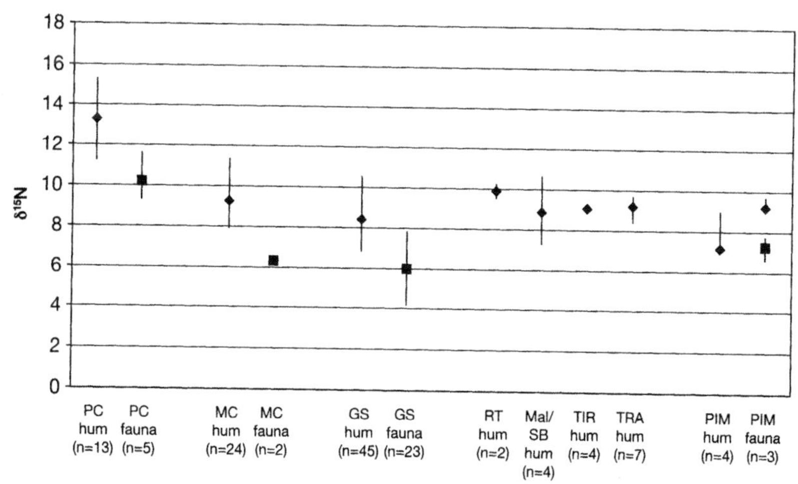

Figure 8.4 Mean stable nitrogen isotopic ratios ($\delta^{15}N$) with ranges for both human and animal specimens at Passo di Corvo (PC), Masseria Candelaro (MC) and Grotta Scaloria (GS), together with mean comparative values from Ripa Tetta (RT)*, Malerba/S. Barbara (Mal/SB)*, Tirlecchia (TIR)*, Trasano (TRA)* and Poggio Imperiale (PIM)*. * (Tafuri *et al.*, unpublished data).

analogous to the ones obtained at Passo di Corvo – were registered by Lelli *et al.* (2012) at Ripa Tetta, in contrast with the data we obtained at the same site, although on different individuals; they interpreted such high nitrogen isotopic values as indicative of elevated protein intake, mostly in consideration of the animal/human offset.

We cannot rule out contamination as the possible cause for high $\delta^{15}N$ in the bone samples at Passo di Corvo; modern nitrates used for farming could have altered the nitrogen composition of the soils in the area (Heaton 1986), though both the areas of Passo di Corvo and Masseria Candelaro – only a few kilometres apart – have been affected by farming activities and would have similarly suffered from soil contamination. The proximity to the sea might also have caused high $\delta^{15}N$ values in plant species, as observed elsewhere (Virginia and Delwiche 1982, Heaton 1987), but again it would be difficult to explain how this had an effect on Passo di Corvo samples but not on those from the other sites considered, all similarly located near the coast.

We tentatively interpret the data from Passo di Corvo as the result of a manuring effect occurring at the site. Animal manure has high $\delta^{15}N$ because of the greater loss of the more volatile nitrogen isotope (^{14}N), with relative enrichment of the heavier one (^{15}N) in the residual ammonia which turns into ^{15}N-enriched nitrates; such nitrates are then taken up by plants, triggering a high ^{15}N cycle (Kreitler and Jones 1975, Kendall 1998), which alters the nitrogen content of the tissues of the plant consumers.

What might have favoured a nitrogen cycle at Passo di Corvo can be linked to the size and the plan of the site. Within the Tavoliere, among the over 500 ditched villages identified (Skeates 2000), Passo di Corvo is the only one that reaches such an extraordinary size; its outer ditch encircles an area of approximately 90–130 ha, while other sites rarely exceed 30–40 ha (Tinè 1983). What is of great interest, moreover, is the plan of the site with a smaller ditch to limit the typical area with the C-compounds, and a larger ditch, which encloses an extraordinarily large portion of land with no evidence of recognisable structures (Figure 8.2). It is unquestionable that such an area had a function connected to the sustenance of the population. Whatever the function of the Tavoliere ditches might have been, they limited a piece of land within a wider landscape; at other villages (Masseria Candelaro, but also Tirlecchia, Trasano, Malerba, Ripa Tetta and Poggio Imperiale), such limited spaces did not allow for the tending of domestic animals, so that daily/seasonal activities would have moved to areas outside the borders of the village. At Passo di Corvo this might not have been necessary, as the large encircled area could guarantee a considerable plot of land (Figure 8.2), and might have been devoted to both herding and farming activities. This is even more convincing if we consider that according to the excavators (Tinè 1983, 184) the creation of the ditches can be ascribed to a single moment (phase IVa1), so that the village reached its maximum extension at a very early stage of its life.

Our isotope data seem to be consistent with Tinè's idea that the community at Passo di Corvo was devoted to both farming and herding, with the C-compounds space, the larger encircled area and the zone immediately near the ditch occupied on a (seasonal) cycle for keeping animals and growing crops (Tinè 1983, 1985). This scenario would correspond to Bogaard's (2012) reconstruction of Neolithic farming on long-established plots of land, which would inevitably require actions to enhance soil productivity, that is, manuring. The practice of manuring would have implied a long-term investment in a portion of territory that in most cases would have been claimed through visible structures (in this case, the outer ditch; Bogaard *et al.* 2013).

Our interpretation seems to agree also with a model recently proposed by Monaco (2011) on the carrying capacity of the Tavoliere, according to which the size of Passo di Corvo outer ditch could have supported a combined farming/herding exploitation system, which did not require the seasonal movement of animals. It is worth considering that, within the Tavoliere, Passo di Corvo lies south of the river Candelaro, which during the sixth millennium cal BC might have created a natural barrier to the highlands of the Gargano, that could instead be easily reached by the groups of Masseria Candelaro and, possibly, Grotta Scaloria (Caldara and Pennetta 2002, Monaco 2011, Fiorentino *et al.* 2013).

If we were to use a cautious estimate of two tons of dung produced by a cow per year (Geden *et al.* 1990), it would require 15 cows and about double this number of sheep/goat or swine to cover an area of about 100 ha on Wulff's (1966, 270) calculation of approximately 1.5 tons per hectare per year required for manuring purposes. This calculation is pessimistic when compared to numbers proposed elsewhere (Tinè 1983, Rowly-Conwy 1981, Monaco 2011). In any case, the permanence of the animals on the site would have reduced the need to transport the dung, although they could have been used as draught animals occasionally.

As Bogaard suggests, 'it appears plausible that a household keeping a few cattle for meat and perhaps milk, as well as a few sheep/goat and pigs, could, by strategic folding of animals on stubble and spreading manure as well as household refuse, manage to replenish nutrients in intensively cultivated plots' (Bogaard 2004, 46). At Passo di Corvo, unlike at other sites, this practice might have been performed in an enclosed area, a place – spatially defined – where work, but also social relations, might have been concentrated. It is difficult, at this stage, to affirm whether the use of animal dung at the village was intentional rather than unintentional; other data (for example, from micromorphology, and isotopic analysis of plant remains) will help to refine our interpretation in the future, but it is suggestive of 'locally adapted' subsistence practices (*sensu* Skeates 2000, 177).

Whether this 'vicious circle' at Passo di Corvo was the result of an intentional action to improve soil productivity, or rather the outcome of a practice that did not entail a strategy, it is difficult to determine at this point. It is however extremely interesting that at a very same moment in the sixth millennium, such diverse practices in what we have hitherto believed to be a homogeneous archaeological landscape (i.e., the ditched villages of the Tavoliere plain) were being followed. Despite the many new perspectives put forward, for many archaeologists the Neolithic has remained a 'package' that only rarely has shown its diversity. In southern Europe this is becoming less convincing, as we are confronted with new scenarios that differ from the traditional view of a monolithic process of Neolithisation. There is no ubiquitous spread of farming communities over increasingly larger territories in Italy, perhaps because of the composite nature of the landscape made of different and often very diverse ecological niches. In the south-eastern regions of the Peninsula, for a very long time, the spread of farming communities remains particularly marked within key areas (Radina 2002). This archaeological landscape demonstrates a great deal of complexity within the larger comprehensive model. To name but a few, along with a local raw material procurement activity, such as that attested at Masseria Candelaro and other coeval sites, stands the impressive system of corridors and chambers of the Defensola flint mine complex (Tarantini and Galiberti 2011). The ritual cave of Scaloria stands out for complexity and longevity, so does the site of Passo di Corvo, as the largest and more complex Neolithic village of south-eastern Italy. This dense network of sites, impressive cult caves, large flint mines falls within a radius of only about 50 km.

It is likely that this set of 'exceptionality' within a rather homogeneous landscape might mirror different forms of adaptations and specific socio-cultural organizations. Within this scenario we might explain the variety in the isotopic signatures at the sites investigated: coeval villages located only a few kilometres apart, found on a homogeneous model that would produce homogeneous archaeological landscapes (i.e., material culture, settlement patterns) but engaging in manifold economic practices within a shared background.

Acknowledgements

We thank Mike Hall and James Rolfe at the Godwin Lab, Department of Earth Sciences, University of Cambridge for help with isotopic analyses. We thank Catherine Kneale, Louise Butterworth, Alex Pryor, and Hazel Reade, McDonald Institute for Archaeological Research, for help in sample preparation and analysis. We are grateful to Fulvio Bartoli for the material from Ripa Tetta, Tirlecchia and Trasano. We thank Sandro Sublimi Saponetti for the material from Malerba and S. Barbara and Rocco Sanseverino for the material from Poggio Imperiale. Many thanks to Emanuele Cancellieri for help with the images. Special thanks to Andrea Monaco for comments and suggestions.

References

Ambrose, S.H. 1990 Preparation and characterization of bone and tooth collagen for isotopic analysis. *Journal of Archaeological Science* 17, 431–51.

Ambrose, S.H. and Norr L. 1993. Experimental evidence for the relationship of the carbon isotope ratios of whole diet and dietary protein to those of bone collagen and carbonate. In J.B. Lambert and G. Grupe (eds), *Prehistoric human bone archaeology at the molecular level*, 1–38. Berlin: Springer.

Antoniazzi, A., Bermond Montanari, G., Giusberti, G. Massi Pasi, M., Mengoli, D. and Prati, L. 1990. Lo scavo preistorico a Fornace Cappuccini in Faenza. *Archeologia a Faenza, Ricerche e scavi dal Neolitico al Rinascimento*, 23–59. Bologna: Nuova Alfa Editore.

Bogaard, A. 2004. *Neolithic farming in Central Europe: an archaeobotanical study of crop husbandry practices*. London: Routledge.

Bogaard, A. 2012. Middening and manuring in Neolithic Europe: issues of plausibility, intensity and archaeological method. In R.L. Jones (ed.), *Manure matters: historical, archaeological and ethnographic perspectives*, 25–39. Farnham: Ashgate.

Bogaard, A., Fraser, R.A., Heaton, T.H.E., Wallace, M., Vaiglova, P., Charles, M., Jones, G., Evershed, R.P., Styring, A.K., Andersen, N.H., Arbogast, R.-M., Bartosiewicz, L., Gardeisen, A., Kanstrup, M., Maier, U., Marinova, E., Ninov, L., Schäfer, M. and Stephan, E. 2013. Crop manuring and intensive land management by Europe's first farmers. *Proceedings of the National Academy of Sciences of the United States of America* 110, 12589–94.

Bradford, J.S.O. 1949. Buried landscapes in southern Italy. *Antiquity* 23, 58.

Bradford, J.S.O. 1950. The Apulian expedition: an interim report. *Antiquity* 24, 84.

Bradford, J.S.O.1957. *Ancient landscapes*. London: G. Bells & Sons.

Brown, T.A., Nelson D.E., Vogel, J.S. and Southon, J.R. 1988. Improved collagen extraction by modified Longin method. *Radiocarbon* 30, 171–7.

Caldara, M. and Pennetta, L. 2002. L'ambiente fisico delle Murge durante il Neolitico. In F. Radina (ed.), *La Preistoria della Puglia. Paesaggi, uomini e tradizioni di 8.000 anni fa*, 21–6. Bari: Mario Adda Editore.

Cassano, S.M. and Manfredini, A. 1983. *Studi sul Neolitico del Tavoliere della Puglia. Indagine territoriale in un'area campione*. Oxford: British Archaeological Reports.

Cassano, S.M. and Manfredini, A. (eds) 2005. *Masseria Candelaro: vita quotidiana e mondo ideologico in una comunità neolitica dell Tavoliere*. Foggia: Grenzi Editore.

DeNiro, M.J. 1985. Postmortem preservation and alteration of in vivo bone collagen isotope ratios in relation to palaeodietary reconstruction. *Nature* 317, 806–9.

Fiorentino, G., Caldara, M., De Santis, V., D'Oronzo, C., Muntoni, I.M., Oronzo, S., Primavera, M. and Radina F. 2013. Climate changes and human–environment interactions in the Apulia region of southeastern Italy during the Neolithic period. *The Holocene* 23, 1297–316.

Geden, C.J., Rutz, D.A. and Bishop, D.R. 1990. Cattle lice (Anoplura, Mallophaga) in New York: seasonal population changes, effects of housing type on infestations of calves, and sampling efficiency. *Journal of Economic Entomology* 83, 1435–8.

Gimbutas, M. 1981. *Grotta Scaloria: resoconto sulle richerche del 1980 relative agli scavi del 1979*. Manfredonia: Amministrazione Comunale.

Halstead, P. 2011. Farming, material culture and ideology: repackaging the Neolithic of Greece (and Europe). In A. Hadjikoumis, E. Robinson and S. Viner (eds), *Dynamics of Neolithisation in Europe: studies in honour of Andrew Sherratt*, 131–51. Oxford: Oxbow Books.

Heaton, T.H.E. 1986. Isotopic studies of nitrogen pollution in the hydrosphere and atmosphere: a review. *Chemical Geology (Isotope Science Section)* 59, 87–102.

Heaton, T.H.E. 1987. The 15N/14N ratios of plants in South Africa and Namibia: relationship to climate and coastal/saline environments. *Oecologia* 74, 236–46.

Hedges, R.E.M. and Reynard, L. 2007. Nitrogen isotopes and the trophic level of humans in archaeology. *Journal of Archaeological Science* 34, 1240–51.

Hedges, R.E.M., Clement, J.G., Thomas C. D. L. and O'Connell T.C. 2007. Collagen turnover in the adult femoral mid-shaft: modeled from anthropogenic radiocarbon tracer measurements. *American Journal of Physical Anthropology* 133, 808–16.

Katzenberg, M.A. and Krouse, H.R. 1989. Application of stable isotope variation in human tissue to problems in identification. *Canadian Society of Forensic Science Journal* 22, 7–19.

Kendall, C. 1998. Tracing nitrogen sources and cycling in catchments. In C. Kendall and J.J. McDonnell (eds), *Isotope tracers in catchment hydrology*, 519–76. Amsterdam: Elsevier.

Knüsel, C., Robb, J. and Tafuri M.A. In press. The humans skeletal remains from Scaloria Cave. In E.E. Elster and J.E. Robb (eds), *The Scaloria Cave*. Los Angeles: Cotsen Institute.

Kreitler, C.W. and Jones, D.C. 1975. Natural soil nitrate: the cause of the nitrate contamination of groundwater in Runnels County, Texas. *Ground Water* 13, 53–61.

Lelli Roberta, Allen R., Biondi R., Calattini M., Conati Barbaro C., Gorgoglione M.A., Manfredini A., Martínez-Labarga C., Radina F., Silvestrini M., Tozzi M., Rickards O. and Craig O.E. 2012. Examining dietary variability of the earliest farmers of south-eastern Italy. *American Journal of Physical Anthropology* 149, 380–90.

Longin, R. 1971. New method of collagen extraction for radiocarbon dating. *Nature* 230, 241–2.

Minagawa, M. and Wada E. 1984. Stepwise enrichment of 15N along food chain: further evidence and the relation between d15N and animal age. *Geochimica et Cosmochimica Acta* 48, 1135.

Monaco, A. 2011. A simulation of farming and breeding activities: comparing the economic strategies in south east Italy Neolithic communities. *Origini* 33, 61–82.

O'Connell, T.C. and Hedges, R.E.M. 1999. Investigations into the effect of diet on modern human hair isotopic values. *American Journal of Physical Anthropology* 108, 409–25.

Pessina, A. and Tinè V. 2008. *Archeologia del Neolitico: L'Italia tra VI e IV millennio a.C.* Roma: Carocci.

Privat, K.L., O'Connell, T.C., Neal, K. and Hedges, R.E.M. 2005. Fermented dairy product analysis and palaeodietary repercussions: is stable isotope analysis not cheesy enough? In J. Mulville and A. Outram (eds), *The archaeology of animal fats, oils and dairying*, 60–6. Oxford: Oxbow Books.

Radina, F. (ed.) 2002. *La Preistoria della Puglia. Paesaggi, uomini e tradizioni di 8.000 anni fa*. Bari: Mario Adda Editore.

Rowley-Conwy, P. 1981. Slash and burn in the temperate European Neolithic. In R. Mercer (ed.), *Farming practices in British prehistory*, 85–96. Edinburgh: Edinburgh University Press.

Salvadei, L. and Santandrea E. 2003. Condizioni di vita e stato di salute nel campione neolitico di Masseria Candelaro (FG). *Atti della Riunione Scientifica dell'IIPP* 35, 829–34.

Schoeninger, M.J., De Niro M.J. and Tauber H. 1983. Stable nitrogen isotope ratios of bone collagen reflect marine and terrestrial components of prehistoric human diet. *Science* 220, 1381–3.

Skeates, R. 2000. The social dynamics of enclosure in the Neolithic of the Tavoliere, south-east Italy. *Journal of Mediterranean Archaeology* 13, 155–88.

Tafuri, M.A., O'Connell, T.C., Souter, E., Libianchi, N. and Robb, J. In press. Diet during life: Paleoeconomic studies of human diet using stable carbon and nitrogen isotopes. In E.E. Elster and J.E. Robb (eds) *The Scaloria Cave*. Los Angeles: Cotsen Institute.

Tarantini, M. and Galiberti, A. (eds) 2011. *Le miniere di selce del Gargano VI–III millennio a.C. Alle origini della storia mineraria europea*. Firenze: All'Insegna del Giglio.

Tinè, S. 1972. Gli scavi nel villaggio neolitico di Passo di Corvo. *Atti XIV Riunione Scientifica Istituto Italiano Preistoria Protostoria*, 313–31.

Tinè, S. 1975. La civiltà neolitica del Tavoliere. *Atti del Colloquio Internazionale di Preistoria e Protostoria della Daunia*, 99–111. Firenze: Istituto Italiano di Preistoria e Protostoria.

Tinè, S. (ed.) 1983. *Passo di Corvo e la civiltà neolitica del Tavoliere*. Genova: Sagep.

Tinè, S. and Isetti, E. 1980a. Culto neolitico delle acque e recenti scavi nella Grotta Scaloria. *Bullettino di Paletnologia Italiana* 82, 31–70.

Tinè, S. and Isetti, E. 1980b. Recenti scavi nella Grotta Scaloria. *Atti del convegno archeologico "Civiltà e culture antiche tra Gargano e Tavoliere"* 1 (ns), 77–96.

Tunzi Sisto, A.M., Danesi, M. and Simonetti, R. 2006. Il grande abitato neolitico di Troia – Monte San Vincenzo. *Atti del 26 Convegno Nazionale sulla Preistoria Protostoria e Storia della Daunia* 1, 39–58.

van Klinken, G.J. 1999. Bone collagen quality indicators for palaeodietary and radiocarbon measurements. *Journal of Archaeological Science* 26, 687–95.

Virginia, R.A. and Delwiche, C.C. 1982. Natural ^{15}N abundance of presumed N_2-fixing and non N_2-fixing plants from selected ecosystems. *Oecologia* 54, 317–25.

Wulff, H.E. 1966. *The traditional crafts of Persia: their development, technology, and influence on Eastern and Western civilizations*. Cambridge: MIT Press.

9

Linear Pottery Culture Household Organisation

An Economic Model

LAMYS HACHEM AND CAROLINE HAMON

General aim of the study

HUSBANDRY, HUNTING AND AGRICULTURE are keys to understanding the economic and ideological aspects of the first Neolithic societies. The colonisation process of the first farmers of continental Europe relies heavily on the diffusion of a new model of production and a new structure of society, implying a real balance between farmers, their plant production and their animals. Through faunal remains and grinding tools, it is then possible to examine the organisation of Linear Pottery culture (LBK) societies.

In the Aisne valley (Picardy, France), such an approach is made possible by the quality of the archaeological data. The faunal assemblage is currently the largest in Europe for the LBK, and it is the first time that grinding tools have been studied at the scale of an entire region. Moreover, these artefacts come from clearly defined contexts, dating approximately to the late LBK, around 5100–4900 cal BC. The low density of structures and the short duration of occupation in comparison to other European sites limit the risk of chronological mixtures. The close relationship between the houses and the refuse pits which constitutes the basic household, together with the importance of the corpus of pottery, provide a fine chronological framework. All these are assets when attempting a comprehensive analysis of the organisational processes of the villages and locating them in the regional network.

A combined analysis of animal remains and macrolithic tools, reflecting meat and plant food consumption, was conducted on six multi-phase settlements dating to the late LBK of Paris basin (*Rubané récent du Bassin parisien* or *RRBP* in French: Constantin 1985, Constantin and Ilett 1997). This study enables us to propose a new socio-economic model for LBK settlement organisation, which sheds light on the definition of the household (Coudart 1998, Moddermann 1988, Soudský and Pavlů 1972), the significance of domestic space and its function (Boelicke 1982,

Coudart 2009, Bickle 2013, Pavlů 2013, Hamon *et al.* 2013, Allard *et al.* 2013), and village organisation (Boelicke *et al.* 1988, Whittle 1996, Lenneis 2010, Pavlů 2010, Wolfram *et al.* 2012).

Presentation of the sites in their regional context

The Aisne valley offers the highest density of LBK settlements in the Paris basin. In the context of a pioneer rescue excavation programme developed since the mid-1970s, 80 km of the valley have been studied by a team composed of members of the Paris 1 University and the former ERA 12 of the CNRS, now known as the UMR 8215 *Trajectoires*. Our study was carried out as part of two collective research projects (an *ACR*: Ilett *et al.* 2006; and a *PCR*: Allard and Hamon 2010). The long-term project in the river Aisne valley – 40 years of archaeological fieldwork – has produced very large quantities of data; 20 sites, 90 houses and 80 graves have been discovered and excavated. Such exceptional archaeological documentation has permitted reflection on the relative chronology of these different occupations, the organisation of the territory and the economy of the LBK people in question (Dubouloz *et al.* 2012, Ilett in press).

Most of the LBK sites are located at the bottom of the alluvial valleys, and are less than 1 km from the nearest water source (Figure 9.1). The domestic units are composed of a long partitioned house oriented east–west, together with their lateral pits used as refuse pits. They respond to specific technical, social and identity rules (Coudart 1998, 2009).

Despite the lack of a cemetery in the Aisne valley, a complex system of burial practices has been highlighted in close relation to the house and village organisation (Thevenet 2009). The smallest occupations are composed of two to three contemporary houses, whereas most important sites show several stages of occupation. The best known and most important of these sites is Cuiry-lès-Chaudardes, with more than 33 houses (Figure 9.2; Ilett 2012).

Adopting different scales of analysis, the pattern of the occupations and their changes through time and space can be approached in relation to their immediate environment, raw material procurement strategies and the configuration of local soils. The pattern indicates the existence of a network of main and secondary sites, distributed along the Aisne valley (Ilett and Plateaux 1995, Dubouloz *et al.* 2012). The distance between theoretical areas of influence is between 4–5 km for secondary sites and between 18–20 km for the main sites.

The radiocarbon dates of the LBK/RRBP occupations of the Aisne valley fall between 5100–4900 cal BC (Dubouloz 2003). Three ceramic stages have been defined by Michael Ilett after a seriation of the ceramic assemblages (Ilett 2012, Blouet *et al.* 2013, 316). These stages are not all represented on each site, except at Cuiry-lès-Chaudardes which shows the longest duration of occupation.

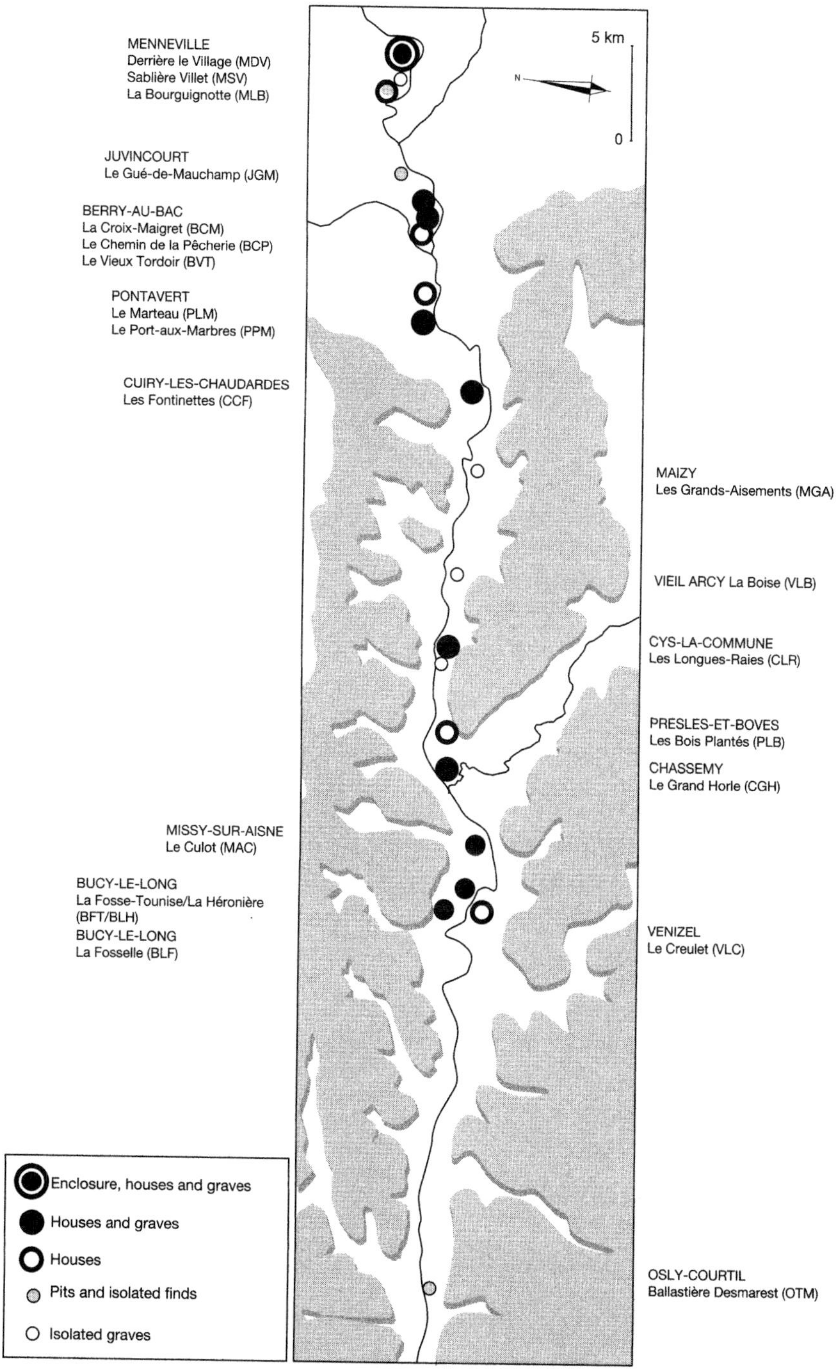

Figure 9.1 Map of the main LBK sites excavated in the Aisne Valley (France). Document ASAVA-UMR 8215 *Trajectoires.*

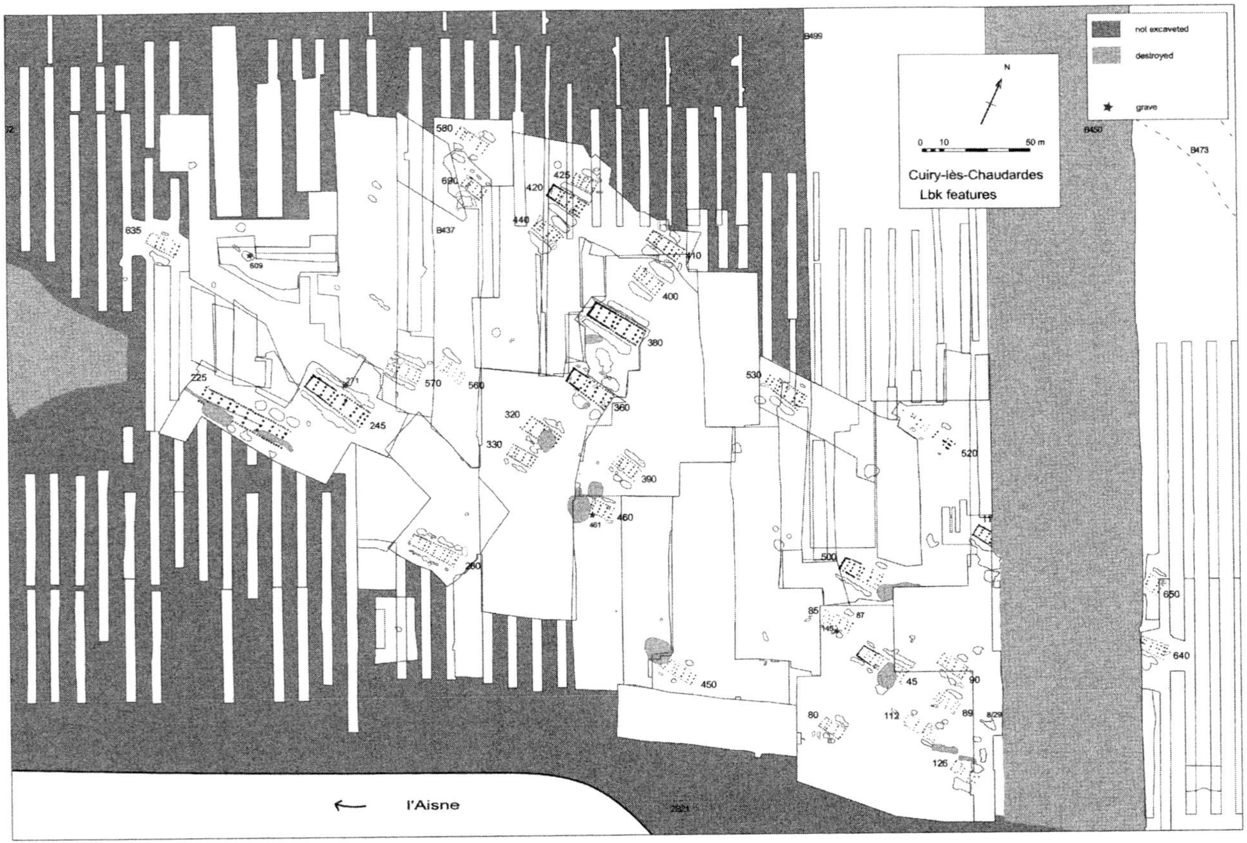

Figure 9.2 The LBK site of Cuiry-lès-Chaudardes (Aisne). Document ASAVA-UMR 8215 *Trajectoires*. CAD: C. Monchablon.

In the Aisne valley, we have a good knowledge of the subsistence economy and technical system of the LBK populations. The economy of Linear Pottery farmers depended on the cultivation of cereals (einkorn, wheat and common barley), legumes (peas and lentils), and a few other rarer plant species such as the poppy and flax (Bakels 1995, 1999, 2008). Cattle, sheep, goat, and pig constitute the majority of the faunal remains, and cattle breeding is geared towards the production of meat (Hachem 2011, Hachem and Bedault 2008). Although domestic animals are abundant, the hunting rate remains relatively high, especially in terms of the weight of meat from red deer and wild boar. Roe deer and aurochs are the two other large game, while wolf, bear and horse are rare. Small game is essentially represented by beaver, but there are many other fur species. Various bird, fish and amphibious species complete the list of animals found in the pits.

The lithic industry is based on different regional flint sources, among which Senonian and Tertiary sources dominate (Allard 2005). Their relative proportions vary from the western to the eastern part of the valley, in relation to different cultural influences. If blade debitage by indirect percussion characterises the assemblages, the proportion of long blades and flake debitage increases at the end of the sequence. The tools mainly consist of scrapers, arrowheads, sickle blades, burins, retouched blades and splintered forms. The macrolithic tools are made from local sandstones of different qualities (Hamon 2006). They include two sizes of grinding tools, used for cereal- and colouring-processing, polishing tools for the production of ornaments, bone industry and polished adzes, and percussion tools used for lithic production. Despite having little chronological significance, they are a good indicator of technological behaviour and give indications of the localisation of food processing in the domestic space.

For the bone industry, the debitage by abrasion during the RRBP is used for the shaping of perforating tools made mainly on the metapodials of sheep/goats and cattle, but also of deer (Sidéra 2000). Ornaments consist of beads in shell, limestone and bone, and rings in shell and stone (Bonnardin 2009).

Methodological approach

In this context, our study of the food practices in LBK households relies on the comparison of data concerning the size and typology of the houses, the characteristics of the faunal assemblage and the main activities involving macrolithic tools.

These criteria have been compared on six sites of the Aisne valley:

- Bucy-le-Long (the precise location is called *La Fosselle*) (Hachem *et al.* 1998a; seven houses selected from ten)
- Missy-sur-Aisne (*Le Culot*) (Farruggia and Constantin 1984; Charier 1986; two houses selected from five)

- Berry-au-Bac (*Le Vieux Tordoir*) (Allard *et al.* 1996)
- Berry-au-Bac (*Le Chemin de la Pêcherie*) (Ilett and Plateaux 1995; all three houses selected)
- Menneville (*Derrière-le-Village*) (Farruggia *et al.* 1996; Hachem *et al.* 1998b; three houses selected from nine)
- Cuiry-lès-Chaudardes (*Les Fontinettes*) (Ilett and Hachem 2001; nine houses selected from 33).

The inclusion of a house in the present study was determined by the visibility of the houseplan. Erosion or destruction of post-holes should not be a barrier to the understanding of the whole house plan, especially as regards the number of units to the west of the rear corridor. Only well preserved houses with rich find assemblages were included in this study. There is no standard amount of discarded animal bone per house; bone quantity varies widely from one house to another. The minimum for a normally preserved house is about 1000 fragments or 10 kg of bone (numbers below this threshold indicate major erosion of features) and the maximum is just under 7000 fragments or 65.5 kg of bone. This heterogeneity is not the result of taphonomy, and is simply reflecting domestic activities. So all houses with fewer than 1000 bone remains (or 100 identified bones) were eliminated from the analysis, with one exception (Missy-sur-Aisne, structure 75).

The same observations can be made of macrolithic tools. As we are not dealing with daily consumption refuse, the number of tools appears much lower in quantity than faunal remains. The houses with less than three tools, and fragments that could not be classed as tools, have been eliminated from the analysis. Importantly, the presence of only two or three grinding tools in the refuse pits must be considered with great care, as the duration of their use can span tens of years. Their deposition is a far from meaningless act in this context.

A thorough analysis was conducted to assess variation between households first on the numerous and well dated data from Cuiry-lès-Chaudardes (Table 9.1), and then on the other selected sites of the Aisne valley. In the end, 26 selected houses presented the most reliable contexts (complete plans, good preservation of the lateral pits and significant assemblages), and when possible, precise chronological attribution. They were chosen in order to compare house features with the different indicators of the subsistence economy.

A study of the house plans in terms of number of back units and chronological stages was conducted to establish clear distinctions between houses (Hachem 2009, 2011).

First, the number of bays (units) to the west of the rear corridor was used as a basis for dividing the houses into three categories (see Figure 9.5 later):

- small houses with one rear unit (house length: 9.5–15 m)
- medium-long houses with two rear units (house length: 15–21 m)
- long houses with three rear units (house length: 21–39 m).

Table 9.1 Number of faunal remains and macrolithic tools by main categories for the 33 households of Cuiry-lès-Chaudardes

Household no	*Number back units*	*Ceramic stage*	*Hunting*	*Herding*	*Bovines*	*Caprines*	*Wild boar*	*Grinding tool fragments*	*Hammering tool fragments*	*Abrading tool fragments*
112	?	1	179	494	291	105	71	6	2	1
500	3	3	8	201	126	39		7	1	
580	1	2	62	304	192	59	23	14	1	2
640	?	1	71	320	215	51	12	11		3
80	1	?	6	87	59	9	1		1	
85	1	?	5	97	74	14	1	1		
89	1	2	79	350	196	95	38	2	1	
90	1	1	153	488	328	114	33		1	1
126	1	1	263	577	295	146	108	1	2	
320	1	?	39	125	64	39	2			
330	1	2	37	196	122	36	4		4	1
390	1	1	38	214	148	28	12	6	3	3
400	1	2	103	420	334	41	44	2	1	5
410	1	2	14	87	65	7		2		
425	1	2	308	441	281	50	80	27	2	2
440	1	2	109	454	331	50	46	13	6	16
530	1	3	89	450	305	72	18	35	6	15
560	1	?	4	27	18	4				2
570	1	2	73	427	245	99	19	30	1	11
635	1	?	32	98	57	8	21	17	2	5
690	1	3	101	353	197	77	12	1	2	7
280	2	3	38	548	275	211	3	13		3
360	2	3	184	995	644	258	58	23	7	6
420	2	3	355	552	352	68	93	7	2	3
450	2	?	7	132	98	25		1	1	
520	2	3	6	70	47	8				2
11	3	2	14	145	98	27		1		
45	3	1	37	194	137	27	9	12		3
225	3	3	98	1804	710	761	12	31	2	1
245	3	3	30	530	221	247	8	14	1	1
380	3	2	281	1926	1411	285	72	30	22	8
TOTAL			2823	13106	7936	3060	800	307	71	101

A recurrent pattern characterises every LBK hamlet in the Aisne valley. Each group of contemporary houses making up a settlement phase is composed of several small houses associated with one (or two) long houses. Secondly, chronological stage was determined from study of the pottery (Ilett 2012; Blouet *et al.* 2013, 316). M. Ilett identified three main chronological stages for the settlement of the whole valley: Aisne 1, Aisne 2 and Aisne 3 (corresponding to early, middle and late RRBP). The first stage, which reflects the first occupation by migrants, involves fewer houses, while the last stage includes a larger number of houses.

Chi-squared tests were carried out to determine whether the domestic or wild fauna and the type of houses were linked or independent, and to ask the same question for the macrolithic tools. A correlation test was also conducted to ask if there was a link between grinding tools and domestic fauna, and also between wild fauna and abraders. The χ^2 test on the variables table of data for domestic animals and querns per household gives the result of a probability of no link between the two variables as almost near zero ($n = 25$, $\chi = 3.7\ 10^{-28}$), so there is a link. The χ^2 test on the variables table data for wild animals/abraders per household gives the result of a probability of no link between the two variables as almost near zero ($n = 25$, $\chi = 0$), so there is a link. However, these results should be affected by a very strong statistical weight of the fauna variable, which has very high data numbers. This leads to high chi-squared values on quern and abrader variables. If the linear correlation coefficient is calculated between wild fauna and abraders, it is weakly positive ($r = 0.297$), and between querns and domestic animals it is almost zero ($r = 0.02$). This weak correlation is probably a consequence of data weight differences.

Faunal analysis

There is indeed significant variation in the distribution of species per house. One highly discriminating factor is the amount of domesticated and hunted fauna assigned to each house (Figure 9.3). Correspondence analysis on Cuiry-lès-Chaudardes data has revealed three clusters of houses grouped around dominant species (Hachem 2011, 114). Three species offer particularly rich distributional variability for patterning: wild boar, domestic cattle and caprines. Two other species, pigs and to a lesser extent red deer, contribute marginally to the overall pattern. Pigs are particularly linked to hunted animals, especially wild boar. Red deer are always strongly linked with roe deer and show an overall opposition to cattle. One also notes that red deer and wild boar are quite well opposed, as the two species are never abundant in the same house. It seems that the houses with a lower hunting presence have more red deer than boar, while conversely the houses with a high rate of hunting have more wild boar than red deer. The aurochs is associated with livestock, particularly domesticated cattle. All these trends are also identified in the other villages of the Aisne valley.

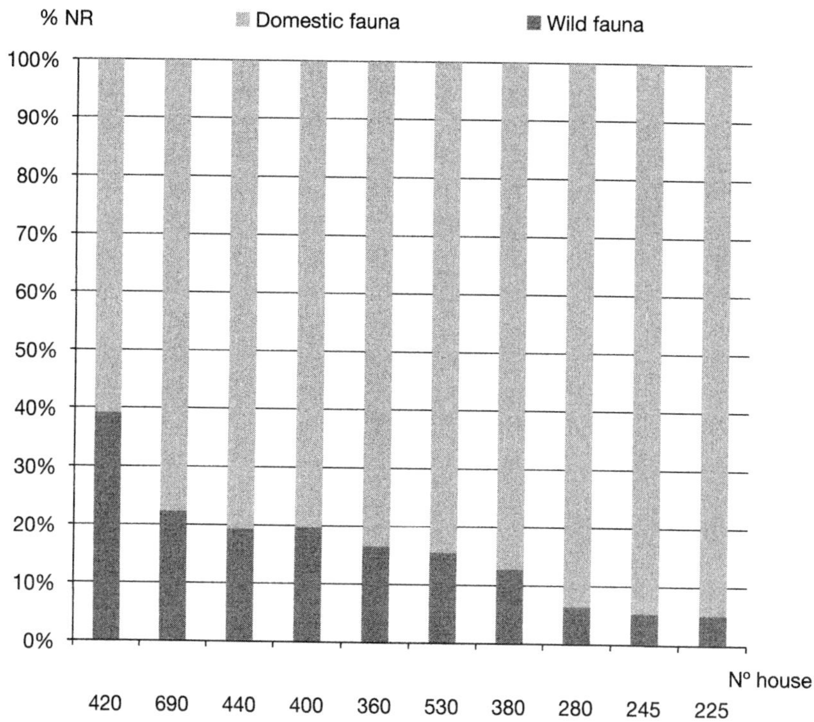

Figure 9.3 Relative proportions of domestic and wild fauna in the best-preserved houses of Cuiry-lès-Chaudardes.

When all the chronological and 'functional' results are combined, a coherent pattern emerges. Thus, three categories of houses are found in each settlement phase:

- houses where hunting is abundant (with more than 23–24% wild animals)
- houses where husbandry is massively predominant (with 91–96% domestic animals)
- houses where neither of these trends are apparent (domestic animals 60–76%).

For houses where the hunting rate is high, wild boar is the predominant large game animal. For houses where the husbandry rate is higher than average, sheep or cattle predominate. With the exception of beaver, the frequency of small game is not correlated with frequency of large game.

Analysis of grinding and abrading tools

Among macrolithic artefacts, the three main categories of tools were taken into account as evidence of the activities of grinding (querns and grinders), percussion (anvils and hammerstones) and abrading (polishers, abraders and so on) (Hamon 2006, 2008a). Grinding tools reveal the existence of food preparation, especially cereal processing. Abrading activities are more directly related to recurrent craft activities, such as the shaping of ornaments and bone tools. Finally, percussion activities were linked to flint debitage and the maintenance of querns.

Only assemblages presenting an acceptable number of complete tools were chosen, in order to avoid distortions caused by excessive fragmentation processes (Figure 9.4). This number of archaeologically complete tools, generally proportional to the total number of tool fragments, is a good indicator of the importance of each activity in the households. Each activity is represented by 1–17 tools, and even 27 abraders in one house at Missy-sur-Aisne. The apparently low number of tools in most of the households hides a more complex interpretation, as not all these tools show the same duration of use and do not have the same economic significance.

Three types of domestic assemblages can be clearly distinguished:

- those with a predominance of abrading tools. These show that specific episodes of craft activities such as bone tool shaping have been realised in the household.

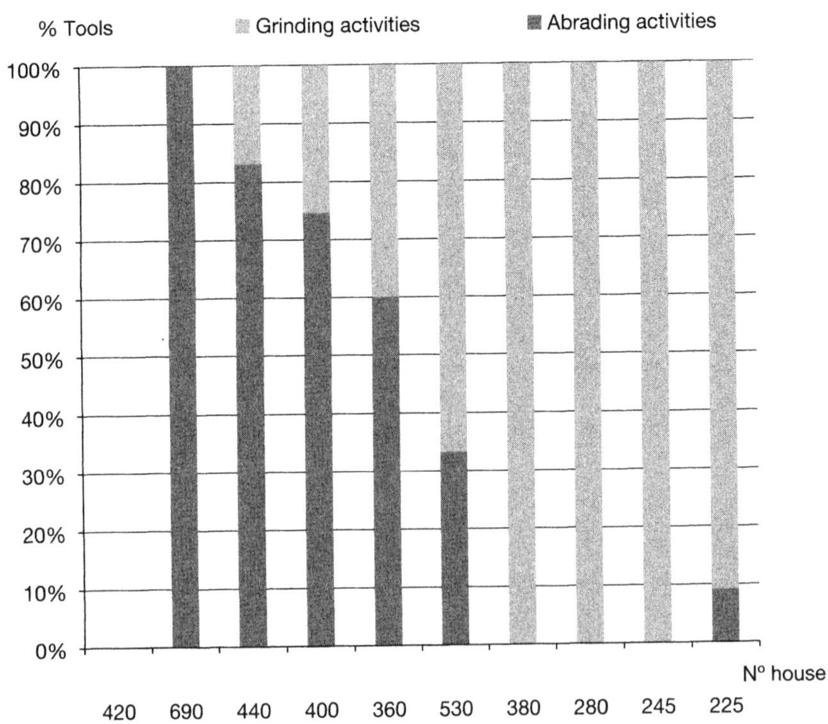

Figure 9.4 Relative proportions of grinding tools and abrading tools in the best-preserved houses of Cuiry-lès-Chaudardes.

- assemblages dominated by grinding tools. These reflect the importance of cereal processing and food preparation in these households.
- mixed assemblages. These show a quite complete range of domestic activities.

The concentration of several abrading tools in a household shows that this activity was practised quite intensively. For grinding tools, considering their long duration of use, sometimes up to tens of years, one must consider that when more than three tools were deposited, cereal processing had a really important place in the economy of the household. This could be linked either to a special organisation of the family unit for meal preparation (e.g. several wives in one house) or to the production of a great quantity of flour, beyond the domestic need. The presence of percussion tools together with abrading and grinding tools corresponds to a quite complete range of activities in the household.

Results

When all the results are combined, a coherent pattern emerges and a typology of the households can be proposed (Tables 9.2 and 9.3). Three groups can be defined, on the basis of the size of the houses, the activities represented by the macrolithic tools and the importance of the domestic fauna in the assemblage:

- In houses with one rear unit, hunting is abundant (with wild boar often predominant) together with abrading tools (Group 1).
- In houses with one or two rear units, where neither of these trends are apparent, husbandry is around 80% and macrolithic tools mix the three main categories (grinding, hammering, abrading) in average proportions (Group 2).
- In houses with three rear units, husbandry is massively predominant (90% sheep or cattle) and grinding tools are over-represented (Group 3).

Comparison of the distribution of faunal remains and macrolithic tools, as well as other artefacts, suggests that the household was an 'autonomous' entity in terms of consumption, with commonly shared rules and economic basis (Sahlins 1974). At one level, a general rule is that, despite the range of species potentially available, the domestic animals trio (cattle, caprines, pigs), red deer and aurochs were systematically eaten by each household. Similarly, it was observed that grinding tool and abrading tool fragments were present, even in very low numbers, in each house. As a result, one can assume that each house was basically autonomous in terms of consumption. However, differential access to agriculture and hunting-collecting products beyond the necessary minimum cannot be ruled out. This phenomenon is apparent in the rather high proportions of some species and tool types, by the recurrence of one of these activities in one area of a site, and by the correspondence of both activities with the different types of house plan. We are thus dealing with a cultural context within which variations are possible. Different

Table 9.2 Correlation between the characteristics of the houses, fauna (% of NR) and functional categories of macrolithic tools (only archaeologically preserved tools, fragments excepted) in Cuiry-lès-Chaudardes

Household no.	*Number back units*	*Ceramic stage*	*Herding*	*Hunting*	*Caprines*	*Bovines*	*Wild boar*	*Grinding tool*	*Hammering tool*	*Abrading tool*	
420	2	3	60.9	39.1	7.5	38.8	10.3		3		GROUP 1
690	1	3	77.3	22.2	17.0	43.3	2.6		1	5	
400	1	2	80.3	19.7	7.8	63.9	8.4	1	1	3	
440	1	2	80.6	19.4	8.9	58.8	8.2	1		5	
530	1	3	83.5	16.5	13.4	56.6	3.3	2	5	1	GROUP 2
360	2	3	84.4	15.6	21.9	54.6	4.9	2	3	3	
380	3	2	87.3	12.7	12.9	63.9	3.3	2	3		
280	2	3	93.5	6.5	36.0	46.9	0.5	4			GROUP 3
245	3	3	94.6	5.4	44.1	39.5	1.4	3	1		
225	3	3	94.8	5.2	40.0	37.3	0.6	10	1	1	

parameters introduce variations in the pattern of finds: chronology, type and function of the households, individual choices and finally position of the houses in the settlement. One has to analyse the data at each scale, to evaluate the significance and meaning of each parameter.

While no real chronological evolution of the macrolithic assemblages can be identified, chronological trends in the fauna are apparent between the beginning and the end of the occupation of the settlements. The first (Aisne 1) and third (Aisne 3) stages mark the two extremes, the former having the highest average rate for hunting, and the latter the lowest. Correspondingly, the proportion of caprines increased sharply, especially during the last settlement stage. Proportions of roe deer increased correspondingly.

Rates of husbandry-cereal processing and hunting-abrading vary according to each type of house within a specific pattern (Figure 9.5). The over-representation of an animal or type of tools, and its absence or rarity, indicate some specificity per household related to the type of house plan. Thus the over-representation of sheep, goat and cattle together with grinding tools is linked to the houses with three rear units. The presence of a high amount of grinding tools (more than five tools) in the refuse pits has important economic consequences, as a grinding set is generally enough to feed a small domestic unit for 20–40 years. On the other hand, the over-representation of wild boar and abrading tools is linked to the houses with one rear unit (and pig is more numerous in these smaller houses).

Table 9.3 Correlation between the characteristics of the houses, fauna (% of NR) and functional categories of macrolithic tools (only archaeologically preserved tools, fragments excepted) on a sample of households from six sites of the Aisne Valley (Cuiry-lès-Chaudardes, Bucy-le-Long, Missy-sur-Aisne, Berry-au-Bac *le Vieux Tordoir* (BVT), Berry-au-Bac *le Chemin de la Pêcherie* (BCP), and Menneville)

Site	*Household no.*	*Ceramic stage*	*Number of back units*	*Grinding tools*	*Hammering tools*	*Abrading tools*	*Macrolithic activities*	*Herding*	*Hunting*	*Number of identified remains*	*Fauna type*	
Cuiry-lès-Chaudardes	420	3	2		3		hammering	60.9	39.1	1535	hunting	GROUP 1
Menneville	140	2	1			10	abrading	76.9	23.1	443	hunting	
Cuiry-lès-Chaudardes	440	2	1	1		5	mixed/abrading	80.6	19.4	568	mixed	
Cuiry-lès-Chaudardes	400	2	1	1	1	3	mixed/abrading	80.3	19.7	528	mixed	
Bucy-le-Long	45	2	3	1	1	3	mixed/abrading	84.2	15.8	164	hunting	
Missy-sur-Aisne	75	3	1	1		3	mixed/abrading	87.9	12.1	33	mixed	
Bucy-le-Long	40	3	3	1	3	3	mixed/abrading	95.4	4.6	112	breeding	
Bucy-le-Long	90	2	1	1	6	5	mixed/abrading	93.7	6.3	198	mixed	
Bucy-le-Long	30	2	3	1	6	11	mixed/abrading	91	9	950	mixed	
Bucy-le-Long	116	?	?	2	5	17	mixed/abrading	92.4	8.2	254	mixed	
Missy-sur-Aisne	60	3	1	1	8	27	mixed/abrading	88.2	11.8	922	mixed	
Bucy-le-Long	10	2	3	6	8	17	mixed assemblage	95.5	4.5	318	breeding	GROUP 2
Cuiry-lès-Chaudardes	380	2	3	2	3	1	mixed assemblage	87.3	12.7	2236	mixed	
Menneville	200	3	2	2	3	1	mixed assemblage	96.1	3.9	951	breeding	
Cuiry-lès-Chaudardes	530	3	1	2	5	1	mixed assemblage	83.5	16.5	544	mixed	
Cuiry-lès-Chaudardes	360	3	2	2	3	3	mixed assemblage	84.4	15.6	1207	mixed	
Bucy-le-Long	20	3	3	9	7	9	mixed assemblage	96.9	3.1	1315	breeding	
Cuiry-lès-Chaudardes	225	3	3	10	1	1	mixed/grinding	94.8	5.2	1923	breeding	GROUP 3
Berry-au-Bac (BCP)	300	1	3	10	7	1	mixed/grinding	94.8	5.2	427	breeding	
Berry-au-Bac (BCP)	195	1	3	13	3	1	mixed/grinding	97.2	2.8	564	breeding	
Berry-au-Bac (BCP)	200	1	3	14	3	7	mixed/grinding	96.8	3.2	877	breeding	
Cuiry-lès-Chaudardes	245	3	3	3	1		grinding	94.6	5.4	562	breeding	
Berry-au-Bac (BVT)	630	3	3	4	1		grinding	78.1	21.9	406	hunting	
Cuiry-lès-Chaudardes	280	3	2	4			grinding	93.5	6.5	589	breeding	
Menneville	10	2	3	6			grinding	86.9	13.1	356	mixed	
Berry-au-Bac (BVT)	590	3	3	9	1		grinding	71.2	28.8	250	hunting	

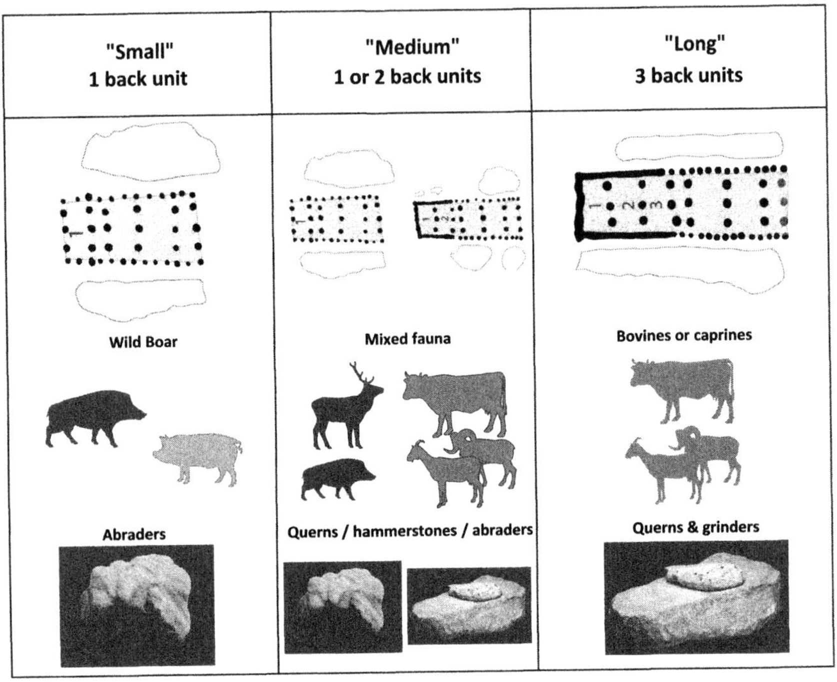

Figure 9.5 Schematic representation of the correlation between the size of the houses, the fauna and the macrolithic tools on the LBK sites of the Aisne valley.

Finally, a level corresponding to 'individual' variability (a term that applies to the household) is involved in the choice of small game, in the presence of birds and possibly fish. The presence of other types of macrolithic tools, such as handstones or burnishers involved in occasional craft activities, corresponds to the same level of individual contribution to the assemblage.

Another important parameter which introduces variation to the finds pattern is the position of the household within the village itself. A specific organisation has been highlighted by the analysis of the domestic fauna through the five stages of occupation of the site which lasted more than 100 years (Hachem 1997, 2000, 2011). This model was tested with the macrolithic tools (Hamon 2006) and has led to positive results. In the north-west of the village there are houses with a high frequency of wild fauna and abrading equipment. In the opposite area, to the south-west, other houses show high sheep frequencies and grinding equipment (Figure 9.6). It thus seems that we are seeing a specific village organisation, with economic, social and symbolic trends linked to household subsistence and craft activities.

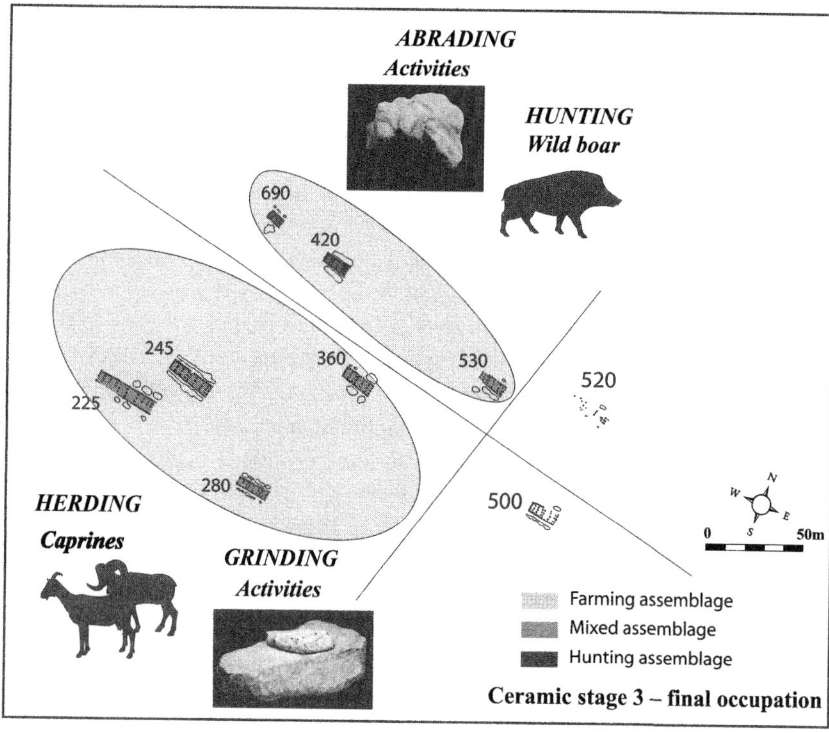

Figure 9.6 Cuiry-lès-Chaudardes, ceramic stage 3, final occupation: main characteristics of the households combining fauna and macrolithic tools.

Discussion

Several questions are raised by these results, and one can suggest different kinds of interpretation. The first is at the core of this study: How can one interpret the links between, on the one hand, querns and domestic fauna, and on the other hand, abraders and wild animals? It seems easy to link animal breeding to the processing of cereals, in a global economic system based on husbandry and agriculture. The link between an over-consumption of wild animals, a high proportion of furred game, and an important level of bone tool shaping in small houses is harder to understand. However, one could interpret such association in terms of the over-exploitation of secondary products from animal exploitation, including fur, bones and skins.

The second question is about the variation of house plans: How can one interpret the links between the type of the house and such associations of subsistence indicators in the domestic refuse? Small houses occur in large numbers within each village. An interpretation would be that each small house could represent a family cell unit. Some of them could be occupied by young families with a small herd and limited land, and could complement their initial farming production with hunting and the exploitation of secondary animal products, for their own consumption or for exchange with other households in the village. Large houses are less numerous. Some of them have considerable quantities of pottery, especially for storage, faunal remains and complete querns. An interpretation would be that each large house could represent an enlarged family unit, or several ones together, with important needs for daily subsistence. A second interpretation could be that such large houses would be used as more collective or meeting houses, for more or less special events in the village, and would reflect the needs of the broader community for cereals. One can even imagine an intermediate interpretation, where large houses with a higher number of inhabitants would receive frequent collective meetings of a large part of the village community, requiring important food production events.

Such hypotheses of the subsistence economy of small and large houses allow us to propose a new model of the social structuring of LBK settlement and the functioning of the Neolithic house (for an overview, see Bickle 2013, Coudart 1998, Allard and Hamon 2010). This also fits with the demographic model proposed for the LBK by Dubouloz (2008). According to this, the population overall was in constant growth; in a period of a century in the Aisne valley, the number of sites and houses multiplied threefold, and the average size of households multiplied by 1.5. This study has suggested the link between the rear portion of houses, the number of spaces in the house and the number of inhabitants. Following this proposal, the minimum number of people in the small houses (one back unit) would be of five to seven persons, and this family cell would grow quite quickly in two to three generations. A minimum of 12–16 persons could live in medium-long houses (with two back-units) and 36–48 persons in long houses (three back units). Such a model of demographic distribution would be in accordance with our proposal of village structure.

The last question is how, knowing the links between the type of house and the nature of refuse, to grasp the economic, social and ideological factors of the house. If 'the house is both a whole entity, (social and cultural), and a separate entity, (the individuals and the household who live therein), so it wears and produces individual and collective sense' (Coudart 2009, 217), we wanted to see if we could identify these different components in our sample. Throughout this chapter, we have been able to determine what was the cultural standard in terms of basic diet, and what was more related to the identity of the house.

A general principle, governing the main aspects of subsistence strategy, is visible. It is reflecting a piece of the LBK economic system. This principle determines the dominance of cultivation over foraging, of breeding over hunting, and in some cases of food preparation over other activities. One can assume that this general principle is the cultural one. A second principle involves variability; the basics of the general principle are respected with a certain margin. One interpretation could be that the LBK community was segmented into symbolic 'farmers' and 'hunters'. One can interpret this as a social factor. Around these poles, some assemblages might reflect more individual choices. A third principle is expressed by the relationship between the subsistence economy and the two main types of house; large houses are related to husbandry and cereal consumption, while small houses are linked to hunting and secondary animal exploitation. The accentuation of these activities could be motivated by a system of complementarity or exchange between households, rather than a strict specialisation of the houses. One can interpret this as a social factor.

Conclusion

How can the answers to the three questions stated above be interpreted in terms of a social anthropological model? The combination of cultural, social, individual and ideological trends leads to a general interpretation. It is based on subsistence economy refuse and on cross-reference to the ritual domain and anthropological examples. Fauna, querns and abraders are clearly part of the symbolic and ritual LBK system, as they are frequently involved in hoards. Bovine bucrania, numerous young domestic animals, and partial remains of female and male aurochs are unusual finds in the lateral pits of the longhouses of the Aisne valley (Hachem 2011). Querns are also found in different types of ritual or foundation deposits (Hamon 2008b, Graefe *et al.* 2009), and fauna and macrolithic tools are also involved in funerary practices. Two out of the three animals which determine Neolithic household diet have considerable symbolic importance for this agro-pastoral society: the cattle and the sheep. They were found in the funerary context of the Menneville enclosure ditch or in graves with child burials (Thevenet 2009, forthcoming; Farruggia *et al.* 1996). There is a strong link between the deposition of sheep and children. In the Hinkelstein graves of the Trebur cemetery, Germany (Spatz 1999, Spatz and von den Driesch 2001), the association of animals and grinding tools with the sex of the individuals also stressed a gendered dimension to their symbolism. When the tables are analysed, pigs or wild boar parts are deposited in the male graves, for example (with one exception), and querns are associated in the majority of cases with women or children. Anthropologically, hunting is known to be a male activity, and socially valued, while grinding is a female one (Godelier 1982).

Taking into account all these elements, the small houses that stood apart in the western part of a village like Cuiry-lès-Chaudardes could be interpreted in the following way. Apart from being a small family unit, they would also shelter young men who could face a dangerous activity like hunting wild boar, as an initiation rite for example. The large houses in the southern part would have gathered women and small children, more directly involved in food preparation for special occasions.

In summary, our model is based on a combined economic and social interpretation. Some small houses could represent a family cell unit more involved in hunting and the exploitation of secondary animal products, while the members of larger family cell units and/or a meeting house would concentrate their subsistence on farming activities. Such organisation could be based on lineage, and this could have involved a clan system. Through the distribution of agricultural and hunting-foraging products, one could assume that after having fulfilled basic dietary needs, some over-production was used as an exchange between households, or clans – such as furs for cereals.

Acknowledgements

We are glad to thank all the researchers of our team who have been working on the Aisne valley project; their results have fed our reflection on LBK society. The Aisne valley archaeological research project has been supported financially for many years by the Conseil général de l'Aisne and the Association pour le Sauvetage Archéologique de la Vallée de l'Aisne (ASAVA). New funding for post-excavation work has been provided recently by the French Ministry of Culture and Communication. We also thank F. Giligny, professor at the University of Paris I, Panthéon-Sorbonne, for his help with the statistical tests.

References

Allard, P. 2005. *L'industrie lithique des populations rubanées du nord-est de la France et de la Belgique*. Rahden: Marie Leidorf.

Allard, P., Dubouloz, J., Farruggia, J.-P., Hachem, L., Ilett, M. and Robert, B. 1996. Berry-au-Bac «le Vieux Tordoir»: la fin d'un grand sauvetage et la fouille d'un nouveau site rubané. *Fouilles Protohistoriques dans la vallée de l'Aisne, Rapport de fouille 1995*, 11–46. Paris: Université de Paris I, Panthéon-Sorbonne,

Allard, P., Hamon, C. (eds), with the collaboration of Bonnardin, S. Cayol, N., Chartier, M., Coudart A., Dubouloz, J., Fronteau, G., Gomart L., Hachem, L., Ilett, M., Meunier K., Monchablon, C. and Thevenet, C. 2010. *Economie et société des populations rubanées de la vallée de l'Aisne*. Premier, deuxième et troisième rapport annuel d'étape 2007–2010. Projet Collectif de Recherche, Service régional de l'Archéologie. Amiens: DRAC Nord-Picardie.

Allard, P., Hamon, C., S. Cayol, N., Chartier, M., Coudart A., Dubouloz, J., Fronteau, G., Gomart L., Hachem, L., Ilett, M., Meunier K., Monchablon, C. and Thevenet, C. 2013. Linear Pottery domestic space: taphonomy, distribution of finds and economy in the Aisne valley settlements. In C. Hamon, P. Allard and M. Ilett (eds), *The domestic space in LBK settlements*, 9–28. Internationale Archäologie 17. Rahden: Marie Leidorf.

Bakels, C. 1995. Late glacial and holocene pollen record from the Aisne and Vesles valleys, Northern France: the pollen diagrams Maizy-Cuiry and Bazoches. In G.F. Herngreen and L. Van der Valk (eds), *Neogene and Quaternary geology of north-west Europe*, 223–34. Medelingen Rijks Geologische Dienst 52.

Bakels, C. 1999. Archeobotanical investigations in the Aisne valley, northern France, from the Neolithic up to the early Middle ages. *Vegetation History and Archaeobotany* 8, 71–7.

Bakels, C. 2008. L'agriculture rubanée/post-rubanée, continuité ou discontinuité ? In L. Burnez-Lanotte, M. Ilett and P. Allard (eds), *Fin des traditions danubiennes dans le Néolithique du Bassin parisien et de la Belgique (5100–4700 av. J.C.). Autour des recherches de Claude Constantin*, 191–6. Paris: Mémoire de la Société Préhistorique Française, 44.

Bickle, P. 2013. Of time and the house: the early neolithic communities of the Paris basin and their domestic architecture. In D. Hofmann and J. Smyth (eds), *Tracking the Neolithic house in Europe: sedentism, architecture and practice*, 151–80. New York: Springer.

Blouet, V., Klag, T., Petitdidier, M.-P. and Thomashausen, L. 2013. *Le Néolithique ancien en Lorraine*. Paris: Mémoire de la Société Préhistorique Française, 55.

Boelicke, U. 1982. Gruben und Häuser: Untersuchungen zur Struktur bandkeramischer Hofplätze. In J. Pavůk (ed.), *Siedlungen der Kultur mit Linearkeramik in Europa. Internationales Kolloquium Nové Vozokany, 17.–20. November 1981*, 17–28. Nitra: Archäologisches Institut der Slowakischen Akademie der Wissenschaften.

Boelicke, U., von Brandt, D., Lüning, J., Stehli, P. and Zimmermann, A. (eds) 1988. *Der bandkeramische Siedlungsplatz Langweiler 8, Gemeinde Aldenhoven, Kreis Düren*. Köln: Rheinland-Verlag.

Bonnardin, S. 2009. *La parure funéraire des premières sociétés agro-pastorales des Bassins parisien et rhénan: Rubané, Hinkelstein et Villeneuve-Saint-Germain*. Paris: Mémoire de la Société Préhistorique Française, 49.

Charier, M.-A. 1986. *Missy-sur-Aisne, le Culot: étude d'un village néolithique danubien*. Mémoire de maîtrise, Université de Paris 1, Panthéon-Sorbonne, Paris.

Constantin, C. 1985. *Fin du Rubané, céramique du Limbourg et Post- Rubané: le Néolithique le plus ancien en Bassin parisien et en Hainaut*. Oxford: British Archaeological Reports, International Series 273.

Constantin, C. and Ilett, M. 1997. Une étape finale dans le Rubané récent du Bassin Parisien. In C. Jeunesse (ed.), *Le Néolithique danubien et ses marges entre Rhin et Seine (Actes du 22e colloque interrégional sur le Néolithique, Strasbourg, 27–29 octobre 1995)*, 281–300. Zimmersheim: Cahiers de l'Association pour la Promotion de la Recherche archéologique en Alsace, supplément n° 3.

Coudart, A. 1998. *Architecture et société néolithique: l'unité et la variance de la maison danubienne*. Paris: Editions de la Maison des Sciences de l'Homme.

Coudart, A. 2009. La maison néolithique: métaphore matérielle, sociale et mentale des petites sociétés sédentaires. In J.-P. Demoule (ed.), *La révolution Néolithique dans le monde*, 215–35. Paris: Inrap/Universcience and Editions CNRS.

Dubouloz, J. 2003. Datation absolue du premier Néolithique du Bassin parisien: complément et relecture des données Rubané Récent du Bassin Parisien et Villeneuve-Saint-Germain. *Bulletin de la Société Préhistorique Française* 100, 671–89.

Dubouloz, J. 2008. Impacts of the Neolithic demographic transition on Linear Pottery culture settlement. In O. Bar Yosef and J.-P. Bocquet-Appel (eds), *The Neolithic demographic transition and its consequences*, 207–35. New York: Springer.

Dubouloz, J., with Chartier, M., Hachem, L. and Ilett, M. 2012. Interdépendance et cohésion des différents niveaux de territorialité au Néolithique Rubané en Bassin parisien. In

V. Charpentier and C. Marcigny (eds), *Des hommes aux champs: pour une archéologie des espaces ruraux du Néolithique au Moyen Age*, 23–34. Rennes: Presses Universitaires de Rennes.

Farruggia, J.-P. and Constantin, C. 1984. Le site néolithique et des âges des métaux de Missy-sur-Aisne (Le Culot). In *Les fouilles protohistoriques dans la vallée de l'Aisne* 12, 61–94. Paris: Université de Paris I, Panthéon-Sorbonne.

Farruggia, J.-P., Hachem, L. and Guichard, Y. 1996. Les ensembles funéraires rubanés de Menneville "Derrière le Village" (Aisne). In P. Duhamel (ed.), *La Bourgogne entre les Bassins rhénan, rhodanien et parisien: carrefour ou frontière? Actes du 18e colloque interrégional sur le Néolithique, Dijon 25–27 octobre 1991*, 119–74. *Revue Archéologique de l'Est*, 14e supplément.

Godelier, M. 1982. *La production des Grands Hommes: pouvoir et domination masculine chez les Baruyas de Nouvelle Guinée*. Paris: Fayard.

Graefe, J., Hamon, C., Lidström-Holmberg, C., Tsoraki, C. and Watts, S. 2009. Subsistence, social and ritual practices: quern deposits in the Neolithic societies of Europe. In S. Bonnardin, C. Hamon, M. Lauwers and B. Quilliec (eds), *Du matériel au spirituel. Réalités archéologiques et historiques des «dépôts » de la préhistoire à nos jours*, 87–96. Antibes: Éditions APDCA.

Hachem, L. 1997. Structuration spatiale d'un village du Rubané Récent, Cuiry-lès-Chaudardes (Aisne). Analyse d'une catégorie de rejets domestiques: la faune. In G. Auxiette, L. Hachem and B. Robert (eds), *Espaces physiques, espaces sociaux, dans l'analyse interne des sites du Néolithique à l'Age du Fer. Actes du 119e Congrès National des Sociétés Historiques et Scientifiques, Amiens, 24–28 octobre 1994*, 245–61. Paris: Comité des Travaux Historiques et Scientifiques.

Hachem, L. 2000. New observations on the Bandkeramik house and social organisation. *Antiquity* 74, 308–12.

Hachem, L. 2009. Elevage, chasse et société dans le Néolithique français: exemples dans le Danubien du Nord de la France. In J.-P. Demoule (ed.), *La révolution Néolithique dans le monde*, 197–213. Paris: Inrap/Universcience and Editions CNRS.

Hachem, L. 2011. *Le site néolithique de Cuiry-lès-Chaudardes – I. De l'analyse de la faune à la structuration sociale*. Internationale Archäologie 120. Rahden: Marie Leidorf.

Hachem, L. and Bedault, L. 2008. Recherches sur les sociétés du Néolithique danubien à partir du Bassin parisien: approche structurelle des données archéozoologiques, In L. Burnez-Lanotte, M. Ilett and P. Allard (eds), *Fin des traditions danubiennes dans le Néolithique du Bassin parisien et de la Belgique (5100–4700 av. J.C.). Autour des recherches de Claude Constantin*, 221–43. Paris: Mémoire de la Société Préhistorique Française, 44.

Hachem, L., Allard, P., Constantin, C., Farrugia, J.-P., Guichard, Y. and Ilett, M. 1998a. Le site néolithique rubané de Bucy-le-Long « la Fosselle » (Aisne). *Internéo* 2, 17–27.

Hachem, L., Guichard, Y., Farruggia, J.-P., Dubouloz, J. and Ilett, M. 1998b. Enclosure and burial in the earliest neolithic of the Aisne Valley. In M. Edmonds and C. Richards (eds), *Understanding the Neolithic of north-western Europe*, 127–40. Glasgow: Cruithne Press.

Hamon, C. 2006. *Broyage et abrasion au Néolithique ancien: caractérisation technique et fonctionnelle des outillages en grès du Bassin parisien*. Oxford: Archaeopress.

Hamon, C. 2008a. Functional analysis of stone grinding and polishing tools from the earliest Neolithic of north-western Europe. *Journal of Archaeological Science* 35, 1502–20.

Hamon, C. 2008b. The symbolic value of grindingstones hoards: technical properties of neolithic examples. In C. Hamon and B. Quilliec (eds), *Hoards from the Neolithic to the Metal Ages. Technical and codified practices, Session of the XIth Annual Meeting of the European Association of Archaeologists*, 19–28. Oxford: Archaeopress.

Hamon, C., Allard, P. and Ilett, M. (eds) 2013. *The domestic space in LBK settlements.* Internationale Archäologie 17. Rahden: Marie Leidorf.

Ilett, M. 2012. Linear Pottery and Blicquy/Villeneuve-Saint-Germain settlement in the Aisne valley and its environs. In S. Wolfram, H. Stäuble, M. Cladders and T. Tischendorf (eds), *Siedlungstruktur und Kulturwandel in der Bandkeramik*, 69–79. Leipzig: Landesamt für Archäologie.

Ilett, M. In press. Research into early Neolithic settlement in the Paris basin. In C. Marcigny and L. McFadyen (eds), *The prehistory of France.* Cambridge: Cambridge University Press.

Ilett, M. and Hachem, L. 2001. Le village néolithique de Cuiry-lès-Chaudardes (Aisne, France). In J. Guilaine (ed.), *Communautés villageoises du Proche Orient à l'Atlantique*, 171–84. Paris: Editions Errance.

Ilett, M. and Plateaux, M. (eds) 1995. *Le site néolithique de Berry-au-Bac « le Chemin de la Pêcherie » (Aisne).* Paris: Editions CNRS.

Ilett, M., Hachem, L. and Coudart, A. (eds), with Allard, P., Auxiette, G., Baillieu, M., Bonnardin, S., Cayol, N., Chartier, M., Hamon, C., Farruggia, J.-P., Meunier, K., Monchablon, C., Peltier, V., Sidéra, I., Tardieu, C., Thevenet, C. and Thouvenot, S. 2006. *L'implantation du Néolithique rubané dans la vallée de l'Aisne.* Premier, deuxième et troisième rapport annuel d'étape. Projet d'Action Collective de Recherche (ACR) 2003-2006. Service régional de l'Archéologie. Amiens: DRAC Nord-Picardie.

Lenneis, E. 2010. *Die bandkeramische Siedlung von Mold bei Horn in Niederösterreich. Teil 1. Naturwissenschaftliche Beiträge und Einzelanalysen.* Rahden: Marie Leidorf.

Luning, J. 1982. Research into the Bandkeramik settlement of the Aldenhovener Platte in the Rhineland. *Analecta Praehistorica Leidensia* 15, 1–30.

Modderman, P.J.R. 1988. The Linear Pottery culture: diversity in uniformity. *Berichten van de Rijksdienst voor het Oudheidkundig Bodemonderzoek* 38, 63–139.

Pavlů, I. 2010. *Činnosti na neolitickém sidlišti Bylany: prostorová analýza keramiký (Activities on a Neolithic site of Bylany: an intrasite spatial analysis of pottery).* Praha: Vydal Archeologický ústav ČR.

Pavlů, I. 2013. The role of Linear Pottery houses in the process of neolithisation, *Documenta Praehistorica* XL, 31-37.

Sahlins, M. 1974. *Stone Age economics.* London: Tavistock.

Sidéra, I. 2000. Animaux domestiques, bêtes sauvages et objets en matières animales du Rubané au Michelsberg. *Gallia Préhistoire* 42, 107–94.

Soudský, B. and Pavlů, I. 1972. The Linear Pottery Culture settlement pattern of central Europe. In P. Ucko, R. Tringham and G. Dimbleby (eds), *Man, settlement and urbanism*, 318–28. London: Duckworth.

Spatz, H. 1999. *Die Mittelneolithische Gräberfeld von Trebur, Kreis Gross-Gerau.* Wiesbaden: Materialen zur Vor- und Frühgeschichte von Hessen, Band 19.

Spatz, H. and von den Driesch, A. 2001. Zu den tierischen Beigaben aus dem Hinkelsteiner und Grossgartacher Gräberfeld von Trebur, Kr. Gross-Gerau (Hesse). In R.-M. Arbogast,

C. Jeunesse and J. Schibler (eds), *Rôle et statut de la chasse dans le Néolithique ancien danubien, 5500–4900 av. J.-C.*, 113–28. Rahden: Marie Leidorf.

Thevenet, C. 2009. Les sépultures rubanées du Bassin parisien: composition de l'échantillon funéraire et implantation sépulcrale. In A. Zeeb-Lanz (ed.), *Krisen–Kulturwandel–Kontinuitäten: zum Ende der Bandkeramik in Mitteleuropa*, 111–27. Rahden: Marie Leidorf.

Thevenet, C. Forthcoming. L'enceinte rubanée de Menneville "Derrière le Village" et les structures associées (Aisne, France): de la diversité du traitement des défunts à la cohérence d'un système. *Gallia Préhistoire*.

Whittle, A. 1996. *Europe in the Neolithic*. Cambridge University Press: Cambridge.

Wolfram, S., Stäuble, H., Cladders, M. and Tischendorf, T. (eds) 2012. *Siedlungstruktur und Kulturwandel in der Bandkeramik*. Leipzig: Landesamt für Archäologie.

10

Framing Farming

A Multi-stranded Approach to Early Agricultural Practice in Europe

AMY BOGAARD

Introduction

MOST OF THE DIRECT INFERENCES WE CAN MAKE about early farmers concern their settlements, monuments and burials. Following Neolithic people into their wider landscapes requires inferences that are more indirect and methodologically challenging, and detailed assessments at this scale are relatively sparse. Nevertheless, opportunities to infer the nature of off-site activities are crucial for apprehending Neolithic life in what Tim Ingold (1993) called a 'dwelling perspective'. The scope of this viewpoint is defined not by the boundaries of the archaeological site, or by physical features of the landscape, but rather by what Ingold terms the 'taskscape': the socially constitutive flow of people and their actions. Attending to what people actually *did* in order to detect the contours of 'their' landscapes depends on our ability to interpret on-site occupation debris as evidence of wider practice. Farming offers a particularly favourable set of conditions for achieving the necessary linkages. It leaves (more or less) abundant material traces of processing, storage and discard; such remains normally derive from cultivation areas external to the immediate structures or features where they are found, hence broadening the scale of enquiry; and, as will be argued below, there are now multiple, complementary ways in which crop remains and associated weeds can inform the nature of early farming practice and the landscapes it shaped. If these methods are integrated, farming becomes one of the more accessible lines of archaeological enquiry that takes us into the wider setting and its social geography.

An increasingly accessible arable 'taskscape' affords new understandings of early farmers, and it suits the purpose of the argument to be developed here that Ingold chose *The Harvesters* by Pieter Bruegel the Elder to illustrate the concept. The painting depicts a type of post-medieval farming that could be observed across much of lowland temperate Europe until the arrival of the combine-harvester in

Proceedings of the British Academy **198**, 181–96.

the 1950s, and 'for most European scholars . . . belongs to their childhood or family memories' (Sigaut 1988, 22). Given the nostalgia that surrounds this image, and its place in a genre of 'timeless' scenes of peasant work and rest, it is useful to consider briefly the particular agricultural practices it depicts, as a way into a very different Neolithic 'taskscape' (Figure 10.1). The scythes wielded by men in the painting had replaced the medieval sickles used by both women and men. The action of the scythe mowed the cereal crop very low to the ground, collecting as much straw as possible for working animals, thatch and so on. Sheaves were loaded onto ox-drawn carts like the one visible in the background, presumably heading for a barn, where the crop would be stored and threshed/winnowed over several weeks or months, typically by men (cf. Sigaut 1988). The uniformity of the crop stand suggests a monoculture of a single cereal landrace. Its density is symptomatic of broadcasting, a method wasteful of seed corn but requiring minimal (though skilled) labour per unit area, and hence effective as part of a system finely tuned to work on an extensive scale, and to be profitable to large-scale landowners.

A Neolithic harvesting scene would have been radically different. The focus of the harvest was the cereal ear, which can be 'plucked' by hand, 'stripped' using wooden clamps or cut high on the straw by sickle or knife; in central Europe, for example, harvesting of ears with little straw is suggested by weed data and a scarcity of straw nodes (Jacomet *et al.* 1989, 155–6; Brombacher and Jacomet 1997; Kreuz

Figure 10.1 The Harvesters by Pieter Bruegel the Elder.

et al. 2005; Bogaard 2012a, 41–2, 158–62). Exceptional organic preservation on lakeshore sites provides particularly clear evidence of harvesting practice. For example, wooden sickle hafts from late sixth millennium cal BC La Draga, Catalonia, were designed for harvesting well above ground level, supporting other evidence for harvesting of cereal ears at the site (Palomo *et al.* 2011; Ferran Antolin, pers. comm.), while unprocessed ears stored in houses at Hornstaad-Hörnle 1A, which burned in *c.* 3910 BC, were cut about a hand's breadth below the ear (Maier 2001, 98–9). Forms of sickle or harvesting knife were regionally and culturally specific (Schlichtherle 1992, Gurova 2005), and precise harvesting methods and gestures would have varied, but ethnographies of ear-harvesting (whether by plucking, stripping or cutting) suggest that it was typically a female task, or undertaken by women and men together (Sigaut 1988, 19). A major advantage of harvesting ears with little to no straw is that the crop is much less bulky and easier to transport; it is an approach adapted to small-scale farming where animal labour is in short supply.

Neolithic cereals were likely sown by dibbling or row-sowing rather than broadcasting, giving rise to a more open crop stand, both to invest seed corn efficiently and to facilitate weeding during the growing season (Halstead 1989, Bogaard 2004, Kreuz 2007; cf. Halstead 1990). It is also increasingly clear that Neolithic farmers cultivated 'mixtures' of cereals (species and/or landraces) together in order to improve the chances of obtaining a good harvest by exploiting the contrasting requirements and potentials of different types (Maier 2001, 189; Marinova 2006; Fraser *et al.* 2013, 510). The 'aesthetics' of a Neolithic field were thus profoundly different from the uniform, dense cereal stands idealised by Bruegel. In terms of storage and processing, there is scattered evidence extending from Bulgaria (Marinova 2006) to central Europe (Maier 2001) that cereals were stored as whole ears within houses and processed piecemeal into grain. Removal of the harvest to domestic spaces at the earliest possible stage, as unthreshed ears, followed by frequent, small-scale processing (Kreuz 2007; Bogaard 2012a, 157) would not only spread the labour costs of processing throughout the year but also facilitate household 'ownership' of the crop.

Thus, the tools, gestures, aesthetics, scale and sociality of a Neolithic harvest would depart radically from Ingold's core illustration of the 'taskscape'. In fact, coming to terms with Neolithic farming involves a conscious process of *unlearning* influential images such as *The Harvesters*. This chapter considers how the integration of different methodological approaches to the inference of crop growing conditions can refine social and ecological interpretation of early farming practice in Europe. The broader aim is to demonstrate how these methods bring us much closer to the perspective advocated by Ingold: reconsidering Neolithic settlements, monuments and burials from the vantage point of wider (here, arable) taskscapes.

An integrated methodological framework

Table 10.1 summarises key aspects of farming that become accessible archaeologically by integrating two methodological approaches: ecological interpretation of arable weed assemblages, and stable isotope analysis of crop material. The charred grains and chaff of cereals and seeds of pulses, together with charred seeds of arable weeds that are sometimes preserved with this material, constitute the most direct archaeological evidence available for farming practice. Ecological interpretation of weed assemblages is a well known approach for inferring crop growing conditions (Wasylikowa 1981, Jones 1992, van der Veen 1992), and functional ecology has proved especially useful for setting up comparisons between present-day weed flora surveyed in 'traditional' or experimental farming regimes and archaeobotanical weed assemblages (Bogaard 2004). By combining functional ecology and modern weed survey studies, for example, it is possible to confirm or exclude the possibility of shifting cultivation (recent clearance of extensively managed plots), to distinguish autumn versus spring sowing and to assess the intensity of management (a function of soil disturbance – tillage and weeding – and productivity). These three variables (permanence, seasonality, intensity: Table 10.1), in turn, are needed to distinguish the major models of early farming: shifting cultivation, extensive ard cultivation, floodplain farming and intensive 'garden' cultivation. The limiting factor for the weed ecological approach is the availability of archaeobotanical weed data; weed seeds often occur at low frequency in Neolithic assemblages, meaning that many sites and samples cannot be assessed (Bogaard 2004, 67).

Stable carbon isotope analysis has been under investigation for some time as a method of inferring crop water status (Araus *et al.* 1997, Ferrio *et al.* 2005), while analysis of stable nitrogen isotope ratios offers a means of inferring soil nitrogen composition and, in an arable context, the practice of manuring (Bogaard *et al.* 2007, Fraser *et al.* 2011). Isotopic study of present-day cereals and pulses grown under a range of experimental and 'traditional' farming regimes (Fraser *et al.* 2011, Wallace *et al.* 2013) has demonstrated the usefulness of stable C and N isotope

Table 10.1 The relevance of weed ecology and crop stable isotopes to key aspects of farming

	Weed ecology	*Crop stable isotopes*
Permanence	x	
Seasonality	x	
Intensity	x	x
Location	x	x
Dietary role		x

compositions for assessing whether soil productivity was manipulated through water management and/or manuring, respectively. This potential complements weed ecological analysis; functional attributes relating to general soil productivity are often the best correlates of manuring *and* irrigation (Charles *et al.* 1997, 2003; Jones *et al.* 2000), and stable isotope determinations of crops provide an independent means of identifying these specific contributions to productivity and intensity. Moreover, stable C and N isotope analysis is conducted *directly* on crop remains, and can be used to compare the growing conditions of different crops, whether or not associated weed data are available. On the other hand, weed data provide useful ecological constraints on the interpretation of stable isotope data, indicate levels of soil disturbance, pH, and so on that are not directly accessible through isotopic analysis, and can identify subtle ecological differences that may not be apparent from isotope data. Ideally, therefore, crop growing conditions would be reconstructed through a combination of weed ecological and stable isotope approaches, in order to exploit their complementary strengths.

Particularly in combination, weed ecology and stable isotope analysis are also useful for inferring *where* cultivation took place (Table 10.1). To some extent this may be a further inference from general crop growing conditions, interpreted in the light of ethnographic observation (for example, that more intensively managed plots tend to be nearer to habitation than less intensively managed areas: Jones *et al.* 1999). More direct inferences are potentially accessible through strontium isotope analysis of charred crop remains (Benson *et al.* 2010, Heier *et al.* 2009, Bogaard *et al.* 2013), but further work is needed on diagenetic effects and cleaning protocols.

Finally, the dietary role of crops (Table 10.1) has fundamental implications for understanding early farming systems, but cannot be inferred from the frequency of crop remains *per se*. The contribution of crops and other plants, however, becomes accessible if palaeodietary studies incorporate stable C and N isotope analysis of well preserved archaeobotanical remains along with faunal and human collagen (Fraser *et al.* 2013). In the absence of direct plant isotope data, palaeodietary studies have made assumptions about the isotopic values of plant foods that tend to minimise the protein contribution of crops to early farming diets.

The discussion below turns to case studies in farming practice from south-east, central and north-west Europe, and seeks to demonstrate how varying opportunities to integrate weed- and isotope-based inferences can 'frame farming' in new ways.

Diversity and convergence in early farming regimes: crop stable isotope results from south-east to north-west Europe

Available weed ecological data from southern and central Europe, Britain and Ireland exclude the possibility that shifting cultivation or slash-and-burn was the predominant mode of farming (Bogaard 2004, Bogaard and Jones 2007, Bogaard and Halstead in press, McClatchie *et al.* 2012). Tending and maintenance of long-lived cultivation plots, therefore, were apparently widespread. More detailed functional ecological analysis of Neolithic weed assemblages in central Europe suggests that management of arable plots was intensive, though the precise social and ecological strategies at work were highly context-dependent (see below; Bogaard 2004, 2012a). A recent project investigating the stable carbon and nitrogen isotopes ratios of crop remains from Neolithic sites in south-eastern, central and north-west Europe found widespread evidence of manuring – an aspect of intensive management – but equally important indications that farming practices were contingent on local factors (Bogaard *et al.* 2013).

A revealing comparison is that between two tell sites of the sixth millennium cal BC in south-eastern Europe: Koufovouno, southern Greece (Cavanagh *et al.* 2007, Bogaard unpublished), and Slatina, western Bulgaria (Docheva 1990, Marinova 2006). The ranges of $\delta^{15}N$ values in cereal grain samples at the two sites (Figure 10.2a–b) are similar, and both also display clustering of cereal $\delta^{15}N$ values into 'high' and 'low' groups. At Koufovouno, the 'high' $\delta^{15}N$ cereals are all free-threshing wheat, while the 'low' values belong to hulled barley. At Slatina, einkorn and emmer wheat, the major cereals, occur in both 'high' and 'low' groups. Thus, the two site assemblages reflect intensive manuring of some but not all cereals, but they contrast in the way manuring levels relate to cereal species. At Koufovouno hulled barley was (relatively) unmanured in contrast to free-threshing wheat, whereas at Slatina einkorn and emmer were grown with or without intensive manuring. Stable isotope analysis of faunal remains at both sites indicates that herbivore forage $\delta^{15}N$ values were similar to the lowest cereal $\delta^{15}N$ values (see 'forage' reference lines in Figure 10.2) and provide a proxy for vegetation largely unaffected by manuring (Bogaard *et al.* 2013).

The pulse crops – pea at Koufovouno, and lentil and grass pea at Slatina – are lower in $\delta^{15}N$ than the cereals, as expected due to atmospheric N-fixation by *Rhizobium* bacteria that infect legume roots, but the values at both sites suggest that they were also manured intensively (cf. Fraser *et al.* 2011, figs 4 and 5). The similarity in manuring treatment between the pulses and cereals with 'high' $\delta^{15}N$ values at both sites suggests that they were grown in the same part of the arable landscape, possibly in rotation. The cereals with 'low' $\delta^{15}N$ values were probably grown in separate areas; this inference is justified not only by the discrete clustering of 'high' and 'low' cereals but also by the fact that enrichment of ^{15}N appears to decrease only gradually after manuring ceases (Fraser *et al.* 2011, fig. 2), meaning that apparently 'manured' and 'unmanured' crops at these sites are unlikely to have been grown within a year or two of each other on the same plot of land. Thus, at both of these sites it appears that there were distinct parts of the arable landscape: one that was intensively manured and another that received less or no manure.

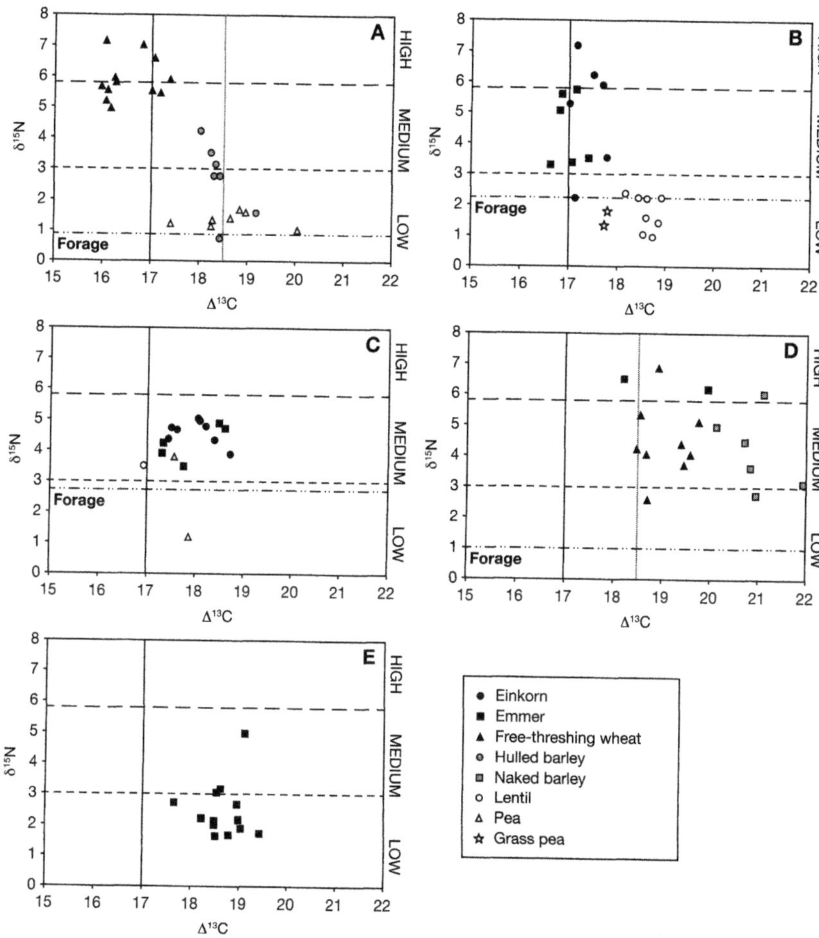

Figure 10.2 $\Delta^{13}C$ and $\delta^{15}N$ values of cereal and pulse samples at five Neolithic sites: (A) Koufovouno, (B) Slatina, (C) Vaihingen, (D) Hornstaad, (E) Sarup. Dash-dot-dot horizontal lines represent estimate of large herbivore forage $\delta^{15}N$ value (by subtracting 4‰ from the mean value for herbivore bone collagen to account for trophic shift). All archaeological data are taken from Bogaard *et al.* 2013 and Fraser *et al.* 2013. Dashed horizontal lines represent thresholds of low, medium, and high manuring rates inferred from modern experiments (Fraser *et al.* 2011). Black vertical lines represent well watered wheat and pulse threshold; grey lines represent well watered barley threshold (Wallace *et al.* 2013).

Carbon isotope discrimination ($\Delta^{13}C$ values, which take into account changes in $\delta^{13}C$ of source CO_2 through time: Ferrio *et al.* 2005) adds another dimension to the story. Some or all of the pulses at both sites have higher values than the cereals, suggesting that they grew under wetter conditions. The high variability among pea samples at Koufovouno is similar to that among present-day hand-watered pulses studied in Evvia (Wallace *et al.* 2013, fig. 3), while the more uniform values at Slatina suggest a different kind of regime that ensured relatively wet conditions. Here again, therefore, diverse practices are apparent alongside convergent tendencies or principles.

A pair of central European sites, LBK Vaihingen and late Neolithic Hornstaad-Hörnle 1A, offers another informative comparison (Figure 10.2C–D). At Vaihingen cereal $\delta^{15}N$ values were on average *c.* 2‰ higher in $\delta^{15}N$ than herbivore forage, assuming an offset of 4‰ between herbivore collagen values and the plant matter they consumed (Fraser *et al.* 2013). Larger trophic offsets of 5–6‰, suggested recently by O'Connell *et al.* (2012) in the context of $\delta^{15}N$ data from human subjects fed on controlled diets, would imply an even greater discrepancy between cereals and herbivore forage. Manuring probably contributed to enrichment in crop ^{15}N. The Vaihingen weed data (see below) exclude other causes of ^{15}N elevation, and the cereal $\delta^{15}N$ values fall within those of experimental cereals grown on long-term manured plots receiving *c.* 10–15 t/ha, while pulse $\delta^{15}N$ values suggest even higher levels of manuring (Fraser *et al.* 2011, figs 4–5). It is difficult to infer precise manuring rates but such levels would imply careful management of animal dung given that herding was probably small in scale (such as a few cattle, pigs, sheep and goats per household: Knipper 2012, 276; cf. Bogaard 2012b, 34). Dung could have been supplemented with household waste, but availability of midden material would be even more restricted, and some household waste was likely fed to pigs at Vaihingen (Bogaard 2012b, 33), contributing to increases in $\delta^{15}N$ over wild boar (Fraser *et al.* 2013). The relative consistency of cereal $\delta^{15}N$ values (Figure 10.2C) is in agreement with a lack of weed ecological evidence for differential productivity levels among cereal samples (Bogaard 2012a, 212). It was not possible to compare cereal grain samples deriving from different house groups at Vaihingen (Bogaard *et al.* 2011) (in isotopic terms, since most of the assemblage is too grain-poor for stable isotope measurements.

Comparison of crop isotope data from Vaihingen with the later Neolithic site of Hornstaad-Hörnle 1A (Maier 2001) on Lake Constance suggests similarities as well as differences in agricultural regime between these two well documented sites. Ecological analysis of the weed assemblage accompanying stored cereals at Hornstaad suggests that cultivators here had established a relatively intensive form of fixed-plot farming by the time the settlement burned down in 3910 BC, ten years after it was built (Maier 2001, 108–9; Bogaard 2004, 156, 168). Stable isotope analysis of stored cereal samples from the burned settlement layer yielded $\delta^{15}N$ values that are markedly more variable than those at Vaihingen, though average cereal values are similar at both sites, and also resemble those at Koufovouno and Slatina (Figure 10.2). It appears that cumulative manuring rates among plots varied appreciably within the harvest of a single year at Hornstaad. The range in cereal $\delta^{15}N$ values is similar to that observed in single-year studies of present-day cereal fields in two traditional farming areas where manuring is practised: Sighishoara, Romania, and Asturias, northern Spain (Fraser *et al.* 2011, fig. 5). In these areas manuring rates vary because farmers spread manure on some plots more heavily than others (for example, due to differential proximity to byres/penning areas), and because they tend to target different plots for manuring each year. At Vaihingen more consistent $\delta^{15}N$ values may be due in part to mixing of grain from different sources in pit fills, though one of the contexts analysed appears to represent a burnt store of multiple glume wheat species, and these exhibit similar values to samples from other contexts (Fraser *et al.* 2013). Relatively even manuring could have been achieved by grazing animals on all plots and/or by the dispersal of pens (for example for pigs) amongst arable areas. Overall, the results from Koufovouno, Slatina, Vaihingen and Hornstaad (Figure 10.2) hint at different modes of manuring practice, representing alternative solutions to the challenge of limited manure resources, on the one hand, and the need to balance cooperation and competition among production groups, on the other.

At the north-western end of Europe, crop stable isotope data from 'ritual' enclosure sites and settlement contexts hint at another possible axis of diversity in farming practice. Isotope analysis of emmer wheat grain samples from Sarup, a TRB causewayed enclosure on Fyn (Andersen 1997), yielded distinctively low $\delta^{15}N$ values in comparison with the sites discussed above (Figure 10.2E) and also with grain from the nearby TRB settlement site of Skaghorn; similarly, emmer wheat grain samples from the Stepleton enclosure at Hambledon Hill, Dorset, are lower in their $\delta^{15}N$ values than those from the burned 'house' site of Lismore Fields, Derbyshire (Bogaard *et al.* 2013, table 2). The dataset is limited, and could be interpreted in different ways, but one explanation is that grain deposits at the enclosure sites derived from the cultivation of land with a different history – and certainly less affected by manuring – than that typically associated with grain at residential sites. Relatively unmanured fields might be located at the enclosures sites themselves, or the Sarup and Hambledon grain samples could represent cereals produced at settlement sites in 'outfield' areas without the usual intensive inputs. Either way, these results raise the possibility that the grain deposited at Sarup and Hambledon Hill represents a distinct form of production.

LBK Vaihingen: integration of functional weed ecology, crop stable isotope analysis and palaeodietary study

Through a functional ecological analysis of weed assemblages across Neolithic central Europe, the models of shifting cultivation, extensive ard cultivation and floodplain farming can be excluded based on an assessment of the permanence, intensity and seasonality of arable land. 'Intensive garden cultivation' emerges as the most plausible model (Bogaard 2004), but farming practice only appears uniform at this general level of inference. Relative differences in growing conditions are apparent at a regional level within the LBK (Bogaard 2004, 118, 147–8; 2012a,

194–6, 214–22), for example, and well sampled individual sites also demonstrate ecological variation, at least some of which has to do with the social configuration of the arable landscape (Bogaard 2004, 145–7; 2012a, 187–93, 212–14). At the LBK site of Vaihingen an der Enz, south-west Germany, Hans-Christoph Strien has identified a series of house groups or 'clans' based on differences in lithic types, pottery decoration and architecture (Strien 2005, 2012; Bogaard *et al.* 2011). These groups, in turn, are linked into regional 'provinces' of the Neckar valley LBK, such that groups of 'local' Middle Neckar people appear to have founded the site along with several groups from the Unterland-Kraichgau region to the north. Exploratory multivariate analysis of archaeobotanical samples from a specific crop type and processing stage (glume wheat dehusking by-products) reveals ecological gradients among samples in soil disturbance levels and perhaps soil pH (Bogaard *et al.* 2011; Bogaard 2012a, 212–14). These trends articulate with the settlement's social geography. The most intensively disturbed plots, probably closest to the settlement, belong to house group A and to a lesser extent D, both representing the 'local' Middle Neckar tradition, while group C, of the Unterland-Kraichgau network, is linked with more extensively managed land, probably at a greater distance from habitation (Bogaard *et al.* 2011; Bogaard 2012a, 187–93). These trends exclude the possibility that house groups cultivated equivalent radial 'slices' of land around the ditched settlement, pointing instead to cultivation of distinct areas that advantaged the Middle Neckar groups at the expense of those from the Unterland-Kraichgau. After the departure of the latter at the end of the Flomborn period, a smaller community of the Middle Neckar tradition at Vaihingen cultivated only the more proximate, intensively managed plots (Bogaard *et al.* 2011; Bogaard 2012a, 212–14).

When combined with Strien's artefactual analysis, therefore, the Vaihingen weed data shed light on the social configuration of the early Neolithic arable taskscape. It is of interest not only that different house groups were (dis)advantaged in their holdings, but also that neighbouring households worked alongside each other in distinct areas. Flomborn house group A contained three subdivisions, distinguished by slight variation in ceramic decorative motifs, each consisting of two to three longhouses that apparently stood together in rows; the weed data suggest clustering at this level within the broader zone of group A, but the patterning is weak (Bogaard *et al.* 2011; Bogaard 2012a, 187). Whether or not members of neighbouring households farmed in a truly cooperative manner, it is evident that *groups* of households constituted 'communities of practice' that worked in adjacent plots, shared detailed knowledge of local conditions and adhered to certain norms of husbandry.

At Vaihingen, assessment of the dietary role of crops, through integrated stable carbon and nitrogen isotope analysis of crops, fauna and humans, further underscores the salience of nested social identities in its farming 'taskscape' (Fraser *et al.* 2013). Palaeodietary studies of Neolithic sites in central and north-west Europe have often inferred a distinct emphasis on animal-based foods from $\delta^{15}N$ values in humans and associated fauna (Dürrwächter *et al.* 2006, Nehlich *et al.* 2009, Richards 2000, Richards and Hedges 1999, Richards *et al.* 2003, Richards *et al.* 2008, Schulting and Richards 2002; but see Oelze *et al.* 2010), with high estimates of animal dietary protein exceeding 60% (Hedges and Reynard 2007). Simple linear mixing models for Vaihingen that take account of the *actual* crop $\delta^{15}N$ values measured all point to higher consumption of plant-derived protein (*c.* 50–70%) than the standard model estimate (40%), which assumes that human plant food $\delta^{15}N$ is the same as that of animal forage (Table 10.2; Fraser *et al.* 2013). Estimates of plant-derived protein are higher still if greater trophic offsets of 5–6‰ are used (cf. O'Connell *et al.* 2012) rather than 4‰ as the mid-point of the conventional 3–5‰ range (Table 10.2D). Interestingly, a higher trophic offset of 5–6‰ would suggest that cereals could not have been the predominant plant protein source, and that (manured) pulses played a significant role (Table 10.2D). The aim here is not to argue for any particular scenario, but rather to point out that direct measurement of crop stable isotope ratios at Vaihingen builds a strong case for their staple role in the diet, and contextualises the finely tuned balance of cooperation and competition in its arable taskscape.

Conclusions

It is hoped that this contribution demonstrates both the advantages of integrating complementary strands of evidence for the nature and role of Neolithic cultivation,

Table 10.2 Linear-mixing model parameters and outputs used for estimating the fraction of animal protein of total dietary protein in human diets at Vaihingen (modified from Fraser *et al.* 2013, table 5)

Model inputs for $\delta^{15}N$ values	*Human dietary model scenarios*			
	A. Standard model	*B. Cereal consumer*	*C. Mixed cereal/pulse consumer*	*D. Mixed cereal/pulse consumer (+5‰)*
Human collagen	9.1‰	9.1‰	9.1‰	9.1‰
Herbivore collagen	6.7‰	6.7‰	6.7‰	6.7‰
Inferred herbivore forage	2.7‰	na	na	na
Vaihingen cereal crops	na	4.5‰	3.7‰	3.7‰
Vaihingen pulse crops	na	na		
Δplant-herbivore	4‰	na	na	na
Δdiet-human	4‰	4‰	4‰	5‰
Estimated % animal protein	~60%	~29%	~46%	~13%

and the necessity of embedding these efforts in the broader study of material culture. Ingold's concept of taskscape is challenging not only in terms of archaeological methodology but also because it calls for accounts of the human past that are at once fully cultural and ecological. None of the case studies discussed here approaches the coherence of a 'dwelling perspective', but it is encouraging that awkward distinctions between artefactual and ecofactual material, or social and economic interpretations, lose relevance once linkages between farming practice and material culture emerge.

Farming in south-east, central and north-west Europe tended towards the creation and maintenance of arable plots that could be cultivated for many generations. Their management often involved manuring in some form, but strategies varied from one site to the next, depending on factors such as herding practices, residential patterns and the range of crops grown. At Vaihingen it is possible to get a sense of the temporality of the farmed landscape; its configuration was bound up with the history and regional affiliations of the families who lived there. Here, arable plots were heritable property, passed down within house groups, and their ecology shaped, and was shaped by, the development of the community. The residential household, within which food was probably stored and systematically shared, embedded itself in broader groupings of houses practising (at least partially) cooperative production. This strategy helped to ensure access to adequate supplies of crops, which were dietary staples according to the emerging palaeodietary reconstruction.

Acknowledgements

This chapter draws on research funded by the Natural Environment Research Council (Standard Grant NE/E003761/1, PI Bogaard), l'Ecole française d'Athènes (Koufovouno), the Landesdenkmalamt Baden-Württemberg (Vaihingen) and the Arts and Humanities Research Council (extended sabbatical leave).

References

Andersen, N.H. 1997. *The Sarup enclosures I.* Aarhus: Jutland Archaeological Society.

Araus, J.L., Febrero, A., Buxó, R., Rodríguez-Ariza, M.O., Molina, F., Camalich, M.D., Martín, D. and Voltas, J. 1997. Identification of ancient irrigation practices based on the carbon isotope discrimination of plant seeds: a case study from the south-east Iberian peninsula. *Journal of Archaeological Science* 24, 729–40.

Benson, L.V., Taylor, H.E., Plowman, T.I., Roth, D.A. and Antweiler, R.C. 2010. The cleaning of burned and contaminated archaeological maize prior to $^{87}Sr/^{86}Sr$ analysis. *Journal of Archaeological Science* 37, 84–91.

Bogaard, A. 2004. *Neolithic farming in Central Europe.* London: Routledge.

Bogaard, A. 2012a. *Plant use and crop husbandry in an early Neolithic village: Vaihingen an der Enz, Baden-Württemberg*. Bonn: Habelt.

Bogaard, A. 2012b. Middening and manuring in Neolithic Europe: issues of plausibility, intensity and archaeological method. In R.L. Jones (ed.), *Manure matters: historical, archaeological and ethnographic perspectives*, 25–39. Farnham: Ashgate.

Bogaard, A. and Halstead, P. In press. Subsistence practices and social routine in Neolithic southern Europe. In press. In C. Fowler, J. Harding and D. Hofmann (eds), *The Oxford handbook of Neolithic Europe*. Oxford: Oxford University Press.

Bogaard, A. and Jones, G. 2007. Neolithic farming in Britain and central Europe: contrast or continuity? In A. Whittle and V. Cummings (eds), *Going over: the Mesolithic–Neolithic transition in north-west Europe*, 357–75. London: Oxford University Press for The British Academy.

Bogaard A., Strien H.-C. and Krause R. 2011. Towards a social geography of cultivation and plant use in an early farming community: Vaihingen an der Enz, south-west Germany. *Antiquity* 85, 395–416.

Bogaard, A., Heaton, T.H.E., Poulton, P. and Merbach, I. 2007. The impact of manuring on nitrogen isotope ratios in cereals: archaeological implications for reconstruction of diet and crop management practices. *Journal of Archaeological Science* 34, 335–43.

Bogaard, A., Fraser, R.A., Heaton, T.H.E., Wallace, M., Vaiglova, P., Charles, M., Jones, G., Evershed, R.P., Styring, A.K., Andersen, N.H., Arbogast, R.-M., Bartosiewicz, L., Gardeisen, A., Kanstrup, M., Maier, U., Marinova, E., Ninov, L., Schäfer, M. and Stephan, E. 2013. Crop manuring and intensive land management by Europe's first farmers. *Proceedings of the National Academy of Sciences of the United States of America* 110, 12589–94.

Brombacher, C. and Jacomet, S. 1997. Ackerbau, Sammelwirtschaft und Umwelt: Ergebnisse archäobotanischer Untersuchungen. In J. Schibler, H. Hüster-Plogmann, S. Jacomet, C. Brombacher, E. Gross-Klee and A. Rast-Eicher (eds), *Ökonomie und Ökologie neolithischer und bronzezeitlicher Ufersiedlungen am Zürichsee*, 220–91. Zurich: Zürich und Egg.

Cavanagh, W., Mee, C. and Renard, J. 2007. Excavations at Kouphovouno, Laconia: Results from the 2001 and 2002 field seasons. *Annual Report of the British School at Athens* 102, 11–101.

Charles, M., Jones, G. and Hodgson, J.G. 1997. FIBS in archaeobotany: functional interpretation of weed floras in relation to husbandry practices. *Journal of Archaeological Science* 24, 1151–61.

Charles, M., Hoppé, C., Bogaard, A., Jones, G. and Hodgson, J.G. 2003. Using weed functional attributes for the identification of irrigation regimes in Jordan. *Journal of Archaeological Science* 30, 1429–41.

Docheva, E. 1990. Plant macrorest research of early Neolithic dwelling in Slatina. *Studia Praehistorica* 10, 86–90.

Dürrwächter, C., Craig, O.E., Collins, M.J., Burger J. and Alt, K.W. 2006. Beyond the grave: variability in Neolithic diets in southern Germany? *Journal of Archaeological Science* 33, 39–48.

Ferrio, J., Araus, J., Buxo, R., Voltas, J. and Bort, J. 2005. Water management practices and climate in ancient agriculture: inferences from the stable isotope composition of archaeobotanical remains. *Vegetation History and Archaeobotany* 14, 510–17.

Fraser, R., Bogaard, A., Heaton, T., Charles, M., Jones, G., Christensen, B.T., Halstead, P., Merbach, I., Poulton, P.R., Sparkes, D. and Styring, A. 2011. Manuring and stable nitrogen isotope ratios in cereals and pulses: towards a new archaeobotanical approach to the inference of land use and dietary practices. *Journal of Archaeological Science* 38, 2790–804.

Fraser, R.A., Bogaard, A., Schäfer, M., Arbogast, R.-M. and Heaton, T.H.E.H. 2013. Integrating botanical, faunal and human stable carbon and nitrogen isotope values to reconstruct land use and palaeodiet at LBK Vaihingen an der Enz, Baden-Württemberg. *World Archaeology* 45, 492–517.

Gurova, M. 2005. Elements de faucilles néolithiques en silex de la Bulgarie: evidence et contexte. *Archaeologia Bulgarica* 9, 1–14.

Halstead, P. 1989. Like rising damp? An ecological approach to the spread of farming in south east and central Europe. In A. Milles, D. Williams and N. Gardner (eds), *The beginnings of agriculture*, 23–53. Oxford: British Archaeological Reports International Series.

Halstead, P. 1990. Quantifying Sumerian agriculture – some seeds of doubt and hope. *Bulletin on Sumerian Agriculture* 5, 187–95.

Hedges, R.E.M. and Reynard, L.M. 2007. Nitrogen isotopes and the trophic level of humans in archaeology. *Journal of Archaeological Science* 34, 1240–51.

Heier, A., Evans, J.A. and Montgomery, J., 2009. The potential of carbonized grain to preserve biogenic $^{87}Sr/^{86}Sr$ signatures within the burial environment. *Archaeometry* 51, 277–91.

Ingold, T. 1993. The temporality of the landscape. *World Archaeology* 25, 152–74.

Jacomet, S., Brombacher, C. and Dick, M. 1989. *Archäobotanik am Zürichsee. Ackerbau, Sammelwirtschaft und Umwelt von neolithischen und bronzezeitlichen Seeufersiedlungen im Raum Zürich*. Zurich: Orell Füssli.

Jones, G. 1992. Weed phytosociology and crop husbandry: identifying a contrast between ancient and modern practice. *Review of Palaeobotany and Palynology* 73, 133–43.

Jones, G., Bogaard, A., Halstead, P., Charles, M. and Smith, H. 1999. Identifying the intensity of crop husbandry practices on the basis of weed floras. *Annual of the British School at Athens* 94, 167–89.

Jones, G., Bogaard, A., Charles, M. and Hodgson, J.G. 2000. Distinguishing the effects of agricultural practices relating to fertility and disturbance: a functional ecological approach in archaeobotany. *Journal of Archaeological Science* 27, 1073–84.

Knipper, C. 2012. *Die räumliche Organisation der linearbandkeramischen Rinderhaltung: naturwissenschaftliche und archäologische Untersuchungen*. Oxford: Archaeopress.

Kreuz, A. 2007. Archaeobotanical perspectives on the beginning of agriculture north of the Alps. In S. Colledge and J. Conolly (eds), *The origins and spread of domestic plants in southwest Asia and Europe*, 259–94. Walnut Creek, CA: Left Coast Press.

Kreuz, A., Marinova, E.M., Schäfer, E. and Wiethold, J. 2005. A comparison of early Neolithic crop and weed assemblages from the Linearbandkeramik and the Bulgarian Neolithic cultures: differences and similarities. *Vegetation History and Archaeobotany* 14, 237–58.

Maier, U. 2001. Archäobotanische Untersuchungen in der neolithischen Ufersiedlung Hornstaad-Hörnle IA am Bodensee. In U. Maier and R. Vogt (eds), *Botanische und pedologische Untersuchungen zur Ufersiedlung Hornstaad-Hörnle IA*, 9–384. Siedlungsarchäologie in Alpenvorland 6. Stuttgart: Konrad Theiss.

Marinova, E. 2006. *Vergleichende paläoethnobotanische Untersuchung zur Vegetationsgeschichte und zur Entwicklung der prähistorischen Landnutzung in Bulgarien*. Dissertationes Botanicae. Berlin/Stuttgart: Cramer.

McClatchie, M., Bogaard, A., Colledge, S., Whitehouse, N.J., Schulting, R.J., Barratt, P. and McLaughlin, T.R. 2012. Neolithic farming in north-western Europe: new archaeobotanical evidence from Ireland. *Journal of Archaeological Science*, online 7 November 2012.

Nehlich, O., Montgomery, J., Evans, J., Schade-Lindig, S., Pichler, S.L., Richards, M. P. and Alt, K.W. 2009. Mobility or migration: a case study from the neolithic settlement of Nieder-Mörlen (Hessen, Germany). *Journal of Archaeological Science* 36, 1791–9.

O'Connell, T.C., Kneale, C.J., Tasevska, N. and Kuhnle, G.G.C. 2012. The diet-body offset in human nitrogen isotopic values: a controlled dietary study. *American Journal of Physical Anthropology* 149, 426–34.

Oelze, V.M., Siebert, A., Nicklisch, N., Meller, H., Dresely, V. and Alt, K.W. 2010. Early Neolithic diet and animal husbandry: stable isotope evidence from three *Linearbandkeramik* (LBK) sites in central Germany. *Journal of Archaeological Science* 38, 270–9.

Palomo, A., Gibaja, J. E., Piqué, R., Bosch, A., Chinchilla, J. and Tarrús, J. 2011. Harvesting cereals and other plants in Neolithic Iberia: the assemblage from the lake settlement of La Draga. *Antiquity* 85, 759–71.

Richards, M.P. 2000. Human consumption of plant foods in the British Neolithic: direct evidence from bone stable isotopes. In A.S. Fairbairn (ed.), *Plants in Neolithic Britain and beyond*, 123–35. Oxford: Oxbow Books.

Richards, M.P. and Hedges, R.E.M. 1999. Stable isotope evidence for similarities in the types of marine foods used by late Mesolithic humans at sites along the Atlantic coast of Europe. *Journal of Archaeological Science* 26, 717–22.

Richards, M.P., Price, T.D. and Koch, E. 2003. Mesolithic and Neolithic subsistence in Denmark: new stable isotope data. *Current Anthropology* 44, 288–95.

Richards, M.P., Montgomery, J., Nehlich, O. and Grimes, V. 2008. Isotopic analysis of humans and animals from Vedrovice. *Anthropologie* 46, 185–94.

Schlichtherle, H. 1992. Jungsteinzeitliche Erntegeräte am Bodensee. *Plattform* 1, 22–44.

Schulting, R.J. and Richards, M.P. 2002. The wet, the wild and the domesticated: the Mesolithic–Neolithic transition on the west coast of Scotland. *European Journal of Archaeology* 5, 147–89.

Sigaut, F. 1988. A method for identifying grain storage techniques and its application for European agricultural history. *Tools and Tillage* 6, 3–32.

Strien, H.-C. 2005. Familientraditionen in der bandkeramischen Siedlung bei Vaihingen/Enz. In J. Lüning, C. Fridrich and A. Zimmerman (eds), *Die Bandkeramik im 21. Jahrhundert: Symposium in der Abtei Brauweiler bei Köln vom 16.9.–19.9.2002*, 189–97. Rahden: Marie Leidorf.

Strien, H.-C. 2012. Chronological and social interpretation of the artefactual assemblage. In A. Bogaard, *Plant use and crop husbandry in an early Neolithic village: Vaihingen an der Enz, Baden-Württemberg*, 19–23. Bonn: Habelt.

van der Veen, M. 1992. *Crop husbandry regimes: an archaeobotanical study of farming in northern England*. Sheffield: J.R. Collis Publications.

Wallace, M., Jones, G., Charles, M., Fraser, Halstead, P., Heaton, T.H.E. and Bogaard, A. 2013. Stable carbon isotope analysis as a direct means of inferring crop water status and water management practices. *World Archaeology* 45, 388–409.
Wasylikowa, K. 1981. The role of fossil weeds for the study of former agriculture. *Zeitschrift für Archäologie* 15, 11–23.

11

Stewing on a Theme of Cuisine

Biomolecular and Interpretive Approaches to Culinary Changes at the Transition to Agriculture

HAYLEY SAUL, AIKATERINI GLYKOU AND OLIVER E. CRAIG

Introduction

THE ACT OF PRODUCING AND CONSUMING FOOD is one of the most intimate modes for the construction of personal and social identity, being both a literal and metaphorical intrusion of the external world on to the internal one. Cuisine studies are an investigation into the processes by which foods are evaluated for perceived properties that inform their role in production and consumption practices. More recently, the value of taste has weighted a notion of 'cuisine' as a refined, often class-related knowledge about acceptable ways to combine and consume foods. Whilst cuisine as the evaluation of foods operates at the social scale, it is not restricted to a single stratum, or represented by culinary exotica alone. Cuisine is the reasoning process and the values behind all food choices. Archaeologically, patterning in the combinations and selective exclusions of food in response to wider socio-economic contexts is the point of departure to enter these past evaluation processes.The idea of cuisine which is advanced here cannot be divorced from the wider socio-economy. Instead, it is suggested that a cuisine perspective is a way of reassessing food studies in the wider prehistoric economy. The detail and spatio-temporal resolution of debates about the value of food are a product of the datasets supporting them. A future that includes biomolecular techniques to study paleodiet and food must also witness a critical review of how archaeologists represent food values, and the way those values are considered to stimulate prehistoric change. In a cuisine perspective, food is part of multiple evaluation processes that cross-cut production and consumption spheres (Parker Pearson 2003). Thus, linear models that posit a singular change, brought about by a dominant food value, such as energy or prestige, are de-emphasised. Foods have many cultural and emotional connotations. There are no cultural prescriptions, and yet coming to terms with why certain foods were valued in certain social contexts in the past is essential to understanding what might have motivated a change in the values of foods, and

Proceedings of the British Academy **198**, 197–213. © The British Academy 2014.

ultimately changes in the foods themselves. This issue is essential to an understanding of the role of food in social complexity, the transition to agriculture, and the relative importance of different foods in wider cultural change.

If food is valued for more than just subsistence, what are its other values? One way to address this question is to consider the micro-contexts for food processing, consumption and deposition. Organic residue analysis of ceramic cooking vessels to reconstruct use provides a unique opportunity to evaluate choices made during food preparation. Here, we evaluate the results of lipid residues (Craig *et al.* 2011) and plant microfossil analyses (Saul *et al.* 2012, 2013) on pottery residues to illustrate how foods were valued as cuisines in the late fifth to early fourth millennia cal BC in the southern Baltic.

The contribution of biomolecular datasets to the study of prehistoric food

Due to acceleration of biomolecular research in archaeology in the last two decades, traditional research agendas, which have relied on conventional macro-remains to investigate the production and consumption of food, are poised for extension. Stable isotope analysis of human bone collagen has already revolutionised the study of diet and more recently dietary change through an individual's lifetime (Beaumont *et al.* 2013). The sequencing of protein fragments by mass spectrometry (ZooMS) renders tiny fragments of bone and shell 'visible' in the archaeological record, allowing them to be identified to genus and sometimes species (Richter *et al.* 2011, Stewart *et al.* 2013, Buckley *et al.* 2009). Lipid residue analysis of pottery can provide direct evidence for the processing of foods such as dairy and insect products (such as beeswax) (Copley *et al.* 2003, Heron *et al.* 1994) that leave no other direct archaeological trace. Advances in the detection of plant microfossils, such as starch granules, associated with artefacts (Crowther 2005, Saul *et al.* 2012), permit new insights into the culinary role of plants, including non-staple plants such as spices (Saul *et al.* 2013). Finally, the analysis of DNA, proteins and plant microfossils in coprolites and dental calculus is beginning to offer high-resolution dietary records towards the scale of single meals (Botella *et al.* 2010, Hardy *et al.* 2009, Henry and Piperno 2008).

With the application of such new methods, the greatest advance for archaeology is that they often allow for direct, and temporally more precise, inference. With methods to identify lipids, proteins and plant microfossils across a range of micro-contexts, such as pottery vessels, dental calculus, coprolites and soil micro-strata (Weiner 2010), the resolution of datasets is altering to meet more highly resolved timescales of post-modern interpretive demands. Whilst we have not yet dis-aggregated these micro-archaeological palimpsests to make individual meals visible (Isaksson *et al.* 2010), the patterns of choice in the way foods were selected for at community scales are a window into the attributes that made them important. The purpose of this chapter is to propose a framework for interpreting the relative value of foods, in the light of these enhanced culinary datasets, and in the context of weaknesses in prevailing theoretical perspectives. One area where the application of biomolecular methods has made, and continues to make, substantial impact is in determining the changes in the acquisition and consumption of foods with the introduction of farming in the European Neolithic, and this will be the focus of this chapter.

Theoretical perspectives of food value and economy at the transition to agriculture in northern Europe

What are the main interpretations about how the transition occurred?

Two main models of economy characterise interpretations of the agricultural transition: the subsistence economy, and the prestige economy. Respectively, they describe different views of what value is, and how it is structured into processes that motivate social change. In the subsistence economy, value is at least partially measurable in the relative quantities of various food macro-remains. For example, in the resource ranking model (Rowley-Conwy 1984), human behaviour is rational and bent towards utility maximisation of energy. In this way, the oyster (*Ostrea edulis*) comes to be considered a dietary linchpin that forced the adoption of a more 'stable', high-energy food base that included domesticated species when their numbers declined (Rowley-Conwy 1984). Yet it is inherently difficult to judge the degree to which alterations to the resource base pushed the adoption of agro-pastoralism, because of differences in the preservation of food remains or their proxies. Wild plant food studies are a casualty of this relatively reduced systematic investigation compared to domesticated edible flora (Zvelebil 1994). As a result of the focus on high energy species, the Late Mesolithic broad spectrum economy is seen in the negative, as a push towards starvation species, rather than a positive period of culinary creativity or experimentation. Ranking resources (Rowley-Conwy 1984) overlooks the way foods were combined, even though the choices made to combine, mix, or exclude food elements are important in understanding the value of foods and food culture. A reinvestigation of oyster seasonality showed that this food was not exclusively a spring 'gap-filler' as had been suspected (Milner 2002). Representation to equitable levels in other seasons showed that oysters were not a singular staple; they were combined with other culinary choices, diffusing primary reliance on them.

In the prestige economy the inclusion of domesticated foods arises because they were 'exotica', and conferred status on to those that consumed them (Fischer 2002, Zvelebil 2006). Food items mark the end stages of a series of other importations of artefact adoptions (Zvelebil and Rowley-Conwy 1984): shoe-last adzes from central Europe (Klassen 2002), green slate artefacts from eastern Finland (Zvelebil 2006), jadeitite and eclogite axes from the Italian Alps (Klassen 2002).

Agro-pastoralism is part of an inflationary socio-economic process (Fischer 2002), where 'prestige' transfers from one class of object to the next as the 'market' becomes gradually more saturated and value through exclusivity is lost. However, a non-local food source does not necessarily confer high-status. South Asian Vedic literature, for example, testifies to food preferences oriented towards local rice staples since the first millennium BC, used in important religious offerings as well as mundane daily consumption, partly because of medicinal connotations which the starchy plant held (Smith 2006).

What is the state of economy and value?

The prestige economy model is a reaction to the subsistence model which assumes value to be solely utilitarian and divorced from contingent social circumstances, yet both models share much in common despite their intended contrast. Formalist economics are expressed in emphases on the intensity of food use as a factor in subsistence change. Means–end relationships are universal and thus explainable as measures of costs, be they time or energy, for example. A reactionary substantivist economics focused on embedding the economy in social relations, the key critique of formalist universalism. 'Prestige' is a problematic concept in value terms, however. In its archaeological use it possesses absolute qualities; it is either ascribed or it is not. What is being ascribed might be considered as much a description of a value state or hierarchical placement, as a value in itself. Although the concept of economy is less socially isolated, the substantivist agenda sacrifices the measurability of value, and thus cannot explain mechanisms of change. Instead, prestige 'proxies', namely feasting or exotica, are used as markers of the presence of value. The prestige model describes when change happens through the sequence of when items gain or lose prestige, but it cannot articulate a measure of that value, or a more detailed characterisation of it. Thus, in substantivist economics the potential for change is always uni-directional, from high to low prestige, whilst in formalism social change is uni-directional towards predictable acts of efficiency.

The key similarity for both models is that the mechanisms of change act repetitively despite variations in the economic commodities. So, the sequence of prestige foods alters, but the process of building value saturation occurs regardless of the food commodity. Likewise, in the subsistence model, food value is more abstractly described by calories, or time, or energy. These categories are common to all the diverse foods implicated in the process allowing the foods themselves to be rendered comparable. Animal has the same value as vegetable, bitter to sweet. The result of both models is that *past* evaluation processes are overlooked, as are *past* motivators to change, and all types of food are considered to be part of a singular, linear process of change towards an agro-pastoral society. Additionally, neither model integrates food data into wider culinary contexts, so that the material culture of cuisine – for example, pottery, quernstones, basketry, hunting equipment and storage pits – has been treated as separate artefact classes. Taken together, however, the ways in which culinary artefacts were structured into evaluation processes are essential to understand why changes to cuisine happened in the late fifth millennium cal BC.

Cuisine and food values

Hunter-gatherer communities must have acted to alter their food base in the late fifth to early fourth millennia cal BC, according to what foods they valued and why they valued them. Whilst cuisine can be defined in a literal sense as the mixing of foods, what it is socially acceptable to combine are actually the values of those foods. Cuisine is understood here as the evaluation of suitable properties or perceived attributes of edible items and related technologies, both singularly and in combination, to fulfil roles in contingent social circumstances. If communities were not living at their ecological limits, they were making choices about consumables. Cuisine studies are a window into choice and decision-making, underpinned by the evaluation of foods. Concepts like health, medicine, disease, consciousness, pollution, danger, the self, aesthetics and virtue are commonly features of cuisines. Since cuisine deals with value systems it is inappropriate to divorce it from the wider economy. Rather, a cuisine perspective on food is one that targets which values informed the mechanisms of economy, instead of presupposing that those values relate to efficiency or a substantive notion of 'prestige'.

Lévi-Strauss (1970, 1993) situates the driving force of cuisine changes at the social, rather than the individual, scale. Food culture is an expression, or microcosm of much wider social processes. Even personal responses in the emic realm of taste, disgust and craving are set within a culturally established framework that makes routine choice and material patterning a reality. In New Guinean society, for example, medicine and food are often considered the same thing. Sugarcane is eaten by men after sexual intercourse, as a tonic against the dangers of female sexuality (Counihan1999). The medicinal value of sugarcane therefore directly affects social structure and demography, making a greater amount of contact between the sexes socially acceptable than would otherwise be considered safe. Up until the eighteenth century such medicinal connotations to food existed in western Humorism cuisines too (Lupton 1996), which was a fashionable topic for the educated mind, even culminating in cookbooks (e.g. Charleton 1659).

Though it is counter-intuitive, cuisine can also be expressed through ideas of poison and pollution. This is the manner in which food is perceived by many people with eating disorders, such as anorexia nervosa and bulimia. Bruch (1982) sees the obsessive pursuit of thinness as resulting from a lack of core identity, and ascription to a media-defined one instead. Conversely, such disorders could be a resistance to the prescriptions of a patriarchal society and a 'feminine' destiny, by identifying with an androgynous or masculine body form (Squire 2003), the seat of greater personal agency (Littlewood 2004). As such, food is rendered a poison to 'the self', or the projected self. For the Ituri forest pygmies of the Republic of Congo, the highly ethical construction of cuisine posits a different sort of pollution risk. In the processes of creating or preparing food the Ituri 'self' is blurred, because of an ontology based around fluid identity. The attributes of animal, plant and human entwine (Turnbull 1993), and these beings are valued in an intimate, ethical sense.

Ceramic cuisine in the western Baltic

Here we consider the use of ceramics at two sites situated on the Baltic coast of Northern Germany (Figire 11.1), to highlight food values during the late fifth to early fourth millennia cal BC. Neustadt (LA 156) is a submerged coastal site spanning the transition to agriculture. Radiocarbon dates fall between 4600–3800 cal BC. Excavations were conducted by the *ArchäologischesLandesamt Schleswig-*

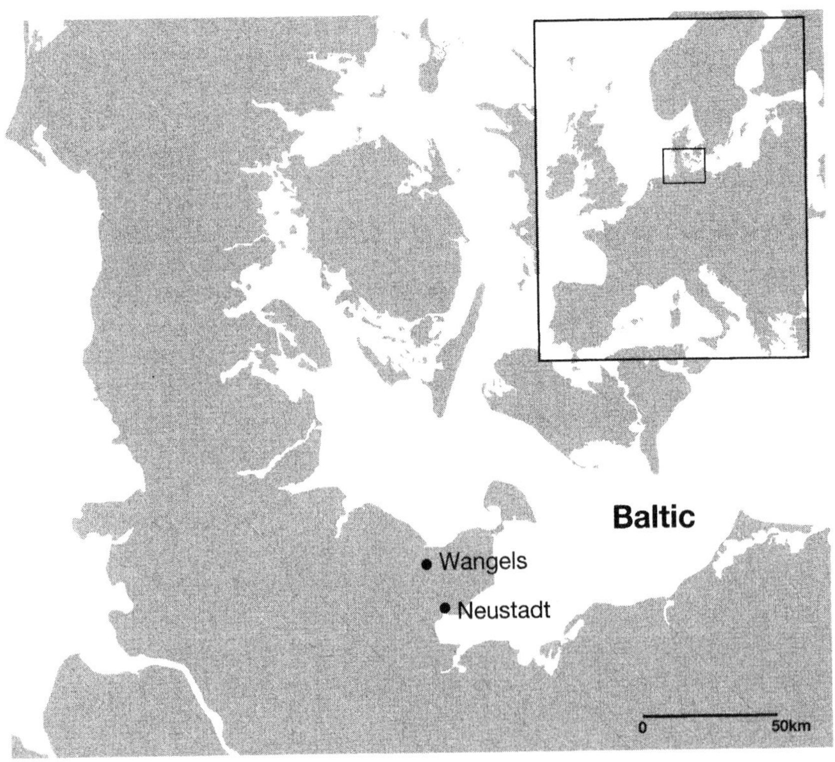

Figure 11.1 A map showing the locations of Neustadt and Wangels in Schleswig-Holstein, the sites from which pottery was sampled for residue analysis of lipids and plant microfossils.

Holstein from 2000–2006. Among the finds are *c.* 7500 pottery sherds including both Ertebølle (EBK) and Funnel Beaker styles (TRB), typically associated with the Late Mesolithic and Early Neolithic respectively (Glykou 2010). Exploitation of marine resources was very important at this site, indicated by thousands of fish bones and a particularly high frequency of marine mammals, accounting for 37% (NISP) of the mammalian faunal assemblage, suggesting that marine mammal hunting was a major activity at this site (Glykou 2013). Conversely, Wangels (LA 505), a submerged settlement on a former island off the coast of Schleswig-Holstein is orientated to the exploitation of cattle. This domesticate is represented by 37% of the skeletal specimens. Radiocarbon dates on animal bone place occupation at Wangels in the 4100–3800 cal BC range, which correlates with a pottery assemblage dominated by Funnel Beakers (Hartz *et al.* 2007).

Organic residue analysis of lipids extracted from ceramic interior surfaces and charred interior deposits or 'foodcrusts' by gas chromatography mass spectrometry (GCMS) and GC combustion isotope ratio MS (GC-c-IRMS) from both sites has been reported by Craig *et al.* (2011). Isotopic analysis of the 'foodcrusts' by elemental analysis-IRMS (EA-IRMS) has been reported by Craig *et al.* (2011). Plant microfossil analysis has been reported by Saul *et al.* (2012, 2013). Here we consider the combined results of these analyses to investigate cuisine. In total, organic residue analysis was carried out on 30 pointed-bottomed vessels from Neustadt, typologically identified as Ertebølle, and 31 round- or flat-bottomed Funnel Beakers (Types 0, I and II; Koch 1998), and a further 14 TRB vessels from Wangels were considered for analysis. Table 11.1 shows the main classes of lipids

Table 11.1 Main classes of lipids identified in pottery vessels from Neustadt and Wangels. The primary criteria for assigning lipids were the GC-c-IRMS data using the ranges shown in Figure 11.2. Where mixtures were present the dominant lipid class is reported, defined by contributing >75% of the fatty acids present (Craig *et al.* 2011). If no GC-c-IRMS data were available, lipid biomarkers were used to identify marine derived residues (Hansel *et al.* 2004).

	Lipid class	*EBK*	*TRB*	*Total*
Neustadt	Marine	5	6	11
	Ruminant	8	11	19
	Porcine	5	0	5
				35
Wangels	Marine	–	2	2
	Ruminant	–	8	8
	Porcine	–	0	0
				10
Total	Marine	5	8	13
	Ruminant	8	19	27
	Porcine	5	0	5
				45

identified from 45 vessels that yielded sufficient lipid. Identifications were made by considering the isotope ratios of individual fatty acids (Figure 11.2) or the presence of specific biomarkers (Craig *et al.* 2011).

Culinary changes at the transition to agriculture

A comparison of lipids recovered from pointed-based EBK and TRB Funnel Beaker pottery from the German Baltic coast shows that marine and ruminant products were a feature in ceramic cuisine during both periods (Table 11.1, Figure 11.2). This mirrors the results of a much larger sample showing continuity in the processing of marine foods beyond the introduction of domesticated animals (Craig *et al.* 2011). Interestingly, there is no evidence for the processing of porcine products in Funnel Beaker pottery at either site despite the fact that porcine lipids, most likely

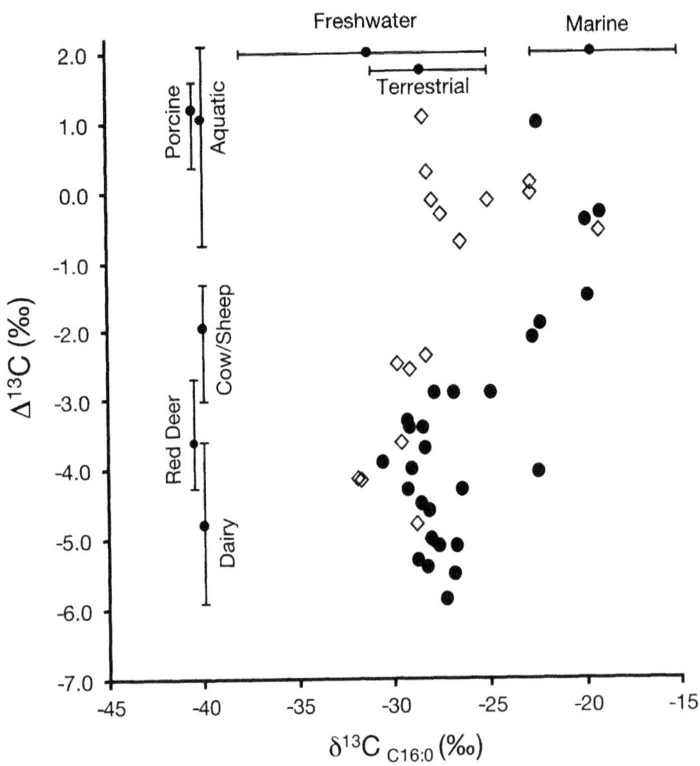

Figure 11.2 GC-c-IRMS analysis of Ertebølle (open diamonds) and Funnel Beaker cooking vessels (closed circles). Different pots are distinguished by their carbon isotope value of their C16 fatty acid (x-axis) and the carbon isotope difference between the C16 and C18 fatty acids (y-axis). Reference ranges (medians and ranges) are from Dudd *et al.* 1999 and Craig *et al.* 2011, 2012.

wild boar, were detected in the Late Mesolithic vessels and domesticated pigs account for 3.3% NISP of the Wangels assemblage. A large increase in vessels with lipids that match the criteria for dairy products (Copley *et al.* 2005) was evident in the Early Neolithic which is attributable to the establishment of dairy economies with the arrival of domesticated ruminants, although lipids from three Ertebølle vessels also match the dairy range (Figure 11.2). Recently, the criteria for identifying dairy products have been revised (Craig *et al.* 2012) to consider carcass fats from wild ruminants which have partly overlapping isotopic ranges (Figure 11.2), which may account for the 'dairy' residues identified on Ertebølle pottery. An alternative, but less likely, explanation, is the exchange of dairy products between coastal foragers and other Early Neolithic farming groups to the south, or perhaps the exploitation of dairy products from wild ruminants following death, or even early domestication, although there is currently no faunal or DNA evidence for domesticated ruminants in Ertebølle assemblages (Scheu *et al.* 2008).

Altogether, over twice as many pots were used for processing predominantly terrestrial residues compared to marine. This was true at both sites and in both periods despite the fact that at Neustadt, marine fish and marine mammals are a dominant feature of the faunal assemblage. Partial analysis of the Neustadt assemblage, where fish and mammals were recorded, revealed that the ratio of marine to terrestrial was 10:1 by individual skeletal parts (NISP). Whilst the relative body mass and differential preservation need to be accounted for, it was notable that bovine species (aurochs and domesticated cattle) count for less than 10% of the NISP at Neustadt but were identified in over half the vessels analysed. Such discrepancies between the composition of faunal assemblages and pottery use are of course entirely to be expected, considering the major cultural transformation processes involved in food preparation and consumption, namely cuisine.

Evidence for the selective separation of foods in different vessels

A consideration of the plant microfossils, bulk isotope composition from charred surface deposits and the composition of lipids present allows insight into a wider range of foods processed in individual pots than previously possible. Samples of foodcrusts from Funnel Beaker pottery which were more enriched in $\delta^{15}N$ and $\delta^{13}C$, indicative of marine sources, also contained more substantial numbers of starch grains (Figure 11.3). These are also more often associated with pots containing predominantly marine lipids, strongly suggesting that plant and marine foods were preferentially processed together whilst terrestrial foods, including dairy products, were processed separately (Figure 11.3). Such a pattern is not evident in the Ertebølle pottery, partly because nearly all the bulk foodcrust values from this period were isotopically enriched. Whilst mixing of foods is evident (Figure 11.2) it is hard to disentangle given the number of potential sources and uncertainties regarding the period of vessel use, we propose that at the minimum the group of Funnel Beakers

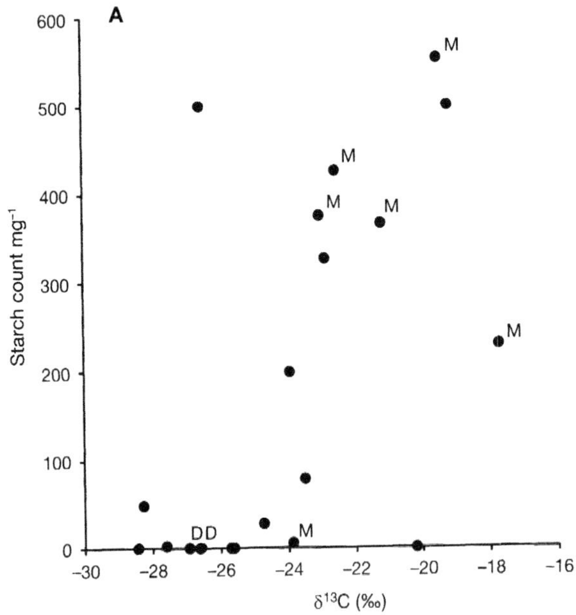

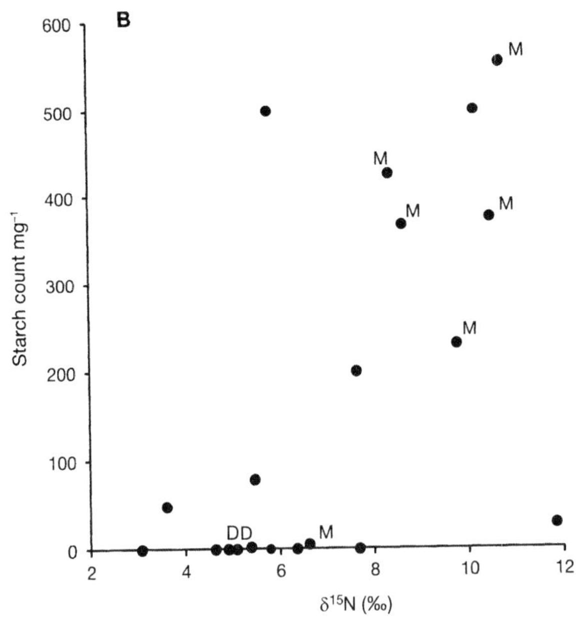

Figure 11.3 Plots of starch counts against bulk $\delta^{13}C$ and $\delta^{15}N$ values in charred surface deposits. M indicates pottery vessels with the majority of lipid derived from a marine source, as determined by GC-c-IRMS. In the absence of GC-c-IRMS data, the presence of marine lipid biomarkers was used to identify a marine source. D indicates pots with evidence for dairy products as determined by GC-c-IRMS.

with very low fatty acid $\Delta^{13}C$ values <–5‰; Figure 11.2) were used exclusively for processing milk, supporting a claim for a tradition of controlling culinary contact of this product.

The relationship between vessel form and use

Changes in both the mode of food preparation and the types of food prepared are evident between vessel forms, suggesting role-specific design in the production of pottery. As well as the clear typological change between pointed-bottomed EBK and round- or flat-bottomed TRB pots, the distribution of carbonised residue on the inside of the vessels also changes. Carbonised deposits extend from rim to base of EBK pots but are more localised on the rim and upper body of TRB pots, when all vessels available were considered (n = 2162). Such patterned variations in the presence of 'foodcrust' must be related to variable preparation techniques, in the context of particular foods.

Within the Funnel Beaker pottery assemblage, there is a preference for pro-

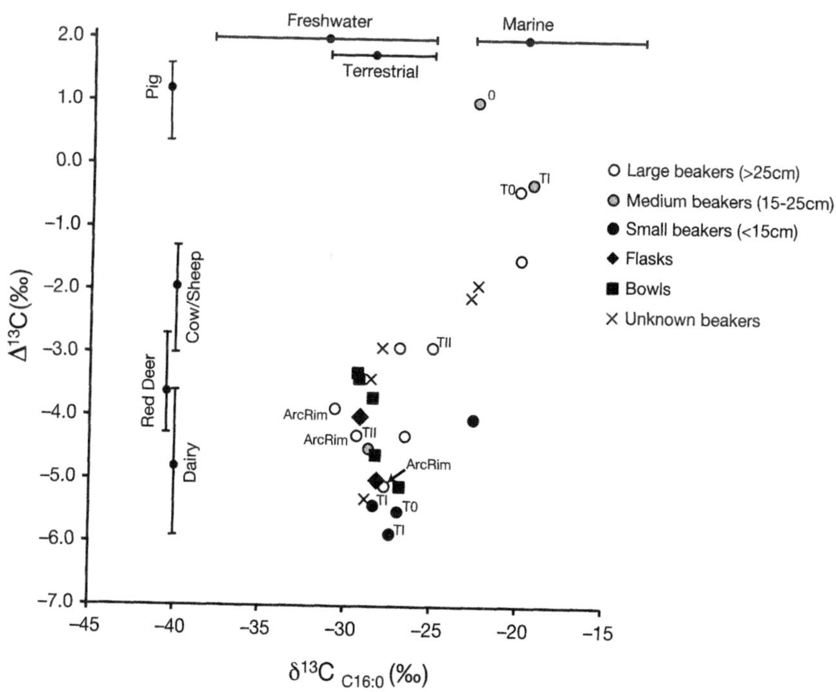

Figure 11.4 GC-c-IRMS analysis of Early Neolithic ceramic vessels distinguished by vessel type. Different pots are distinguished by the carbon isotope value of their C16 fatty acid (x-axis) and the carbon isotope difference between the C16 and C18 fatty acids (y-axis). Reference ranges (medians and ranges) are from Dudd *et al.* 1999, Craig *et al.* 2011, Craig *et al.* 2012. ArcRim = arcade rim; Tx indicates TRB vessel typology after Koch 1998.

cessing dairy products in small cup-sized beakers (rim diameter <15cm; Figure 11.4), as well as in bowls and flasks. Larger and medium Funnel Beakers have a more varied range of uses including the processing of marine and plant foods. Perhaps unsurprisingly, bowls, flasks and small cups have features designed for the storage or transport of liquids, entrenching a notion of their association with the manipulation of milk. The open orifice of a Funnel bowl, for example, is suitable for the skimming of cream, whilst narrow necked jars/flasks allow the restriction of oxygen to fermentation bacteria, slowing and controlling souring, that could lead to the production of low-lactose cheeses and yoghurts. Of the four large vessels that could be associated with dairy products, all from Neustadt, three are decorated with very unusual 'arcade' rims (or *Tupfenleist*), a notched cordon of clay applied to the rim exterior. This decorative feature is found on Michelsberg-style ceramics produced by farming groups across west-central Europe, perhaps suggesting some exchange with established pastoral communities to the south.

Changes in food values and the transition to agriculture in the Baltic

In the Early Neolithic period, we suggest based on the residue evidence that dairy foods were selectively excluded from other foods (Figures 11.2–11.3), such as starchy plants and marine products, and separately processed in small beakers, flasks and bowls. The acceptability of the use of dairy was therefore subject to limitations. The presence of dairying at these maritime-focused sites, particularly Neustadt, is in itself interesting and is more likely to reflect a specialised practice than the wider economy. Apparent control was being exerted over the processing and use of dairy, as if a toxic evaluation of dairy lactose had been recognised. Regardless of the timing for the introduction of the lactase persistence allele into northern Europe (Gerbault *et al.* 2009, Malmström *et al.* 2010), the absence of the allele in the *c.* late fifth to early fourth millennium cal BC (Burger *et al.* 2007) indicates that lactase persistence was rare in Early Neolithic farming groups. For some therefore, dairy food could be associated with negative physical effects, potential toxicity. Despite this, dairy still clearly makes a significant contribution to pottery cuisine. The element of risk, and the responsibility afforded to those that made low-lactose products like cheese and yoghurt, may have situated milk at the boundary between the edible and inedible at the transition. Arguably the proliferation of Funnel Beaker pottery forms conducive to liquid food processing – most likely related to handling dairy products – was in response to a desire for fine control over the almost proto-alchemical transformations of milk to solid curds and liquid whey.

Conversely, arguments can be made for the nutritional and health benefits of milk, for example, as infant weaning foods, and/or by transformation into low-lactose products such as yoghurts and cheese. The demographic portion of society for which dairy foods pose the greatest benefit, and least toxic risk, are infants, where the value of dairy may have been the antithesis of a toxic one. From the predominance of arranged child burials, it has been suggested that fertility and population viability were important issues within the group consciousness of the Late Mesolithic (Jensen and Hansen 1999, Petersen 1990, Petersen *et al.* 1993) and it is likely that infant feeding practices and weaning were carefully controlled. In this context, the introduction of animal milk, as a replacement or supplement to human milk, may have been highly valued for its perceived medicinal benefits, even if the actual health benefits were negligible (Howcroft *et al.* 2012).

The earliest inclusion of a domestic-based cuisine in evidence at Neustadt and Wangels is in the form of dairy foods. Further, the mode by which these novel cuisines were introduced seems to be through exchange, with the value of managing dangerous culinary elements playing a potentially pivotal role. Three Funnel Beakers with potential dairy signatures (Figure 11.4) are decorated with an applied clay cordon, or the Michelsberg 'arcade rim'. Exchange with agro-pastoral groups in the Early Neolithic was perhaps spearheaded by dairy cuisine. It may be that dairy foods were valuable exchange items because their consumption was enabled by their correct preparation; they embodied the value of trust necessary for strong inter-group communications. Additionally, however, the transportation of dairy products would have introduced new bacterial strains, culminating in innovative flavours, and enhancing the clear investment being made during the Early Neolithic in exploring transformation processes.

The floral assemblage from Neustadt is overwhelmingly dominated by acorn casings (*Quercus* sp.), the toxic tannins of which could have been leached using the pottery. Starch analyses of the Neustadt residues certainly support an intensive association with starchy foods (Saul *et al.* 2012). In both periods, vessels with these high starch counts are associated with marine or ruminant foods, and are often to be found in large-sized pottery (up to 40cm rim diameter) with a volume suitable for up to 60 consumers. Where rim sherds were available from these large vessels, carbonised deposits high on their profile visibly demonstrate that they were used to capacity. Communal consumption that persisted from the EBK to the TRB period was therefore targeted on traditional food sources, such as locally available plants and coastal animals, perhaps a means of engendering an ethic of cohesion in the face of non-local exchange and contact.

Conclusion

Acknowledging that all datasets are designed to answer certain questions, the challenge for a mature and theoretically informed scientific archaeology becomes to design a dataset that can interrogate fine-scale social processes which are proposed by interpretive questions. Proceeding from this, the development of scientific techniques will adopt a greater synergy with wider archaeological trajectories, rather than the two working in relative isolation. The biomolecular approaches offered here act upon small samples of novel material, such as pottery residues, which are otherwise not amenable to macroscopic investigation. A cuisine perspective emphasises patterning in culturally sanctioned choices of what was important, and contextualises the reasons for this decision making by matching culinary attributes to the social arenas where they are deployed. Such an approach is proposed by cuisine studies, which emphasise neither a formalist subsistence view of economy, nor a substantive characterisation of 'prestige' consumption economy. Instead, economy can start to be disaggregated into sub-processes of food evaluation. Significantly, this 'value'-based approach is in contrast to 'meaning' studies that have tended to adopt post-structuralist conceptions of a manifold plurality that is caught in a state of flux, always in a 'signifying' verb state and never as static, motionless noun (Harland 1987). The effect of an applied version in archaeology has been detrimental to studies of change, such as the transition to agriculture, since change is appreciable through contrast, and the insubstantial essence conceived for 'meaning' negates the possibility of contrast. If there is neither form nor substance to anchor the flow of relations motionless and substantial, there is no hope of being able to reconstruct past concepts of 'meaning' so that they can be ascribed with value content.

Selection and creativity are inherent in the practice of cuisine, and therefore scientific analyses of food and diet need to pursue sampling strategies and techniques that add detail to culinary processes. The greater identification resolution that is developing through the recovery of ancient biomolecules, alongside their relatively refined temporal anchoring, means a cuisine perspective on food is increasingly possible. The values, or attributes, of foods to fulfil certain social roles motivate change along particular trajectories, so it is only by engaging with past evaluations that it is possible to represent the palimpsest of multiple changes at the transition to agriculture, rather than simplistic unilinear models, or the polar opposite, post-modern fractalised accounts that lack a trajectory to change.

Acknowledgements

The authors gratefully acknowledge the financial support of the Arts and Humanities Research Council (AH/E008232/1). Samples of pottery sherds and carbonised foodcrust were kindly donated by Sönke Hartz of Schleswig-Holstein Landesmuseen. Thanks to Carl Heron and Val Steele for their contributions to the production of the original lipid residue data.

References

Beaumont, J., Gledhill, A., Lee-Thorp, J. and Montgomery, J. 2013. Childhood diet: a closer examination of the evidence from dental tissues using stable isotope analysis of incremental human dentine. *Archaeometry* 55, 277–95.

Botella, H.G., Vargas, J.A.A., Arnay de la Rosa, M., Leles, D., Gonzále Reimers, E., Vicente, A.C. and Iñiguez, A.M. 2010. Paleoparasitologie, paleogenetic and paleobotanic analysis of XVIII century coprolites from the church La Conceptiòn in Santa Cruz de Tenerife, Canary Islands, Spain. *Memróias Instituto Oswaldo Cruz, Rio de Janeiro* 105, 1054–6.

Bruch, H. 1982. Anorexia nervosa: therapy and theory. *American Journal of Psychiatry* 139, 1531–8.

Buckley, M., Collins, M., Thomas-Oates, J. and Wilson, J.C. 2009. Species identification by analysis of bone collagen using matrix-assisted laser desorption/ionisation time-of-flight mass spectrometry. *Rapid Communications in Mass Spectrometry* 23, 3843–54.

Burger, J., Kirchner, M., Bramanti, B., Haak, W. and Thomas, M.G. 2007. Absence of the lactase-persistence-associated allele in early Neolithic Europeans. *Proceedings of the National Academy of Sciences of the United States of America* 104, 3736–41.

Charleton, W. 1659. *Natural history of nutrition, life and voluntary motion*. London: Henry Herringman.

Copley, M.S., Berstan, R., Dudd, S.N., Docherty, G., Murkherjee, A.J., Straker, V., Payne, S. and Evershed, R.P. 2003. Direct chemical evidence for widespread dairying in prehistoric Britain. *Proceedings of the National Academy of Sciences of the United States of America* 100, 1524–9.

Copley, M.S., Berstan, R., Mukherjee, A.J., Dudd, S.N., Straker, V., Payne, S. and Evershed, R.P. 2005. Dairying in antiquity. III. Evidence from absorbed lipid residues dating to the British Neolithic. *Journal of Archaeological Science* 32, 523–46.

Counihan, C.M. 1999. *The anthropology of food and body: gender, meaning and power*. New York: Routledge.

Craig, O.E., Steele, V., Fischer, A., Andersen, S., Donohoe, P., Glykou, A., Saul, H., Jones, M., Koch, E. and Heron, C. 2011. Ancient lipids reveal continuity in culinary practices across the transition to agriculture in Northern Europe. *Proceedings of the National Academy of Sciences of the United States of America* 108, 17910–15.

Craig, O.E., Allen, R.B., Thompson, A., Stevens, R.E., Steele, V.J. and Heron, C. 2012. Distinguishing wild ruminant lipids by gas chromatography/combustion/isotope ratio mass spectrometry. *Rapid Communications in Mass Spectrometry* 26, 2359–64.

Crowther, A. 2005. Starch residues on undecorated Lapita pottery from Anir, New Ireland. *Oceania* 40, 62–6.

Dudd, S.N., Evershed, R.P. and Gibson, A.M. 1999. Evidence for varying patterns of exploitation of animal products in different prehistoric pottery traditions based on lipids preserved in surface and absorbed residues. *Journal of Archaeological Science* 26, 1473–82.

Fischer, A. 2002. Food for feasting? An evaluation of explanations of the Neolithisation of Denmark and southern Sweden. In A. Fischer and K. Kristiansen (eds), *The Neolithisation of Denmark: 150 years of debate*, 343–93. Sheffield: J.R. Collis Publications.

Gerbault, P., Moret, C., Currat, M. and Sanchez-Mazas, A. 2009. Impact of selection and demography on the diffusion of lactase persistence. *PLoS ONE* 4, 6369–79.

Glykou, A. 2010.Technological and typological analysis of Ertebølle and early Funnel Beaker pottery from Neustadt LA 156 and contemporary sites in northern Germany. In B. Vanmontfort, L. Louwe Koojimans, L. Amkreutz and L. Verhart (eds), *Pots, farmers and foragers: pottery traditions and social interaction in the earliest Neolithic of the lower Rhine area*, 177–88. Amsterdam: Amsterdam University/Leiden University.

Glykou, A. 2013. Seal hunting at the Baltic Sea coast – a case study from the late Mesolithic and earliest Neolithic Neustadt in Holstein, Germany. In O. Grimm and U. Schmölcke (eds), *Hunting in northern Europe until 1500 AD – old traditions and regional developments, continental sources and continental influences. Papers presented at a workshop organised by the Centre for Baltic and Scandinavian Archaeology (ZBSA), Schleswig, June 16th and 17th, 2011*, 101–112. Neumünster: Schriften des ArchäologischenLandesmuseums, Ergänzungsreihe, Band 7.

Hansel, F.A., Copley, M.S., Madureira, L.A.S. and Evershed, R.P. 2004. Thermally produced [omega]-(o-alkylphenyl) alkanoic acids provide evidence for the processing of marine products in archaeological pottery vessels. *Tetrahedron Letters* 45, 2999–3002.

Hardy, K., Blakeney, T., Copeland, L., Kirkham, J., Wrangham, R. and Collins, M. 2009. Starch granules, dental calculus and new perspectives on ancient diet. *Journal of Archaeological Science* 36, 248–55.

Harland, R. 1987. *Superstructuralism: the philosophy of structuralism and post-structuralism*. London and New York: Methuen.

Hartz, S., Lübke, H. and Terberger, T. 2007. From fish and seal to sheep and cattle: new research into the process of neolithisation in northern Germany. In A. Whittle and V. Cummings (eds), *Going over: the Mesolithic–Neolithic transition in north-west Europe*, 567–94. London: Oxford University Press for the British Academy.

Henry, A. and Piperno, D. 2008. Using plant microfossils from dental calculus to recover human diet: a case study from Tell al-Raqā'I, Syria. *Journal of Archaeological Science* 35, 1943–50.

Heron, C., Nemeek, N., Bonfield, K.M., Dixon, D. and Ottaway, B.S. 1994.The chemistry of Neolithic beeswax. *Naturwissenschaften* 81, 266–9.

Howcroft, R., Eriksson, G. and Lidén, K. 2012. The Milky Way: the implications of using animal milk products in infant weaning. *Anthropozoologica* 47, 31–43.

Isaksson, S., Karlsson, C. and Eriksson, T. 2010. Ergosterol (5, 7, 22-ergostatrien-3β-ol) as a potential biomarker for alcohol fermentation in lipid residues from prehistoric pottery. *Journal of Archaeological Science* 37, 3263–8.

Jensen, O.L. and Hansen, K.M. 1999. BarnegravenfraNivågård. *Hørsholm Egns Museum* 1998, 7–21.

Klassen, L. 2002. The Ertebølle culture and Neolithic continental Europe: traces of contact and interaction.In A. Fischer and K. Kristiansen (eds), *The Neolithisation of Denmark: 150 years of debate*. Sheffield: J.R. Collis Publications.

Koch, E. 1998. *Neolithic bog pots from Zealand, Møn, Lolland and Falster*. København: Fortidsminder.

Lévi-Strauss, C. 1970. *The raw and the cooked: introduction to a science of mythology I*. London: Cape.

Lévi-Strauss, C. 1993. *Structural anthropology*. London: Penguin.

Littlewood, R. 2004. Commentary: globalisation, culture, body image, and eating disorders. *Culture, Medicine and Psychiatry* 139, 1531–8.

Lupton, D. 1996. *Food, the body and the self.* London: Sage Publications.
Malmström, H., Linderholm, A., Lidèn, K., Srora, J., Molnar, P., Hlomund, G., Jakobsson, M. and Götherström, A. 2010. High frequency of lactose intolerance in a prehistoric hunter-gatherer population in northern Europe. *BMC Evolutionary Biology* 10, 89–95.
Milner, N. 2002. Oysters, cockles, and kitchen middens: changing practices at the Mesolithic/Neolithic transition. In N. Milner and P. Miracle (eds), *Consuming passions and patterns of consumption*, 89–96. Cambridge: McDonald Institute.
Parker Pearson, M. 2003. Food, culture and identity: an introduction and overview. In M. Parker Pearson (ed.), *Food, culture and identity in the Neolithic and Early Bronze Age,* 1–30. Oxford: Archaeopress.
Petersen, E.B. 1990. Nye grave fra Jægerstenalderen. *Nationalmuseets Arbejdsmark* 1990, 19–31.
Petersen, E.B., Alexandersen, V. and Meiklejohn, C. 1993. Vedbæk, graven midtibyen. *Nationalmuseets Arbejdsmark* 1993, 61–9.
Richter, K.K., Wilson, J., Jones, A.K.G., Buckley, M., van Doorn, N. and Collins, M.J. 2011. Fish n' chips: ZooMS peptide mass fingerprinting in a 96 well plate format to identify fish bone fragments. *Journal of Archaeological Science* 38, 1502–10.
Rowley-Conwy, P. 1984. The laziness of the short-distance hunter: the origins of agriculture in western Denmark. *Journal of Anthropological Archaeology* 3, 300–24.
Saul, H., Wilson, J., Heron, C.P., Glykou, A., Hartz, S. and Craig, O. 2012. A systematic approach to the recovery and identification of starches from carbonised deposits on ceramic vessels. *Journal of Archaeological Science* 39, 3483–92.
Saul, H., Madella, M., Fischer, A., Glykou, A., Hartz, S. and Craig, O.E. 2013. Phytoliths in pottery reveal the use of spice in European prehistoric cuisine. *PLoS ONE* 8, e70583.
Scheu, A., Hartz, S., Schmölcke, U., Tresset, A., Burger, J. and Bollongino, R. 2008. Ancient DNA provides no evidence for independent domestication of cattle in Mesolithic Rosenhof, Northern Germany. *Journal of Archaeological Science* 35, 1257–64.
Smith, M. L. 2006.The archaeology of food preferences. *American Anthropologist* 108, 480–93.
Squire, S. 2003. Anorexia and bulimia: purity and danger. *Australian Feminist Studies* 18, 17–26.
Stewart, J.R.M., Allen, R.B., Jones, A.K.G., Penkman, K.E.H. and Collins, M.J. 2013. 'ZooMS': making eggshell visible in the archaeological record. *Journal of Archaeological Science* 40, 1797–1804.
Turnbull, C. 1993.*The forest people.* New York: Touchstone.
Weiner, S. 2010. *Microarchaeology: beyond the visible archaeological record.* Cambridge: Cambridge University Press.
Zvelebil, M. 1994. Plant use in the Mesolithic and its role in the transition to farming. *Proceedings of the Prehistoric Society* 60, 35–74.
Zvelebil, M. 2006. Mobility, contact, and exchange in the Baltic Sea Basin 6000–2000 BC. *Journal of Anthropological Archaeology* 25, 178–92.
Zvelebil, M. and Rowley-Conwy, P. 1984. Transition to farming in northern Europe: a hunter-gatherer perspective. *Norwegian Archaeological Review* 17, 104–28.

12

Life Conditions and Health in Early Farmers

A Global Perspective on Costs and Consequences of a Fundamental Transition

CLARK SPENCER LARSEN

Health in early farmers

Introduction

THERE HAS LONG BEEN AN INTEREST in the shift from a lifeway based exclusively on the hunting, gathering and collecting of food to a modern lifeway in which, with no exception, all food is derived from domesticated sources. This complete transformation in how people acquired food and the kinds of foods eaten began around 10–12,000 years ago, and it took most of that time to get to where we are today – complete dependence on domesticated plants and animals. To many authorities, a time frame of thousands of years represents a long, slow process of domestication and change in diet (e.g. Smith 2006). However, when one considers the fact that humans and human-like ancestors have been around for six to eight million years, the change in food practices is just a drop in the bucket of time. Regardless whether this fundamental transition took a short or long time, all authorities agree that it is a remarkable change, whereby subsistence strategies and economic practices have become so dramatically different from those of our pre-Holocene ancestors. Arguably, this is one of the most important adaptive events in human evolution. It comes as little surprise, then, that there is a pervasive scholarly and scientific interest in the foraging-to-farming transition, asking a central question: What were the costs and benefits for our species once farming was adopted, especially involving the adoption of a subsistence regime based on the production and consumption of domesticated plants from the Poaceae family, the cereals?

The bioarchaeological record of prehistoric human remains best informs our understanding of impacts and outcomes of this transition (Larsen 2006). In this discussion, I present an overview of new findings through which human bio-archaeologists, based on their study of skeletal remains from archaeological settings,

Proceedings of the British Academy **198**, 215–32. © The British Academy 2014.

have contributed to the discussion of early farming. I first present the historical context for this body of research, and then a comparative perspective, drawing upon recent analyses from different settings around the globe.

The context for human bioarchaeology: reconstructing and interpreting the farming transition in prehistory

Human remains are central to any discussion of economic transitions, primarily because it is the skeletons that represent the actual people who experienced the transitions. For several reasons, human remains offer insights into the biological signatures of costs and outcomes. First, human remains provide a record of foods consumed, such as from stable isotope analysis (Pearson and Meskell, this volume, Chapter 13) and inferences about the nutritional value of domesticated plants. Secondly, there is an enormous sample of human skeletal remains representing early farmers and their pre-farming ancestors, collectively on a global scale, regionally with focus on specific kinds of plants consumed, locally with focus on individual communities, and individually with focus on a person's life history. Finally, when viewed in the larger archaeological context reconstructed from various lines of evidence (described in many of the other contributions to this volume), human remains provide insights into social behaviour, such as relating to social inequality, access (or not) to food and other resources, and the lived social experience in diversely structured societies.

Working in different areas of the world, a handful of human bioarchaeologists in the 1970s began to recognise through a series of independent studies, a similar pattern of health and/or lifestyle (activity) outcomes in populations that had made the transition from foraging to farming or from less intensified to more intensified farming. These studies addressed questions and tested hypotheses about the impacts of increased dependence on cereals for health and lifestyle within the larger context of post-Pleistocene human evolution. These investigations provided regional and comparative contexts involving temporal continuity and foraging-to-farming transitions, for example, on the North American Atlantic coast (Larsen 1982, 1984), Midwest (Cook 1984, Goodman *et al.* 1984), the eastern Mediterranean basin and south-west Asia (Angel 1984), the Nile valley (Armelagos *et al.* 1984), and other settings in western Asia, Europe and South America (various in Cohen and Armelagos 1984). The recognition of similar patterns of health outcomes led to the first synthesis of the bioarchaeology of the foraging-to-farming transition, presented in *Paleopathology at the origins of agriculture* (Cohen and Armelagos 1984). The general picture that emerged was the presence of increased morbidity virtually whenever and wherever farming was adopted or intensified. Importantly, although in the minority, some settings revealed no change or even signs of improvement. As a collective, this group of studies supported the hypothesis that the adoption of foodways involving production and consumption of plant carbohydrates and living in sedentary circumstances with more crowded populations occasioned negative outcomes. These findings motivated a series of debates and discussions and generated new ideas and approaches to this important problem.

This newly invigorated bioarchaeology of the foraging-to-farming transition motivated a number of major research programmes, including the History of Health in the Western Hemisphere Project, directed by Richard Steckel and Jerome Rose (Steckel and Rose 2002, Steckel *et al.* 2002). This project focused on health and lifestyle outcomes generally, including the study of the foraging to farming transition. Unlike previous comparisons that relied on a disparate literature using varied methods of analysis, the Western Hemisphere Project involved a large collaboration of researchers who coded a suite of health indicators in the same manner. This collaboration of several dozen human bioarchaeologists, economic and medical historians, and demographers and their study of some 12,500 skeletons covering the last 5000 years revealed a general decline in quality of life with the transition to and intensification of farming, mostly pertaining to the introduction, adoption, and intensification of maize agriculture (see also Steckel *et al.* 2002). Similarly, other large-scale comparative analyses emerged (e.g. Cohen and Crane-Kramer 2007, Pinhasi and Stock 2011, Roberts and Cox 2003) that largely confirmed these patterns of health outcomes, but also pointed to a record of variation strongly suggesting that while there are indications of health decline and lifestyle/activity variation, it is nonetheless a variable record of change. Moreover, this record confirmed earlier findings that while maize farming in the Americas had a profound impact on health as it is measured in human remains, other localities in the world where other grains were produced and consumed also provided circumstances for increased morbidity.

All of the aforementioned work provided a remarkable wealth of skeletal data on morbidity derived from excellent archaeological contexts, allowing for inferences about health changes that accompanied the foraging-to-farming transition and agricultural intensification around the globe. This record for early farmers is now especially strong for eastern North America, but it has considerably expanded for Europe and Asia. This expanded record is important because it allows us to reassess conclusions about general trends in health from earlier analyses begun more than three decades ago (e.g. those from Cohen and Armelagos 1984).

Impact of early farming on the human condition

In the last decades especially, there has been an explosion in research and publication on human remains derived from archaeological contexts, especially in relation to the foraging-to-farming transition. It is now clear that plant domestication occurred in at least ten independent centres in Old World contexts – wheat, barley and rye in west Asia and rice and millet in east Asia – and New World contexts – maize, beans and squash in the Americas (Smith 1995, Staller *et al.* 2006). From primary centres of domestication, the plants and people spread, such that by the later Holocene, full-blown use of and dependence on cultigens had spread throughout Asia, Europe and the Americas.

For purposes of this discussion, seven settings having especially strong documentation of use of domesticated grains, temporal and cultural context, and abundant samples of human skeletal remains with clearly described data on paleopathology and health indicators are presented (Table 12.1). This represents just a sampling of results presented in the literature on the links between health outcomes and farming. The completeness of the archaeological and human bioarchaeological data sets makes these settings ideally suited for comparative purposes. All seven settings have a common profile of nutritional quality of domesticated cereals:

1. All are deficient in one or more essential amino acids (lysine for maize, millet and wheat);
2. None of the common domesticated cereals have adequate iron;
3. All are deficient in one or more vitamins: B_1 (thiamine), B_2 (riboflavin), C (ascorbic acid);
4. All have links with malnutrition, immunosuppression and susceptibility to a variety of pathogens rendering the individual prone to infection;
5. All are carbohydrates, creating circumstances that promote a cariogenic oral environment and dental caries, oral infection and susceptibility to degenerative conditions in later life (e.g. cardiovascular disease and diabetes).

Table 12.1 Comparative bioarchaeological investigations of the foraging to farming transition

Location	*Setting*	*Domesticated plant(s)*
West Asia	Catalhoyuk, Turkey[1]	wheat, barley, rye
	Near East/Levant[2]	wheat, barley, rye
Europe	Britain[3]	wheat, barley, rye
	Scandinavia[4]	wheat, barley, rye
East Asia	China (Central Plains)[5]	millet
	Thailand[6]	rice
North America	Georgia Bight[7]	maize

Sources: [1]Boz (2005), Hillson *et al.* (2013); [2]Eshed *et al.* (2006); [3]Roberts and Cox (2007); [4]Bennike and Alexandersen (2007); [5]Pechenkina *et al.* (2007, 2012, 2013); [6]Douglas and Pietrusewsky (2007); [7]Larsen *et al.* (2001, 2007).

The consumption of domesticated cereals as a focus of diet predicts poor health

Given the above nutritional circumstances surrounding consumption of cereal grains, it is possible to predict specific patterns of health outcomes, especially with respect to oral health, skeletal health, and the impact on growth and development. In this regard, this overview considers evidence from oral health, including dental caries (tooth decay) and ante-mortem tooth loss. Dental caries is a disease process characterised by the focal demineralisation of dental hard tissues (enamel and other constituents of tooth crowns and roots; Figure 12.1). Ante-mortem tooth loss is the loss of teeth while the person is living. Under circumstances of no or minimal oral hygiene and consumption of soft, sticky carbohydrates, the tissues surrounding the teeth harbour micro-organisms that promote caries and periodontal disease, resulting in tooth decay and loss during the person's lifetime (Figure 12.2).

While the foods consumed by members of a population contributed to poor oral health in a direct way, there are also indirect ways in which foods consumed

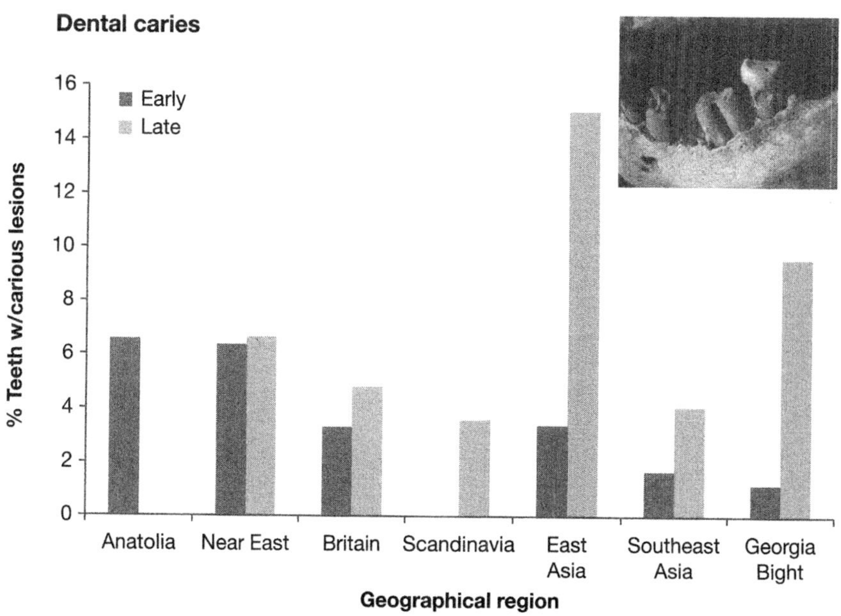

Figure 12.1 Dental caries prevalence in settings in east Asia, Europe, west Asia, and North America (data from Boz 2005; Hillson, Larsen *et al.* 2013; Eshed *et al.* 2006; Roberts and Cox 2003, 2007; Bennike and Alexandersen 2007; Pechenkina *et al.* 2002, 2012, 2013; Douglas and Pietrusewsky 2007; Larsen 1984; Larsen *et al.* 2007). Settings compare late foragers versus early farmers or earlier farmers versus later farmers; therefore 'Early' refers to either late foragers or early farmers; 'Late' refers to earlier farmers or later farmers (see text for discussion) (image courtesy of Leslie Lea Williams).

will affect health. Although not always the case, there is a strong tendency for farmers to live in more permanent settlements than foragers, and population size among farmers tends to be larger and more densely crowded than among foragers. These conditions promote the spread of infectious disease as well as reduce cleanliness. Under these living conditions, wounds and abrasions are far more prone to infection, which can spread to bone, resulting in osteoperiostitis or an inflammatory response generally. It is also the case that osteoperiostitis can be caused by a trauma to the subperiosteum, resulting in proliferative response (Weston 2012). However, given the living conditions involving sedentism and crowding, it seems most likely that much of the bony response observed in past societies was due to infectious disease and/or localised bone infection originating in the surrounding soft tissue.

Under circumstances of poor diets, growth and development are attenuated during key developmental periods, such as in the first several years of life and during the adolescent growth spurt. In the absence or reduction of key nutrients, including consumption of a poor protein diet (domesticated cereals), growth is reduced, leading to short stature in adulthood (Bogin 1999, Bogin and Keep 1999). In

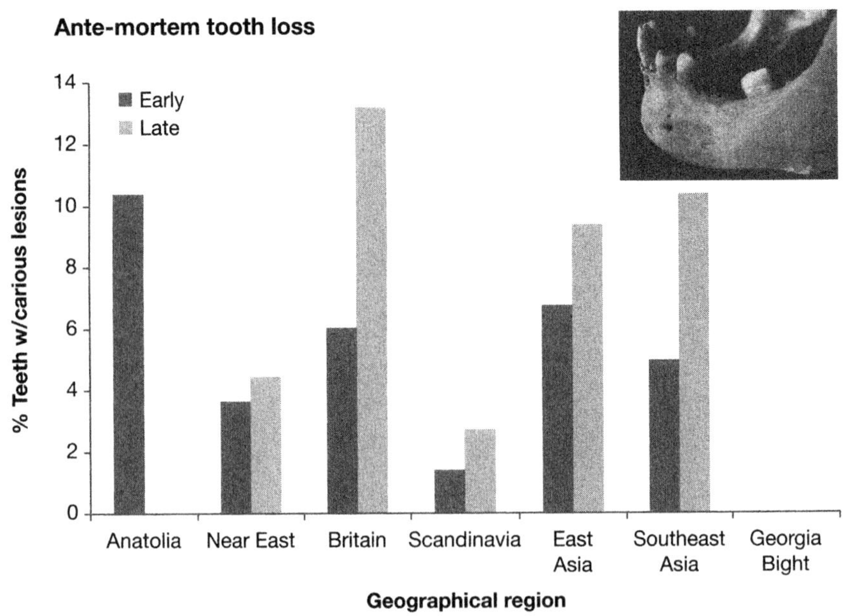

Figure 12.2 Ante-mortem tooth loss prevalence in settings in east Asia, Europe, west Asia, and North America (data from Boz 2005; Hillson *et al.* 2013; Eshed *et al.* 2006; Roberts and Cox 2003, 2007; Bennike and Alexandersen 2007; Pechenkina *et al.* 2002, 2012, 2013; Douglas and Pietrusewsky 2007). Settings compare late foragers versus early farmers or earlier farmers versus later farmers; therefore 'Early' refers to either late foragers or early farmers; 'Late' refers to earlier farmers or later farmers (see text for discussion; no data available for the Georgia Bight; image courtesy of Leslie Lea Williams).

particular, four expectations emerge with regard to the shift from foraging to farming: (1) increase in dental caries, (2) increase in ante-mortem tooth loss, (3) increase in osteoperiostitis, and (4) growth disruption leading to reduced adult height. Although not a complete one by any means, these four attributes provide a profile of negative health outcomes and consequences of the foraging-to-farming transition.

The bioarchaeological record matches the expectation of health decline with farming

The results of this comparative analysis are drawn from the published literature and are presented in Figures 12.1–12.4, representing respectively, prevalence changes of dental caries, ante-mortem tooth loss, osteoperiostitis, and reduction in adult stature. This record focuses on regions that adopted or intensified their use of domesticated crops: wheat, barley and rye (Neolithic west Asia and Europe), rice and millet (Neolithic south-east and east Asia), and maize (North America). Specifically, early and late comparisons (foraging versus farming) are presented from the Near East, Scandinavia and the Georgia Bight, whereas intensification of farming is documented via data sets from Britain and east and south-east Asia (China, Thailand). The pattern is reversed for the Anatolian site of Çatalhöyük, where a period of intensified farming and heavy dependence on domesticated plants was followed by population decline, dispersal, and ultimately, abandonment of the site.

Virtually all temporal comparisons reveal trends consistent with the expectations described above. The results from these regions reveal a general trend of increasing frequency of dental caries (Figure 12.1). The increase is only slight in the earliest farmers in the Near East (Levant) setting, where Natufian foragers are compared with early farmers. This only marginal increase suggests that the earliest farmers were incipient farmers, deriving significant resources from collected, hunted, or gathered foods in addition to domesticated cereals (Eshed *et al.* 2006). The prevalence is similar to Anatolia (Çatalhöyük), where caries prevalence is relatively low, but certainly high enough to reflect a diet that includes a significant presence of cultigens (Hillson *et al.* 2013). The other data sets reveal significant increase in dental caries, showing the increased commitment to cultigens (wheat, barley and rye in Britain and Scandinavia, millet in north China, rice in Thailand, and maize on the US Atlantic coast). These comparisons likely conceal the presence of considerable variation in especially east and south-east Asia. That is, the increases in Thailand are primarily associated with Non Nok Tha where Douglas and Pietrusewsky (2007) report an increase in caries with intensification of rice agriculture (and see Temple and Larsen 2007, for Japan), whereas others working in Thailand and Vietnam report no or little increase in caries (Domett and Tayles

2007, Oxenham 2006, Tayles *et al.* 2000). Moreover, Oxenham and Domett (2010) report a high prevalence of dental caries (11%) in a Neolithic population in Vietnam (Man Bac) where there is no rice agriculture. Oxenham (2000, 2006) suggests the possibility that low prevalence of caries in some of these settings may be due to local populations raising rice for export to China, and not for local consumption. Furthermore, Tayles and co-workers argue that rice although a starchy carbohydrate is not especially cariogenic but that there is 'no reason why one dietary component would be solely responsible for the development of caries or why all carbohydrates should have the same effect' (Tayles *et al.* 2009, 162). Thus, rice agriculture may not have led to increased cariogenesis and declines in oral health, at least not on the same scale as the changes observed in populations subsisting on other domesticated staple crops. Indeed, sucrose, a simple sugar that is favoured by caries-causing oral bacteria (e.g. *Streptococcus mutans*) is highest in barley and rye, closely followed by millet and maize, and relatively low in milled rice (Shelton and Lee 2000). What may be more important than its starch content in interpreting the cariogenecity of rice is the manner in which it is prepared for consumption. In past Asian societies where rice figured prominently in diet, it was cooked and eaten whole rather than milled into a flour and prepared into a soft, sticky substance, such as a bread. Certainly, the softer consistency and texture of bread rendered it far more cariogenic than whole rice.

Despite these variable results, the overwhelming record presented here and in the literature in general for both living and past populations makes a very strong case that the consumption of starchy carbohydrates does indeed contribute to increased caries prevalence. While rice may not have been as cariogenic as other starches, when viewed as a whole, each of the regions discussed in this chapter show broadly a reduction in oral health in early farmers, including in east and south-east Asia, where rice figured so prominently in past diets (various in Oxenham 2006, Pechenkina and Oxenham 2013).

Comparisons for ante-mortem tooth loss, a condition caused by a poor oral environment, usually involving periodontal disease, reveals trends for every setting that are similar to dental caries (Figure 12.2). This similarity in pattern (and see Larsen 1997) suggests a common set of factors in the etiology of both dental caries and ante-mortem tooth loss, namely elevated levels of oral bacteria leading to decay and periodontal disease. In combination with dental caries, the oral pathology reflecting declining circumstances and the shift to carbohydrates, the farming transition profoundly affected health, especially related to diseased teeth and elevated loss of teeth prior to death.

The record presented here suggests that farming was also deleterious for skeletal health, but at a considerably lower level. That is, most of the osteoperiostitis prevalences are low (e.g. south-east Asia, Britain, Near East), but all show slight to moderate increases with the transition to and intensification of farming (Figure 12.3). The highest prevalence in this comparison is reported for Çatalhöyük in

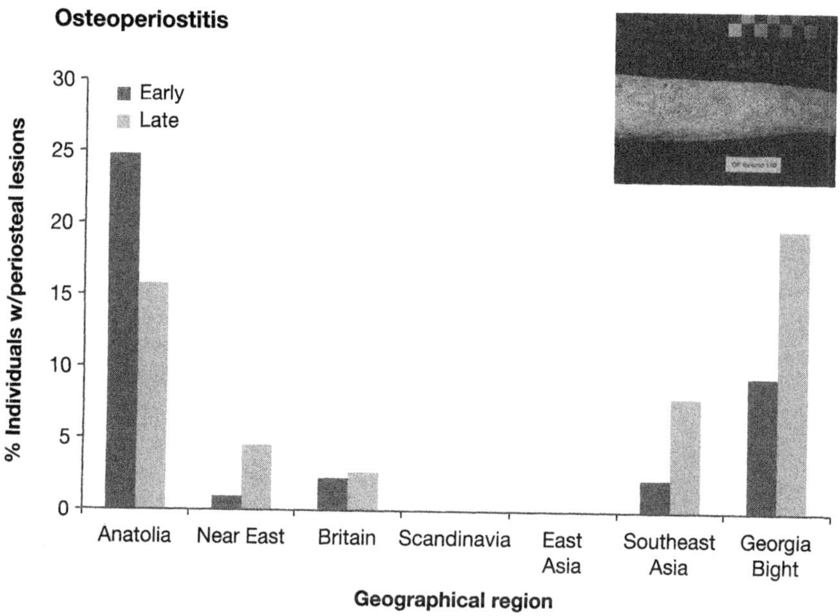

Figure 12.3 Osteoperiostitis prevalence in settings in east Asia, Europe, west Asia, and North America (data from Boz 2005; Hillson *et al.* 2013; Eshed *et al.* 2006; Roberts and Cox 2003, 2007; Douglas and Pietrusewsky 2007; Larsen 1984; Larsen *et al.* 2007). Settings compare late foragers versus early farmers or earlier farmers versus later farmers; therefore 'Early' refers to either late foragers or early farmers; 'Late' refers to earlier farmers or later farmers (see text for discussion; no data available for Scandinavia and east Asia; image courtesy of Leslie Lea Williams).

south-central Anatolia. Overall, the setting reveals a relatively high prevalence of osteoperiostitis, which likely reflects the very close, crowded living conditions (Hillson *et al.* 2013). It is also one of the few settings globally showing a decline in prevalence, which may be related to a decline in population, interpreted to be associated with increasing aridity and resource stress in the latest phase of the site's occupation (Hillson *et al.* 2013; cf. Temple and Larsen 2013). The findings are consistent with the record in general for osteoperiostitis in a variety of settings, largely reflecting increased sedentism, increased population size and density, and the declining circumstances revolving around poor hygiene and exposure to pathogens (Larsen 1995, 2006).

Stature is the only variable discussed in this chapter that does not follow a consistent pattern of temporal change (Figure 12.4). On the one hand, in accordance with expectations associated with a shift to a nutritionally deficient diet and broadly poor living conditions, there are settings showing a decrease in stature, including in the Near East, east Asia, south-east Asia (females only), and the Georgia Bight. Overall, these trends mirror stature change in historical and recent societies during times of hardship and deprivations (e.g. Bogin and Keep 1999, Kemkes-Grottenthaler 2005, Steegmann 1985). In contrast to expectations, however, are the increases in stature for Anatolia, Britain and south-east Asia (males only). The increase in stature in these settings may reflect an improvement in living conditions. For example, at Çatalhöyük, the decrease in population size and crowding in the final period of the occupation of the community may have contributed to improved living standards, as also reflected in the decreasing frequency of osteoperiostitis noted above. In addition, $\delta^{15}N$ values, a measure of meat consumption, increase in this setting, suggesting increased access to animal sources of protein (Hillson *et al.* 2013, Pearson 2013). In south-east Asia, a number of settings show increased stature (e.g. Non Nok Tha, Khok Phanom Di, Nong Nor, Noen U-Loke; Pechenkina *et al.* 2013, Domett 2001, Domett and Tayles 2007, Douglas and Pietrusewsky 2007). Domett and Tayles (2007) suggest that health conditions improved in this setting perhaps owing to the manner in which rice was prepared for consumption. Rice in prehistoric contexts was unpolished and home-pounded with higher nutritional values than today's processed, boiled rice. Similarly, body size increases are also documented in later Yoyoi populations compared to earlier Jomon populations in Japan (Okazaki 2013, Pechenkina *et al.* 2013), but various authorities argue that this increase in stature reflects possibly evolutionary circumstances whereby later taller populations emigrated from mainland China. In summary, the

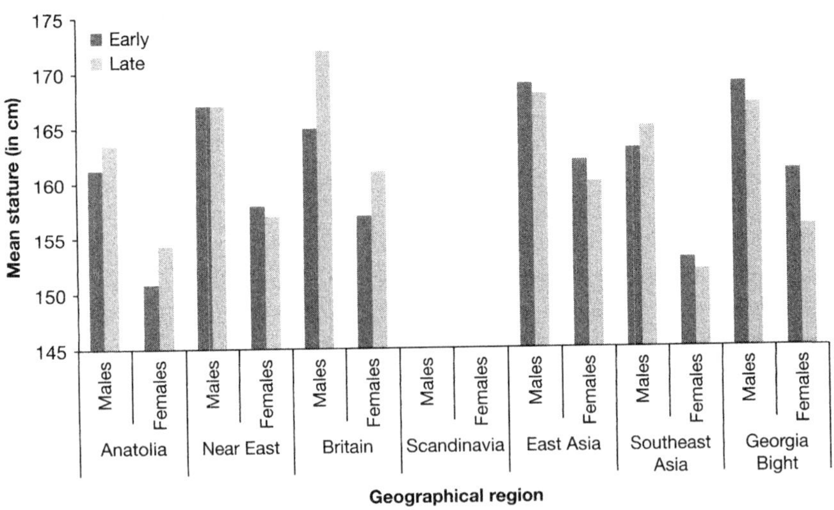

Figure 12.4 Adult height (stature) estimates from settings in east Asia, Europe, west Asia, and North America (data from Boz 2005; Hillson *et al.* 2013; Roberts and Cox 2003, 2007; Pechenkina *et al.* 2002, 2012, 2013; Douglas and Pietrusewsky 2007; Larsen 1984; Larsen *et al.* 2007; Smith *et al.* 1984; Smith and Horwitz 2007). Each setting compares either late foragers versus early farmers or earlier farmers versus later farmers; therefore 'Early' refers to either late foragers or early farmers; 'Late' refers to earlier farmers or later farmers (see text for discussion; no data available for Scandinavia).

foraging-to-farming transition shows a general pattern of stature reduction (Mummert *et al.* 2011), but stature reductions identified in archaeological contexts are not universal as various settings globally show no change or increased stature (see also various in Cohen and Armelagos 1984, Steckel and Rose 2002).

Assessment of the chronology for each of these settings reveals that these changes in health indicators were relatively rapid, within scores to hundreds of years, suggesting that once agriculture was adopted it became the fundamental driver of diet and subsistence practices. The human bioarchaeological record discussed here also reveals a consistent pattern of change in indicators that collectively serve to characterise a pattern of declining health with the shift to farming or increased intensification of farming. Importantly, it is this record that serves as the context for understanding later health changes that accompanied further economic intensification.

In order to document health trends, a collaboration of 75 European, Canadian, and American biological anthropologists began a long-term data collection and analysis project, focusing on data previously collected and newly collected data. This European History of Health Project was an extension of the History of Health in the Western Hemisphere Project, producing new data for approximately 17,500 skeletons across Europe and dating from Classical Antiquity (*c.* 800 cal BC–AD 500) to the Industrial/Modern era (*c.* AD 1800–1950; Steckel *et al.* 2009). The analysis suggests that at the early end of the temporal series caries, ante-mortem tooth loss and osteoperiostitis in Europe remained at prevalence levels comparable to the early farmers described in this chapter until the Late Middle Ages when there was a marked increase in oral and skeletal infection, especially in females (Larsen *et al.* 2012, Wittwer-Backofen *et al.* 2009, Figure 12.5). Thus, the path of declining oral and skeletal health was set into motion from the earliest adoption of agriculture, but with highly intensified farming and increased emphasis on domesticated grains, there was a dramatic spike in poor oral health.

Of considerable importance is the implication that this change has for other health parameters, especially degenerative conditions such as cardiovascular disease. Periodontal disease is a tissue-destructive disease leading to degradation of the attachment tissues of teeth, resulting in tooth loss. A large and growing body of clinical and epidemiological research has examined the relationship between poor oral environments and general health, especially in relation to the links between poor oral health, adult mortality, and chronic, systemic health conditions such as cardiovascular disease, respiratory infections, and other degenerative diseases (Buhlin *et al.* 2003, Irwin *et al.* 2008, Johnson *et al.* 2006, Koren *et al.* 2011, Slots 2004, Williams *et al.* 2008). Periodontal disease and cardiovascular disease show the clearest association with poor oral health (e.g. DeStefano *et al.* 1993, Li *et al.* 2000). It may be the case that poor oral health beginning early in life is the root cause of systemic infection and attendant inflammatory response, which may in turn promote chronic inflammation and

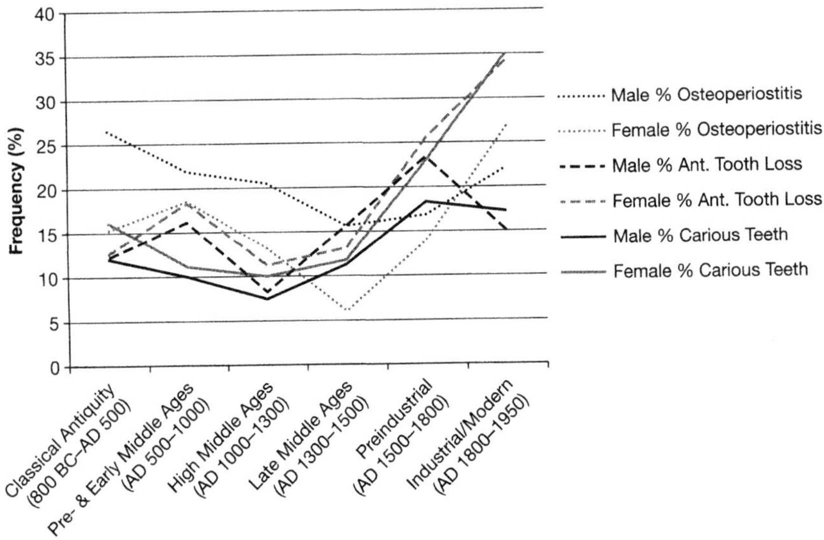

Figure 12.5 Temporal trends in prevalence of dental caries, ante-mortem tooth loss, and osteoperiostitis in Classical Antiquity through Industrial/Modern periods in Europe (Larsen *et al.* 2012, Wittwer-Backofen *et al.* 2009).

cardiovascular disease later in life. The oral environment is a clear target for developing an increased understanding of adult systemic conditions both today and in the past, as poor oral health is a risk factor for increased mortality linked to cardiovascular conditions, and pulmonary disease, as well as other degenerative conditions and early death.

The finding of links between oral and systemic health has strong implications for interpreting general health in past populations, especially in regard to insights into the potential relationship between risk of death and poor oral environment as expressed by elevated prevalence of periodontal disease (alveolar bone resorption and ante-mortem tooth loss) and dental caries. The links between inflammation associated with oral health and elevated mortality from cardiovascular disease in living populations suggest the possibility of links between oral health status and mortality in past populations. Indeed, the link between poor oral health and increased risk of death (based on mortality analysis) is clearly illustrated in a large series of remains from the medieval St. Mary Graces cemetery in London (DeWitte and Bekvalac 2010). In this setting, there is a significant relationship between oral health and mortality; individuals with dental caries and periodontal disease were at greater risk of dying than those without these conditions. Similarly, individuals having poor oral health were far more likely to have osteoperiostitis (DeWitte and Bekvalac 2010). While it is speculative to say that the origins of cardiovascular disease can be traced to the Neolithic and the adoption of a subsistence and dietary regime based at least in part on farming, it is an important implication for the history of human health reaching far into the past and informing the present.

Conclusions

Beginning some 10–12,000 years ago, human beings underwent a process of dietary change involving the shift from foraging to farming. This process of domestication has culminated in today's experience where virtually every human on the planet is dependent to one extent or another on domesticated plants. For the earliest farmers and continuing to the present, this has meant a general decline in nutritional quality, leading to a decrease in oral and general health as measured by dental caries, antemortem tooth loss, osteoperiostitis, and altered growth patterns, sometimes with reduced height and body size. The outcomes for health and well-being were not just a North American phenomenon relating to maize agriculture. Rather, this shift and its associated negative health outcomes were part of a broader global event. The ubiquitous presence of degenerative conditions (e.g. cardiovascular disease) across modern populations worldwide in today's world may be related to declines in oral and systemic health set into motion by the shift from foraging to farming in the ancient past.

Acknowledgements

The research presented in this chapter reflects the results of collaborations with many individuals on a range of projects involving the study of human remains from archaeological contexts. I especially acknowledge collaborations with David Thomas, Mark Cohen, George Armelagos, Ian Hodder, Simon Hillson, Richard Steckel, Jerry Rose, Dale Hutchinson, Phillip Walker, Charlotte Roberts, Kim Williams, Christopher Ruff, Scott Simpson, Patricia O'Brien, Robert Kelly, Giuseppe Vercellotti, and Gino Fornaciari. I acknowledge the generous support of the US National Science Foundation, the National Geographic Society, the Edward John Noble Foundation, and the St. Catherines Island Foundation for the support of research programmes. Joshua Sadvari provided useful comments on an earlier draft.

References

Angel, J.L. 1984. Health as a crucial factor in the changes from hunting to developed farming in the eastern Mediterranean. In M.N. Cohen and G.J. Armelagos (eds), *Paleopathology at the origins of agriculture*, 51–73. Orlando: Academic Press.

Armelagos, G.J., Van Gerven, D.P., Martin, D.L. and Huss-Ashmore, R. 1984. Effects of nutritional change on the skeletal biology of northeast African (Sudanese Nubian) populations. In J.D. Clark and S.A. Brandt (eds), *From hunters to farmers: the cause and consequences of food production in Africa*, 132–46. Berkeley: University of California Press.

Bennike, P. and Alexandersen, V. 2007. Population plasticity in southern Scandinavia: from oysters and fish to gruel and meat. In M.N. Cohen and G.M.M. Crane-Kramer (eds), *Ancient health: skeletal indicators of agricultural and economic intensification*, 130–48. Gainesville: University Press of Florida.

Bogin, B. 1999. *Patterns of human growth*. Cambridge: Cambridge University Press.

Bogin, B. and Keep, R. 1999. Eight thousand years of economic and political history in Latin America revealed by anthropometry. *Annals of Human Biology* 26, 333–51.

Boz, B. 2005. The oral health of Çatalhöyük Neolithic people. In I. Hodder (ed.), *Inhabiting Çatalhöyük: reports from the 1995–99 seasons*, 587–92. Cambridge: McDonald Institute Monographs.

Buhlin, K., Gustafsson, A., Pockley, A.G., Frostegard, J. and Klinge, B. 2003. Risk factors for cardiovascular disease in patients with periodontitis. *European Heart Journal* 24, 2009–2107.

Cohen, M.N. and Armelagos, G.J. (eds) 1984. *Paleopathology at the origins of agriculture*. Orlando: Academic Press.

Cohen, M.N. and Crane-Kramer, G.M.M. (eds) 2007. *Ancient health: skeletal indicators of agricultural and economic intensification*. Gainesville: University Press of Florida.

Cook, D.C. 1984. Subsistence and health in the lower Illinois valley: osteological evidence. In M.N. Cohen and G.J. Armelagos (eds), *Paleopathology at the origins of agriculture*, 235–69. Orlando: Academic Press.

DeStefano, F., Anda, R.F., Kahn, H.S., Williamson, D.F. and Russell, C.M. 1993. Dental disease and risk of coronary heart disease and mortality. *British Medical Journal* 306, 688–91.

DeWitte, S.N. and Bekvalac, J. 2010. Oral health and frailty in the medieval English cemetery of St. Mary Graces. *American Journal of Physical Anthropology* 142, 341–54.

Domett, K.M. 2001. *Health in late Prehistoric Thailand*. Oxford: Archaeopress.

Domett, K.M. and Tayles, N. 2007. Population health from the Bronze to the Iron Age in the Mun River valley, Northeastern Thailand. In M.N. Cohen and G.M.M. Crane-Kramer (eds), *Ancient health: skeletal indicators of agricultural and economic intensification*, 286–99. Gainesville: University Press of Florida.

Douglas, M.T. and Pietrusewsky, M. 2007. Biological consequences of sedentism: agricultural intensification in north-eastern Thailand. In M.N. Cohen and G.M.M. Crane-Kramer (eds), *Ancient health: skeletal indicators of agricultural and economic intensification*, 300–19. Gainesville: University Press of Florida.

Eshed, V. Gopher, A. and Hershkovitz, I. 2006. Tooth wear and dental pathology at the advent of agriculture: new evidence from the Levant. *American Journal of Physical Anthropology* 130, 145–59.

Eshed, V., Gopher, A., Pinhasi, R. and Hershkovitz, I. 2010. Paleopathology and the origin of agriculture in the Levant. *American Journal of Physical Anthropology* 143, 121–33.

Goodman, A.H., Lallo, J., Armelagos, G.J. and Rose, J.C. 1984. Health changes at Dickson Mounds, Illinois (A.D. 950–1300). In M.N. Cohen and G.J. Armelagos (eds), *Paleopathology at the origins of agriculture*, 271–305. Orlando: Academic Press.

Hillson, S.W., Larsen, C.S., Boz, B., Pilloud, M.A., Sadvari, J.W., Agarwal, S.C., Glencross, B., Beauchesne, P., Pearson, J., Ruff, C.B., Garofalo, E.M., Hager, L. and Haddow, S.D. 2013. The human remains I: Interpreting community structure, health, and diet in Neolithic Çatalhöyük. In I. Hodder (ed.), *Humans and landscapes of Çatalhöyük*, 339–96. Los Angeles: Cotsen Institute of Archaeology Press.

Irwin, C., Mullally, B., Ziada, H. and Allen, E. 2008. Periodontics: 9. Periodontitis and systemic conditions – is there a link? *Dental Update* 35, 92–4, 97–8, 100–1.

Johnson, N.W., Glick, M. and Mbuguye, T.N. 2006. Oral health and general health. *Advances in Dental Research* 19, 118–21.

Kemkes-Grottenthaler, A. 2005. The short die young: the interrelationship between stature and longevity – evidence from skeletal remains. *American Journal of Physical Anthropology* 128, 340–7.

Koren, O., Spor, A., Felin, J., Fak, F., Stombaugh, J., Tremaroli, V., Behre, C.J., Knight, R., Fagerberg, B., Ley, R.E. and Backhed, F. 2011. Human oral, gut, and plaque microbiota in patients with atherosclerosis. *Proceedings of the National Academy of Sciences* 108 (Supplement 1), 4592–8.

Larsen, C.S. 1982. The anthropology of St. Catherines Island: 3. Prehistoric human biological adaptation. *Anthropological Papers of the American Museum of Natural History* 57, 157–276.

Larsen, C.S. 1984. Health and disease in prehistoric Georgia: the transition to agriculture. In M.N. Cohen and G.J. Armelagos (eds), *Paleopathology at the origins of agriculture*, 367–92. Orlando: Academic Press.

Larsen, C.S. 1995. Biological changes in human populations with agriculture. *Annual Review of Anthropology* 24, 185–213.

Larsen, C.S. 1997. *Bioarchaeology: interpreting behavior from the human skeleton*. Cambridge: Cambridge University Press.

Larsen, C.S. 2006. The agricultural revolution as environmental catastrophe: implications for health and lifestyle in the Holocene. *Quaternary International* 150, 12–20.

Larsen, C.S., Hutchinson, D.L., Stojanowski, C.M., Williamson, M.A., Griffin, M.C., Simpson, S.W., Ruff, C.B., Schoeninger, M.J., Norr, L., Teaford, M.F., Driscoll, E.M., Schmidt, C.W. and Tung, T.A. 2007. Health and lifestyle in Georgia and Florida: Agricultural origins and intensification in regional perspective. In M.N. Cohen and G.M.M. Crane-Kramer (eds), *Ancient health: skeletal indicators of agricultural and economic intensification*, 20–34. Gainesville: University Press of Florida.

Larsen, C.S., Wittwer-Backofen, U., Steckel, R.H. and Williams, K.D. 2012. The global history of dental caries and links with systemic health. Paper presented, *American Association for the Advancement of Science, Vancouver, British Columbia.*

Li, X., Kolltveit, K.M., Tronstad, L. and Olsen, I. 2000. Systemic diseases caused by oral infection. *Clinical Microbiology Reviews* 13, 547–58.

Mummert, A., Esche, E., Robinson, J. and Armelagos, G.J. 2011. Stature and robusticity during the agricultural transition: evidence from the bioarchaeological record. *Economics and Human Biology* 9, 284–301.

Okazaki, K. 2013. Change in the linear growth of long bones with the adoption of wet-rice agriculture in Japan. In E. Pechenkina and M. Oxenham (eds), *Bioarchaeological perspectives on migration and health in ancient east Asia*, 368–98. Gainesville: University Press of Florida.

Oxenham, M.F. 2000. *Health and behaviour during the mid-Holocene and Metal Period in northern Viet Nam*. Unpublished PhD dissertation, Northern Territory University, Darwin.

Oxenham, M. 2006. Biological responses to change in prehistoric Viet Nam. *Asian Perspectives* 45, 212–39.

Oxenham, M.F. and Domett, K.M. 2010. Palaeohealth at Man Bac, in Mac Bac: the excavation of a Neolithic site. In M.F. Oxenham, H. Matsumura and N.K. Dung (eds), *Northern Vietnam: the biology*, 77–93. Canberra: ANU E Press.

Oxenham, M. and Tayles, N. (eds) 2006. *Bioarchaeology of southeast Asia*. Cambridge: Cambridge University Press.

Pearson, J.A. 2013. Human and animal diet as evidenced by stable carbon and nitrogen isotope analysis. In I. Hodder (ed.), *Humans and Landscapes of Çatalhöyük*, 271–98. Los Angeles: Cotsen Institute of Archaeology.

Pechenkina. E. and Oxenham, M. (eds) 2013. *Bioarchaeological perspectives on migration and health in ancient east Asia*. Gainesville: University Press of Florida.

Pechenkina, E., Benfer, R.A., Jr. and Wang, Z. 2002. Diet and health changes at the end of the Chinese Neolithic: the Yangshao/Longshan transition in Shaanxi province. *American Journal of Physical Anthropology* 117, 15–36.

Pechenkina, E., Fan, W. and Ma, X. 2012. Male-female oral health differences in early Chinese communities. *American Journal of Physical Anthropology* 147 (Supplement 54), 235.

Pechenkina, E., Ma, X. and Fan, W. 2013. Trajectories of health in early farming communities of east Asia. In E. Pechenkina and M. Oxenham (eds), *Bioarchaeological perspectives on migration and health in ancient east Asia*, 288–322. Gainesville: University Press of Florida.

Pechenkina, E.A., Benfer, R.A., Jr. and Ma, X. 2007. Diet and health in the Neolithic of the Wei and Middle Yellow River basins, northern China. In M.N. Cohen and G.M.M. Crane-Kramer (eds), *Ancient health: skeletal indicators of agricultural and economic intensification*, 255–72. Gainesville: University Press of Florida.

Pinhasi, R. and Stock, J.T. (eds) 2011. *Human bioarchaeology of the transition to agriculture*. Chichester: Wiley-Blackwell.

Roberts, C. and Cox, M. 2003. *Health and disease in Britain: from prehistory to the present day*. Phoenix Mill: Sutton Publishing.

Roberts, C. and Cox, M. 2007. The impact of economic intensification and social complexity on human health in Britain from 6000 BP (Neolithic) and the introduction of farming to the mid-nineteenth century A.D. In M.N. Cohen and G.M.M. Crane-Kramer (eds), *Ancient health: skeletal indicators of agricultural and economic intensification*, 149–63. Gainesville: University Press of Florida.

Shelton, D.R. and Lee, W.J. 2000. Cereal carbohydrates. In K. Kulp and J.G. Ponte (eds), *Handbook of cereal science and technology*, 385–416. New York: Marcel Dekker.

Slots, J. 2004. Update on human cytomegalovirus in destructive periodontal disease. *Oral Microbiology and Immunology* 19, 217–23.

Smith, B.D. 1995. *The emergence of agriculture*. New York: Scientific American Library.

Smith, B.D. 2006. Documenting domesticated plants in the archaeological record. In M.A. Zeder, D.G. Bradley, E. Emshwiller and B.D. Smith (eds), *Documenting domestication: new genetic and archaeological paradigms*, 15–24. Berkeley: University of California Press.

Smith, P. and Horwitz, L.K. 2007. Ancestors and inheritors: a bio-cultural perspective on the transition to agro-pastoralism in the southern Levant. In M.N. Cohen and G.M.M. Crane-Kramer (eds), *Ancient health: skeletal indicators of agricultural and economic intensification*, 207–22. Gainesville: University Press of Florida.

Smith, P., Bar-Yosef, O. and Sillen, A. 1984. Archaeological and skeletal evidence for dietary change during the late Pleistocene/Early Holocene in the Levant. In M.N. Cohen and G.J. Armelagos (eds), *Paleopathology at the origins of agriculture*, 101–36. Orlando: Academic Press.

Staller, J., Tykot, R. and Benz, B. (eds) 2006. *Histories of maize: multidisciplinary approaches to the prehistory, linguistics, biogeography, domestication and evolution of maize*. Burlington: Academic Press.

Steckel, R.H. and Rose, J.C. (eds) 2002. *The backbone of history: long-term trends in health and nutrition in the Americas*. New York: Cambridge University Press.

Steckel, R.H., Rose, J.C., Larsen, C.S. and Walker, P.L. 2002. Skeletal health in the western hemisphere from 4000 BC to the present. *Evolutionary Anthropology* 11, 142–55.

Steckel, R.H., Larsen, C.S., Roberts, C. and Walker, P.L. 2009. *Reconstructing health and disease in Europe: the Middle Ages through the Industrial Period*. Symposium organised for the annual meetings of the American Association of Physical Anthropologists, Chicago, Illinois.

Steegmann, A.T., Jr. 1985. 18th century British military stature: growth cessation, selective recruiting, secular trends, nutrition at birth, cold, and occupation. *Human Biology* 57, 77–95.

Tayles, N., Domett, K. and Halcrow, S. 2009. Can dental caries be interpreted as evidence of farming? The Asian perspective. *Frontiers of Oral Biology* 13, 162–6.

Tayles, N., Domett, K. and Nelsen, K. 2000. Agriculture and dental caries? The case of rice in prehistoric south-east Asia. *World Archaeology* 32, 68–83.

Temple, D.H. and Larsen, C.S. 2007. Dental caries prevalence as evidence for agriculture and subsistence variation among the prehistoric Yayoi of Japan: biocultural interpretations of an economy in transition. *American Journal of Physical Anthropology* 134, 501–12.

Temple, D.H. and Larsen, C.S. 2013. Bioarchaeological perspectives on systemic stress during the agricultural transition in prehistoric Japan. In E. Pechenkina and M. Oxenham (eds), *Bioarchaeological perspectives on migration and health in ancient east Asia*, 344–67. Gainesville: University Press of Florida.

Weston, D.A. 2012. Nonspecific infection in paleopathology: interpreting periosteal reactions. In A. Grauer (ed.), *A companion to paleopathology*, 492–512. Chichester: Wiley-Blackwell.

Williams, R.C., Barnett, A.H., Claffey, N., Davis, M., Gadsby, R., Kellett, M., Lip, G.Y. and Thackray, S. 2008. The potential impact of periodontal disease on general health: a consensus view. *Current Medical Research and Opinion* 24, 1635–43.

Wittwer-Backofen, U., Coppa, A., Larsen, C.S., Steckel, R.H., Walker, P.L., Blondiaux, Grupe, G., Jankauskas, R., Maat, G., McGlynn, G., Papathanasiou, A. Roberts, C., Teschler-Nicola, M., Agnew, A., Assis, S., Bereczki, Z., Bertrand, B., Betsinger, T.K., Binder, M., Boulter, S., Bourbou, C., Boylston, A., Brickley, M., Burli, L., Cooper, C., Coughlan, J., Dropzd, A., During, E., Elikopoulos, C., Eng, J. Engel, F., Fox, S., Furtado, M., Guntis, G., Groves, S., Harkins, K., Holck, P., Holst, M., Hotz, G., Ives, R., Jakob, T., Jennings, J., Justus, H., Kaminska, K., Kjellström, A., Knüsel, C., Kozlowski, T.,

Lagia, A., Lopes, C., Manolis, S., Marcsik, A., Marques, C., Moenke, C., Niel, C., Novak, S.A., Novotny, F., Peck, J., Potiekhina, I., Rega, B., Richman, R., Rijpma, F., Rose, J., Ruiz, J., Sannen, P., Sciulli, P., Smith, M., Soficaru, A., Spannagl, M., Storm, R., Stroud, G., Subira, E., Swales, D., Tristaroli, V., Tyler, E., Ulrich-Bochsler, S., Vatteoni, S., Villar, V., Wiggins, R. and Williams, L.L. 2009. The history of European oral health: evidence from dental caries, dental abscesses, ante-mortem tooth loss. *American Journal of Physical Anthropology* Supplement 48, 429.

13

Biographical Bodies

Flesh and Food at Çatalhöyük

JESSICA PEARSON AND LYNN MESKELL

Introduction

IT HAS LONG BEEN ARGUED that the body is simultaneously a physical and social entity (Douglas 1978, 70). The social body can be been read as a symbolic representation and that representational reality 'constrains the way the physical body is perceived'. We suggest here that these two realms, the physical body and the representational body, while distinct, need to be considered in tandem by archaeologists. These two types of bodies constitute different nodes of experience; the physical body is interpolated into social experience while the symbolic dimension of embodiment is understood via bodily physicality (Van Wolputte 2004). In this chapter we attempt to move back and forth between individual datasets that relate to the body at the Neolithic site of Çatalhöyük (7400–6000 cal BC) and their tacit interpretive strengths and limitations: carbon and nitrogen isotope analysis as evidence of diet, and figurines, plastered features and the burial assemblage as evidence of bodily representation.

Çatalhöyük was discovered in the 1950s and excavated between 1961 and 1965 by James Mellaart (1962, 1963, 1964, 1966) and more recently by Ian Hodder (2005). The site, which was inscribed as a UNESCO World Heritage Site in 2012, is particularly well known for its agglomerated architecture with rooftop entry, elaborate wall paintings, plastered reliefs and bucrania (Figure 13.1). A large number of figurines have been recovered including those that depict corpulent individuals. Once interpreted as evidence for a Mother Goddess cult (Mellaart 1967, Gimbutas 1989), new studies suggest other possible readings of the significance of flesh, ageing, maturity and longevity (Nakamura and Meskell 2009). Such concerns would have been long-lived and would have required regular maintenance and reinforcement in social settings, including household activities and commensality. Stable isotope evidence of diet at the site provides a direct link between individuals and their bodies by cataloguing the long-term regulations about food consumption through which individuals and groups invested in bodies.

Proceedings of the British Academy **198**, 233–50. © The British Academy 2014.

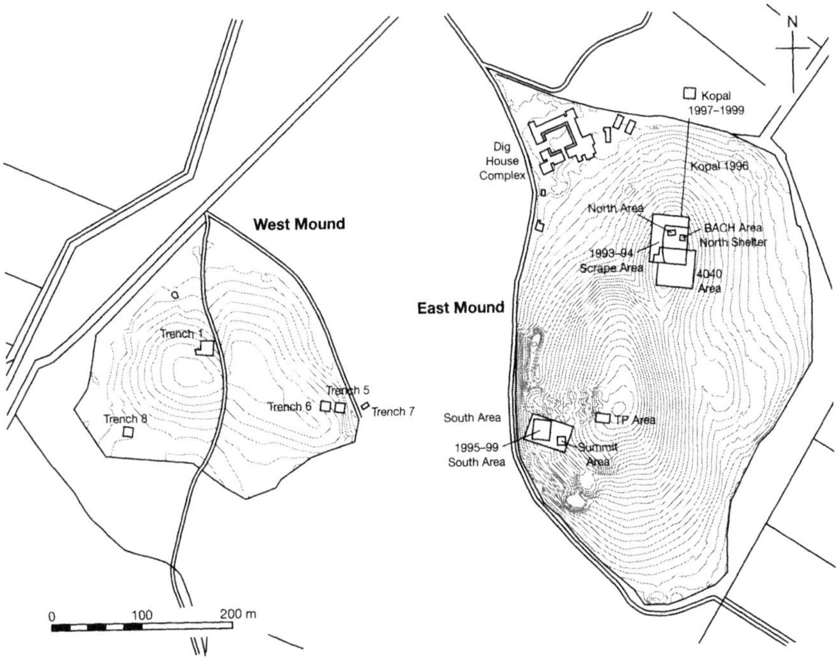

Figure 13.1 Plan of Çatalhöyük, Turkey. Courtesy of the Çatalhöyük Research Project, 2012.

Food, flesh and social structures

The consumption of food at Çatalhöyük would have represented a significant investment in the body, one which can be identified by stable isotope analysis of physical bodies, and reaches far beyond the biological status such as diet or health of the skeleton (e.g. Fuller *et al.* 2010, Lösch *et al.* 2006). Food is an everyday necessity but the choices made concerning which meat or plants to eat, their methods of preparation and how these are distributed within a group mean that this necessity is also a carefully structured system of communication (Barthes 1997). Such practices form part of an embodied *habitus* (Bourdieu 1998), a 'socially constituted system of cognitive and motivating structures' which can be self-perpetuating and tends towards the reproduction of existing social systems (Shilling 1996, 129). This suggests that the adoption of the adage 'you are what you eat' by practitioners of stable isotope analysis relates not just to the biochemical makeup of the Çatalhöyük body, but also to the nuanced perspective that Brillat-Savarin (1825) and Feuerbach (1841) originally intended when this phrase was coined. Stable isotope measurements of human bones should provide a summary of the implementation of social structures over the period of bone formation (Pearson *et al.* 2013; see also White 2005), which represents perhaps a decade or more and only reveals the most repetitious of food habits, such as daily meals, rather than less frequent episodes such as feasting.

The lived body is built from food. Protein in food, which is otherwise a poor source of energy (Speth 1983, Stefansson 1932), is used to build proteinaceous body tissues such as muscle, hair and nails but also bones, which can be sampled by archaeologists. Lipids and carbohydrates in the diet are metabolised for energy allowing the body to perform activities. If diets contain adequate amounts of protein and energy, extra protein can also be used to build up muscle. Adipose fat is deposited on the body when the energy ingested from food exceeds the energy spent on activities. Since protein is not a source of energy, the bodies at Çatalhöyük could only be enlarged beyond their basic musculoskeletal framework by consuming fats and sugars. Eating particular foods, therefore, creates specific physical body shapes and illustrates how food plays a central role in the constitution of corporeality. It is not clear at what point in history humans understood the connection between food and corpulence or that people consumed particular foods to deliberately generate a particular physical image. Therefore, we suggest that the inherent vulnerability and precariousness of the body as lived (Van Wolputte 2004) might be mitigated, ameliorated, or overcome through visual and material repertoires. At Çatalhöyük we suggest there was a preoccupation with rendering flesh and that this was inextricably linked with concomitant ideas of ageing, maturity and longevity. We suggest that this was not only materialised in the figural record (Nakamura and Meskell 2009), but also reflected in the burial assemblage (Nakamura and Meskell 2013) and the human remains themselves.

Fleshing out bodies

Instances of this recognition of bodily vulnerability are found in the treatment of particular bodies after death at Çatalhöyük. Given the practice of intramural burial at the site and the particular type of generational circulation and manipulation of bodies, we posit that the inhabitants were very familiar with bodily decay, physical partibility and the fragility of human remains. Various cultural strategies were employed to ameliorate these physical realities, the most obvious being the enhancement of the dead body through substances like plaster. Just as walls were repeatedly plastered and built up in layers to give them a new skin, so too were skeletal remains. In Building 49 a middle-aged female (sk. 14441) was buried with plaster applied to the lower legs, both feet, the lower left arm and right hand. Some of these bones were entirely encased in plaster. In the same building a child was also buried with plaster on the legs and feet. But the most dramatic example of this technique is the plastered skull (Hodder 2006) from Building 42 showing multiple layers of plaster applied to flesh out the life-like appearance of the head as a living, not deceased, person (Hodder 2007). Given the number of plasterings, this skull was probably in circulation for a lengthy time. This concern with flesh as a living substance, mimicked by smoothed plasters, was a preoccupation that crossed the species divide as well. For example, in Building 77 (Figure 13.2) there is a burial platform with two pillars which have plastered horns extending from them, and above the platform there is a niche and a plastered calf skull embedded into the wall.

Both clay and plaster could have symbolised flesh, the former specifically for figurines and the latter for house installations and the walls or 'skins' of houses, as well as animal and human re-fleshing and revivifying. The colour, texture, softness, sheen, plasticity and ability to layer and smooth must have made plaster an evocative material. Given the qualities of plaster – that it protects, transforms and fortifies an underlying substructure – it is tempting to view the practice of plastering in terms of maintaining, building up, and indeed 'enfleshing' (Meskell 2008). Plaster provides the possibility to transform an individual beyond recognition and yet the use of plaster on the skull at Çatalhöyük is modest, suggesting a focus on reconstruction rather than transformation, a point also made by Fletcher *et al.* (2008) with respect to the plastered skulls of Jericho.

Figurines and plastered bucrania, animal remains and skulls all underwrite the tension between fleshed and skeletal bodies, which are mediated by practices such as plastering bucrania, human skulls and figurine production. This tension is materialised in a headless figurine (12401.X7; Figure 13.3) depicting a skeleton

Figure 13.2 Building 77 showing plastered platform with horns and plastered calf skull embedded in the wall. Photo courtesy of L. Meskell.

Figure 13.3 Figurine 12401.X7, showing a fleshed front and skeletonised back. Photo courtesy of the Çatalhöyük Research Project.

on the back and a corpulent female on the front. This figurine can be interpreted as representing that tension between flesh and bone and their attendant, complex associations with life, survival and vitality, and as emphasising that these figural bodies are indeed made, modified and unmade. Figurine makers sought to reconstitute the living body through plastering and painting, thus improving upon the bony scaffolding of bodies after death (Meskell 2008). These practices may be testament to a material concern for co-producing and rendering permanent ancestors by again improving upon the frailties of flesh.

Flesh, therefore, seemed to play a pivotal role in the life history of bodies at Çatalhöyük, whether human or animal. Because flesh is built from the food we eat, isotope analysis allow us to reconstruct which foods may have enhanced, or attempted to enhance, corpulence and whether diet varied according to sex, age or other social structure (Bentley *et al.* 2007, Barrett and Richards 2004). Gaining dietary information on males, females, younger adults and older members of the community enables us to address the issue of how the imaged individual and the lived individual might relate to one another. A central question has always been whether the examples of obese figurines could have in fact depicted real people. Nitrogen isotope data provide information about the varying proportions of meat and plants which people consumed as part of their regular diet, allowing us to identify which foods were involved (DeNiro and Epstein 1981, Hedges and Reynard 2007). Relatively high nitrogen isotope values would suggest greater consumption of higher trophic level foods (animal protein), whereas lower nitrogen isotope values would indicate greater consumption of lower trophic level foods (plants). Since corpulence cannot be generated through lean animal protein consumption (Speth 1983, Stefansson 1932), it would be those individuals with relatively low nitrogen isotope values that were more likely to have eaten significant amounts of plant resources and subsequently have had the potential for the deposition of unused calories as adipose fat, whether this was deliberate or not. The degree to which different individuals consumed higher and lower trophic level resources can be inferred by their isotope profiles.

The stable carbon and nitrogen isotope analysis at Çatalhöyük now constitutes a large corpus of data about human diet and where possible uses fauna from middens directly associated with the houses beneath which the dead were buried (Pearson 2013). The mean isotope values for females (−18.8‰ and 12.6‰) and males (−18.6‰ and 12.7‰) are virtually identical, indicating that their diets were basically the same. This would suggest that differentiation in diet was not normally organised along gender lines, although there may have been exceptions. One individual, a young adult female (sk. 11665) from Building 44, has nitrogen isotope values lower than the mean of most inhabitants, suggesting that she may have eaten less meat and instead had a diet focused more on plant foods. Alternatively, she could have consumed meat from animals with lower mean nitrogen isotope values such as wild equids and boar (since sheep and aurochs/cattle have the highest nitrogen isotope values at the site) (Pearson 2013). In Building 56 (Figure 13.4), the skeleton of a middle-aged woman (sk. 12863) has a carbon isotope value of −18.3‰ (similar to the mean for all of the inhabitants), and a nitrogen isotope value of 13.3‰, which is 0.7‰ higher than average. Comparing these to the fauna from the associated midden, which has a combined mean nitrogen isotope value of 10.7‰, indicates that she probably had access to more meat from sheep and aurochs/cattle than the average inhabitant at the site.

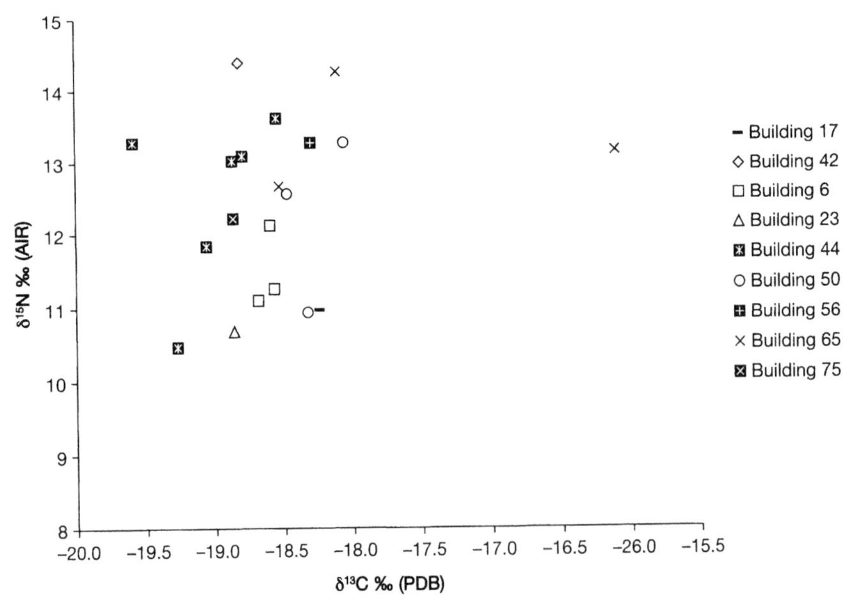

Figure 13.4 Isotope data from adults buried beneath buildings from the South Area (after Pearson 2013).

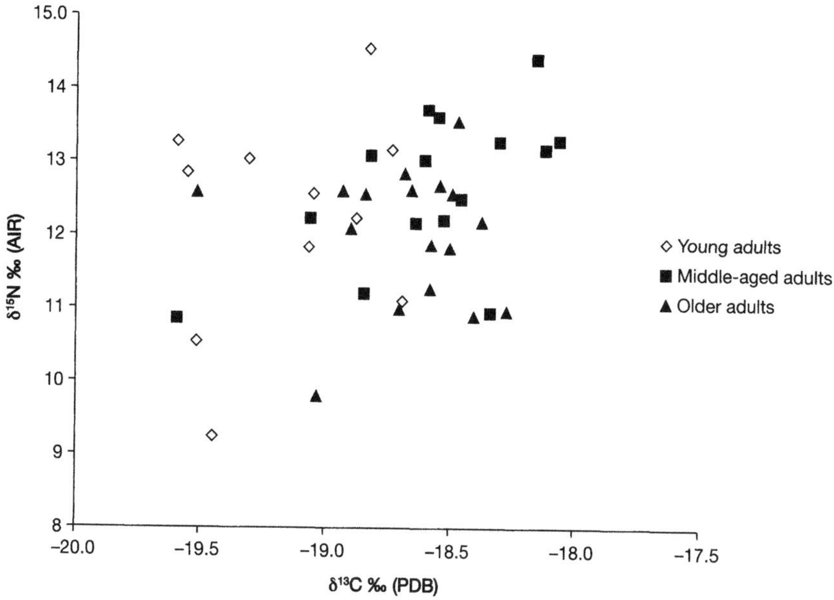

Figure 13.5 Human isotope data according to age stage (young adults: 20–30 years; middle-aged adults: 30–40 years; older adults: 50 years +).

We also examined the possibility of dietary distinctions between individuals based on age (Figure 13.5). Comparing the isotope values of adults by age (younger adults: n = 11; middle-aged adults: n = 15; older adults: n = 17), our data show that younger adults had different diets compared to middle-aged and older adults. An independent samples 2 tailed Student's t-test comparing the three groups shows the following levels of significance:

	$\delta^{13}C$	$\delta^{15}N$
Younger adults versus middle-aged adults	p = <0.001	p = 0.419
Younger adults versus older adults	p = 0.001	p = 0.663
Middle-aged adults versus older adults	p = 0.450	p = 0.076

These data suggest that the carbon isotope values of the younger adults are significantly different from both the middle-aged and older adult individuals, while the values of middle-aged and older adults were not significantly different. No significant difference was observed in nitrogen isotope values in any age group. The carbon isotope data for the different age groups (younger adults: −19.1‰; middle-aged adults: −18.6‰; and older adults: −18.7‰) suggest that younger adults of both sexes had access to plants or animals from areas of the landscape with lower amounts of C_4 plants (Pearson *et al.* 2007). Isotope characterisation of the faunal assemblage indicates that wild equids and boar had lower nitrogen isotope values as well as lower concentrations of C_4 plants in their diet than domestic animals. If the younger adults had the same amount of meat in their diet as middle-aged and older adults but from wild equids and boar, then we would expect the nitrogen isotope values of the younger adults to reflect the lower values in the animals they eat and have lower values themselves as a consequence. Since there is no significant difference in the nitrogen isotope values across the adult age groups, this suggests that younger adults consumed more wild meat than the middle-aged and older adults did from domestic animals. The evidence from dental pathologies indicates that increasing numbers of dental caries occurred in older individuals (more than would be expected by age alone), as the diet shifted from starch-rich to sugar-rich foods in later life (Hillson *et al.* 2013). This would support the hypothesis that younger adults consumed greater amounts of protein and thus help to explain the transition seen in the isotope data between younger and middle-aged/older adults. In terms of body weight, middle-aged and older adults of both sexes could thus be expected to be larger in body size, assuming they did not burn off the extra carbohydrates through activity.

The stable isotope and skeletal data suggest that age rather than sex was the primary axis of difference. The mean isotope values for females and males provide strong evidence that their diets were virtually identical. Larsen *et al.* (2013) and Hillson *et al.* (2013) also confirm that there is great similarity in both sexes in terms of overall health, physical development, stress, injury, other non-isotope indicators of diet, and both were generally robust and active, becoming more sedentary later in life. Such data render questionable older ideas, founded on representational readings of figurines, about obesity at the site, and at other Neolithic sites, predominantly focused on large women, and arguing, by extension, for matriarchy and Mother Goddesses. Instead, we argue that flesh was a culturally marked category that pertained to the entwined concepts of age, maturity and longevity, rather than gender, hierarchy and religion.

Flesh may serve as a material sign of longevity, good health, reliable access to food, sedentary lifestyles, and the ability to provide. The explicit roundness of numerous figurines may have tangibly rendered an ideal visual metaphor for abundance. We suggest that examination of the anthropomorphic figurines provides another avenue to explore the cultural significance of corporeality. Prior analysis of a subset of 455 anthropomorphic figurines and their bodily traits (Nakamura and Meskell 2009) reveals how Neolithic people marked their own preoccupations with bodily form. Nakamura and Meskell argue for a strong tendency for delineating and exaggerating the buttock and stomach regions in the non-gendered and female figurines. The emphasis of the buttocks and stomachs was typically at the expense of other bodily characteristics such as limbs and sometimes even breasts. While breasts were the trait most commonly depicted (since both males and females have them), the stomach and buttocks received the most exaggeration. This phenomenon was characterised as the Three Bs: breasts, buttocks and bellies (Table 13.1). These

Table 13.1 Groups of bodily trait occurrences (after Nakamura and Meskell 2008)

Grouped traits	*Number*	*Grouped traits*	*Number*
3Bs	**17**	**Breasts + other traits**	**24**
Br-Bu-Be	4	Br	15
Br-Bu-Be-F-N	4	Br-\He	3
Br-Bu-Be-\He-N	2	Br-\He-F	1
Br-Bu-Be-\He	2	Br-F-C/A	1
Br-Bu-Be-F	1	Br-N	1
Br-Bu-Be-N	1	Br-C/A	1
Br-Bu-Be-H/C-PT	1	Br-H/C	1
Br-Bu-Be-C/A	1	Br-Ph	1
2Bs	**16**	**Head elaboration**	**8**
Bu-Be	5	H/C	4
Br-Bu	3	PM/Ha	3
Br-Bu-N	2	Pha	1
Br-Be-N	2	H/C-C/A	1
Br-Be	1		
Br-Be-\He-F-n	1	**Sex traits**	**2**
Br-Bu-H/C	1	Ph-PT	1
		PT	1
Buttocks + other traits	**7**		
Bu	4	**Body adornment**	**2**
Bu-\He	3		
		Headless	**2**
Belly + other traits	**3**		
Be	1	**Fingers**	**2**
Be-Pha	1		
Be-\He-N	1	**Beard**	**2**

Key to codes

Code	Feature	Code	Feature
Br	Breasts	C/A	Clothing/body adornment
Bu	Buttocks	PM/Ha	Puncture marks/hair
Be	Belly	H/C	Hair/cap
N	Navel	FHE	Folded head element
PT	Pubic triangle	PHe	Pointed head
F	Fingers	Bd	Beard
PHa	Parted hair	Ph	Phallus
\He	Headless		

are the fleshiest parts of the body, where excess energy from the diet accumulates as fat and where the body can manifest distinctive visual signs of ageing or maturity.

The combined visual emphasis on breasts, buttocks and stomachs has prompted scholars to interpret these figurines as emblems of fertility or pregnancy. However, these features are depicted in a manner that is not suggestive of fertility, but rather of the ageing body (Nakamura and Meskell 2009; see also Voigt 2007, Meskell 2007). When depicted, these bodily features are typically flattened, drooping and angular rather than robust and rounded in shape, as one might expect of a healthy pregnant female (Figure 13.6). This led us to consider the non-generative aspect of figurines (Nakamura and Meskell 2009, 226), and since so many anthropomorphic figurines were intentionally left un-gendered, it would be incorrect to assume that all corpulent figurines are female.

Mary Voigt (2007) has similarly drawn attention to this preoccupation at Hacılar (*c.* 6000 cal BC) in regard to 76 clay and stone figurines from Level VI. She too notes the predilection for drooping stomachs and accentuated buttocks. She argues that they represent bodies worn by work and childbirth, and as such these were ordinary women that served as models for adult roles within the society. Indeed elders might have supervised and safeguarded the transmission of socio-economic skills: some individuals gaining more renown than others and thus attaining social position. Yet as they aged they may have become increasingly confined to the house both in a practical sense and in the sense of becoming guardians of the goods, skills, capacities and identities stored therein (Hodder and Pels 2010, 183).

We should also consider what Neolithic makers chose *not* to represent. For example, images of pregnancy, childbirth, or the portrayal of children in either two or three dimensions are all but absent across the site. If fertility were the main fixation for image production, we might expect to see children play a much more visible role. In terms of physical evidence for fertility, it is difficult to determine the average number of births that occurred at Çatalhöyük. We can look to evidence

Figure 13.6 Assemblage of figurines showing emphasised buttocks, flattened or drooping breasts and stomachs. Photo courtesy of the Çatalhöyük Research Project and Jason Quinlan.

of weaning to provide an indirect measure, since the transition from exclusive breastfeeding to the first introduction of weaning foods provides us with an idea of when breastfeeding may have dropped sufficiently to enable ovulation. Evidence from nitrogen and carbon isotope data from females and infants suggests that women at Çatalhöyük (Pearson 2013) practised a programme of exclusive breastfeeding and weaning which was similar to other Neolithic sites in Anatolia (Pearson *et al.* 2010) and in later periods in other parts of the world (e.g. Dupras and Tocheri 2007, Waters-Rist *et al.* 2011). This suggests that there was nothing remarkable about the practice of breastfeeding and weaning that would have caused fewer births and thereby resulted in concerns about fertility that may have been emphasised in the figural imagery.

Biographical burials: age and maturity

The burial assemblage is another dataset that reinforces the idea that ageing and maturity were valued and that biographical elements were accrued through the lifecourse and constituted part of individual bodies and identities. Just as isotope values from bone reveal a cumulative biography of individual life choices and corporeal history, since isotope values in human bone are generated through repetitive practice and thus literally written on the body as a diet history, so too does the burial assemblage. For the people of Çatalhöyük, the objects placed with them at burial also reveal their biographies and are testament to their ability to survive and accumulate over their lifecourse.

During excavations from 1995 to 2008, some 456 objects and 6252 beads from 244 Neolithic burial features were recovered. Objects found directly associated with individuals include over 40 kinds of tools, personal adornments and other objects; these include jewellery, incised tusks, claws, shells, chipped stone, clay balls, ground stone, baskets, lumps of pigment, textiles, wood, lumps of plaster and worked bone (Nakamura and Meskell 2013). Most individuals, however, received no burial goods whatsoever during the Neolithic occupation of the site and those that did were typically meagre. Our analysis reveals that when burial goods are included with an individual they are drawn from life, rather than being a suite of objects specifically directed towards death or any notion of an afterlife.

Like many domains of life at Çatalhöyük, gender differentiation was not strongly marked in the burial assemblage. Both men and women are found with about 30 different types of artefact both directly and indirectly associated with the body. Of the most common occurrences we find beads, pigment, and worked bone with both male and females, although beads and pigment are found more frequently with females. From an extensive analysis of burial artefacts we suggest that age, not gender, may in fact be the most salient structuring principle. Neonates and infants were buried with matting, baskets and occasionally burial goods. The objects gifted to children were indeed similar to those placed with mature and older adults. Adolescents, on the other hand, rarely received burial items and when they did so it was only beads and bone pins. It was the adults, specifically older individuals, who acquired the most complex and biographically rich burials (Nakamura and Meskell 2013). This may extend beyond a simple expression of their technical skill, to encompass ritual or ancestral prowess, and to reference wider connections in the landscape and even human–animal relationships. Significantly, many of these objects interred with older individuals have an accumulated history of use.

Similar to the figural evidence, the burial assemblage also hints at the salience of maturity as a concept and life stage. Longevity and survival were probably markers of status and this is bolstered by the few burials that contain the most diverse, elaborated and biographical objects such as those from Building 50, especially two older individuals (sks. 10829 and 10813) who lived beyond 50 years of age. Skeleton 10829 (Figure 13.7) is an older female, buried with her legs drawn up to the chest and her arms extended along the body towards the pelvis. Three incised boar tusks lay on the upper body, similar to one which Mellaart found in another female burial in house VII.12 (Mellaart 1967, Pl. 98). These tusks may have been worn as jewellery or attached to a garment (Russell *et al.* 2004). The fact that the only other example has been found with an adult woman, at roughly the same time period (South mound), and strikingly, in a directly adjacent building, suggests a marked connection. This co-occurrence might signify a shared identity, age cohort, ritual affiliation or other grouping. Lastly, a string of bone and stone beads was placed on her upper chest and she wore an anklet made of mock deer canine beads. In the same building, an older man (sk. 10813, Figure 13.8) was buried on his back with his arms and legs falling to the right side. The legs are drawn up against the chest, and the arms folded upon the right shoulder. There are a number of directly associated artifacts such as a bone hook that was placed on his chest (10813.x1). The hook was made by shaping and perforating the caudal end of an otherwise unmodified left aurochs premaxilla (Russell *et al.* 2004). It is very similar to another item described by Mellaart (1962, Pl. VI) as being 'carved in the form of a stork's head'. Beneath the left leg and above the lower right ribs was a cluster of five flint tools (10813.x2–5) and one antler tool (10813.x6). Below the skeleton, reddish brown discolorations were observed that might be interpreted as residues of textiles. Taken together, this unique concentration of tools and equipment may hint at the man's activities and skills acquired during his lifetime. Longevity, as Caspari and Lee argue (2004, 10895), may be necessary for the transgenerational accumulation and transfer of information which allow for complex social networks that are uniquely human. There is no reason to suppose that such a hypothesis does not extend to the transmission of important objects, some of which are eventually included in burials.

Stable carbon and nitrogen isotope values from human bone collagen have the ability, as we have shown here, to reveal whether food was used in daily life to

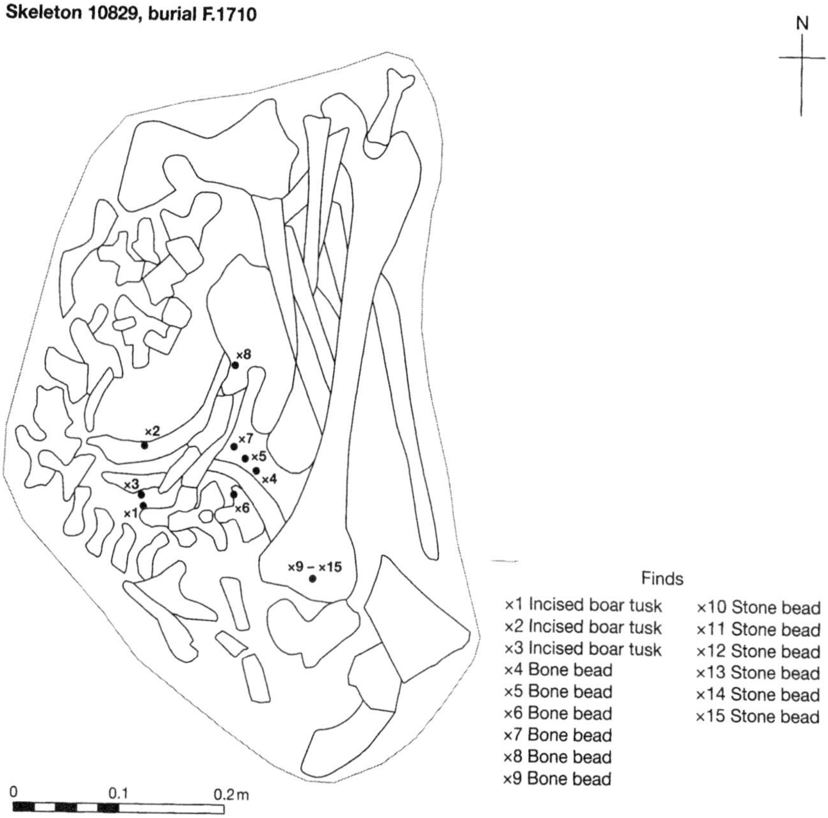

Figure 13.7 Skeleton 10829 with associated finds, courtesy of Scott Haddow and Camilla Mazzucato.

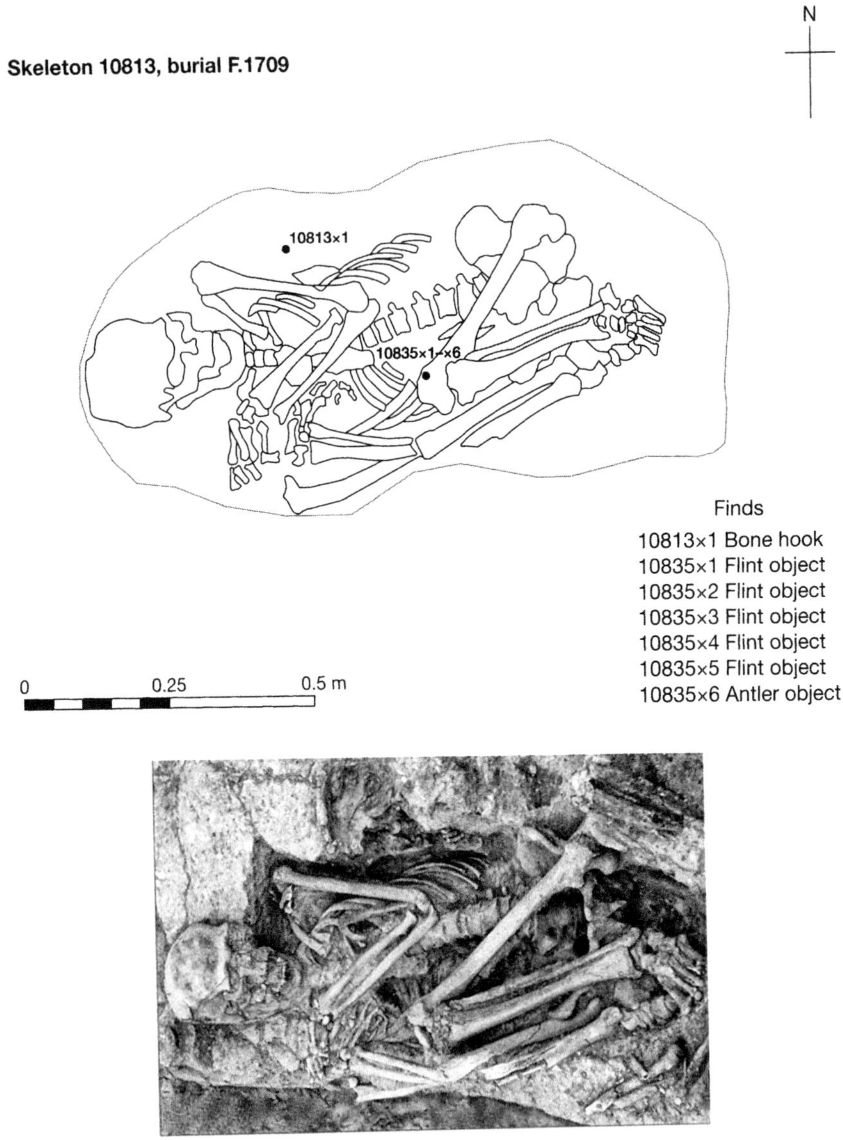

Figure 13.8 Skeleton 10813 with associated finds, courtesy of Scott Haddow and Camilla Mazzucato.

reinforce social structures, because bone collagen isotope values are compiled from diets consumed over many years. The variations in carbon and nitrogen isotope values observed at Çatalhöyük, which indicate different diets between younger and middle-aged/older adults, could only have been achieved through repeatedly eating particular foods on a regular, probably daily, basis. These data are not evidence of one-off feasts and neither do they represent the objects chosen by the community to include with the dead. Instead bone collagen preserves evidence of the persistent nature with which some foods were reserved and consumed by certain people in the community while they were alive. Thus, these isotope data provide a complementary source of biographical information about the body. This suggests that food played both a nutritive and symbolic role in the lives of people at Çatalhöyük through a daily repetitious performance that reinforced long-established social identities, which were later confirmed in death through inclusion of increasing numbers of grave objects accumulated by older individuals.

Conclusion

We have shown here how the seemingly disparate archaeological evidence from figurines, plastered installations, burials and diet can be woven together to provide a deeper understanding of both the social and the physical realms of the body. We see repeatedly that age and maturity rather than gender and fertility were the structuring principles at the site. This lack of differentiation is a notable feature throughout the site, whether one is addressing diet and injury, or burial treatment such as head removal or even burial goods. Biological anthropology shows that the inhabitants were probably not obese as is often presumed based on the figural record. Instead these people were generally robust and active, becoming more sedentary with age. The isotope data show us that food sharing was divided based on age, with middle-aged and older individuals probably consuming less meat from domestic animals, than younger adults whose diet, it seems, contained more meat and was sourced from wild species. This pattern is also borne out in the burial assemblages by age cohort at the site; older individuals accrued the most diverse and biographical materials that were included at death. A particular attention to age, ageing and flesh is also observed in the representational sphere. Flesh and the practice of enfleshing were a preoccupation we see repeatedly throughout the life of the site in the building installations, plastered features, plastered skulls, figurines and other material culture.

Therefore, just as isotope profiles can reveal the biography of an individual's life choices and circumstances, so too corporeal histories can be gleaned from material culture that circulated through the spheres of life and death at Çatalhöyük. Flesh was a material fact of life, particularly for the site's elders, imbued with qualities of endurance and maturity, possibly even with associations of knowledge and skill. Flesh was obviously a bodily necessity during life and similarly needed to be materially sustained after death. Important individuals, both human and animal, were subject to these special acts of enfleshing. Figurines too reflected these bodily preoccupations and priorities, regardless of gender categories. Taken together this new perspective challenges older notions about matriarchy, gender hierarchies and the privileging of female fertility. While we have only hinted at the possibilities here, future collaboration and interdisciplinary work continue to weave together and interpret these diverse threads of evidence.

Acknowledgements

Many people have been instrumental in helping with our research. We would particularly like to thank Ian Hodder, Carrie Nakamura, Shahina Farid, Clark Larsen, Simon Hillson, Sabrina Agarwal, Scott Haddow, Camilla Mazzucato, Jason Quinlan, Nerissa Russell, Kathy Twiss and Amy Bogaard.

References

Barrett, J.H. and Richards, M.P. 2004. Identity, gender, religion and economy: new isotope and radiocarbon evidence for marine resource identification in early historic Orkney, Scotland, UK. *European Journal of Archaeology* 7, 249–71.

Barthes, R. 1997. Toward a psychosociology of contemporary food consumption. In C. Counihan and P. Van Esterik (eds), *Food and culture: a reader*, 20–7. London: Routledge.

Bentley, A., Tayles, N., Higham, C., Macpherson, C. and Atkinson, T.C. 2007. Shifting gender relations at Khok Phanom Di, Thailand: isotopic evidence from the skeletons. *Current Anthropology* 48, 301–14.

Bourdieu, P. 1998. *Distinction: a social critique of the judgement of taste*. London: Routledge.

Brillat-Savarin, J.A. 1825. *Physiologie du gout, ou Méditations de Gastronomie Transcendante; ouvrage théorique, historique et à l'ordre du jour, dédié aux Gastronomes parisiens, par un Professeur, membre de plusieurs sociétés littéraires et savantes*. Bruxelles: J.P. Meline.

Caspari, R. and Lee, S.-H. 2004. Older age becomes common late in human evolution. *Proceedings of the National Academy of Sciences of the United States of America* 101, 10895–900.

DeNiro, M.J. and Epstein, S. 1981. Influence of diet on the distribution of nitrogen isotopes in animals. *Geochimica et Cosmochimica Acta* 45, 341–51.

Douglas, M. 1978. *Natural symbols*. New York: Praeger.

Dupras, T.L. and Tocheri, M.W. 2007. Reconstructing infant weaning histories at Roman period Kellis, Egypt, using stable isotope analysis of dentition. *American Journal of Physical Anthropology* 134, 63–74.

Feuerbach, L.A. 1841. *Das Wesen des Christentums*. Leipzig: Verlag Otto Wigand.

Fletcher, A., Pearson J. and Ambers, J. 2008. The manipulation of social and physical identity in the Pre-Pottery Neolithic. *Cambridge Archaeological Journal* 18, 309–25.

Fuller, B.T., Marquez-Grant N. and Richards, M.P. 2010. Investigation of diachronic dietary patterns on the islands of Ibiza and Formentera, Spain: evidence from carbon and nitrogen stable isotope ratio analysis. *American Journal of Physical Anthropology* 143, 512–22.

Gimbutas, M. 1989. *The language of the goddess: unearthing hidden symbols of western civilisation*. London: Thames and Hudson.

Hedges, R.E.M. and Reynard, L.M. 2007. Nitrogen isotopes and the trophic level of humans in archaeology. *Journal of Archaeological Science* 34, 1240–51.

Hillson, S.W., Larsen, C.S., Boz, B., Pilloud, M.A., Sadvari, J.W., Agarwal, S.C., Glencross, B., Beauchesne, P., Pearson, J.A., Ruff, C.B., Garofalo, E.M., Hager, L.D. and Haddow, S.D. 2013. The human remains I: interpreting community structure, health, and diet in Neolithic Çatalhöyük. In I. Hodder (ed.), *Humans and Landscapes of Çatalhöyük: Reports from the 2000-2008 Seasons*, 339–96. Los Angeles: Cotsen Institute.

Hodder, I. (ed.) 2005. *Inhabiting Çatalhöyük: reports from the 1995–1999 seasons.* Cambridge: McDonald Institute for Archaeological Research.

Hodder, I. 2006. *The leopard's tale: revealing the mysteries of Çatalhöyük.* London: Thames and Hudson.

Hodder, I. 2007. Çatalhöyük in the context of the Middle East Neolithic. *Annual Review of Anthropology* 36, 105–20.

Hodder, I. and Pels, P. 2010. History houses: a new interpretation of architectural elaboration at Çatalhöyük. In I. Hodder (ed.), *Religion in the emergence of civilization: Çatalhöyük as a case study*, 163–86. Cambridge: Cambridge University Press.

Larsen, C.S., Hillson, S.W., Ruff, C.B., Sadvari, J.W. and Garofalo, E.M. 2013. The human remains II: interpreting lifestyle and activity in Neolithic Çatalhöyük. In I. Hodder (ed.), *Humans and landscapes of Çatalhöyük: reports from the 2000–2008 seasons*, 397–412. Los Angeles: Cotsen Institute.

Lösch, S., Grupe, G. and Peters, J. 2006. Stable isotopes and dietary adaptations in human and animals at pre-pottery Neolithic Nevalı Çori, Southeast Anatolia. *American Journal of Physical Anthropology* 131, 181–93.

Mellaart, J. 1962. Excavations at Çatal Hüyük. First preliminary report, 1961. *Anatolian Studies* 12, 41–65.

Mellaart, J. 1963. Excavations at Çatal Hüyük. Second preliminary report, 1962. *Anatolian Studies* 13, 43–103.

Mellaart, J. 1964. Excavations at Çatal Hüyük. Third preliminary report, 1963. *Anatolian Studies* 14, 39–119.

Mellaart, J. 1966. Excavations at Çatal Hüyük, 1965. Fourth preliminary report. *Anatolian Studies* 16, 165–91.

Mellaart, J. 1967. *Çatal Hüyük: a Neolithic town in Anatolia.* London: Thames and Hudson.

Meskell, L.M. 2007. Refiguring the corpus at Çatalhöyük. In A.C. Renfrew and I. Morley (eds), *Material beginnings: a global prehistory of figurative representation*, 137–49. Cambridge: McDonald Institute Monographs.

Meskell, L.M. 2008. The nature of the beast: curating animals and ancestors at Çatalhöyük. *World Archaeology* 40, 373–89.

Nakamura, C. and Meskell, L.M. 2009. Articulate bodies: forms and figures at Çatalhöyük. *Journal of Archaeological Method and Theory* 16, 205–30.

Nakamura, C. and Meskell, L.M. 2013. The Çatalhöyük burial assemblage. In I. Hodder (ed.), *Humans and landscapes of Çatalhöyük: reports from the 2000–2008 seasons*, 441–66. Los Angeles: Cotsen Institute.

Pearson, J.A. 2013. Human and animal diet as evidenced by stable carbon and nitrogen isotope analysis. In I. Hodder (ed.), *Humans and landscapes of Çatalhöyük: reports from the 2000–2008 seasons*, 271–98. Los Angeles: Cotsen Institute.

Pearson, J.A., Buitenhuis, H., Hedges, R.E.M, Martin, L., Russell, N. and Twiss, K.C. 2007. New light on early caprine herding strategies from isotope analysis: a case study from Neolithic Anatolia. *Journal of Archaeological Science* 34, 2170–9.

Pearson, J.A., Hedges, R.E.M., Molleson, T.I. and Özbek, M. 2010. Exploring the relationship between weaning and infant mortality: an isotope case study from Aşıklı Höyük and Çayönü Tepesi. *American Journal of Physical Anthropology* 143, 448–57.

Pearson, J.A., Grove, M., Özbek, M. and Hongo, H. 2013. Food and social complexity at Çayönü Tepesi, southeastern Anatolia: Stable isotope evidence of differentiation in diet according to burial practice and sex in the early Neolithic. *Journal of Anthropological Archaeology* http://dx.doi.org/10.1016/j.jaa.2013.01.002.

Russell, N., Pawlowska, K., Twiss, K. C. with a contribution by E. Jenkins and R. Daly. 2004. Animal bone report in *Çatalhöyük 2004 Archive Report,* www.catalhoyuk.com.

Shilling, C. 1996. *The body and social theory.* London: Sage.

Speth, J.D. 1983. *Bison kills and bone counts.* Chicago: University of Chicago Press.

Stefansson, V. 1932. *The friendly Arctic.* New York: Macmillan.

Van Wolputte, S. 2004. Hang onto yourselves: of bodies, embodiment, and selves. *Annual Review of Anthropology* 33, 251–69.

Voigt, M. 2007. The splendour of women: Late Neolithic images from Central Anatolia. In A.C. Renfrew and I. Morley (eds), *Material beginnings: a global prehistory of figurative representation*, 151–69. Cambridge: McDonald Institute Monographs.

Waters-Rist, A.L., Bazaliiskii, V.I., Weber, A.W. and Katzenberg, M.A. 2011. Infant and child diet in Neolithic hunter-fisher-gatherers from Cis-Baikal, Siberia: intra-long bone stable nitrogen and carbon isotope ratios. *American Journal of Physical Anthropology* 146, 225–41.

White, C.D. 2005. Gendered food behaviour among the Maya: time, place, status and ritual. *Journal of Social Archaeology* 5, 356–82.

14

Neolithic Lifeways

Microstratigraphic Traces within Houses, Animal Pens and Settlements

WENDY MATTHEWS, LISA-MARIE SHILLITO,
SARAH ELLIOTT, IAN D. BULL AND JAMES WILLIAMS

Introduction

RECENT RESEARCH ON EARLY FARMING GLOBALLY has identified considerable local variation in community lifeways and relations with plants and animals (Barker 2006). For the Near East, Willcox (2005) argues that there were multiple local centres of domestication within the heartland of wild progenitor species of wheat, barley, sheep and goats. There is also, however, evidence for remarkable contact between communities across this region (Kozlowski and Aurenche 2005). Asouti and Fuller (2013) have recently argued that to understand these local variations and pathways in early farming we need to develop more contextual approaches that consider a wide range of evidence of different spheres of life and to integrate interdisciplinary analyses and ecological and social approaches.

One of the challenges in integrating different data sets is that during routine excavation and bulk sampling, aspects of the diversity, context and association of bioarchaeological, artefactual and sediment residues from activities are irreversibly lost, bulked together or separated. In this process, finely stratified lenses, too thin and numerous to excavate separately, may be bulked together and the specific actions and timescales that they represent are irreversibly amalgamated and homogenised. Even when excavated and sampled as single depositional units, only selected materials are recovered in many analyses during processes of excavation, dry or wet sieving, flotation and extractions from spot or bulk samples of deposits. Each material is also often and necessarily studied separately. In addition, crucial information may also frequently be lost on the environment and history of deposition and post-depositional alterations that are discernible from geoarchaeological analysis of sedimentary context.

This loss of information in composition and of contextual relations has a direct impact on methodological and research issues in the study of early farming. Many

Proceedings of the British Academy **198**, 251–79. © The British Academy 2014.

studies of early plant management and domestication are based principally on the study of charred plant remains recovered from bulk sampling and water-flotation and wet- or dry-sieving. Charred plant remains, however, only represent plants that have been burnt, and generally only those that have been burnt at low temperatures <500°C, which are exceeded in many domestic or other fires (Boardman and Jones 1990, Van der Veen 2007, W. Matthews 2010). To widen the range of plant materials analysed, other plant materials in archaeological deposits are increasingly being recovered, even in these semi-arid environments. Pollen has been recovered from on-site Epi-Palaeolithic-Neolithic deposits in the Zagros (Leroi-Gourhan 1969) and could be more widely studied. Plant silica phytoliths are increasingly being recovered by extraction from spot samples of deposits to provide information on non-burnt as well as burnt plant remains (Rosen 2005, Ryan 2011). Interpretation of the ecological and social significance of these separate plant materials, however, is not straightforward. First, these plant remains specimens are often disarticulated during extraction, making identification of plant anatomy, species and original configuration when deposited problematic, particularly in the study of phytoliths (Shillito 2013). Secondly, dissociation of these plant remains from their precise depositional context and associations makes it more difficult to interpret their diverse depositional and taphonomic pathways and thereby their ecological and social significance (Van der Veen 2007).

The earliest stages in animal management, furthermore, may be problematic to detect in zooarchaeological assemblages as changes in bone morphology indicative of domestication may be delayed by 500–1,000 years, and studies of kill-off profiles are dependent on identification of indicators of sex and age, which are influenced by a range of factors including environment (Zeder 2005). New indications of environment, vegetation, animal diet and management practices are being provided by analysis of a range of stable isotopes in animal and human bone more widely (this volume). These analyses, however, do not provide information on specific plant species, critical to in-depth study of ecological niches and wild and domestic resources, and often represent bulk seasonal or indeterminate longer-term time fluxes.

The presence of dung on archaeological sites is one potential independent marker of greater human proximity to animals and early management, as traces of dung collected for fuel or from penning, for example (W. Matthews 2005a, 2010; Bull *et al.* 2005; Shahack-Gross 2011; Portillo *et al.* 2009; Shillito *et al.* 2011). Studies of dung and plant and microfossil content provide indicators of wider environment and vegetation (Ghosh *et al.* 2008) as well as animal diet and management practices at timescales of one to two days (Shahack-Gross 2011). These integrated studies are particularly important in the investigation of interrelations between environment, early plant and animal management and sedentism. In addition, dung burnt as fuel is one major routeway for the presence of charred plant remains on archaeological sites, but remains difficult to identify (Charles 1998, Valamoti 2013). Charred dung pellets may be recovered by flotation, but many are difficult to identify due to fragmentation either in antiquity during trampling in pens or dung-cake manufacture, for example, or during recovery in flotation. It is currently uncertain whether an increase in the diversity of charred plant species in the Epi-Palaeolithic to Early Neolithic sites represents a 'broad spectrum revolution' in human diet or plants consumed by animals and burnt as dung fuel (Miller 1996, Jones 1998).

This chapter briefly reviews ways in which integrated approaches that include micro-contextual analysis of materials *in situ* within their microstratigraphic sequence in large resin-impregnated thin-sections may contribute to more precise data on the diversity and contextual significance of materials in early farming sites. It also examines ways in which micromorphological approaches can be linked to geochemical and phytolith analyses.

The aim of this research is to develop integrated high-resolution micro-contextual approaches and to apply these first to evaluate how *in situ* analysis of diverse plant materials within their precise depositional context can contribute to understanding plant taphonomy and thereby their ecological and social significance. A second aim is to study dung as an indicator of animal management, especially in its earliest stages. The third aim is to examine the interrelationship between changes in plant and animal management and changes in activities, roles and relations at the scale of individuals/households and communities.

This chapter begins with a review of the case studies and methodology. It then examines climate and environment as a key context for local and regional variation in early farming and sedentism, and then considers micro-contextual data on plant taphonomy and use; dung as an indicator of early management; and the nature and organisation of Neolithic activities, roles and relations.

Case studies

The case studies are selected from one of the key heartlands of early plant and animal management, in the central Zagros in the east of the Fertile Crescent. Selective comparison is made to Neolithic sites in central Anatolia, over 1,000km to the west where similar analyses have been conducted, in order briefly to examine local and regional variation in early farming strategies and lifeways.

In the Zagros, many key interdisciplinary approaches and theories in the study of early farming were forged during field work in the 1950s–60s (Braidwood *et al.* 1961, Flannery 1969, Hole *et al.* 1969). Although subsequent research has identified earlier Neolithic sites and domesticated species in Cyprus, Anatolia and the Levant (Vigne *et al.* 2012), new research in the Zagros is re-emphasising the importance of this region in studies of local, regional and global options and pathways in the development of agriculture and more sedentary lifeways and early stages in this (Charles 2008, Zeder 2009, Riehl *et al.* 2013). A wide range of early Holocene sites have been identified by new surveys and excavations in the Iranian and Iraqi Zagros

(R. Matthews and Fazeli-Nasheli 2013). In addition, recent analysis of modern DNA suggests that the Zagros may have been one area where barley and goat were domesticated (Morell and Clegg 2007, Naderi *et al.* 2008).

This chapter examines results from new excavations and interdisciplinary research by the Central Zagros Archaeological Project to investigate local variation in ecology and lifeways at sites on a transect through different ecozones from the high to low Zagros mountains spanning 290km (Figure 14.1; R. Matthews *et al.* 2013). The two principal sites examined in this chapter are Sheikh-e Abad (9800–7600 cal BC) and Jani (*c.* 8000 cal BC) 90km apart in the high Zagros in Iran. Both are settlement mounds, *c.* 1 hectare in size, characteristic of many Neolithic sites (Baird 2005), and 10 and 8m high respectively. These sites were excavated, recorded and sampled in 2008 (R. Matthews *et al.* 2013). At Sheikh-e Abad, three trenches were excavated. The first trench revealed a sequence of very early accumulations of ashy deposits and possible surfaces, *c.*10100–9140 cal BC (Beta-258647, Beta-267509), on natural deposits. Trench 2 revealed a sequence of *in situ* burnt deposits, fire-cracked stones, architecture and midden deposits, at 6m above the base of the mound, *c.* 8230–7730 cal BC (Beta-258646). Trench 3, 13 × 10m uncovered a wide range of open areas and two buildings: Building 1 with at least four small rooms in a linear arrangement, and Building 2, a ritual building with four wild goat and one wild sheep skull, *c.* 7640–7580 cal BC (Beta-258648). At Jani, a 45m long section cut by a stream and track through the south-east of the mound was cleaned, photographed and sampled, with a date 1.3m above natural of *c.* 8240–7740 cal BC. Brief comparative reference is also made to ongoing excavations and analyses at Bestansur, *c.* 8000 cal BC, which is up to 2.5ha in size and Shimshara *c.* 7450–7080 cal BC, 0.8ha in size in the lower Zagros in Iraq.

In central Anatolia, survey and excavation of a cluster of sites within 45km of each other on the Konya Plain and in Cappadocia are enabling study of the ecology and lifeways of a range of local communities, from 9000–6000 cal BC (Baird 2005, 2008; Baird *et al.* 2011). This chapter focuses on selected results from the mega-site of Çatalhöyük (7400–6000 cal BC), excavated in the 1960s by Mellaart (1967) and since 1993 by Hodder (2006).

Methodology

Field and micromorphology sampling and analysis

In the field at all sites, sequences of occupation surfaces and deposits were analysed, recorded and sampled, at 1–2m intervals within interior spaces and more distant in larger exterior areas during excavation and in section profiles, adapting methods in soil science and archaeology (Hodgson 1976, Courty *et al.* 1989). Profiles were selected from sections in strategic plinths, baulks or cross-sections of features,

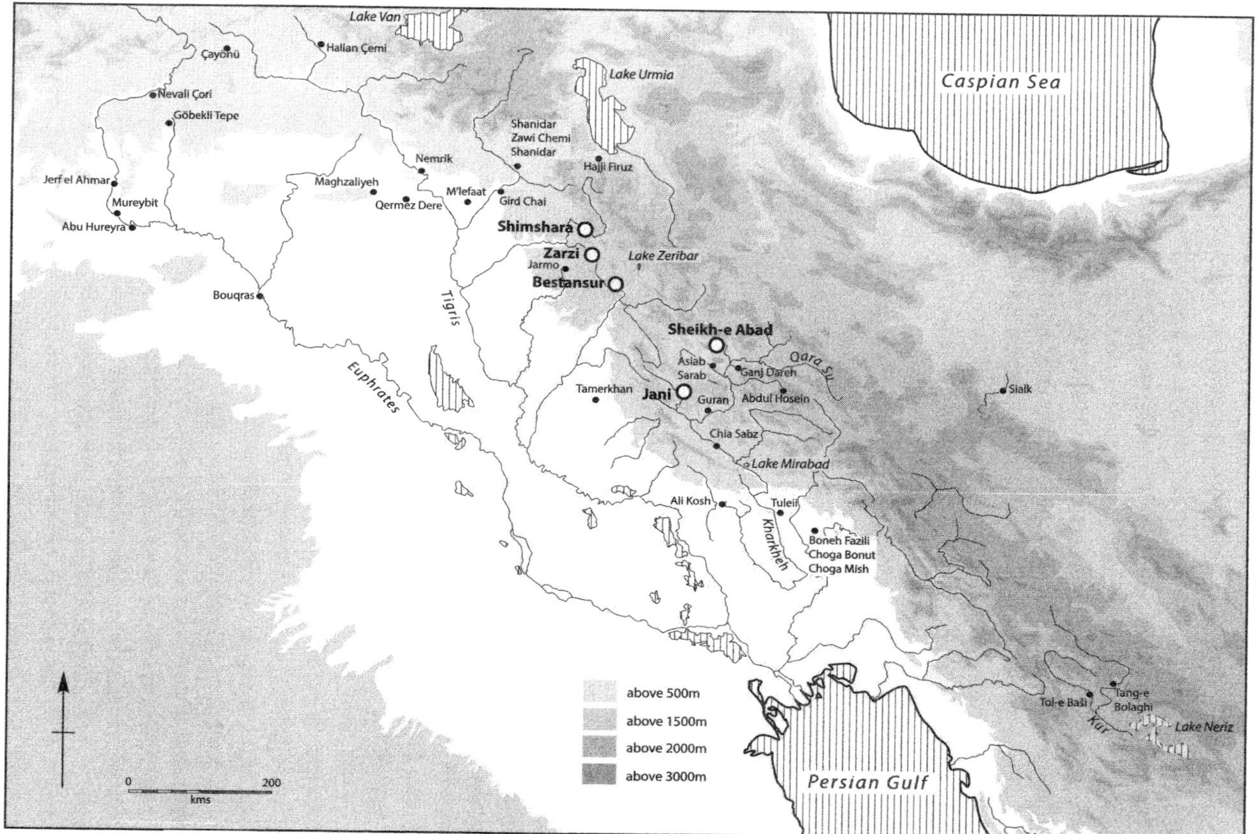

Figure 14.1 Location of Neolithic sites in the Central Zagros. White circles mark the five sites investigated by the Central Zagros Archaeological Project.

and pit or trench edges. At Bestansur, deposits and sequences have been examined *in situ* in the field using a Niton XL3t GOLDD+ portable X-Ray fluorescence analyser to identify deposits high in phosphorus that may indicate areas of organic residues and dung, and to inform excavation and sampling strategies (Elliott n.d.). Spot samples of deposits were analysed in smear slides in field laboratories at Çatalhöyük (W. Matthews 2005a), Sheikh-e Abad (R. Matthews *et al.* 2013, Shillito and Elliott 2013) and Bestansur (Elliott n.d.) to assess deposit composition and plant phytolith and calcareous dung spherulites content in particular. Intact block samples of sequences, *c.* 14 × 7 × 8cm were cut from section-profiles from all sites, overlapping where necessary and feasible to study entire sequences within buildings and open areas. In the laboratory in Reading, spot samples of deposits, *c.* 2–20g in weight, were collected from each stratum in the block samples to enable high-precision correlation of results from *in situ* micromorphological analyses, with spot samples for geochemical and phytolith analyses. The blocks were then impregnated with resin and large thin-sections 14 × 7cm in size, 25–30µm thick, cut ground and polished (Colour Plate 8; Courty *et al.* 1989). Samples analysed to date include: 11 from Sheikh-e Abad and six from Jani (W. Matthews *et al.* 2013); 30 from Bestansur and six from Shimshara. More than 350 have been analysed from the comparative site of Çatalhöyük (W. Matthews 2005a, Shillito 2011).

In thin-section, the type, abundance, size and microstratigraphic context of deposit components and features were analysed at magnifications of ×25–400 using an optical polarising microscope and internationally standardised protocols (Bullock *et al.* 1985, Courty *et al.* 1989, Stoops 2003). Abundance of components was measured as a percentage by area in thin-section by comparison to visual charts, with an error range of ± 5–10% (W. Matthews 2010). The size of components was measured by graticule and Leica software. The autofluorescence of components was examined as presence and intensity of autofluorescence is one potential characteristic of components of biological origin and of animal dung (Courty *et al.* 1989, Altemüller and Van Vliet-Lanoe 1990), using incident fluorescent light (Leica Filter system AS, Filter system N2.1S: wavelength excitation 515–560nm, transmitting >590 nm).

Plant remains in thin-section were identified by reference to key atlases and reference collections (Schweingruber 1990, Piperno 2006, Rosen 1992). Identification of plant family, genus and species in thin-section was dependent on plant type, part, size, articulation, orientation and preservation (W. Matthews 2010).

Animal dung was identified in thin-section by analysis of: pellet morphology; fine fabric composition; inclusion type, comminution, orientation and distribution; and the presence of calcareous dung spherulites, 5–20µm in size, that form in the guts of animals during digestion (see Figure 14.2; Courty *et al.* 1989, Brochier 1992, Macphail *et al.* 1997, Canti 1999, W. Matthews 2010). Ruminant dung was identified by the presence of abundant spherulites (Canti 1999) and finely comminuted plant remains. Omnivore coprolites were identified by the presence of: a

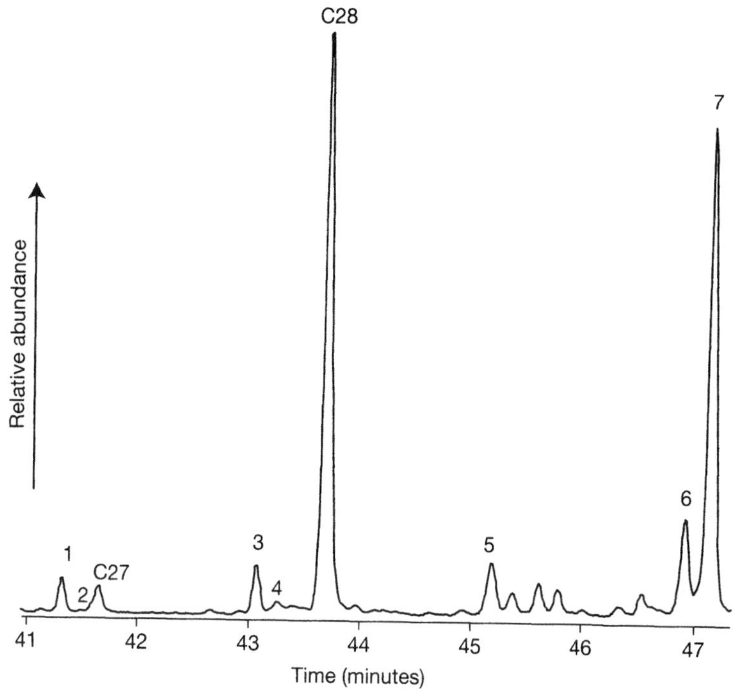

Figure 14.2 GC trace showing sterols in C804 S1.2. Ratio 3 indicates that this is ruminant faeces, supported by the lack of LC with small quantities of DOC bile acid. 1.coprostanol 2. epi-coprostanol 3. cholesterol 4. 5α-cholestanol 5. 5β-stigmastanol 6. sitoserol 7. 5α-stigmastanol.

dense organic fine fabric, which is yellowish-orange in plane-polarised light (PPL) and isotropic in cross-polarised light (XPL); partially digested bone or tooth remains and lower density of plant remains and spherulites (Canti 1999). Only cases where all of the characteristics listed above are present are discussed in this chapter, as the presence and abundance of each characteristic may vary according to genesis and taphonomic processes. Dung spherulites presence and abundance, for example, are variously influenced by animal species, age, diet and environment and by pH as spherulites dissolve in a pH of <6–7.7 (Canti 1999). Similarly the type and abundance of plant phytoliths produced may vary according to plant species and part, and phytoliths are dissolved in a pH>8–8.5 (Schiegl *et al.* 1996, Tsartsidou *et al.* 2007).

Phytolith analyses

To enhance and test the identification and quantification of plant opal silica phytoliths in micromorphological thin-sections, plant opal silica phytoliths were extracted from the correlated spot samples following the method outlined in Rosen

(Rosen 2005, Shillito and Elliott 2013). Identification to genus where possible was conducted by comparison to reference collections at the University of Reading and to published material (Rosen 1992, Wang and Lyu 1992, Jenkins 2009). Nineteen high-precision spot samples have been analysed for Sheikh-e Abad and seven from Jani (Shillito and Elliott 2013).

Biomolecular analysis of faecal remains

To enhance and test the identification of ruminant dung and omnivore coprolites in thin-section, correlated spot samples were analysed by gas chromatography mass spectrometry (GC/MS), 16 from Sheikh-e Abad and five from Jani (Shillito *et al.* 2013), based on approaches developed at Çatalhöyük and other sites (Bull *et al.* 2005, Shillito *et al.* 2011, Bull and Evershed 2012).

Comparative archaeobotanical analyses

For the Zagros sites, the study of charred plant remains from flotation is currently in progress as part of PhD research conducted by Jade Whitlam, supervised by Dr Amy Bogaard and Dr Mike Charles. Comparisons are therefore based on preliminary data only (Whitlam *et al.* 2013). The samples come from Sheikh-e Abad. Five pilot samples collected from Jani will be studied by Hengami Ilkhani in the future.

Dating

AMS radiocarbon age determinations for the Central Zagros Archaeological sites are provided as calibrated date estimates at 2 sigma (95% probability) and were dated by Beta Analytic.

Identifications of plant remains and dung: material and contextual variation

Plant remains in thin-section

In thin-section, a diverse range of non-burnt as well as burnt plant materials, parts and families and occasionally species have been identified in occupation deposits from Sheikh-e Abad and Jani in thin-section (Colour Plates 8, 10–12); W. Matthews 2005a, 2010; R. Matthews *et al.* 2013). These include: impressions of plant remains that have since decayed in fine sediments; monocotyledonous and dicotyledonous plant silica opal phytoliths – non-burnt, burnt with occluded carbon from charring, as well as melted at temperatures >850°C (Canti 2003); charred and partially charred wood, stem, leaf and seeds; and calcitic ashes. Pollen and starch have not yet been positively identified as they may be masked by fine sediments, but a number of fluorescent micro-fossils are currently being investigated.

Whilst a much greater range of species have been identified by analysis of charred plant remains from bulk flotation samples (Whitlam *et al.* 2013) than in thin-section, there is a general correspondence in the relative densities of charred plant remains densities as well as fragmentation in both data sets, which we are currently documenting as in previous studies (R. Matthews and Postgate 2001).

Up to >50–70% of deposits may comprise non-charred plant remains (W. Matthews *et al.* 2013). The lowest concentrations of plant remains are in natural and architectural materials, <2–5%, the highest in midden-like deposits and oxidised burnt layers, up to >50–70%.

In spot samples from Sheikh-e Abad and Jani the extracted phytolith types and their abundance varied significantly according to context, as discussed in greater detail below (Colour Plate 10, TJ8.2; Shillito and Elliott 2013).

Identification of faecal deposits

A range of ruminant and omnivore coprolites have been identified in thin-section from all four Zagros sites, based on the criteria above, and discussed below (Figure 14.2 and Colour Plate 12). GC/MS analyses were conducted on a sub-set of 21 spot samples from Sheikh-e Abad and Jani (Colour Plates 9 and 12). Coprostanols and bile acids were absent from two control samples from natural sediments from both sites, as expected. Faecal characteristics were positively identified in 18 of the suspected 19 faecal deposits, with trace amounts only in three of these (Shillito *et al.* 2013). Of the seven faecal deposits identified as ruminant in thin-section, six were positively identified as ruminant by GC/MS (Colour Plate 9). Of the six omnivore coprolites identified in thin-section, four were positively identified as omnivore. Furthermore, two of the omnivore coprolites were identified as human at Sheikh-e Abad in open area deposits in Trench 2 and a probable pen/latrine in Trench 3. Ruminant, omnivore and human faecal deposits have also been identified by GC/MS for examples from Çatalhöyük (Bull *et al.* 2005, Shillito *et al.* 2011).

Post-depositional alterations

Post-depositional alterations observed in thin-section from the four Zagros sites include: compaction; decay of organic tissues and traces of organic staining and microbial action including framboids; bioturbation and the presence of modern plant roots, soil microfauna excremental pellets; precipitation of gypsum salts and pseudomorphic impressions of these; localised translocation and in-washing of sediments; expansion and contraction of clays and cracking; trampling; secondary burning and secondary discard. Although post-depositional alterations are more marked within *c*. 40cm of the surfaces, no thin-sections comprise totally reworked sediments.

Environment and use of wild and domesticated plants: microcontextual analysis of diverse plant materials and parts

Introduction

There is increasing evidence for local and regional variation in climate, environment and the availability of plants and animals during rapid global warming after the Younger Dryas, *c*. 9600 cal BC, across the Near East (Mithen 2003, Willcox 2005, Zeder 2009, Conolly *et al*. 2011). The environmental contexts of early farming in both the central Zagros as well as at Çatalhöyük, however, remain disputed. In this section we examine micro-contextual evidence for spatial and temporal variability in biomes and the implications of this for the nature and viability of early sedentism and interrelations between humans, plants and animals in particular locales. Studies of plant remains on archaeological sites potentially provide crucial indicators of the intersection between humans and the environment for comparison to off-site records.

Plant ecology and resources

All four sites were situated on fertile intermontane plains in the central Zagros mountain belt, close to water-sources (Figure 14.1). The site of Sheikh-e Abad is located in the high Zagros at an elevation of 1430m, surrounded by mountain peaks over 3,000m. It is situated in a small fertile plain by a modern spring that connects with the principal river system on which many Neolithic sites are located, including Asiab, Sarab and Ganj Dareh within 42km. Jani, 90km to the southwest, at 1,280m asl (above sea level) is close to a local stream and surrounded by lower peaks, 1,500m. At much lower elevations, the site of Bestansur at 550m asl is located by a major spring at the head of the large Sharizor plain, with local peaks up to 1,500m. Shimshara, 490m asl, was located at the head of the Rania plain by a major pass along the banks of the Lesser Zab, with local peaks 1,000–1,500m.

There are major discrepancies in interpretation of climate and environment proxies for the Early Holocene in the Zagros from the principal lake core for this region, Lake Zeribar, at 1,290m asl, between these two clusters of sites (Wasylikowa and Witkowski 2008, W. Matthews 2013). It has been widely argued that the central Zagros was colder, drier and less habitable than other regions of the Near East, based on: the scarcity of tree pollen (<15%) notably oak (van Zeist and Bottema 1977, van Zeist 2008); a rapid and sustained reduction in $\delta^{18}O$ stable isotopes of carbonate-rich sediments from Lake Zeribar, from *c*. 9900–6200 cal BC (Stevens *et al.* 2001); and an apparent absence of sites from *c.* 10500–8500 cal BC (Hole 1996). Trees, however, are likely to have been more abundant as pistachio are insect-pollinated and almond poor pollen dispersers, and likely to be under-represented in the Lake Zeribar core (Asouti 2005, Asouti and Austin 2005). In addition, the delay in oak was widespread in many continental interior regions of south-west Asia and may have been due to the impact of human activity and animal browsing in these areas, particularly by goat and deer (Roberts 2002, Turner *et al.* 2010). It is also argued that the reduced $\delta^{18}O$ stable isotope values in the Zagros may be due to a number of factors, including distance from source or a change in storm tracks (Jones and Roberts 2008). That the Zagros was habitable in the Early Holocene is supported by new and increasing evidence for occupation in this region, not only at Sheikh-e Abad, 10100–7580 cal BC, but at lower altitudes at Chogha Golan and East Chia Sabz (Riehl *et al.* 2013, R. Matthews and Fazeli-Nashli 2013).

The presence of at least parkland with pistachio and almond, herbaceous cover and swards of grass in the high Zagros is confirmed by independent identification of these species in charred plant assemblages on archaeological sites recovered by flotation, which comprise up to 85–98% charred wood (Hubbard 1990, Willcox 1990). In thin-sections from Sheikh-e Abad, we have identified diverse remains of trees and nuts that confirm the importance of this resource base for local communities from the earliest levels at both sites. These remains include: charred dicotyledonous wood (Colour Plate 11), including Pistachio and Chenopodeaceae shrubs [804 S4]; charred, mineralised and calcitic ash nut pericarps (Colour Plate 10); and dicotyledonous phytoliths (R. Matthews *et al.* 2013). Dicotyledonous wood and leaf phytoliths are also present in the majority of field spot and laboratory extraction samples, comprising up to 80% of the assemblage in some samples (Shillito and Elliott 2013).

By contrast, the sparsity of charred wood in occupation deposits at the site of Bestansur in the lower Zagros in both thin-section samples and flotation assemblages (W. Matthews n.d., Whitlam n.d.) suggests that trees may have been sparser, more distant or conserved in the lower Zagros, as also observed at other sites in in the piedmont and north-eastern Mesopotamian steppe and plains (Miller 2003a and b).

Grass pollen rapidly increased in the Early Holocene at Lake Zeribar from 15% in *c.* 10000 cal BC to up to 50% by *c.* 8500 cal BC (van Zeist and Bottema 1977, van Zeist 2008). Its decline after this is associated with an increase in *Plantego lanceolata* pollen which is often associated with disturbance, including agriculture. Abundant charred grass seeds have been identified in flotation samples from Sheikh-e Abad (Whitlam *et al.* 2013). In thin-sections from all four archaeological sites both charred and siliceous phytolith remains of Poaceae and Cyperaceae stems/leaves have been identified in many contexts and in densities of up to 50–70%, notably reeds, grasses and sedges (Colour Plate 8). Densities in Sheikh-e Abad Trench 1, however, are lower, corresponding with lower counts of grass pollen in the early tenth millennium cal BC.

In the high Zagros, at Sheikh-e Abad and Jani the dominant short cells in laboratory-extracted phytoliths are rondels, which indicate the presence of C_3 grasses that prefer moist and wet conditions in well-watered habitats (Colour Plate 9; Piperno 2006, Vincentini *et al.* 2008, Shillito and Elliott 2013). This evidence of abundant reeds and wetland grasses supports Helbaek's (1969) and Rosen's (2003) arguments that wetlands were more abundant than today and an important focus and resource for communities in the foundation of sites not only at Chogha Bonut in Khuzistan and Ali Kosh in the piedmont but also the low and high Zagros, based on on-site charred and phytolith assemblages. It is questionable whether these wetlands and extensive grasslands could have been supported if there was little or no spring/summer precipitation as suggested by Stevens *et al.* (2001), unless there was sufficient moisture from snow melt (Stevens *et al.* 2008, 300) and/or high sub-surface water tables in the Early Holocene (Hole *et al.* 1969).

Plant use

With regard to use of plants remains, integrated thin-section and phytolith analysis is contributing to identification of a greater range and higher densities of plant materials and parts than is possible from study of charred plants alone. It is also aiding identification of the taphonomic pathways of plants and assessment of the effects of combustion in particular on biases in assemblages.

Construction materials and craft activities

Impressions of non-burnt plants that have since decayed have been identified in a range of fine-grained early architectural construction materials that pre-date *c.* 8240–7740 cal BC at Jani, as well as from at least *c.* 8230–7730 cal BC at Sheikhe-Abad (Colour Plate 12; R. Matthews *et al.* 2013). These linear and curvilinear impressions are most likely from grasses that were incorporated as stabilisers to provide tensile strength and flexibility and reduce cracking (Houben and Guillaud 1989). Cereal husks have been identified in spot phytolith sample TJ 10.4 in deposits at Jani *c.* 8240–7740 cal BC, and other samples higher in the sequence. The presence of plant stabilisers in mudbricks, plasters and diverse secondary aggregates attests widespread knowledge of sediment properties and construction material requirements in the Early Neolithic.

Some dense layers of partially burnt articulated phytoliths in middens and on floors, may represent *in situ* burning/discard of thatch roofing or litter, as in the thick layers of articulated reed phytoliths in open area Jani TJ S9.8 (Colour Plate 8), and on floors in Shimshara, as argued for other Zagros sites (Savard *et al.* 2006, 189; van Zeist *et al.* 1986).

Fuel

Mixed fuel sources have been identified in thin-section and spot phytolith samples. In thin-section, *in situ* burnt fuel and as well as discarded lenses of fuel rake-out includes: grass and reed stems and leaves; dicotyledonous wood and nut shells; and from *c.* 8230–7730 cal BC herbivore dung – with calcareous faecal spherulites; all preserved variously as charred remains, phytoliths and calcitic ashes, as in Sheikh-e Abad in Trench 2, C619 S3 (Colour Plate 10). In extracted phytolith samples, all ashy deposits examined from both sites include both grass phytoliths as well as 2–28% dicot phytoliths, which represent significant quantities of wood fuel considering the low phytolith production rates of dicots (Shillito and Elliott 2013).

Food and diet

It is increasingly recognised that early sedentism was facilitated by access to and significant use of local wild, or managed wild food resources, both prior to and during early agriculture, and that there was significant local variation in the availability of these resources (Willcox 2005).

Well preserved fragile calcitic ash remains of nut shells have been identified as articulated specimens in thin-section in a range of burnt deposits *c.* 8240–7580 cal BC at Sheikh-e Abad and Jani, and *c.* 7450–7080 cal BC at Shimshara, constituting *c.* 10–20% of these deposits (Colour Plate 10). These remains confirm that nuts are an under-represented resource in charred plant remains assemblages as suggested by Savard *et al.* (2006) for multiple Neolithic sites across the northern Fertile Crescent, as they are highly combustible and frequently burnt to ash and are lost during flotation.

The Zagros is one of the heartlands of wild stands of barley (Wilcox 2005, Charles 2008), and genetic analysis of DNA haplotype frequencies of modern wild and traditionally cultivated barley suggests it was one of the centres of barley domestication (Morrell and Clegg 2007). Wild and domesticated barley were recovered from Ganj Dareh, 42km to the southeast of Sheikh-e Abad. The earliest domesticated barley here is dated to *c.* 7950 cal BC (van Zeist *et al.* 1986, Zeder 2009). Interestingly, no distinction could be made in the use of wild and domesticated barley at Ganj Dareh (van Zeist *et al.* 1986, 219), suggesting they were of similar food and cultural value.

Domestic type charred barley grain has been identified in flotation samples at Sheikh-e Abad, from *c.* 8230–7730 cal BC, based on large grain size. Its domestic status, however, cannot currently be confirmed due to the sparsity of chaff material and consequent lack of data on whether this barley had a tough rachis to prevent seed heads shattering when harvested, one of the key markers of domestication (Whitlam *et al.* 2013, Wilcox *et al.* 2008).

As chaff, including husks, is one of the first plant parts from which carbon is combusted (Boardman and Jones 1990), it is currently uncertain from the study of charred plants alone whether the absence of chaff is due to differential preservation due to combustion or to sparse use of cereals; crop processing and/or discard of chaff off-site; or use of chaff as fodder, for example. Micro-contextual study of the plant silica phytoliths that remain from both burnt as well as non-burnt wheat and barley chaff, therefore, is of particular importance in addressing these questions. In addition, wild and domesticated emmer wheat husks can potentially in some cases be distinguished by analysis of papillae morphometrics (Rosen 1992, Piperno 2006). At Sheikh-e Abad and Jani, few silica phytoliths from cereal husks or awns have been identified in thin-section or extracted phytolith samples to date. Some multi-celled cereal husks only had two of the three distinguishing criteria (Rosen 1992). Phytolith husks resembling barley have been identified in low percentages: at Jani in a range of contexts from *c.* 8210–7730 cal BC, including ash, floor sequences and a mudbrick; and at Sheikh-e Abad in ash in an open area in Trench 3, *c.* 7640–7580 cal BC. Phytolith husks resembling wheat have been identified at Jani in finely stratified midden deposits and in a mudbrick, *c.* 0.5m above a date of 8240–7740 cal BC. Additional samples and further micro-contextual analyses in thin-section are needed to examine the specific context of these cereal phytoliths. Their presence in mudbrick samples from Jani suggests scarcity on site may be due to the use of the by-products of cereal processing for a range of additional uses, including temper.

The richest and most diverse food remains identified in thin-section to date are from an exterior area of *in situ* burning, probably cooking, at Sheikh-e Abad Trench 2 (Colour Plate 10). In addition to the calcitic ash remains of nuts, this sequence of deposits in thin-section included charred seeds and possible tuber, 20–30% mollusc shells either from *Unio tigridis* or *Helix salomonica*, both of which have been identified at the site, and 2% burnt bone including fish bone. These diverse remains, together with zooarchaeological and archaeobotanical data, do support suggestions of a broad spectrum diet irrespective of the use of dung fuel, discussed below (R. Matthews *et al.* 2013).

The content of human coprolites identified in thin-section and by GC/MS is currently being studied and will add additional data on diet.

Early animal management (penning, fodder, fuel): integrated analysis of animal dung

The Zagros mountains are a preferred natural habitat for wild goats as they can browse on wood perennials and were hunted here from at least the Middle Palaeolithic (Zeder 2009). Based on sex-specific demographic patterns, the earliest zooarchaeological evidence for selective culling of morphologically wild goat for herding and herd propagation in the central Zagros is from the upland site of Ganj Dareh, *c.* 8100–7800 cal BC, where there are few male goats >2 years old. It is currently uncertain from preliminary zooarchaeological analyses whether goats at

Sheikh-e Abad were wild or managed; the four large goat skulls placed in a ritual building are certainly morphologically wild (Bendrey *et al.* 2013).

In investigating dung as a potential independent indicator of early animal management, it is particularly significant that widespread traces of burnt and non-burnt dung have been identified in thin-section from all four Zagros sites studied, and from at least *c.* 8230–7730 cal BC at Sheikh-e Abad and Jani, contemporary with indicators of management of morphologically wild populations at Ganj Dareh (W. Matthews *et al.* 2013). These identifications in thin-section have been confirmed by GC/MS analyses of six out of the seven corresponding spot samples analysed (Shillito *et al.* 2013), indicating a high degree of accuracy in thin-section characterisations, as discussed above.

At Jani, burnt ruminant dung has been identified in mixed-source fuel rake out on a surface, dated to *c.* 8240–7740 cal BC. The dung comprises 2–10% of the deposit, principally as calcitic ashes with traces of faecal spherulites (Colour Plates 10 and 12). One partially charred fragment, 1.2cm in size, resembles ethnoarchaeological samples of pen deposits, with few or no discernible pellet edges and compacted parallel oriented plant remains and plant impressions (W. Matthews 2005a). At Jani, this use of dung fuel and possible indicator of proximate pens/corrals coincides with a marked increase in the intensity of occupation in this area of the site, discussed below.

At Sheikh-e Abad, the earliest ruminant dung is also from dung burnt as fuel in levels dated to *c.* 8230–7730 cal BC, at the base of Trench 2 in thin-section sample C619 S3, *c.* 6m above natural (Colour Plate 12). Much of this dung fuel is present as calcitic ashes with little residual carbon, suggesting burning at moderately high temperatures >750°C, as observed at Jani and increasingly at Bestansur. In ethnoarchaeological research Sillar (1998) observed that dung is a preferred source of fuel as it retains its internal structure and thereby sustains oxidising conditions during combustion. One major explanation, therefore, for low densities of charred plant remains in some contexts at these sites is that dung was one of the principal sources of fuel. As it is highly combustible carbon is frequently fully burnt off, leaving traces predominantly of calcitic ash that is routinely lost during water flotation and wet-sieving.

Non-burnt ruminant dung has been discovered in many open areas and in thick compacted layers that resemble ethnoarchaeological samples from a pen and penning deposits at Çatalhöyük *c.* 7000 cal BC (W. Matthews 2005a) and in a small room *c.* 1.6 × 1.6m, 2.56m^2, in Building 1 at Sheikh-e Abad, *c.* 7590 cal BC, with human latrine waste. Preliminary micromorphological studies indicate that the diet of these penned animals included reeds, as at Çatalhöyük (W. Matthews 2005a), as well as dicot leaves, with some evidence of periodic perhaps seasonal variation diet (Shillito and Elliott 2013).

This widespread distribution of dung and the identification of a probable pen strongly suggest that herbivores were brought into the settlement at Sheikh-e Abad and may have lived for extended periods of time in close interdependent relationships with humans, speculatively as household members in Building 1, as suggested for the Neolithic more widely (Orton 2010).

Household and community activities, roles and relations: microstratigraphic insights into seasonal, annual and life cycles and histories

In this final section we briefly review the microstratigraphic evidence for how the transition to agriculture was shaped by and impacted on particular activities, roles and relations within households and communities at Jani. Maurice Bloch (2010) has argued that the location, settings and fixtures for particular actions provide an enduring representation of the roles and relations that make specific actions and lifeways possible and transcend the flux of everyday, seasonal and life-cycle changes. In analysis of the microstratigraphic sequences within particular areas, buildings and features and the geographic relations between them, we aim to examine continuity and change in particular actions and the roles, relations and lifeways that they represent by study of surfaces as settings and residues as evidence of particular actions and roles.

One of the most striking examples of how greater management of plants and animals was related to and impacted on activities, roles and relations is emerging at the site of Jani (Colour Plate 11). Here one of the largest cross-sections through a Neolithic settlement mound provides remarkable insight into continuity and change in practices across more than 45m of the south-eastern sector of the site, through 8m of occupation from the foundation of the site on a low natural mound. In this sequence we observed, recorded and sampled four major phases of microstratigraphic continuity across the entire length of the section, each of which was marked by an abrupt change in deposition and practices (R. Matthews *et al.* 2013).

The earliest sequence, Phase 1, comprises a series of massive bands of natural deposits mixed with comparatively sparse traces of anthropogenic activities, 1.3m deep, with fire-cracked stones, bone and lithics. Activity residues in thin-section include: charred plants and dicotyledonous wood, <20%, <0.7mm; phytoliths, <2%; 2–5% burnt aggregates and construction materials; and burnt and non-burnt bone, <2%, <6mm, some cracked and weathered (Colour Plate 11).

Phase 2 is marked by an abrupt change to deposition of multiple layers of well-preserved occupation deposits, *c.* 1.7–2.2m deep (Colour Plate 11). Phase 2a, 30–50cm thick, comprises alternating layers of discontinuous orange silty clay surfaces, *c.* 6m in length, 3–8cm thick, and bands of ash. In thin-section, the silty clay bands represent well-prepared constructed working surfaces with added vegetal stabilisers. The accumulated burnt deposits included more and better preserved plant remains than Phase 1, comprising charred reeds and grasses, shrubs and dicotyledonous wood, at >2–20%, <7–9mm, 10% phytoliths from reeds and grasses, calcitic/mineralised nut shells and calcitic ash *c.* 20% (Colour Plate 12). Also present are 2–5% burnt aggregates and bone >1.3cm in size, and fragments of omnivore coprolites, 2–5%, and herbivore dung, 2–10%, identified by their morphology and faecal spherulites inclusions (Colour Plate 12; Canti 1999, W. Matthews 2010) and confirmed by GC/MS analyses (Shillito *et al.* 2013). One layer is more fragmented and may be partially wind-sorted. Phase 2b, 1.2–1.7m thick, comprises multiple lenses of diverse plant remains in an open area/midden, often <2–20mm thick (Colour Plate 8). In thin-section these deposits include lenses of: a) up to 60% calcitic ashes with 5–30% charred wood, seeds, Poaceae and Cyperaceae stem/leaf and nut shells; 2–5% phytoliths also from Poaceae, 2–10% ruminant dung and 2–5% burnt aggregates; b) 50–70% well preserved articulated Poaceae phytoliths, confirmed by phytolith analysis as including reeds, some with occluded carbon, interbedded with ash and rounded aggregates of plasters, one with flecks of red ochre (Colour Plate 8); c) discarded aggregates, including diverse construction materials; and d) mixed deposits.

Phase 3 is represented by a horizon, 30–50cm thick, of at least four constructed fire-installations that were repeatedly plastered up to ten times (Colour Plate 11). Some include lenses of ash and fire-cracked stones. These were regrettably not sampled due to time constraints in the field. The activity surfaces associated with these were truncated and removed by levelling for Phase 4 in the Neolithic.

Phase 4 is marked by extensive construction of well-built mudbrick buildings with abutting walls and repeated lenses of thin white clean plaster floors, in sequences up to 10–25cm deep (Colour Plate 11). In thin-section these calcareous plasters are <1–2mm thick and only occasionally covered by thin grey brown ash with charred flecks, <1.2mm thick.

Discussion

Phase 1 is interpreted as a period of low intensity/frequency activities and discard associated with food preparation and cooking, mixed with natural deposits. The massive bedding, random orientation of inclusions, notably charred plant remains, and weathering of some bone fragments resemble both a) eroded residues of periodic hearths and camps observed by Mallol *et al.* (2007) in ethnoarchaeological and micromorphological studies of mobile hunter-gatherers, as well as b) periodically discarded residues at the edge of the mound at Çatalhöyük (W. Matthews 2005a), perhaps from a small settlement/camp farther in the core of the mound.

The abrupt change from these infrequent re-worked activity residues in Phase 1, to multiple accumulations of well-preserved finely stratified surfaces and ashy layers in Phase 2 suggests there was a sudden shift to greater sedentism, *c.*

8240–7740 cal BC. The repeated but shifting construction of well-prepared surfaces, *c.* 6m in extent, indicates that particular areas and settings were demarcated for specific activities, and that the activities and roles within these were thereby more visibly bounded. The accumulated occupation deposits on these surfaces in thin-section attest a range of domestic activities within these settings, including food preparation and cooking and consumption of nuts.

It is particularly significant that this sudden increase in the extent and continuity of activities in this area is associated with the first occurrence of herbivore dung burnt as fuel at the site and perhaps collected from a pen, discussed above. This correlation suggests that at Jani greater sedentism was associated with increased management of animals, in this case ruminants. These changes are contemporary with zooarchaeological evidence for selective culling of goat for herd management at the upland site of Ganj Dareh, *c.* 7900 cal BC (Zeder 2005). Zeder (2009) suggests goat were moved from this natural upland habitat zone to lowland foothill sites, by 7500 cal BC when they are present at Ali Kosh. The much earlier date of *c.* 7950 cal BC for herbivore dung at Jani has implications for studies of geographies of domestication and east–west movements, as Jani is at the boundary between the high and low Zagros. It remains to be established, however, whether this dung is from sheep or goat.

It is also significant that the start of this intensification of activity corresponds with a marked increase in thin-section and phytolith evidence for the use of grasses, reeds and sedges from <2–10%, *c.* 8240–7740 cal BC, as well as cereals as attested by husks resembling barley, discussed above. This intensification of activity coincides with the availability of abundant grass parkland as well as probable evidence for human impact on the environment. Grass pollen from Lake Zeribar peaks at 50%, *c.* 8500 cal BC, but is reduced by *c.* 8000 cal BC to *c.* 40% coincident with the introduction of *Plantago lanceolata*, indicative of vegetation disturbance and often associated with cultivation (van Zeist and Bottema 1977, van Zeist 2008). The identification of calcitic ash remains of nuts as well as possible tubers attests additional use of local wild or managed wild resources at this juncture in the development of sedentism, as suggested by Savard *et al.* (2006) for other regions. The presence of red ochre in some of these deposits attests the significance of ritual practices in this region, as argued more widely for the Near East (Asouti and Fuller 2013).

The four large fire-installations that represent Phase 3 resemble the 'baked-in place basin' features at Jarmo, which Braidwood *et al.* (1983) suggest are hearths/ovens. The construction of many early fire-installations in open areas rather than within buildings at Jarmo and other sites in the Zagros suggest communal engagement in food preparation and cooking (Pollock *et al.* 2010, Matthews 2012). As at Jani, the repeated plastering of these suggests increasing permanence of demarcation of features for activities and roles associated with food processing and cooking.

PLATE 1

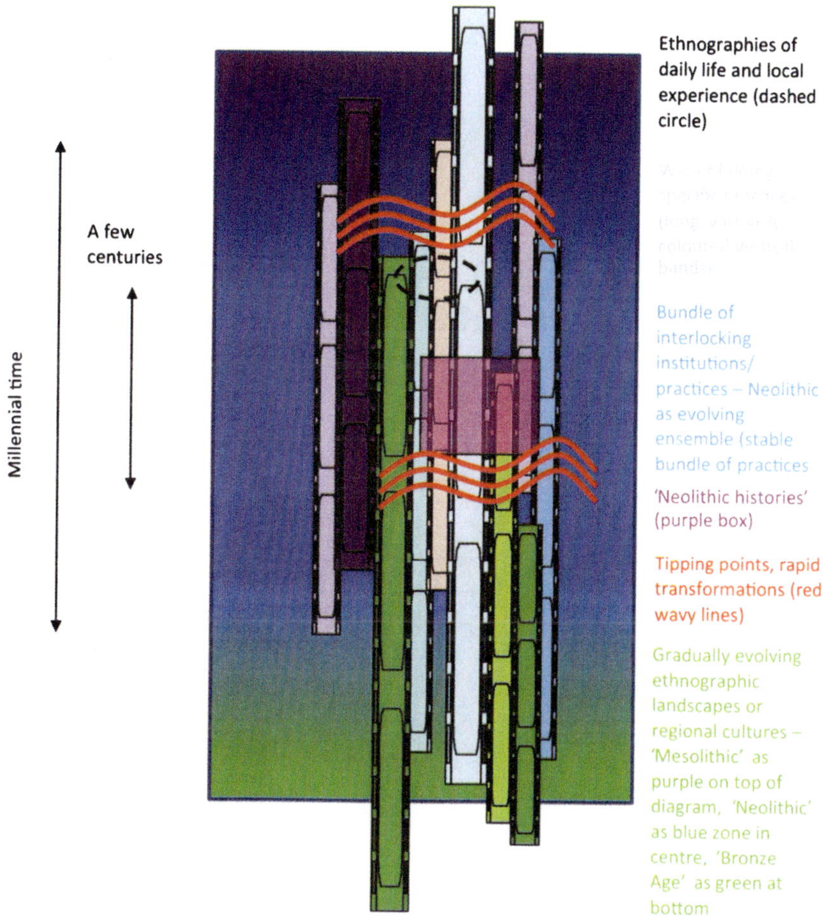

Narrative elements in the European Neolithic (schematic diagram).

PLATE 2

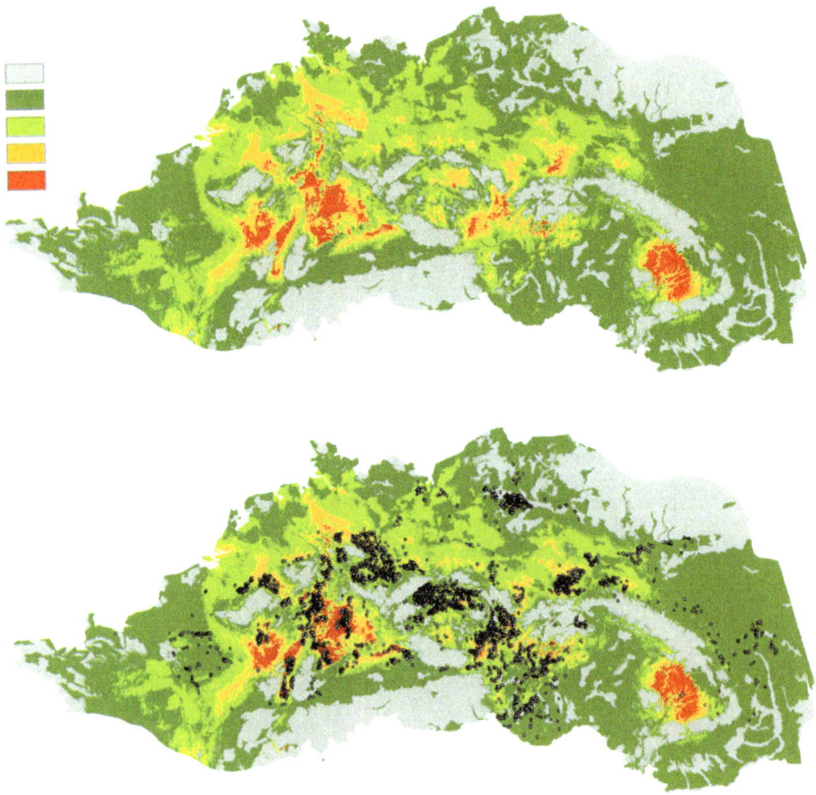

Map of best patch scores (top, red highest to grey lowest) and with the 6543 site locations (bottom).

PLATE 3

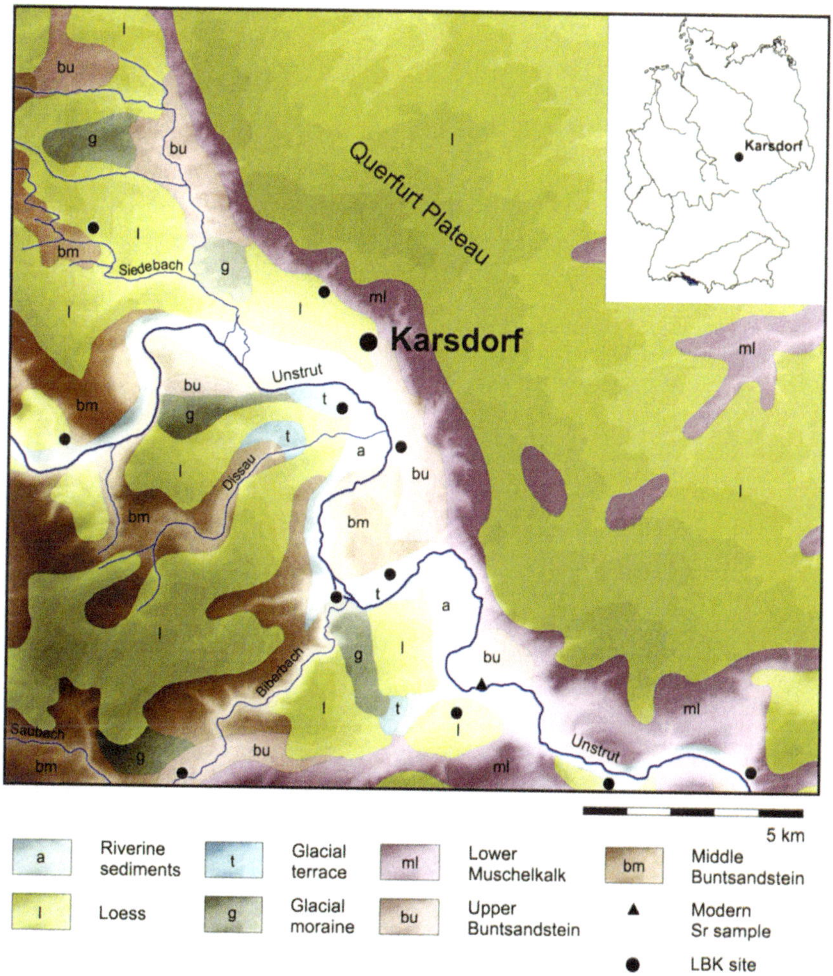

Geological map of the vicinity of the LBK settlement of Karsdorf with locations of known LBK sites. (Graphic: C. Knipper; Elevations after topographical map 1:50,000. Online: www.lvermgeo.sachsen-anhalt.de. Geological units after Geologische Übersichtskarte 1:400.000 (GÜK400). Online: http://webs.idu.de/lagb/lagb default.asp?thm=guek400&tk=C4734).

PLATE 4

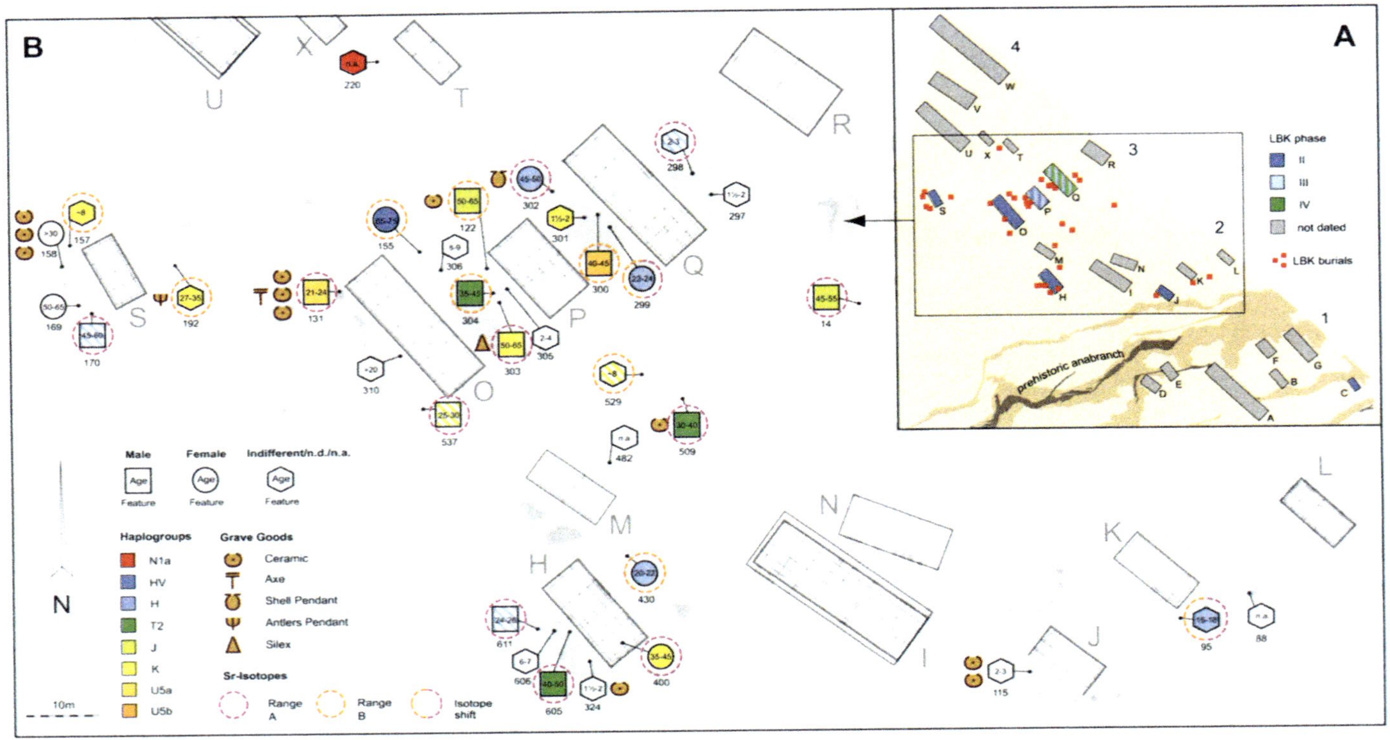

Map of the Linearbandkeramik settlement of Karsdorf. The symbols indicate age, sex, mtDNA haplogroups, grave goods, and strontium isotope ranges of the house-associated burials. Hatched symbols indicate individuals with shared haplotypes (Graphic: G. Brandt, R. Ganslmeier, C. Knipper).

PLATE 5

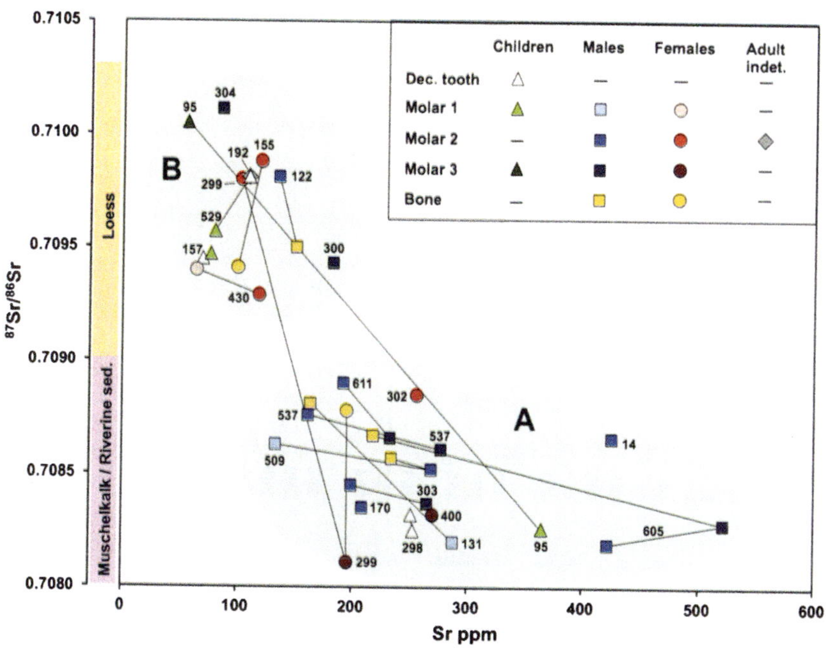

Bivariate plot of $^{87}Sr/^{86}Sr$ ratios and Sr concentrations of human enamel and bone samples. Connected data points indicate teeth from the same individual. The grey shades highlight the two data groups. The Sr isotope ratios can be associated with calcareous weathering products from Muschelkalk limestone, riverine sediments or gypsum of the upper Buntsandstein (group A) or loess (group B) (Data ranges after Maurer *et al.* 2012). (Graphic: C. Knipper).

PLATE 6

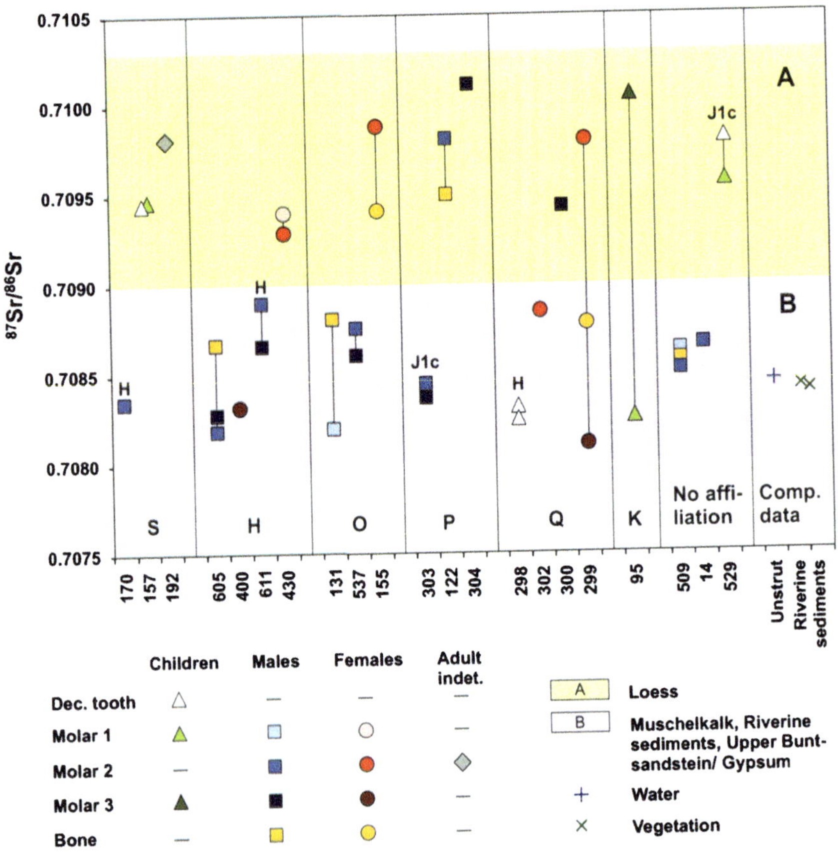

$^{87}Sr/^{86}Sr$ ratios of enamel and bone samples of the Karsdorf settlement burials sorted by houses. Connected data points are derived from the same individual. Identical mtDNA haplotypes of two (J1c), resp. three (H) individuals are indicated above the Sr isotope data points. Comparative data (water of the river Unstrut [blue cross], and a snail shell and a grass sample [green crosses]) as well as Sr isotope ranges after Maurer *et al.* 2012. (Graphic: C. Knipper).

PLATE 7

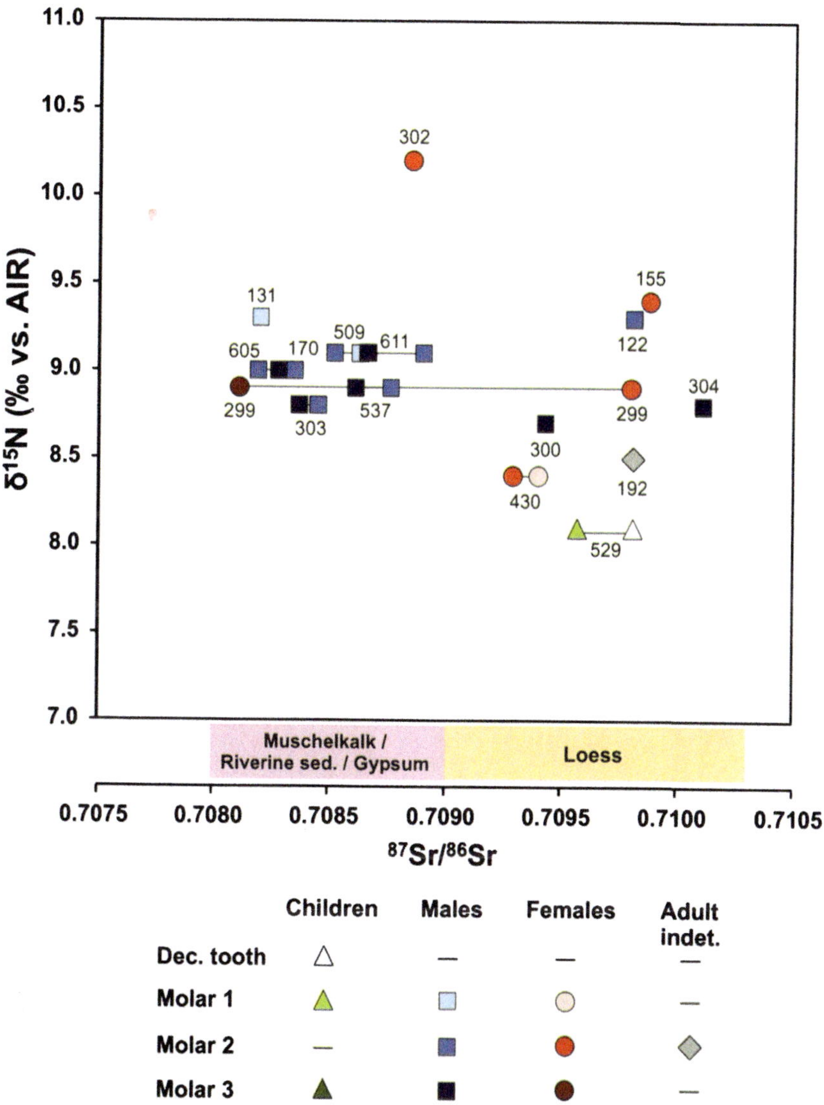

Bivariate plot of $^{87}Sr/^{86}Sr$ ratios of enamel and $\delta^{15}N$ values of bone collagen of the Karsdorf settlement burials. (Individual 95 is not included because of an erroneous association of excavation feature and analytical result in Oelze *et al.* 2011.) (Graphic: C. Knipper; $\delta^{15}N$ data after Oelze *et al.* 2011; Sr isotope ranges after Maurer *et al.* 2012).

PLATE 8

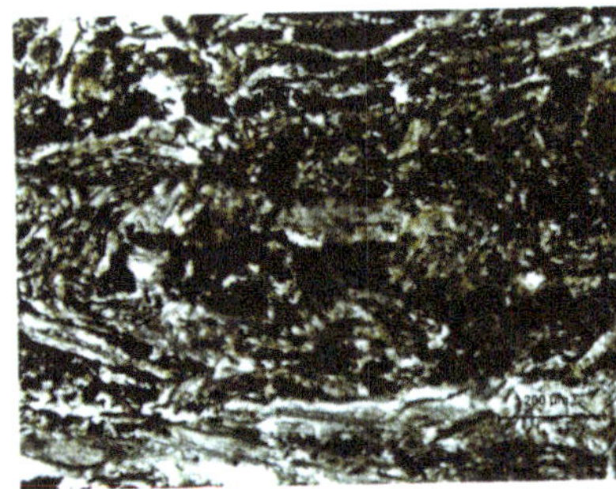

Articulated Poaceae with occluded carbon

Red ochre flecks with aggregates of white plaster

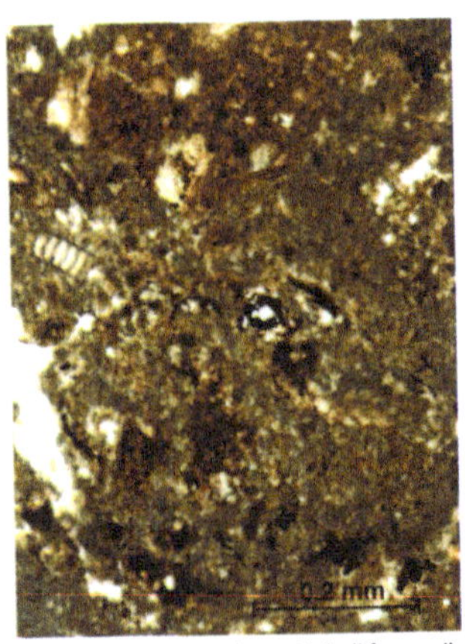

Calcitcic ashes with bulliform cells

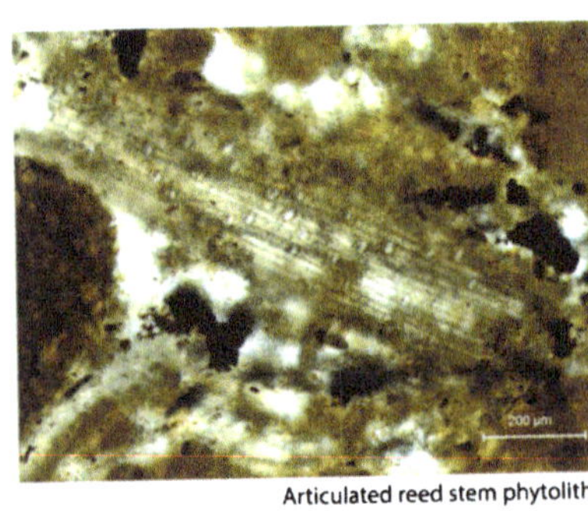

Articulated reed stem phytoliths

Micromorphological thin-section with multiple layers of midden-like deposits, including: burnt articulated Poaceae phytoliths with occluded carbon some reed stems; lens of red ochre flecks with white plaster aggregates; calcitic ashes with bulliform reed cells. (Adapted from W. Matthews *et al.* 2013, Fig. 7.8)

PLATE 9

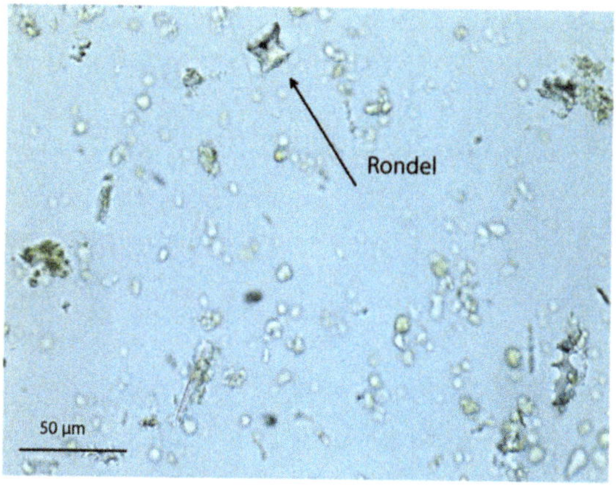

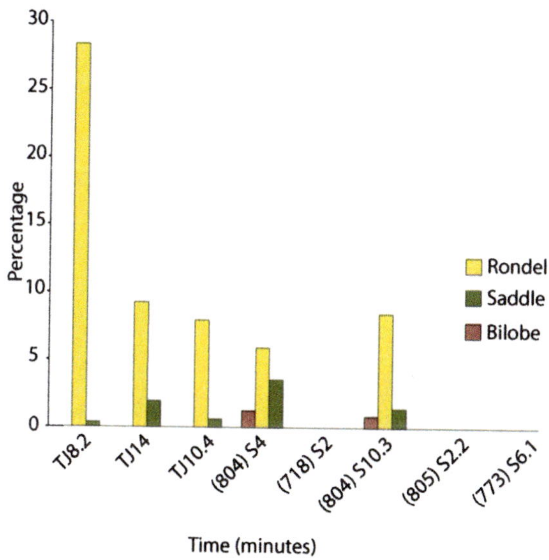

Phytolith analysis of densities of short cells as a percentage by total number counted. (Adapted from Shillito and Elliott 2013, Fig. 16.2)

PLATE 10

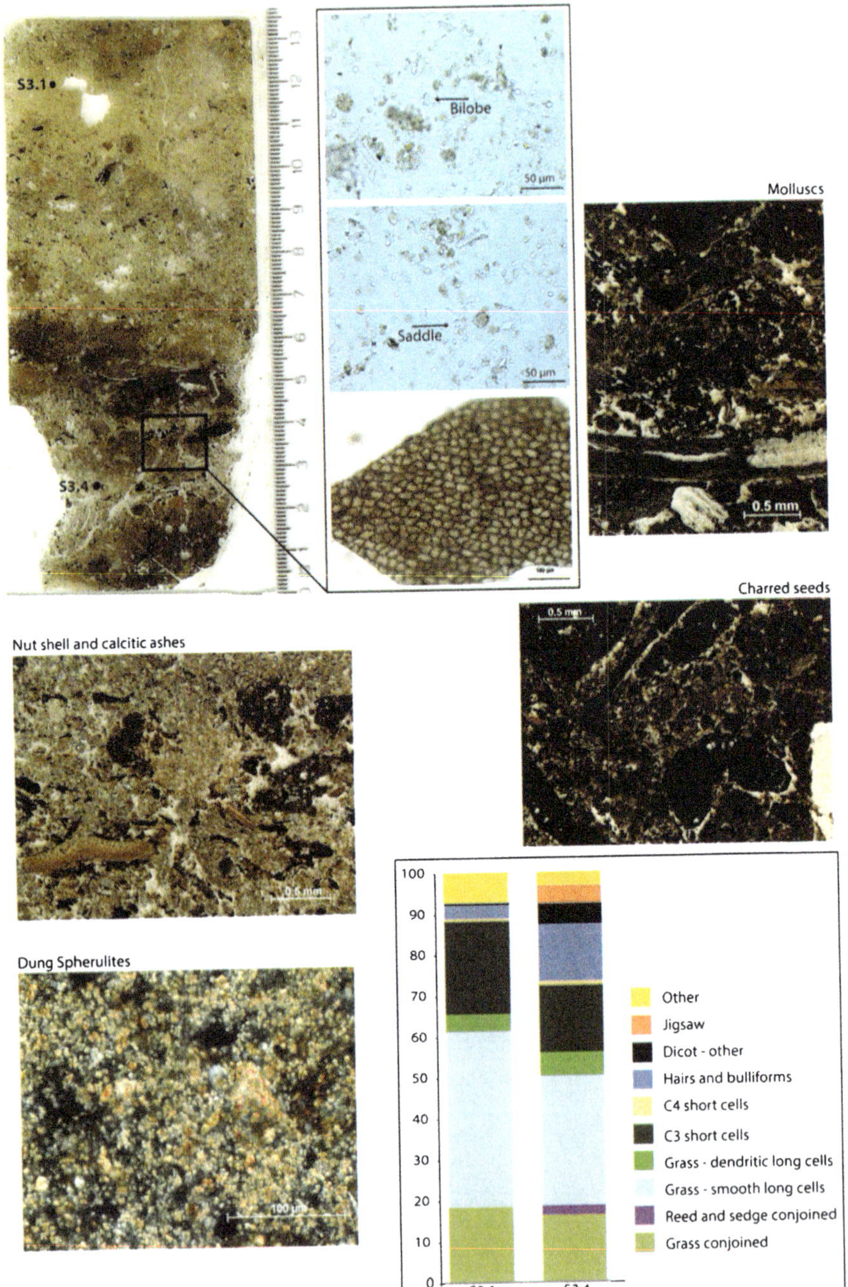

Sheikh-e Abad Trench 2, ShA 619.03: integrated micromorphology and phytolith analyses. (Adapted from W. Matthews *et al*. 2013, Fig. 7.3 and Shillito and Elliott 2013, Fig. 16.2)

PLATE 11

Jani: continuity and change in microstratigraphy, Phases 1–4. (Adapted from W. Matthews *et al.* 2013, Fig. 7.1 and Fig. 5.3)

PLATE 12

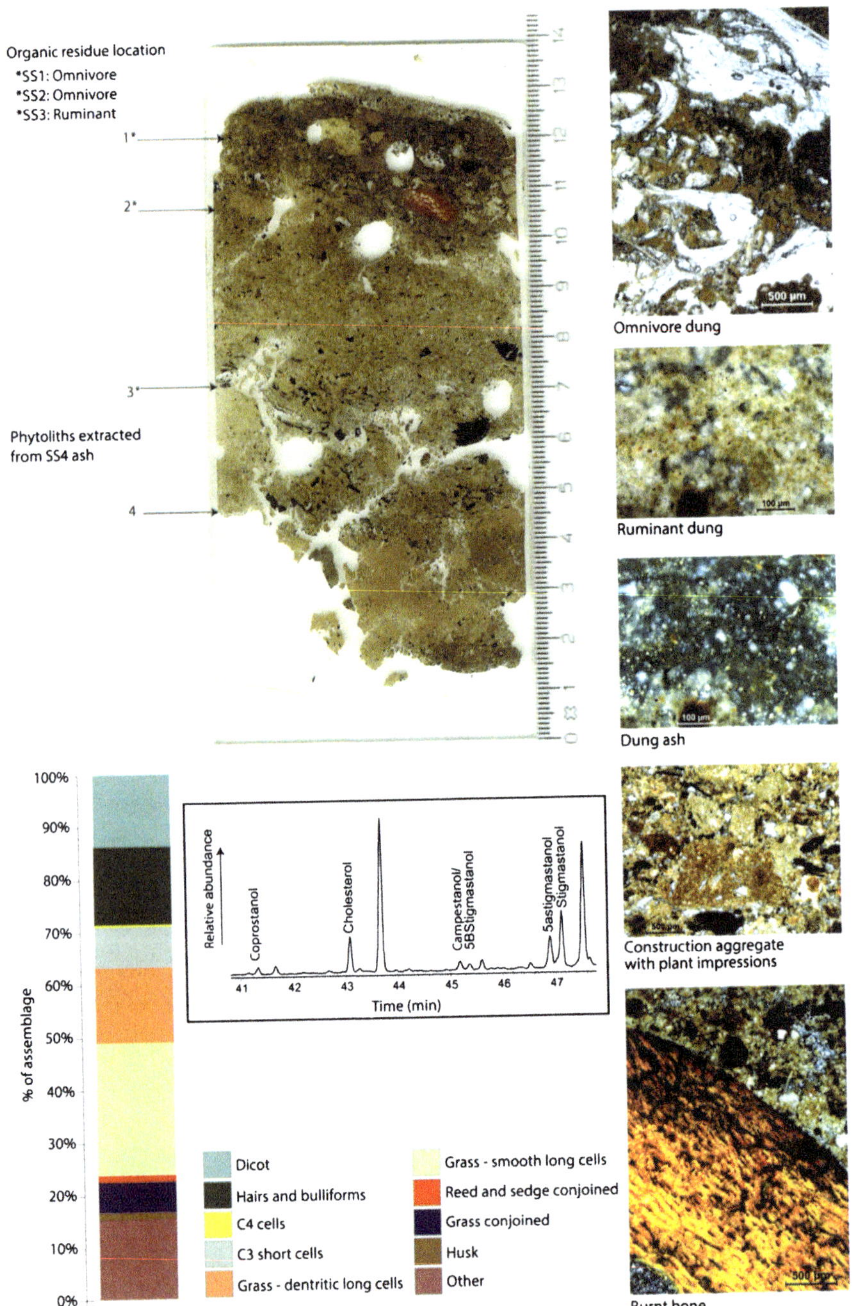

Jani Phase 2, TJ S10: integrated micromorphology and GC trace of ash layers with burnt ruminant dung and omnivore coprolites with partially digested bone. (Adapted from W. Matthews *et al.* 2013, Fig. 7.7)

The widespread levelling and construction of architecture across this area in Phase 4 suggest a major transformation in settlement and lifeways in this sector of the site. This architecture is remarkably well constructed, but regrettably not yet dated. The abutting walls suggest this architecture is rectilinear and agglutinative as also known from other sites from the early eighth to mid-seventh millennium cal BC. The density and conjoining of these buildings suggests an increasingly large as well as closely bound community. Some of the mudbricks are rectilinear, others are boat-shaped, which is characteristic of early eighth millennium cal BC architecture across the Near East, including at Ganj Dareh (Smith 1990). Variation between buildings in the material and shape of the mudbricks, suggests that there was some independent household access to land and selection of architectural materials and technologies. The repeated replastering of floors with multiple layers of white plaster, <1–2mm thick, and maintenance of these, attest greater focus on the house as an important social arena, as observed elsewhere in the Near East (Baird 2005). If the plasters were applied regularly and we assume that the building stood for 40–100 years based on ethnographic examples from the region and the dated life-span of Neolithic buildings at Çatalhöyük, we can estimate that the time interval represented by the *c*. 20 plasters in the West Building sequence, 8–10cm deep, may range from *c*. 2–5 years. The greater depth of floors in the East Building, at 23–25cm, may suggest this house was longer lived, or that the occupants observed different rhythms of renewal (Matthews 2005b).

At Shimshara, the presence of similar highly maintained buildings is suggested by the identification in thin-section of aggregates of multiple layers of whitewash coated in soot probably from the wall of a building that were re-used in a thick floor plaster, *c*. 1m above a date of *c*. 7450–7080 cal BC. These materials and practices most closely resemble those at Çatalhöyük where it has been possible to identify seasonal rhythms in soot accumulations on walls, with more in the winter months, and use of ovens on roofs (Matthews 2005b).

Conclusions: micro-contextual analyses of early farming

Developing integrated contextual approaches

Integration of microstratigraphic analyses in thin-section with high-resolution phytolith analysis of spot samples has significantly enhanced identification of the range of plants in archaeological deposits and interpretation of their significance. Integration with biomolecular analyses by GC/MS has confirmed the identification of many ruminant and omnivore coprolites in thin-section and established that at least two of these are human. Studies of their contents and implications for diet are currently in progress. The omnivore coprostanols at Sheikh-e Abad are currently some of the oldest archaeological examples identified to date in the world, at *c*. 10100–9450 cal BC (Shillito *et al.* 2013). Omnivore coprolites are widespread on Neolithic sites in the Zagros as also reported by Braidwood (1960) and the Near East more widely, with considerable potential for future analyses.

Whilst a much wider range of species have been identified in charred plant assemblages from flotation than in thin-section, there is a general correspondence in the densities and fragmentation of charred plants, which we are seeking statistically to demonstrate, as full analyses are completed. Thin-section analysis, however, has enabled identification of a wide range of other plant remains, not recoverable by flotation, including impressions of plants, abundant silica phytoliths from monocots and dicots, and calcic ashes, that may represent up to >50–70% of deposits, overturning arguments for low densities based on study of charred plants alone. In addition, it has provided significant information on a range of key taphonomic issues in the study of charred plant remains, especially in the study of the effects of combustion on biases in assemblages. For the Neolithic of the Zagros this research has established that nuts are underrepresented as a resource as they are highly combustible, but are widely preserved in articulated calcitic ashes in thin-section. Although cereal chaff is one of the first plant parts to combust (Boardman and Jones 1990), they are well preserved as phytoliths in spot samples and in thin-section. Herbivore dung was widely used as fuel, from at least *c*. 8000 cal BC, and is a potential major pathway for carbonised plant remains if burnt at low temperatures, and has a major bearing on considerations of whether a broad spectrum of plants represents human diet or dung burnt as fuel, as argued by Miller (2003a). Dung, however, is a good combustible source of fuel, and often burns at moderate to high temperatures at which carbon burns off, and is significantly underrepresented in charred plant remains assemblage. Many burnt ruminant dung remains are preserved as calcitic ashes with phytoliths and faecal spherulite inclusions.

Plant ecology and use

Studies of plant remains on archaeological sites are providing crucial indicators of the intersection between humans and the environment for comparison to off-site records. With regard to plant ecology and use, analyses of diverse plant materials on archaeological sites in integrated studies of plants in thin-section, phytolith samples and flotation assemblages, support recent interpretations of multiple proxies from Lake Zeribar that trees and wetlands were more extensive than previously argued. This research has confirmed that nuts are underrepresented, and were an important local resource that contributed to the viability of early sedentism, as argued by Savard *et al.* (2006). The scarcity of cereal husks in charred plant remains, whilst potentially due to combustion, has been confirmed by current phytolith analyses to be valid, and may be due in part to secondary uses, as cereal husk phytoliths have been identified in architectural materials at Jani, for example. Use of chaff as fodder is currently being investigated by study of ruminant dung contents.

Micro-contextual study of the associations of diverse plant remains with other food sources has confirmed that human diet at the cusp of early plant and animal management was broad-spectrum, with integrated zooarchaeological and archaeobotanical analyses. Remains of food resources and diet in one thin-section from an area of *in situ* burning included: charred seeds and tubers; nuts as calcitic ashes; molluscs, and animal bone and fish, *c.* 8000 cal BC.

Ruminant dung and early animal management

Ruminant dung has been identified at all four sites studied from at least as early as *c.* 8000 cal BC, contemporary with zooarchaeological identification of early herd management at Ganj Dareh, based on sex-demographic kill-off patterns. The earliest dung is predominantly for examples of dung burnt as fuel, with one aggregate that suggests dung may have been collected from a proximate pen/corral at Jani. Dung fuel was especially important in lowland Zagros sites, where fewer charred plants and remains of trees have been identified, but was widely used a mixed fuel resource throughout the Zagros.

Widespread non-burnt dung has been identified in open areas and a probable small pen within the settlement at Sheikh-e Abad by 7640–7580 cal BC, suggesting very close proximity and co-habitation in early sedentary settlements. Analyses of dung contents have identified possible seasonal variation in the diet of these penned animals, with evidence for some alternation between dicot leaves and reed- and grass-based diet.

Household and community activities, roles and relations

There is considerable evidence for coincidence of early animal management and greater sedentism at Jani and Sheikh-e Abad alongside use of local wild/managed plant resources including nuts and probably cereals, *c.* 8230–7730 cal BC, pending future excavation and analyses of earlier levels and other areas of these sites. Whilst the apparent abruptness of changes at Jani may be partially due to the location of the study area at the eastern edge of the site, these developments nevertheless mark particular tide-marks in the transition to early farming and greater sedentism.

Local and regional variation in early farming lifeways

Research at these and other sites in Iran and Iraqi Kurdistan has significantly filled a previous apparent gap of 1500 years in occupation in the Zagros from *c.* 10000–8500 cal BC, and is re-emphasising the importance of Zagros in understanding local and regional pathways, as well as early developments in these. The location of the study sites on a transect through the Zagros is enabling study of east–west movements and sharing of knowledge at local scale. That ritual was an important in Neolithic lifeways, as suggested most recently by Asouti and Fuller (2013) in the Zagros, is attested at the microscale by traces of red ochre in a range of contexts, and at the macro by ritual building at Sheikh-e abad, *c*. 7000 cal BC, for example.

Future research

Future research requires more excavation of these and other sites to understand more fully the spatial and temporal variation in ecological and social strategies of communities in the Zagros, and an intensive programme of systematic dating. With regard to integrated contextual approaches, it is apparent from these studies that integration is a multi-stage process, with results from previous analyses providing important new avenues for research and ways of incorporating and managing these. In these particular studies, there will be further comparison of micro-contextual analyses with zooarchaeologocal and charred plant remains analyses from flotation assemblages when these are complete. Current analyses in progress include collaboration with plant anatomists to aid identification of the diverse plant remains in thin-section, further phytolith and GC/MS analyses of new samples and thin-section investigations of the micro-context of cereal husks and omnivore and ruminant faecal contents in particular, as well as experiment and ethnoarchaeology. A range of other microfossils are also examined including pollen, starch, parasites, fungi, and plant lipids. These approaches are being developed by a range of research across the Near East, including Mentzer (Mallol *et al*. 2009) and Portillo *et al*. (2009), and will further enhance contextual understanding of early farming. These approaches are also applicable to studies of later plant and animal ecology management and diet, especially pastoralism and nomadism.

Acknowledgements

We wish to thank the many individuals and institutions who have kindly supported this research including: Iran's Cultural Heritage, Handicrafts and Tourism Organisation, the Iranian Centre for Archaeological Research and its former Director, Dr Hassan Fazeli, the Sulaimaniyah and Erbil Directorates of Antiquities and Heritage, State Board of Antiquities and Heritage, Baghdad, CZAP Co-Directors Prof Roger Matthews, Dr Yahgoub Mohammadifar, Dr Abbass Motarjem and Kamal Rashid Raheem and team members, the British Academy (BARDA-48993), NERC Life Sciences Mass Spectrometry Facility (LSMBRIS038), AHRC (AH/H034315/1), Universities of Reading, UCL and Bu Ali Sina, the Directorate General of Antiquities and Heritage, Turkey, the Çatalhöyük Research Project, John Jack for manufacture of thin-sections, and Sarah Lucas, Lisa Kennard and Margaret Mathews for the illustrations.

References

Altemüller, H.J. and Van Vliet-Lanoe, B. 1990. Soil thin section fluorescence microscopy. In L.A. Douglas (ed.), *Soil micromorphology: a basic and applied science*, 565–79. Amsterdam: Elsevier.

Asouti, E. 2005. Woodland vegetation and the exploitation of fuel and timber at Neolithic Çatalhöyük: report on the wood-charcoal macro-remains. In I. Hodder (ed.), *Inhabiting Çatalhöyük: reports from the 1995–99 seasons*, 213–60. Cambridge: The McDonald Institute for Archaeological Research and the British Institute of Archaeology at Ankara.

Asouti, E. and Austin, P. 2005. Reconstructing woodland vegetation and its exploitation by past societies, based on the analysis and interpretation of archaeological wood charcoal macro-remains. *Environmental Archaeology* 10, 1–18.

Asouti, E. and Fuller, D.Q. 2013. A contextual approach to the emergence of agriculture in southwest Asia: reconstructing early Neolithic plant-food production. *Current Anthropology* 54, 299–345.

Baird, D. 2005. The history of settlement and social landscapes in the early Holocene in the Çatalhöyük area. In I. Hodder (ed.), *Çatalhöyük perspectives. Themes from the 1995–99 Seasons*, 55–74. Cambridge: McDonald Institute for Archaeological Research and British Institute of Archaeology at Ankara.

Baird, D. 2008. The Boncuklu Project; the origins of sedentism, cultivation and herding in central Anatolia. *Anatolian Archaeology* 14, 11–13.

Baird, D., Carruthers, D., Fairbairn, A. and Pearson J. 2011. Ritual in the landscape: evidence from Pinarbasi in the seventh-millennium cal BC Konya Plain. *Antiquity* 85, 380–94.

Barker, G. 2006. *The agricultural revolution in prehistory: why did foragers become farmers?* Oxford: Oxford University Press.

Bendrey, R., Cole, R. and Tvetmarken, C.L. 2013. Zooarchaeology: preliminary assessment of the animal bones. In R.J. Matthews, W. Matthews, and Y. Mohammadifar (eds), *Central Zagros Archaeological Project: the earliest Neolithic of Iran: 2008 Excavations at Sheikh-e Abad and Jani*, 147–58. Oxford: British Institute of Persian Studies and Oxbow Books.

Bloch, M. 2010. Is there religion at Çatalhöyük...or are there just houses? In I. Hodder (ed.), *Religion and the emergence of civilization: Çatalhöyük as a case study*, 146–62. Cambridge: Cambridge University Press.

Boardman, S. and Jones, G. 1990. Experiments on the effects of charring on cereal plant components. *Journal of Archaeological Science* 17, 1–11.

Braidwood, R.J. 1960. Seeking the world's first farmers in Persian Kurdistan. *Illustrated London News* 237, 695–7.

Braidwood, R.J., Howe, B. and Reed, C.A. 1961. The Iranian prehistoric project. *Science* 133, 2008–10.

Braidwood, L.S., Braidwood, R.J., Howe, B., Reed, C.A. and Watson, P.J. (eds) 1983. *Prehistoric archaeology along the Zagros flanks* (Publication 105). Chicago: The University of Chicago Oriental Institute.

Brochier, J.E. 1992. Shepherds and sediments: geo-ethnoarchaeology of pastoral sites. *Journal of Anthropological Archaeology* 11, 47–102.

Bull, I.D., Elhmmali, M.M., Perret, V., Matthews, W., Roberts, D.F. and Evershed, R.P. 2005. Biomarker evidence of faecal deposition in archaeological sediments. In I. Hodder (ed.), *Inhabiting Çatalhöyük: reports from the 1995–99 seasons*, 415–20. Cambridge: The McDonald Institute for Archaeological Research and the British Institute of Archaeology at Anakara.

Bull, I.D. and Evershed, R.P. 2012. Organic geochemical signatures of ancient manure use. In R. Jones (ed.), *Manure matters: historical, archaeological and ethnographic perspectives*, 31–77. Farnham: Ashgate Publishing.

Bullock, P. 1985. *Handbook for soil thin section description*. Wolverhampton: Waine Research.

Canti, M. 1999. The production and preservation of faecal spherulites: animals, environment and taphonomy. *Journal of Archaeological Science* 26, 251–8.

Canti, M. 2003. Aspects of chemical and microscopic characteristics of plant ashes found in archaeological soils. *Catena* 54, 339–61.

Charles, M. 1998. Fodder from dung: the recognition and interpretation of dung-derived plant material from archaeological sites. *Environmental Archaeology: the Journal of Human Palaeoecology* 1, 1111–22.

Charles, M. 2008. East of Eden? A consideration of Neolithic crop spectra in the eastern Fertile Crescent and beyond. In S. Colledge and J. Conolly (eds), *The origins and spread of domestic plants in southwest Asia and Europe*, 37–52. Walnut Creek: Left Coast Press.

Conolly, J., Colledge, S., Dobney , K., Vigne, J-D., Peters, J., Stopp , B. Manning, K. and Shennan, S. 2011. Meta-analysis of zooarchaeological data from SW Asia and SE Europe provides insight into the origins and spread of animal husbandry. *Journal of Archaeological Science* 38, 538–45.

Courty, M.A., Goldberg, P. and Macphail, R.I. 1989. *Soils and micromorphology in archaeology*. Cambridge: Cambridge University Press.

Elliott, S. no date. Investigating early animal management, diet and ecology: portable X-ray fluorescence, spot-sampling and micromorphological sampling. In R.J. Matthews, W. Matthews and K. Rasheed (eds), *Central Zagros Archaeological Project archive report. Excavations at Bestansur and Shimshara, Sulaimaniyah province, Kurdistan Regional Government, Republic of Iraq 18th August–27th September 2012. Survey in Zarzi region, January 2013*, 53–64.

Flannery, K.V. 1969. Origins and ecological effects of early domestication in Iran and the Near East. In P.J. Ucko and G.W. Dimbleby (eds), *The domestication and exploitation of animals*, 73–100. London: Duckworth.

Ghosh, R., Gupta, S., Bera, S., Jiang, H.E., Li, X. and Li, C.S. 2008. Ovi-caprid dung as an indicator of paleovegetation and paleoclimate in northwestern China. *Quaternary Research* 70, 149–57.

Helbaek, H. 1969. Plant collecting, dry-farming and irrigation in prehistoric Deh Luran. In F. Hole, K.V. Flannery and J.A. Neeley (eds), *Prehistory and human ecology of the Deh Luran plain*, 383–426. Michigan: University of Michigan Museum of Anthropology.

Hodder, I. 2006. *Çatalhöyük: the leopard's tale. Revealing the mysteries of Turkey's ancient town*. London: Thames and Hudson.

Hodgson, J.M. 1976. *Soil Survey field handbook*. Harpenden: Soil Survey.

Hole, F. 1996. The context of caprine domestication in the Zagros region. In D. Harris (ed.), *The origins and spread of agriculture and pastoralism in Eurasia*, 263–81. Washington: Smithsonian Institution Press.

Hole, F., Flannery, K.V. and Neely, J.A. 1969. *Prehistory and human ecology of the Deh Luran plain: an early village sequence from Khuzistan, Iran*. Ann Abor: University of Michigan.

Houben, H. and Guillaud, H. 1989. *Earth construction: a comprehensive guide*. Chippenham: Intermediate Technology Development Group.

Hubbard, R.N.L.B. 1990. The carbonised seeds from Tepe Abdul Hosein: results of preliminary analyses. In J. Pullar (ed.), *Tepe Abdul Hosein: a Neolithic site in western Iran, excavations 1978*, 217–22. Oxford: Archaeopress.

Jenkins, E. 2009. Phytolith taphonomy: a comparison of dry ashing and acid extraction on the breakdown of conjoined phytoliths formed in Triticum durum. *Journal of Archaeological Science* 36, 2402–07.

Jones, G. 1998. Distinguishing food from fodder. *Environmental Archaeology: The Journal of Human Palaeoecology* 1, 95–8.

Jones, M.D. and Roberts, C.N. 2008. Interpreting lake isotope records of Holocene environmental change in the Eastern Mediterranean. *Quaternary International* 181, 32–8.

Kozlowski, S.K. and Aurenche, O. 2005. *Territories, boundaries and cultures in the Neolithic Near East*. Lyon: Jean Pouilloux.

Leroi-Gourhan, A. 1969. Pollen grains of Gramineae and Cerealia from Shanidar and Zawi Chemi. In P.J. Ucko and G.W. Dimbleby (eds), *The domestication and exploitation of plants and animals*, 143–8. London: Duckworth.

Macphail, R.I., Courty, M.A., Hather, J., Wattez, J., Ryder, M., Cameron, N.P., Branch, P. 1997. The soil micromorphological evidence of domestic occupation and stabling activities. *Memorie dell'Istituto Italiano di Paleontologia Umana* 5, 53–85.

Mallol, C., Mentzer, S. M. and Wrinn, P.J. 2009. A micromorphological and mineralogical study of site formation processes at the Late Pleistocene site of Obi-Rakhmat, Uzbekistan. *Geoarchaeology* 24, 548–75.

Mallol, C., Marlowe, F.W., Wood, B.M. and Porter, C.C. 2007. Earth, wind, and fire: ethnoarchaeological signals of Hadza fires. *Journal of Archaeological Science* 34, 2035–52.

Matthews, R. and Postgate, J.N. 2001. Contextual analysis of the use of space at two Near Eastern Bronze Age sites: Tell Brak (north-eastern Syria) and Kilise Tepe (southern Turkey). Archaeology Data Service http://ads.ahds.ac.uk/catalogue/projArch/TellBrak/.

Matthews, R.J. and Fazeli-Nashli, H. (eds) 2013. *The Neolithisation of Iran: the formation of new societies*. Oxford: British Association of Near Eastern Archaeology and Oxbow Books.

Matthews, R.J., Matthews, W. and Mohammadifar, Y. 2013. *Central Zagros Archaeological Project: the earliest Neolithic of Iran. 2008 Excavations at Sheikh-e Abad and Jani*. Oxford: British Institute of Persian Studies and Oxbow Books.

Matthews, W. 2005a. Micromorphological and microstratigraphic traces of uses of space. In I. Hodder (ed.), *Inhabiting Çatalhöyük: reports from the 1995–99 seasons*, 355–98, 553–72. Cambridge: McDonald Institute for Archaeological Research and British Institute of Archaeology.

Matthews, W. 2005b. Life-cycle and life-course of buildings. In I. Hodder (ed.), *Çatalhöyük perspectives. Themes from the 1995–99 seasons*, 125–51. Cambridge: McDonald Institute for Archaeological Research and British Institute of Archaeology at Ankara.

Matthews, W. 2010. Geoarchaeology and taphonomy of plant remains and microarchaeological residues in early urban environments in the Ancient Near East. *Quaternary International* 214, 98–113.

Matthews, W. 2012. Defining households: micro-contextual analysis of early Neolithic households in the Zagros, Iran. In B.J. Parker and C.P. Foster (eds), *Household archaeology: new perspectives from the Near East and beyond*, 183–216. Winona Lake: Eisenbrauns.

Matthews, W. 2013. Contexts of Neolithic interaction: geography, palaeoclimate and palaeoenvironment of the Central Zagros. In R.J. Matthews, W. Matthews and Y. Mohammadifar (eds), *Central Zagros Archaeological Project: the earliest Neolithic of Iran: 2008 excavations at Sheikh-e Abad and Jani*. Oxford: British Institute for Persian Studies and Oxbow Books.

Matthews, W. no date. Architecture, traces of activities and site formation processes. In R.J. Matthews, W. Matthews and K. Rasheed (eds), *Central Zagros Archaeological Project archive report. Excavations at Bestansur and Shimshara, Sulaimaniyah Province, Kurdistan Regional Government, Republic of Iraq 18th August-27th September 2012. Survey in Zarzi region, January 2013*, 19–28.

Matthews, W., Shillito, L.-M. and Elliott, S. 2013. Investigating Early Neolithic materials, ecology and sedentism: micromorphology and microstratigraphy. In R.J. Matthews, W. Matthews and Y. Mohammadifar (eds), *Central Zagros Archaeological Project: the earliest Neolithic of Iran: 2008 excavations at Sheikh-e Abad and Jani*, 67–104. Oxford: British Institute for Persian Studies and Oxbow Books.

Mellaart, J. 1967. *Çatal Hüyük: a Neolithic town in Anatolia*. London: Thames and Hudson.

Miller, N.F. 1996. Seed eaters of the ancient Near East: human or herbivore. *Current Anthropology* 37, 521–8.

Miller, N.F. 2003a. Archaeobotany in Iran, past and future. In N.F. Miller and K. Abdi (eds), *Yeki bud, yeki nabud. Essays on the archaeology of Iran in honor of William M. Sumner*, 8–15. Los Angeles: Cotsen Institute of Archaeology, University of California.

Miller, N.F. 2003b. Plant remains from the 1996 Excavation. In Alizadeh (ed.), *Excavations at the prehistoric mound of Chogha Bonut, Khuzestan, Iran*, 123–8. Chicago: Oriental Institute of Chicago Publications.

Mithen, S. 2003. *After the ice: a global human history, 20,000–5,000 BC*. London: Weidenfeld and Nicolson.

Morell, P.L. and Clegg, M.T. 2007. Genetic evidence for a second domestication of barley (Hordeum vulgare) east of the Fertile Crescent. *Proceedings of the National Academy of Sciences of the USA* 104, 3289–94.

Naderi, S., Rezaiei, H.-R., Pompanon, F., Blum, M.G.B., Negrini, R., Naghash, H.-R., Balkiz, O., Mashkour, M., Gaggiotti, O.E., Ajmone-Marsan, P., Kence, A., Vigne, J.-D. and Taberlet, P. 2008. The goat domestication process inferred from large-scale mitochondrial DNA analysis of wild and domestic individuals. *Proceedings of the National Academy of Sciences of the USA* 105, 17659–64.

Orton, D. 2010. Both subject and object: domestication, inalienability, and sentient property in prehistory. *World Archaeology* 42, 188–200.

Piperno, D. 2006. *Phytoliths: a comprehensive guide for archaeologists and paleoecologists*. Oxford: AltaMira Press.

Pollock, S., Bernbeck, R. and Abdi, K. 2010. *The 2003 excavations at Tol-e Basi, Iran: social life in a Neolithic village*. Mainz: Philipp von Zabern.

Portillo, M., Albert, R.M. and Henry, D.O. 2009. Domestic activities and spatial distribution in Ain Abu Nukhayla (Wadi Rum, Southern Jordan): the use of phytoliths and spherulites studies. *Quaternary International* 193, 174–83.

Riehl, S., Zeidi, M. and Conard, N.J. 2013. Emergence of agriculture in the foothills of the Zagros Mountains of Iran. *Science* 341, 65–7.

Roberts, N. 2002. Did prehistoric landscape management retard the post-glacial spread of woodland in Southwest Asia? *Antiquity* 76, 1002–10.

Rosen, A.M. 1992. Preliminary identification of silica skeletons from Near Eastern archaeological sites: an anatomical approach. In G. Rapp and S.C. Mulholland (eds), *Phytolith systematics: emerging issues*, 129–47. New York: Plenum Press.

Rosen, A.M. 2003. Preliminary phytolith analyses. In A. Alizadeh (ed.), *Excavations at the prehistoric mound of Choga Bonut, Khuzestan: seasons 1976/77, 1877/78 and 1996*, 129–35. Chicago: The Orietal Institute of the University of Chicago in association with the Iranian Cultural Heritage Organisation.

Rosen, A.M. 2005. Phytolith indicators of plant and land use at Çatalhöyük. In I. Hodder (ed.), *Inhabiting Çatalhöyük: report from the 1995–99 seasons*, 485–9. Cambridge: The McDonald Institute for Archaeological Research and the British Institute at Ankara.

Ryan, P. 2011. Plants as material culture in the Near Eastern Neolithic: perspectives from the silica skeleton artefactual remains at Çatalhöyük. *Journal of Anthropological Archaeology* 30, 292–305.

Savard, M., Nesbitt, M. and Jones, M.K. 2006. The role of wild grasses in subsistence and sedentism: new evidence from the northern fertile Crescent. *World Archaeology* 38, 179–96.

Schiegl, S., Goldberg, P., Bar-Yosef, O. and Weiner, S. 1996. Ash deposits in Hayonim and Kebara caves, Israel: macroscopic, microscopic and mineralogical observations, and their archaeological implications. *Journal of Archaeological Science* 23, 763–81.

Schweingruber, F.H. 1990. *Anatomy of European woods*. Stuttgart: Haupt.

Shahack-Gross, R. 2011. Herbivorous livestock dung: formation, taphonomy, methods for identification, and archaeological significance. *Journal of Archaeological Science* 38, 205–18.

Shillito, L.-M. 2011. *Daily activities, diet and resource use at Neolithic Çatalhöyük : microstratigraphic and biomolecular evidence from middens*. Oxford: Archaeopress.

Shillito, L.-M. 2013. Grains of truth or transparent blindfolds? A review of current debates in archaeological phytolith analysis. *Vegetation History and Archaeobotany* 22, 71–82.

Shillito, L.-M. and Elliott, S. 2013. Phytolith indicators of plant resource use at Sheikh-e Abad and Jani. In R.J. Matthews, W. Matthews and Y. Mohammadifar (eds), *Central Zagros Archaeological Project: the earliest Neolithic of Iran: 2008 excavations at Sheikh-e Abad and Jani*, 185–200. Oxford: British Institute of Persian Studies and Oxbow Books.

Shillito, L.-M., Bull, I.D., Matthews, W., Almond, M., Williams, J. and Evershed, R.P. 2011. Biomolecular and micromorphological analysis of suspected faecal deposits at Neolithic Çatalhöyük, Turkey. *Journal of Archaeological Science* 38, 1869–77.

Shillito, L.-M., Matthews, W., Bull, I.D. and Williams, J. 2013. Biomolecular investigations of faecal biomarkers at Sheikh-e Abad and Jani. In R.J. Matthews, W. Matthews, and Y. Mohammadifar (eds), *Central Zagros Archaeological Project: the earliest Neolithic of Iran: 2008 excavations at Sheikh-e Abad and Jani*, 105–16. Oxford: British Institute of Persian Studies and Oxbow Books.

Sillar, B. 1998. Dung by preference: the choice of fuel as an example of how Andean pottery production is embedded within wider technical, social and economic practices. *Archaeometry* 42, 43–60.

Smith, P.E.L. 1990. Architectural innovation and experimentation at Ganj Dareh, Iran. *World Archaeology* 21, 323–35.

Stevens, L.R., Ito, E. and Wright Jr., H.E. 2008. Variations in effective moisture at Lake Zeribar, Iran during the last glacial period and Holocene, inferred from the delta O-18 values of authigenic calcite. In K. Wasylikowa and A. Witkowski (eds), *Diatron monographs volume 08: the palaeoecology of Lake Zeribar and surrounding areas, western Iran, during the last 48,000 years*, 283–302. Koenigstein: Koeltz Scientific Books.

Stevens, L.R., Wright, H.E. and Ito, E. 2001. Proposed changes in seasonality of climate during the Late Glacial and Holocene at Lake Zeribar, Iran. *The Holocene* 11, 747–55.

Stoops, G. 2003. *Guidelines for analysis and description of soil and regolith thin-sections*. Madison: Soil Science Society of America.

Tsartsidou, G., Lev-Yadun, S., Albert, R.-M., Miller-Rosen, A., Efstratiou, N. and Weiner, S. 2007. The phytolith archaeological record: strengths and weaknesses evaluated based on a quantitative modern reference collection from Greece. *Journal of Archaeological Science* 34, 1262–75.

Turner, R., Roberts, N., Eastwood, W. J., Jenkins, E. Rosen, A. 2010. Fire, climate and the origins of agriculture: micro-charcoal records of biomass burning during the last glacial-interglacial transition in Southwest Asia. *Journal of Quaternary Science* 25, 371–86.

Valamoti, S.M. 2013. Towards a distinction between digested and undigested glume bases in the archaeobotanical record from Neolithic northern Greece: a prelimimary experimental investigation. *Environmental Archaeology* 18, 31–42.

Van der Veen, M. 2007. Formation processes of desiccated and carbonized plant remains – the identification of routine practice. *Journal of Archaeological Science* 34, 968–90.

Van Zeist, W. 2008. Late Pleistocene and Holocene vegetation at Zeribar. *Diatom Monographs*, 53–104.

Van Zeist, W. and Bottema, S. 1977. Palynological investigations in western Iran. *Paleohistoria* 19, 19–95.

Van Zeist, W., Smith, P.E.L., Palfenier-Vegter, R.M., Suwijn, M. and Casparie, W.A. 1986. An archaeobotanical study of Ganj Dareh. *Palaeohistoria* 26, 201–24.

Vigne, J.-D., Carrère, I., Briois, F. and Guilaine, J. 2011. The early process of mammal domestication in the Near East: new evidence from the Pre-Neolithic and Pre-Pottery Neolithic in Cyprus. *Current Anthropology* 52, S255–71.

Vigne, J.D., Briois, F., Zazzo, A., Willcox G., Cucchi, T., Thiébault, S., Carrère, I., Franel, Y., Touquet, R., Moreau, C., Comby, C. and Guilaine, J. 2012. The first wave of cultivators spread to Cyprus at least 10,600 years ago. *Proceedings of the National Academy of Sciences of the United States of America* 109, 8445–9.

Vincentini, A., Baber, J.C., Aliscioni, S.S., Giussani, L.M. and Kellogg, E.A. 2008. The age of the grasses and clusters of origins of C_4 photosynthesis. *Global Change Biology* 14, 2963–77.

Wang, Y. and Lyu, H. 1992. *The study of phytoliths and its application*. Beijing: Ocean Press.

Wasylikowa, K. and Witkowski, A. 2008. *The palaeoecology of Lake Zeribar and surrounding areas, Western Iran, during the last 48,000 years*. Ruggell: A.R.G. Gantner Verlag K.G.

Whitlam, J. no date. Archaeobotany. In R.J. Matthews and W. Matthews (eds), *Central Zagros Archaeological Project Archive Report. Excavations at Bestansur and Shimshara, Sulaimaniyah Province, Kurdistan Regional Government, Republic of Iraq 18th August–27th September 2012. Survey in Zarzi Region, January 2013*, 75–84.

Whitlam, J., Ilkhani, H., Bogaard, A. and Charles, M. 2013. The plant macrofossil evidence from Sheikh-e Abad: first impressions. In R.J. Matthews, W. Matthews and Y. Mohammadifar (eds), *Central Zagros Archaeological Project: The Earliest Neolithic of Iran: 2008 Excavations at Sheikh-e Abad and Jani*, 175–84. Oxford: British Institute of Persian Studies and Oxbow Books.

Willcox, G. 1990. Charcoal remains from Tepe Abdul Hosein. In J. Pullar (ed.), *Tepe Abdul Hosein: A Neolithic site in Western Iran, excavations 1978*, 227–33. Oxford: Archaeopress.

Willcox, G. 2005. The distribution, natural habitats and availability of wild cereals in relation to their domestication in the Near East: multiple events, multiple centres. *Vegetation History and Archaeobotany* 14, 534–41.

Willcox, G. Fornite S. Herveux L. 2008. Early Holocene cultivation before domestication in northern Syria. *Vegetation History and Archaeobotany* 17, 313–25.

Zeder, M.A. 2005. A view from the Zagros: new perspectives on livestock domestication in the Fertile Crescent. In J.-D. Vigne, J. Peters, and D. Helmer (eds), *First steps of animal domestication*, 125–46. Oxford: Oxbow Books.

Zeder, M. 2009. The Neolithic macro-(r)evolution: macroevolutionary theory and the study of culture change. *Journal of Archaeological Research* 17, 1–63.

15

Violence in Neolithic North-west Europe

A Population Perspective

RICK J. SCHULTING AND LINDA FIBIGER

Introduction

THE PAST FEW DECADES HAVE SEEN A SIGNIFICANT INCREASE in the reporting of prehistoric violence, and a consideration of its role in past societies (Bowles 2012, Keeley 1996). This, in turn, has engendered renewed debate over the timing of the appearance of what can be termed 'war' and its importance in socio-political change. The onset of farming is seen in some discussions as playing a crucial role in this debate, with some denying the existence of warfare prior to this watershed in human history (Beyneix 2007, Peter-Röcher 2007). The European Neolithic provides a rich – if still fragmented – database to bring to bear on these issues (Armit *et al.* 2007, Schulting and Fibiger 2012). Here, we present an overview of selected results from a recently completed research project aimed at re-analysing a large number of Neolithic crania from north-west Europe. In total some 2000 crania were examined from institutions in Denmark, Sweden, Germany, Britain, Ireland and France. A central emphasis in the project was placed on putting the evidence for interpersonal violence into a population perspective, rather than focusing on individual or site-specific cases of trauma, important as these unquestionably are (Meyer *et al.* 2009, Teschler-Nicola 2012, Teschler-Nicola *et al.* 1996, Wahl and König 1987, Wahl and Trautmann 2012). This provides a baseline for the prevalence of at least cranial trauma, enabling comparisons at various levels, from different contemporary contexts (such as caves and monuments) and demographic (age/sex) groups, to broader comparisons between different regions – not just within Europe, but globally – and time periods. The findings presented here are ultimately aimed at this goal.

For this contribution we focus on a comparison of material from southern Scandinavia (Denmark and Sweden) and Britain and Ireland. These areas share some elements of cultural similarity during the early and middle Neolithic (Rowley-Conwy 2004). Both are on the fringes of the spread of farming through central Europe, and both saw a relatively late appearance, probably from *c.* 4000 cal BC onwards (Sørensen and Karg in press, Whittle *et al.* 2011), of the set of traits and behaviours that together make up the 'Neolithic' (though Sheridan (2010; this volume Chapter 19) raises the possibility of an earlier inception for Britain and Ireland). These include first and foremost the domesticated plants and animals lying at the core of a mixed farming/herding way of life, along with (to varying degrees) novel material culture, and new forms of domestic and mortuary architecture. The middle Neolithic in both regions sees the introduction of passage tombs, from which a number of the individuals in the present study derive.

A significant proportion of the extant material was examined for Neolithic Britain and Ireland (*c.* 3800–3000 cal BC), and for middle and late Neolithic Denmark (*c.* 3300–2000 cal BC). For Britain and Ireland in particular, this includes a large number of individuals belonging to communities situated within the first few centuries of the appearance of farming, with coverage from southern England and Wales to Orkney in northern Scotland. Eastern Denmark (mainly Zealand, Lolland and Falster) contributes the majority of the southern Scandinavian material examined, as a result of the generally good preservation conditions there (Fibiger *et al.* 2013). Most individuals from this region considered here belong to a period when farming was already well established. The early Neolithic material from Denmark has previously been analysed (Bennike 1985), and is currently being re-assessed (Pia Bennike, pers. comm.). Thus, while there is some overlap, the southern Scandinavian collections tend to extend rather later in time than the British and Irish material, for which the appearance of Beaker pottery *c.* 2500 cal BC was used as a cut-off point, though late Neolithic (3000–2500 cal BC) human remains are notoriously rare in Britain and feature only peripherally.

Methodology

The focus on cranial trauma is, in part, dictated by the diverse nature of the available material across the study area. Many of the collections are the result of nineteenth- and early twentieth-century excavations, in which interest in post-cranial skeletal remains was often limited, such that they were not always retained. Crania, by contrast, were of special interest given their central role in the study of race in the fledging discipline of physical anthropology (Ripley 1899). But equally, in the absence of edged weapons, and excluding projectile injuries, post-cranial trauma is usually interpreted as reflecting accidents rather than interpersonal violence. The exceptions to this are so-called parry fractures of the ulna, and fractures of the metacarpals and hand phalanges (Lovell 1997; but see Judd 2008). While some parry fractures were observed, and others have been reported in the literature, post-crania were not systematically examined in this study.

A major advantage of the project is that the same criteria for the identification of cranial trauma were applied. A subset of crania were examined by both researchers, with close agreement on the presence or absence of ante-/peri-mortem injury. Crania were examined visually with careful attention to fracture edges, and a digital photographic archive created. Potential cutmarks were examined by hand lens. Crania were scored for condition and completeness. Condition refers to the external cortical surfaces, and is most relevant for assessing the visibility of cutmarks, since eroded or exfoliated surfaces would not preserve such modifications. In a few cases, erosion was so extreme as to obscure any potential small, healed injuries; such specimens have been excluded from the analysis. In this overview, we do not further discuss cutmarks, but focus on the evidence for cranial trauma.

The degree of completeness was scored both in 10% and 25% increments for individual elements of the cranial vault (left and right sides of the frontal, parietals, occipital and temporals), the maxillary region (left and right nasals, malars, and maxilla) and, where present, the left and right halves of the mandible. This is necessary in order to control for differential completeness between collections, which is very much a feature of the material examined. The Danish crania, for example, tended to be both in very good condition and relatively complete, often including the presence of the fragile nasal bones (and indeed a number of healed injuries thereon). Unfortunately, the same cannot be said for most of the other collections, including those from Britain and Ireland. Obviously any comparison of the overall prevalence of trauma – and even more so of nasal trauma specifically – would need to take this into account (cf. Robb 1997, Waldron 1991). For the purposes of this chapter, we include only those elements ≥75% complete in the calculation of injury prevalence for individual cranial elements. As detailed below, this excludes a number of both healed and unhealed injuries in both study areas, on elements not meeting this criterion. This problem is particularly noteworthy in a number of collections from specific sites, in which the material is highly fragmentary. It is possible that some of this material might be reconstructed in future, which would help to both confirm the presence of trauma and allow its inclusion in the kinds of summary calculations presented here. Re-fitting is time-consuming, however, and is further complicated by the division of skeletal material from some sites between different institutions (such as West Tump chambered tomb, Gloucestershire).

Small depressions were treated as probable healed injuries if they were at least 5mm in any one dimension (the majority were larger than this) and were smooth-walled, as opposed to exhibiting lytic reaction, such as seen for example in epidermoid and other cysts (Bikmaz *et al.* 2005) (Figure 15.1). While some of these small defects may of course represent accidental injuries, depressed fractures of the cranial vault are usually attributed to interpersonal violence, with falls more likely to result in linear fractures (Given and Williams 2002). Very few examples of linear fractures were observed in the present study, an exception being an adult individual from Dinnington long barrow, Yorkshire (Schulting 2012, fig. 13.3).

Figure 15.1 Adult left parietal with active lytic lesion indicating a probable cyst (Penywyrlod 74.23H, 9.01, Wales, National Museum of Wales). Photo by Rick J. Schulting.

This is excluded from further consideration in the analysis due to uncertainties over whether it is the result of interpersonal violence.

Peri-mortem injuries were identified using the following criteria: smooth, oblique fracture margins; rounded fracture arcs on the external table (a particular feature of blunt force injuries); overall shape of the defect; the presence of depressed, adhering bone around the injury site; concentric and/or radiating oblique fracture lines; and, crucially for archaeological examples, patinated fracture margins (Berryman and Haun 1996, Berryman and Symes 1998, Delannoy *et al.* 2012, Knight 1991). Although other contexts are represented, many of the crania examined derived from long barrows (in the case of eastern and south-central England) and chambered tombs, in which the skeletal remains were usually highly disarticulated, incomplete and fractured. In the great majority of cases, dry bone breakages could be clearly distinguished by their irregular, rough fracture margins. Given the nature of these contexts, however, a certain degree of ambiguity is introduced by the possibility of post-mortem breakage occurring while the cranium retains sufficient collagen to respond as fresh bone (Jordana *et al.* 2012). The potential for post-depositional damage due to trampling when the tombs or caves were re-entered, and for rockfall, must be considered. It might be argued that such damage, if it occurred while the bone was still in a relatively fresh state, could result in breakages that would be indistinguishable from intentional injuries inflicted during life (cf. Meyer *et al.* 2008). We are confident, however, that this is not a major factor affecting the present analysis. First, crania are arguably less likely than other skeletal elements to be affected by trampling in tomb or cave contexts, nor would a complete cranium be likely to fracture in a way that could be confused with an injury resulting from a directed blow. Secondly, chamber collapse or rockfall would need to occur within the limited period during which the cranium retained sufficient collagen to respond in the same way as fresh bone.

Given the added difficulties of working with archaeological material, we advocate the use of three probability levels – high, medium and low – for the recognition of ante- and peri-mortem trauma (though in practice the issue arises far more frequently with the latter; Schulting and Wysocki 2005). An injury site clearly meeting all or most of the aforementioned criteria would be considered as a 'high probability' case of peri-mortem trauma. If, usually through damage to the injury site and/or incompleteness, some of the criteria were not satisfactorily met, the trauma would be considered as being of medium probability. Finally, if too few criteria were satisfactorily met, or the specimen was too incomplete to adequately assess, the injury site would be classed as low probability and excluded from summary statistics. The advantage of formally noting cases of low probability trauma, rather than simply not recording them, is that it highlights specimens that might be worth future reconsideration (an example of which is noted below).

The cranial injuries

Denmark and Sweden

The combined sample for 378 individuals from Denmark and Sweden provides a crude (i.e. not taking cranial completeness into account) prevalence for cranial trauma of 14.0% (Table 15.1). Considering the good preservation in this sample, this figure, while still presenting a minimum, is probably very close to what can be skeletally diagnosed in this material (see below). The majority of injuries, 10.6%, are healed (Figures 15.2–15.3), with peri-mortem injuries observed on 4.0% of

Table 15.1 Crude prevalence of healed and unhealed injuries by age/sex class for the southern Scandinavian material (male individual with both healed and unhealed injuries is counted only once in total)

	N	*ante-*	*%*	*peri-*	*%*	*Total*	*%*
Adult male	197	31	15.7	7	3.6	37	18.8
Adult female	132	9	6.8	6	4.5	15	11.4
Indet. adol/adult	29	0	0.0	1	3.4	1	3.4
Child/infant	20	0	0.0	0	0.0	0	0.0
Total	378	40	10.6	14	3.7	53	14.0

Figure 15.2 Adult male with large healed injury ('pond fracture') to lower right parietal (Sønderup I, Denmark, Panum Institute, University of Copenhagen). Photo by Linda Fibiger.

Figure 15.3 Adult male with healed fracture to left zygomatic arch (Falshøj II, Denmark, Panum Institute, University of Copenhagen). Photo by Linda Fibiger.

the sample (the discrepancy in the totals being due to a single cranium from Denmark exhibiting both healed and unhealed trauma, counted as only one case in the combined total).

Significant differences between the sexes were noted with regard to the presence of healed cranial injuries, affecting 15.7% of males (31/197) compared with 6.8% of females (9/132; $\chi^2 = 5.886$; df = 1; $p = 0.015$). In contrast, no significant differences between males and females were found with regard to unhealed (presumed fatal) trauma, and in fact a higher percentage of females were affected: 4.5% (6/132) compared to 3.6% of males (7/197) (Figure 15.4). Driven entirely by the healed injuries, overall skull trauma differences between the sexes in the Scandinavian sample can only be said to be suggestive, affecting 18.8% of males (37/197) and 11.4% of females (15/132) ($\chi^2 = 3.268$; df = 1; $p = 0.071$). There are hints of patterning in terms of injury location (right side, left side, anterior) in relation to sex (Fibiger *et al.* 2013, Tables 7 and 8) with women mostly affected by injuries to the right side and anterior aspect of the cranium, while males predominantly suffered injuries to the front of the head. While the results were not statistically significant ($\chi^2 = 3.22$, df = 2, $p = 0.200$) this patterning offers some hints regarding potential gendered differences in attacker–victim constellation and the contexts of violent encounters.

Considering the proportions of individual cranial bones affected (>75% complete), there were no significant differences between the sexes. Generally, more

Figure 15.4 Adult female with unhealed perforations to parietals (Hammer I, Denmark, Panum Institute, University of Copenhagen). Photo by Linda Fibiger.

injuries were recorded in males, and only the left frontal approached significance when comparing males and females (Fisher's exact test, $p = 0.055$), with more males (10/183) than females (1/121) exhibiting injuries. Three injuries – affecting a female right parietal and right temporal (unhealed) and a male right parietal (healed) – are excluded from the individual cranial element calculations, as they occur on elements less than 75% complete (Table 15.2).

Britain and Ireland

The combined British/Irish sample is represented by 573 crania, though many of these are incomplete. The great majority derive from Britain, with Ireland currently contributing only 28 specimens, due to poor preservation conditions over much of that island as a result of acidic soils, as well as a seemingly greater emphasis on cremation in the early and middle Neolithic (Schulting *et al.* 2012). Excluding a number in the 'low probability' category, a total of 74 injuries were identified on 63 individuals, for a crude prevalence of 11.0%, of which 5.1% are healed and 5.9% unhealed (Table 15.3). Individuals with more than one injury are counted only once. These values differ somewhat from those presented in Schulting and Wysocki (2005), where an overall prevalence of 8.9% was cited, with 6.0% healed, and 2.9% unhealed (calculated on the basis of 'composite crania', taking incompleteness into account, albeit informally). It is the figure for peri-mortem trauma that differs most substantially. This is due in part to the incorporation of new material in the present study (Figures 15.4–15.7), but also to a re-assessment of some specimens previously placed in the 'low probability' category but now accepted as probable peri-mortem injuries (such as Bole's Barrow C2; Schulting and Wysocki 2005, fig. 12; for a discussion of other injuries at the same site, see Smith and Brickley 2007).

There are no significant differences between the sexes for either category of injury: healed injuries were recorded on 8.2% (19 of 233) of male crania and 4.2% (six of 144) female crania, while unhealed injuries were recorded on 6.0% (14 of 233) male crania and 6.3% (9 of 144) female crania. Note, however, that this leaves a number of injuries on adolescent and adult crania of indeterminate sex (8 of 132) largely as a result of their incompleteness. An additional seven injuries (out of 64) were observed on the crania of children. This was unexpected, as few such injuries were noted in the earlier study (Schulting and Wysocki 2005). In this case, the difference does lie almost entirely in the inclusion of new material, with the exception of the reclassification of cranium D-4 from Belas Knap as an older child (previously placed in the adolescent category; see Schulting 2012, fig. 13.11).

There is a tendency for both healed (18 of 32) and unhealed (21 of 30) injuries to occur more frequently on the left side of the head, though only the latter figure is statistically significant ($\alpha = 0.05$, two-tailed test, binomial $p = 0.043$: note that this analysis includes a small number of cases of multiple injuries to the same

Table 15.2 Comparison of healed and unhealed trauma to individual elements >75% complete in the southern Scandinavian sample

Element	*Healed*				*Unhealed*				*Total*			*Total*			*Combined*		
	Male		*Female*		*Male*		*Female*		*Male*			*Female*			*all injuries*		
	k	%	*k*	%	*k*	%	*k*	%	*k*	%	*n*	*k*	%	*n*	*k*	%	*N*
R frontal	10	5.5	4	3.3	0	0.0	1	0.8	10	5.5	182	5	4.1	123	15	4.9	305
L frontal	7	3.8	1	0.8	3	1.6	0	0.0	10	5.5	183	1	0.8	131	11	3.5	314
R parietal	7	4.0	3	2.6	3	1.7	4	3.4	10	5.7	175	7	6.0	116	17	5.8	291
L parietal	5	2.8	1	0.9	4	2.2	2	1.8	9	5.0	179	3	2.6	114	12	4.1	293
R temporal	1	0.7	0	0.0	0	0.0	1	1.1	1	0.7	146	1	1.1	95	2	0.8	241
L temporal	1	0.7	2	2.4	2	1.4	0	0.0	3	2.1	141	2	2.4	82	5	2.2	223
occipital	1	0.6	0	0.0	0	0.0	1	1.0	1	0.6	167	1	1.0	105	2	0.7	272
R malar	0	0.0	0	0.0	0	0.0	0	0.0	0	0.0	110	0	0.0	62	0	0.0	172
L malar	0	0.0	0	0.0	0	0.0	0	0.0	0	0.0	112	0	0.0	58	0	0.0	170
R nasal	2	2.3	1	2.2	0	0.0	0	0.0	2	2.3	86	1	2.2	46	3	2.3	132
L nasal	3	3.8	1	2.3	0	0.0	0	0.0	3	3.8	79	1	2.3	43	4	3.3	122
R maxilla	0	0.0	0	0.0	0	0.0	0	0.0	0	0.0	120	0	0.0	65	0	0.0	185
L maxilla	1	0.9	0	0.0	0	0.0	0	0.0	1	0.9	117	0	0.0	56	1	0.6	173
R mandible	0	0.0	0	0.0	0	0.0	0	0.0	0	0.0	68	0	0.0	23	0	0.0	91
L mandible	1	1.3	0	0.0	0	0.0	0	0.0	1	1.3	75	0	0.0	27	1	1.0	102
Total	39	2.0	13	1.1	12	0.6	9	0.8	51	2.6	1940	22	1.9	1146	73	2.4	3086

Table 15.3 Crude prevalence of healed and unhealed injuries by age/sex class for the British and Irish material

Age/Sex	*N*	*ante-*	*%*	*peri-*	*%*	*Total*	*%*
Adult male	233	19	8.2	14	6.0	33	14.2
Adult female	144	6	4.2	9	6.3	15	10.4
Indet. adol/adult	132	3	2.3	5	3.8	8	6.1
Child/infant	64	1	1.6	6	9.4	7	10.9
Total	573	29	5.1	33	5.8	63	11.0

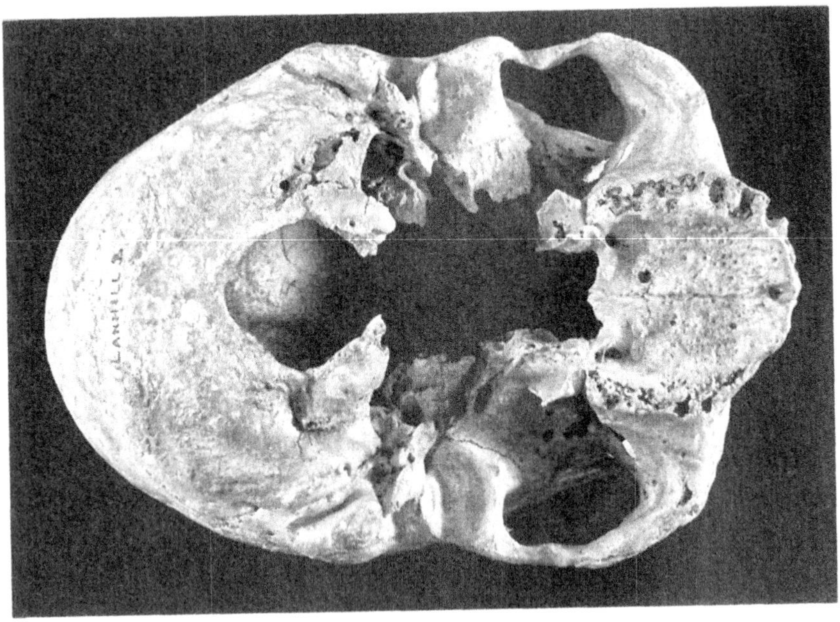

Figure 15.5 Adult ?female with healed fracture of right zygomatic arch (Lanhill, Eu.1.5.105, England, Duckworth Laboratory, University of Cambridge). Photo by Rick J. Schulting.

individual). This excludes one definite and two possible projectile injuries, for which location does not carry the same implications of face-to-face encounters between (usually) right-handed individuals. Given this expectation, an argument could be made for applying a one-tailed test, in which case both the peri-mortem and combined injuries would be significantly biased towards the left side of the cranium (binomial $p = 0.031$ and 0.021, respectively). There are a further four injuries to the rear of the head (occipital). There is no clear evidence for any relationship between sex and injury location.

Taking element completeness into account (Table 15.4), a number of examples of both healed and unhealed trauma in the British material are excluded for failing

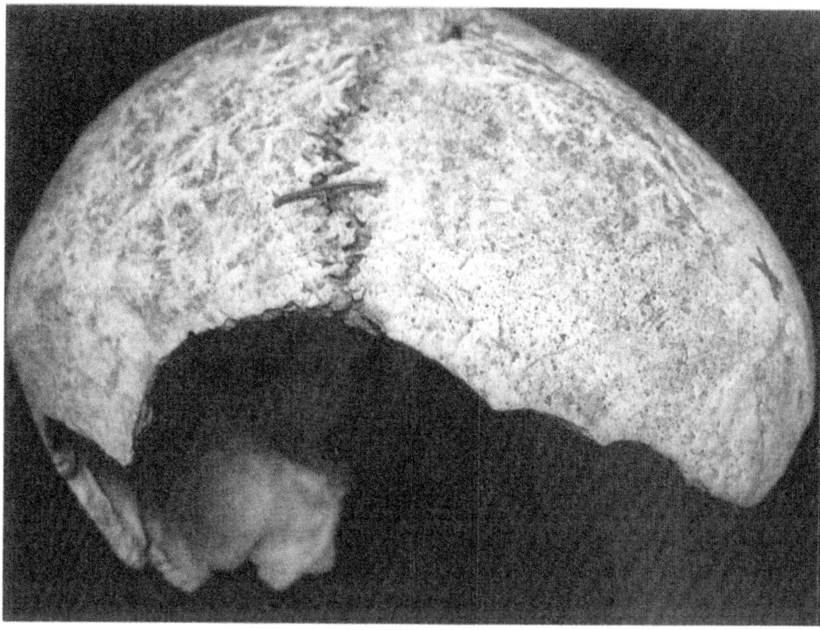

Figure 15.6a Adult male with probable unhealed blunt force traumata to right parietal (Dinnington Sk 1806, England, Natural History Museum, London). Photo by Rick J. Schulting.

Figure 15.6b Internal view, showing patinated bevels.

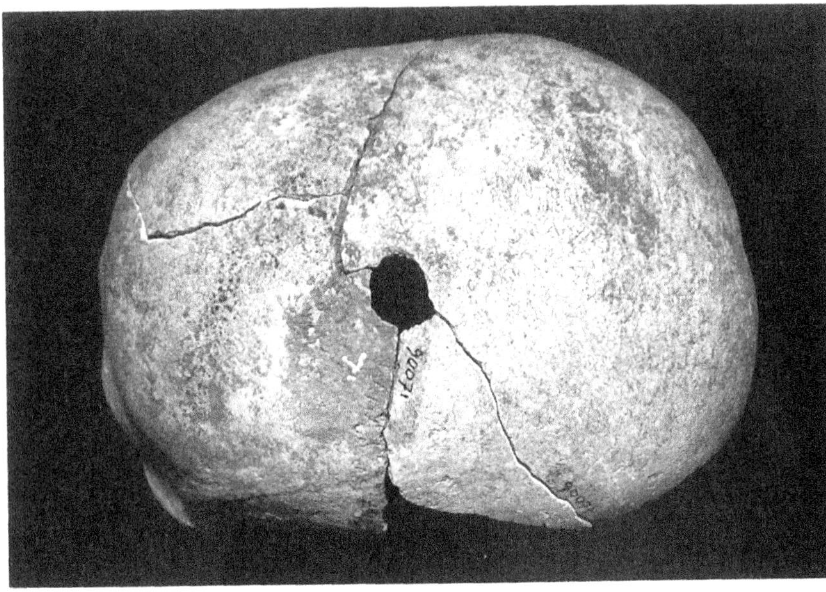

Figure 15.7 Adult male with unhealed circular perforation (13 × 16mm) to left superior fronto-parietal region (Rowiegar F/T1, Scotland, University of Aberdeen Museums). Photo by Rick J. Schulting.

to meet the >75% criterion. A good example is the West Tump chambered tomb, Gloucestershire, where a number of cranial fragments of varying size exhibit suggestive peri-mortem fractures, sometimes on very small fragments (Figure 15.8; see also Smith and Brickley 2009). In fact, four such cases are excluded from the study, including a good candidate for a perforating arrow injury to the right parietal of an adult (?)female (cf. Smith *et al.* 2007, fig. 4), as well as an unhealed probable stone axe injury (Figure 15.9). Other excluded specimens include a female with two healed injuries to the occipital, a male left frontal with two small healed depressions, another male left frontal with a single healed depression, an unhealed injury to a child's right parietal, and an unhealed injury to an adult female left parietal (a similar injury on the right parietal of the same individual meets the criterion).

Southern Scandinavia and Britain and Ireland compared

It must be emphasised that the crude prevalence figures for cranial trauma in both the southern Scandinavian and British/Irish material are minimum estimates, since many crania were incomplete, reducing their chances of recording injuries. Moreover, the collections analysed in the two regions show differential completeness, with crania in the sample from Denmark and Sweden being in general

Table 15.4 Comparison of healed and unhealed trauma to individual elements >75% complete in the British/Irish sample

Element	*Healed*								*Unhealed*								*Total*												*Combined*		
	Male		*Female*		*Indet.*		*Child*		*Male*		*Female*		*Indet.*		*Child*		*Male*			*Female*			*Indet.*			*Child*			*all injuries*		
	k	*%*	*k*	*%*	*k*	*%*	*k*	*%*	*k*	*%*	*k*	*%*	*k*	*%*	*k*	*%*	*k*	*%*	*n*	*k*	*%*	*n*	*k*	*%*	*n*	*k*	*%*	*n*	*k*	*%*	*n*
R frontal	3	1.7	1	1.0	0	0.0	0	0.0	0	0.0	2	1.9	0	0.0	1	2.7	3	1.7	181	3	2.9	105	0	0.0	44	1	2.7	37	7	1.9	367
L frontal	5	2.7	1	0.9	1	2.1	1	3.3	2	1.1	3	2.8	0	0.0	1	3.3	7	3.8	184	4	3.7	109	1	2.1	48	2	6.7	30	14	3.8	371
R parietal	8	4.9	1	1.1	1	2.2	0	0.0	3	1.8	2	2.2	2	4.4	2	7.4	11	6.7	163	3	3.3	92	3	6.7	45	2	7.4	27	19	5.8	327
L parietal	4	2.6	3	3.3	0	0.0	0	0.0	10	6.4	2	2.2	0	0.0	1	3.4	14	9.0	156	5	5.4	92	0	0.0	41	1	3.4	29	20	6.3	318
R temporal	0	0.0	0	0.0	0	0.0	0	0.0	0	0.0	0	0.0	0	0.0	0	0.0	0	0.0	116	0	0.0	62	0	0.0	21	0	0.0	17	0	0.0	216
L temporal	0	0.0	0	0.0	0	0.0	0	0.0	0	0.0	0	0.0	0	0.0	0	0.0	0	0.0	124	0	0.0	63	0	0.0	23	0	0.0	20	0	0.0	230
occipital	1	0.7	0	0.0	0	0.0	0	0.0	2	1.4	0	0.0	1	2.8	1	4.3	3	2.1	143	0	0.0	70	1	2.8	36	1	4.3	23	5	1.8	272
R malar	0	0.0	1	3.1	0	0.0	0	0.0	0	0.0	0	0.0	0	0.0	0	0.0	0	0.0	83	1	3.1	32	0	0.0	11	0	0.0	16	1	0.7	142
L malar	0	0.0	0	0.0	0	0.0	0	0.0	0	0.0	0	0.0	0	0.0	0	0.0	0	0.0	78	0	0.0	39	0	0.0	11	0	0.0	16	0	0.0	144
R nasal	0	0.0	0	0.0	0	0.0	0	0.0	0	0.0	0	0.0	0	0.0	0	0.0	0	0.0	45	0	0.0	18	0	0.0	6	0	0.0	7	0	0.0	76
L nasal	0	0.0	0	0.0	0	0.0	0	0.0	0	0.0	0	0.0	0	0.0	0	0.0	0	0.0	44	0	0.0	18	0	0.0	7	0	0.0	8	0	0.0	77
R maxilla	0	0.0	0	0.0	0	0.0	0	0.0	0	0.0	0	0.0	0	0.0	0	0.0	0	0.0	79	0	0.0	41	0	0.0	19	0	0.0	17	0	0.0	156
L maxilla	0	0.0	0	0.0	0	0.0	0	0.0	0	0.0	0	0.0	0	0.0	0	0.0	0	0.0	75	0	0.0	42	0	0.0	15	0	0.0	18	0	0.0	150
R mandible	0	0.0	0	0.0	0	0.0	0	0.0	0	0.0	0	0.0	0	0.0	0	0.0	0	0.0	57	0	0.0	35	0	0.0	20	0	0.0	10	0	0.0	122
L mandible	0	0.0	0	0.0	1	5.6	0	0.0	0	0.0	0	0.0	0	0.0	0	0.0	0	0.0	63	0	0.0	34	1	5.6	18	0	0.0	10	1	0.8	125
Total	21	1.3	7	0.8	3	0.8	1	0.4	16	1.0	9	1.1	3	0.8	6	2.1	38	2.3	1591	16	1.9	852	6	1.6	365	7	2.5	285	67	2.2	3093

Figure 15.8 Adult? frontal fragment with probable peri-mortem injury (arc on right side of photo) (West Tump 1913/33/1519, England, Cheltenham Museum and Art Gallery). Photo by Rick J. Schulting.

considerably more complete than those from Britain and Ireland. Thus, for example, approximately 80% of the 378 crania in the southern Scandinavian sample had frontal bones that were at least three-quarters complete, whereas this figure drops to *c*. 65% for the British/Irish sample. A similar discrepancy applies to all of the individual elements of the cranium (with the exception of the mandible, which shows equally low representation in both study regions; Figure 15.10). Another way of making this comparison is to note that, while some 573 individuals are represented in the British and Irish sample, compared with 378 in the southern Scandinavian sample, they are very nearly equal in terms of the number of individual cranial elements that are more than 75% complete (Table 15.5).

Looking first at crude prevalence figures, the southern Scandinavian sample has a significantly greater proportion of males affected by healed trauma than the British/Irish sample (15.7% versus 8.2%, $z = 2.44$, $p = 0.0145$; Figure 15.11). There is no significant difference in unhealed trauma to males, nor in either healed or unhealed trauma to females. The high prevalence of healed male trauma is a notable feature of the Danish material in particular, though again this can be at least partly related to the excellent preservation and completeness of this collection (Fibiger *et al*. 2013). In fact, when differential completeness is taken into account, the prevalence of healed and unhealed injuries to adult males and females shifts. Formal evaluation of this comparison is not straightforward, and there are a number of

Figure 15.9a Adult with probable peri-mortem injury to upper posterior left parietal (West Tump 1913/33/3758, England, Cheltenham Museum and Art Gallery). Photo by Rick J. Schulting.

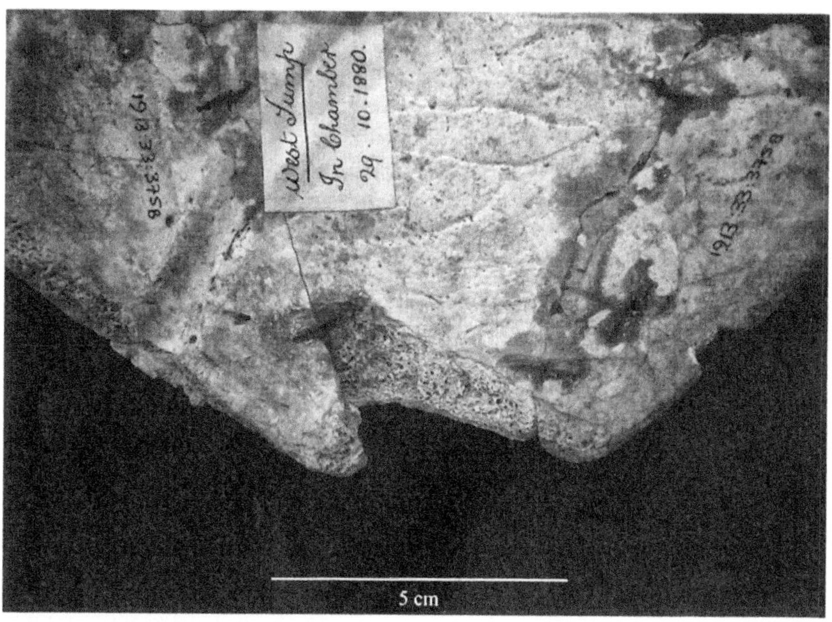

Figure 15.9b Internal view, showing patinated bevels.

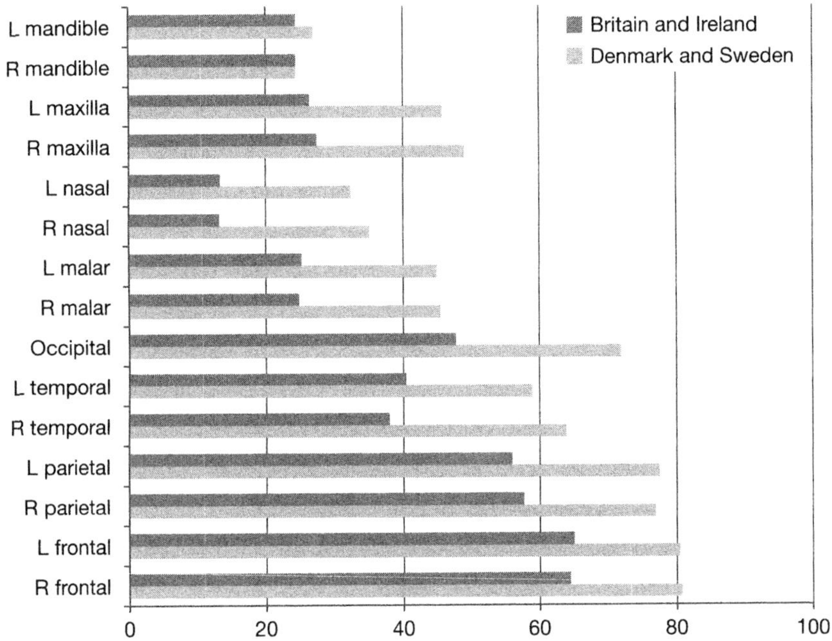

Figure 15.10 Comparison of percentage of elements >75% complete for southern Scandinavia (n = 378) and Britain/Ireland (n = 573).

Table 15.5 Comparison of the number of individual cranial elements >75% complete for southern Scandinavia and Britain and Ireland

Element	*S Scandinavia*	*% of N*	*Britain & Ireland*	*% of N*
R frontal	305	80.7	369	64.4
L frontal	304	80.4	372	64.9
R parietal	291	77.0	330	57.6
L parietal	293	77.5	320	55.8
R temporal	241	63.8	217	37.9
L temporal	223	59.0	231	40.3
occipital	272	72.0	273	47.6
R malar	172	45.5	142	24.8
L malar	170	45.0	144	25.1
R nasal	132	34.9	76	13.3
L nasal	122	32.3	77	13.4
R maxilla	185	48.9	157	27.4
L maxilla	173	45.8	151	26.4
R mandible	91	24.1	138	24.1
L mandible	102	27.0	139	24.3
N	378		573	

possible approaches currently being investigated. A simple first approximation is to compare prevalence figures for the part of the cranial vault most frequently affected by trauma (frontal and left and right parietals), using element counts (>75% complete). Revisiting the comparison of healed trauma to males in the two study areas, the difference highlighted in the crude prevalence figures disappears; 29 of 719 elements (4.03%) are affected in the southern Scandinavian sample, compared with 20 of 684 elements (2.92%) in the British/Irish sample ($z = 1.13, p = 0.258$). A similar result is obtained comparing the total number of healed injuries on all elements (>75%).

The British data show a higher proportion of cranial injuries to children (7 of 64, or 10.9% crude prevalence, compared to 0 of 20 for the smaller Danish sample; no children's crania were examined from Ireland or Sweden). The great majority of these cases (6 of the 7) are unhealed. Given the small number of children in the southern Scandinavian sample, this difference is not significant (binomial $p = 0.200$).

In terms of injury locations, not surprisingly the majority (both in terms of absolute number and relative proportions) of healed and unhealed injuries are found on the frontal and the left and right parietal bones, since together these comprise most of the cranial vault, and are most at risk from blows aimed at the head. In the

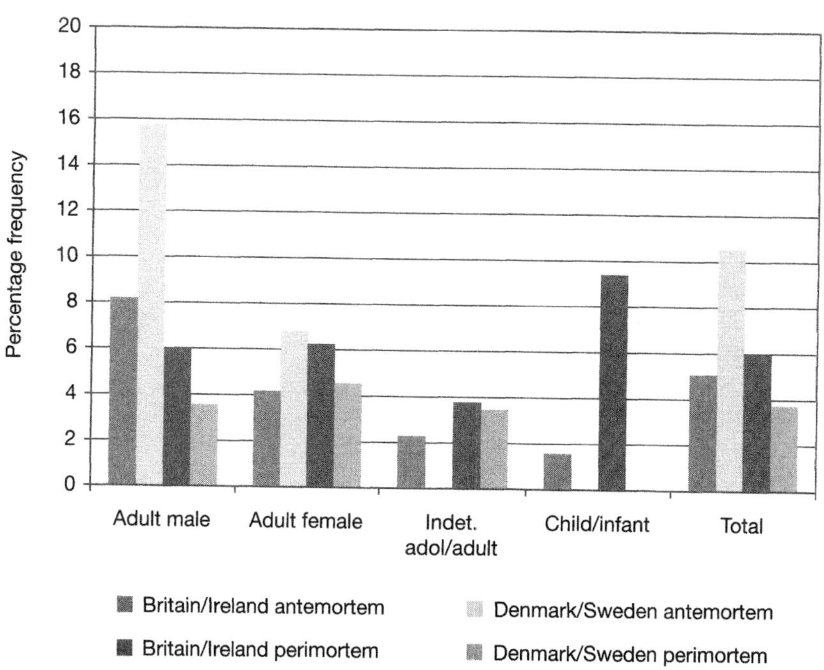

Figure 15.11 Comparison of crude cranial injury prevalence data for southern Scandinavia (n = 378) and Britain/Ireland (n = 573). Note that this does not take differential completeness into account, and so underestimates the true prevalence in both regions, but more so for Britain/Ireland.

southern Scandinavian sample this extends onto the temporals, while no clear injuries were observed on these elements in the British and Irish material. The other difference is in the number of healed breaks to the nasal bones observed on the Danish material (Fibiger *et al.* 2013, fig. 4); as noted above, this is no doubt partly a reflection of its better state of preservation. With this exception, and in marked contrast to modern domestic and unarmed street violence (Le *et al.* 2001, Rogers 2004, Shermis 1984, Walker 1997), injuries to the facial bones (maxilla, zygomatics and mandible) appear to have been relatively rare in the Neolithic. While these elements are certainly less well represented (Table 15.5), this alone is insufficient to account for the difference. Instead, it seems probable that it does relate to the nature of interpersonal violence. Specifically, and with the exception of nasal injuries, the majority of healed trauma observed on the archaeological material are the result of armed rather than unarmed conflict (leaving aside for the moment the question of the nature of the 'arms'; cf. Chapman 1999).

The southern Scandinavian data fail to show even the hint of a tendency for left-side preference, while this is observed to some extent in the British and Irish material. This could suggest different contexts for conflict in the two areas, and/or the use of different tactics, weapons, and so on. Further work is underway on comparisons with the corpus of material from Germany and France also collected as part of the wider study.

The great majority of the recorded headed injuries are small, generally circular depressions averaging about 10mm in diameter. Such injuries could be caused, for example, by blunted antler tines used as clubs. Others are substantially larger – though of course large injuries are less likely to have been survived – and are probably the result of blows with wooden clubs or stones (either hand-held, thrown, or incorporated into a club fixed to a wooden handle). Unhealed injuries generally take similar forms, with the addition of a small number of lenticular fractures for which a stone axe can be suggested (Figures 15.4 and 15.9). Finally, a small number of small lozenge-shaped perforations with marked internal bevels can be suggested to represent arrow injuries.

Discussion

What can be said concerning the nature of Neolithic conflict in north-west Europe on the basis of the findings presented here? First, it is important to recognise the limitations of both the material and of our approach. In terms of the material, the majority of the skeletal remains from the two study areas derive from normative (or at least the most commonly found) mortuary contexts: long barrows (for eastern and south-central England) and chambered tombs. Wherever and however these individuals died – and some died violently – their remains were brought to these (usually) communal places for interment. There is clearly the potential for this to be a biased sample, though whether this bias would be for or against individuals who died violently is unclear, and in any case perhaps no general pattern in this regard should be expected over such a large area and time period in any case (cf. Schulting and Wysocki 2005, Schulting 2012). A further limitation given these mortuary contexts is that, at least for the most part, it is not possible to address the scale of any individual violent encounters. As argued elsewhere, the small number of southern British enclosure sites that appear to have been attacked in force offer greater insights into the existence of large-scale battles and hence the scale of Neolithic conflict ('warfare') (such as Hambledon Hill, Carn Brea, Crickley Hill; Dixon 1988; Mercer 1989, 1999; Schulting 2006, 2012; cf. Whittle *et al.* 2011, chapter 14). Massacre sites such as Talheim in south-west Germany (Wahl and König 1987, Wahl and Trautmann 2012) provide another indication of the scale and ferocity of interpersonal violence sometimes realised, but no comparable sites have as yet been found in the Neolithic of southern Scandinavia or Britain and Ireland. However, some of the sites examined in these regions present with multiple injured individuals, hinting at either repeated episodes of interpersonal violence or the occurrence of violent events that go beyond one-to-one encounters. Examples include Borreby in Zealand and perhaps Whitehawk causewayed enclosure and West Tump chambered tomb in England, though consideration of the latter is hampered by heavy fragmentation. The first phase of the Wayland's Smithy chambered tomb – with a recently confirmed embedded projectile fragment, and two other arrowheads in close association with skeletons – presents another potential case, particularly in the light of a Bayesian analysis of the radiocarbon-dated human remains from this phase, which allows the possibility that all 14 individuals were interred simultaneously in the period 3590–3550 cal BC (Whittle *et al.* 2007).

In terms of our approach, an obvious limitation is that we have systematically examined only crania. While a small number of possible arrow injuries have been identified on the British cranial material (Schulting 2012, fig. 13.14; Schulting and Wysocki 2005, fig. 15; Smith *et al.* 2007, fig. 4) and on a mandible from Kilgreany, Co. Waterford, Ireland (Dowd *et al.* 2006; the embedded bone point in the nasal cavity of the Early Neolithic Porsmose skeleton, Denmark, can also be mentioned here (Becker 1952)), the great majority of such wounds will have impacted on the post-cranial skeleton. A number of such examples are known from Britain (Schulting 2006, 2012; Whittle *et al.* 2007), Ireland (Lynch and Ó Donnabháin 1994) and southern Scandinavia (Becker 1952, Bennike 1985). Arrow injuries in turn will only hit bone in approximately one-third of the cases in which the body was struck (Milner 2005), and not all of these will leave traces in a form recognised by osteologists. In restricting our study to crania, then, we are biasing the injuries towards those received in close personal encounters. This no doubt encompasses a range of situations, including domestic violence and other intra-group disputes as well as inter-community violence. Some cranial injuries might result from conflict at a distance, involving the use of slings or simply hand-thrown stones, though evidence for the former remains elusive for the Neolithic of north-west Europe. This could be a factor in the general lack of strong locational patterning on the cranium, especially for healed injuries (Fibiger 2014, Schulting and Wysocki 2005). If these were mainly the result of face-to-face encounters, a predominance of injuries to the left side of the head should result, as *c.* 90% of most human populations are right-handed, and would naturally tend to strike the left side of a facing opponent. This kind of encounter may itself have been rare, though it may have featured in the kinds of ritualised combat that are seen in a number of societies ethnographically (Abbink 2001, Chagnon 1997, Lizot 1985, Otterbein 1968, Riefenstahl 1973), and are often drawn upon in interpretations of healed blunt force cranial injuries found archaeologically (Papathanasiou 2012, Schulting and Wysocki 2005, Walker 1989), though others have rightly expressed reservations (Jurmain and Bellifemine 1997).

The end result of our focus on cranial trauma is that we are presenting only minimum estimates for the prevalence of interpersonal violence in the Neolithic. Although it has been noted previously, it is worth commenting on the essentially equal prevalence of peri-mortem (presumably lethal) cranial injuries on both men and women in the two study areas. This is counter to what some might expect when dealing with 'warfare', which supposedly directly affects mainly young adult males (Archer 1994). But such an expectation misconstrues the nature of conflict in small-scale societies, which tends to avoid pitched battles between groups of predominantly male warriors, in favour of raids and ambush. There is a wealth of ethnographic evidence to this effect, to the extent that it can be proffered as a central feature of inter-group conflict in small-scale societies (Burch 2007, Das 2006, Gat 1999, Maschner and Reedy-Maschner 1998, Otterbein 1999). Similarly, the strongest motivation to organising raids is revenge for past killings or other grievances, and this may be on the inter-generational level (Schröder and Schmidt 2001). The important point about revenge killings is that the specific identity of the victim often matters little, as long as they are a member of the offending group (Kelly 2000). The outcome of these two observations is that lethal violence against women (and children) is to be expected to feature as or nearly as strongly as against adult men (Schulting 2013, Fibiger 2014). The death of any member of an enemy group (a designation that may change often as alliances are created and broken), then, can satisfy the desire for blood vengeance. In some circumstances, young women may themselves be the objects of raids (Otterbein 2000), and indeed this has been suggested as a possible explanation for the dearth of this demographic group among the skeletons featuring peri-mortem trauma in the ditches of an Early Neolithic enclosure of Asparn-Schletz, Austria (Teschler-Nicola 2012, Teschler-Nicola *et al.* 1996). Children may also be taken as captives on raids, and this and the capture of women could offer an alternative view on the nature of mobility that is becoming an increasing focus of discussion in the European Neolithic, as a result of the application of strontium and oxygen isotope analyses to human remains (Bentley 2007, Bentley *et al.* 2003, Bickle and Whittle 2013): not all individual mobility may have been voluntary.

Conclusions

As has long been recognised, the Neolithic was not an entirely peaceful time. Nor, on the other hand, was it overly violent, at least in the sense that comparisons can be made with other places and times with much higher incidences of cranial trauma (Redfern 2009, Standen and Arriaza 2000, Torres-Rouff and Junqueira 2006, Tung 2008). That being said, the inferred levels of interpersonal violence, while not atypical of small-scale 'tribal' societies, are far beyond what most of the modern Western world has experienced outside of the last century's two world wars and a number of inner city areas (Keeley 1996, Pinker 2011). The aim of the present study has been to provide suitable prevalence data for healed and unhealed trauma from Neolithic north-west Europe that enable just these kinds of comparisons. While there are various caveats associated with estimating actual population prevalence figures (relating largely to issues of representativeness and completeness, concerning which further work is underway), it can be suggested that *c.* 8–15% of the Neolithic population in both southern Scandinavia and in Britain and Ireland experienced blows to their heads sufficient to cause bone breakage at some point or points in their lives. Males may have been at somewhat greater risk, though this may have varied regionally. Rather more critical for those involved, something on the order of 4–8% of the population experienced peri-mortem trauma, that were in most cases probably either themselves the cause of death, or part of a suite of injuries received at the same time.

But it must be acknowledged that these figures in themselves say relatively little about how violence was perceived by people at the time, whether they felt themselves to be living in 'dangerous times', or whether this level of interpersonal violence was taken for granted. One suspects the latter, though it is worth noting that in our own society, the actual incidence of violence may have little bearing to how it is perceived (Lewis and Salem 1986). Understanding the multi-faceted impact of interpersonal violence as experienced in the Neolithic thus presents considerable challenges.

Acknowledgements

Our thanks go to the many institutions and individuals who facilitated our visits to their collections. Meg Hutchinson was of great assistance during research on the collections of the University of Aberdeen Museums, and identified a number of the injuries included in the study. David Lawrence kindly shared his observations on Isbister. The research presented here was supported by a Leverhulme Trust Research Grant to RJS (F/08 735/E) and by a Wenner-Gren Foundation Hunt Postdoctoral Fellowship to LF (Grant No. 8202).

References

Abbink, J. 2001. Violence and culture: anthropological and evolutionary-psychological reflections on inter-group conflict in southern Ethiopia. In B.E. Schmidt and I.W. Schröder (eds), *Anthropology of violence and conflict*, 123–42. London: Routledge.

Archer, J. (ed.) 1994. *Male violence*. London: Routledge.

Armit, I., Knüsel, C., Robb, J. and Schulting, R.J. 2007. Warfare and violence in prehistoric Europe: an introduction. In T. Pollard and I. Banks (eds), *War and sacrifice*, 1–11. Leiden: Brill.

Becker, C.J. 1952. Skeletfundet fra Porsmose ved Næstved. *Fra Nationalmuseets Arbejdsmark* 1952, 23–30.

Bennike, P. 1985. *Palaeopathology of Danish skeletons*. Copenhagen: Akademisk Forlag.

Bentley, R.A. 2007. Mobility, specialisation and community diversity in the Linearbandkeramik: isotopic evidence from the skeletons. In A. Whittle and V. Cummings (eds), *Going over: the Mesolithic–Neolithic transition in north-west Europe*, 117–40. London: Oxford University Press for The British Academy.

Bentley, R.A., Krause, R., Price, T.D. and Kaufmann, B. 2003. Human mobility at the early Neolithic settlement of Vaihingen, Germany: evidence from strontium isotope analysis. *Archaeometry* 45, 471–86.

Berryman, H.E. and Haun, S.J. 1996. Applying forensic techniques to Interpret cranial fracture patterns in an archaeological specimen. *International Journal of Osteoarchaeology* 6, 2–9.

Berryman, H.E. and Symes, S.A. 1998. Recognising gunshot and blunt cranial trauma through fracture interpretation. In K.J. Reichs (ed.), *Forensic osteology: advances in the identification of human remains*, 333–52. Springfield: Charles C. Thomas.

Beyneix, A. 2007. Réflexions sur les débuts de la guerre au Néolithique en Europe occidentale. *L'Anthropologie* 111, 79–95.

Bickle, P. and Whittle, A. (eds) 2013. *The first farmers of central Europe: diversity in LBK lifeways*. Oxford: Oxbow Books.

Bikmaz, K., Cosar, M., Bek, S., Gokduman, C.A., Arslan, M. and Iplikcioglu, A.C. 2005. Intradiploic epidermoid cysts of the skull: a report of four cases. *Clinical Neurology and Neurosurgery* 107, 262–7.

Bowles, S. 2012. Warriors, levelers and the role of conflict in the evolution of social behavior. *Science* 336, 876–9.

Burch, E.S., Jr. 2007. Traditional native warfare in western Alaska. In R.J. Chacon and R.G. Mendoza (eds), *North American indigenous warfare and ritual violence*, 11–29. Tucson: University of Arizona Press.

Chagnon, N.A. 1997. *Yanomamö* (5th edition). Orlando: Harcourt Brace.

Chapman, J. 1999. The origins of warfare in the prehistory of central and eastern Europe. In J. Carman and A. Harding (eds), *Ancient warfare*, 101–42. Stroud: Sutton.

Das, N.K. 2006. Head hunting and warfare: a critical reappraisal of Naga political systems. In S. Dutta and B. Tripathy (eds), *Martial traditions of North East India*, 1–27. New Delhi: Concept Publishing.

Delannoy, Y., Becart, A., Colard, T., Delille, R., Tournel, G., Hedouin, V. and Gosset, D. 2012. Skull wounds linked with blunt trauma (hammer example). A report of two depressed skull fractures – elements of biomechanical explanation. *Legal Medicine* 14, 258–62.

Dixon, P. 1988. The Neolithic settlements on Crickley Hill. In C. Burgess, P. Topping, C. Mordant and M. Maddison (eds), *Enclosures and defences in the Neolithic of western Europe*, 75–87. Oxford: British Archaeological Reports.

Dowd, M., Fibiger, L. and Lynch, L.G. 2006. The human remains from Irish caves project. *Archaeology Ireland* 77, 16–19.

Fibiger, L. 2014. Misplaced childhood? Interpersonal violence and children in Neolithic Europe. In C.J. Knüsel and M. Smith (eds), *The Routledge handbook of the bioarchaeology of human conflict*, 127–45 London: Routledge.

Fibiger, L., Ahlström, T., Bennike, P. and Schulting, R.J. 2013. Patterns of violence-related head trauma in Neolithic southern Scandinavia. *American Journal of Physical Anthropology* 150, 190–202.

Gat, A. 1999. The pattern of fighting in simple, small-scale, prestate societies. *Journal of Anthropological Research* 55, 563–83.

Given, C.A. and Williams, D.W. 2002. The skull. In L.F. Rogers (ed.), *Radiology of skeletal trauma*, 272–313. New York: Churchill Livingstone.

Jordana, F., Colat-Parros, J. and Bénézech, M. 2012. Diagnosis of skull fractures according to postmortem interval: an experimental approach in a porcine model. *Journal of Forensic Science* 58, S156–62.

Judd, M.A. 2008. The parry problem. *Journal of Archaeological Science* 35, 1658–66.

Jurmain, R. and Bellifemine, V.I. 1997. Patterns of cranial trauma in a prehistoric population from central California. *International Journal of Osteoarchaeology* 7, 43–50.

Keeley, L.H. 1996. *War before civilization: the myth of the peaceful savage*. Oxford: Oxford University Press.

Kelly, R.C. 2000. *Warless societies and the origins of war*. Ann Arbor: University of Michigan Press.

Knight, B. 1991. *Forensic pathology*. London: Edward Arnold.

Le, B.T., Dierks, E.J., Ueeck, B.A., Homer, L.D. and Potter, B.F. 2001. Maxillofacial injuries associated with domestic violence. *Journal of Oral and Maxillofacial Surgery* 59, 1277–83.

Lewis, D.A. and Salem, G. 1986. *Fear of crime*. New Brunswick, New Jersey: Transaction.

Lizot, J. 1985. *Tales of the Yanomami: daily life in the Venezuelan forest*. Cambridge: Cambridge University Press.

Lovell, N.C. 1997. Trauma analysis in paleopathology. *Yearbook of Physical Anthropology* 40, 139–70.

Lynch, A. and Ó Donnabháin, B. 1994. Poulnabrone portal tomb. *The Other Clare* 18, 5–7.

Maschner, H.D.G. and Reedy-Maschner, K.L. 1998. Raid, retreat, defend (repeat): the archaeology and ethnohistory of warfare on the North Pacific Rim. *Journal of Anthropological Archaeology* 17, 19–51.

Mercer, R.J. 1989. The earliest defences in western Europe. Part 2. *Fortress* 3, 2–11.
Mercer, R.J. 1999. The origins of warfare in the British Isles. In J. Carman and A. Harding (eds), *Ancient warfare*, 143–56. Stroud: Sutton.
Meyer, C., Kranzbühler, J., Drings, S., Bramanti, B., Nehlich, O., Richards, M.P. and Alt, K.W. 2008. Die menschlichen Skelettfunde aus der neolithischen Totenhütte von Benzingerode. Anthropologische Untersuchungen an den Bestattungen eines Kollektivgrabs der Bernburger Kultur. *Archäologie in Sachsen-Anhalt Sonderband* 7, 107–51.
Meyer, C., Haak, W., Ganslmeier, R.A., Brandt, G., Meller, H. and Alt, K.W. 2009. The Eulau eulogy: bioarchaeological interpretation of lethal violence in Corded Ware multiple burials from Saxony-Anhalt, Germany. *Journal of Anthropological Archaeology* 28, 412–23.
Milner, G.R. 2005. Nineteeth-century arrow wounds and perceptions of prehistoric warfare. *American Antiquity* 70, 144–56.
Otterbein, K.F. 1968. Higi armed combat. *Southwestern Journal of Anthropology* 24, 195–213.
Otterbein, K.F. 1999. A history of research on warfare in anthropology. *American Anthropologist* 101, 794–805.
Otterbein, K.F. 2000. Killing of captured enemies: a cross-cultural study. *Current Anthropology* 41, 439–43.
Papathanasiou, A. 2012. Evidence of trauma in Neolithic Greece. In R.J. Schulting and L. Fibiger (eds), *Sticks, stones and broken bones: Neolithic violence in a European perspective*, 249–63. Oxford: Oxford University Press.
Peter-Röcher, H. 2007. *Gewalt und Krieg im prähistorischen Europa. Beiträge zur Konfliktforschung auf der Grundlage archäologischer, anthropologischer und ethnologischer Quellen*. Bonn: Universitätsforschungen zur Prähistorischen Archäologie, Band 143.
Pinker, S. 2011. *The better angels of our nature: why violence has declined*. New York: Viking Books.
Redfern, R. 2009. Does cranial trauma provide evidence for projectile weaponry in Late Iron Age Dorset? *Oxford Journal of Archaeology* 28, 399–424.
Riefenstahl, L. 1973. *The last of the Nuba*. New York: Harper and Row.
Ripley, W.Z. 1899. *The races of Europe: a sociological study*. New York: D. Appleton and Co.
Robb, J. 1997. Violence and gender in Italy. In D.L. Martin and D.W. Frayer (eds), *Troubled times: violence and warfare in the past*, 111–44. Amsterdam: Gordon and Breach.
Rogers, T. 2004. Recognizing inter-personal violence: a forensic perspective. In M. Roksandić (ed.), *Violent interactions in the Mesolithic*, 9–21. Oxford: British Archaeological Reports, International Series 1237.
Rowley-Conwy, P. 2004. How the west was lost: a reconsideration of agricultural origins in Britain, Ireland, and southern Scandinavia. *Current Anthropology* 45, S83–113.
Rowley-Conwy, P. 2011. Westward Ho! The spread of agriculturalism from central Europe to the Atlantic. *Current Anthropology* 52, S431–51.
Schröder, I.W. and Schmidt, B.E. 2001. Introduction: violent imaginaries and violent practices. In B.E. Schmidt and I.W. Schröder (eds), *Anthropology of violence and conflict*, 1–24. London: Routledge.

Schulting, R.J. 2006. Skeletal evidence and contexts of violence in the European Mesolithic and Neolithic. In R. Gowland and C. Knüsel (eds), *The social archaeology of funerary remains*, 224–37. Oxford: Oxbow Books.

Schulting, R.J. 2012. Skeletal evidence for interpersonal violence: beyond mortuary monuments in southern Britain. In R.J. Schulting and L. Fibiger (eds), *Sticks, stones and broken bones: Neolithic violence in a European perspective*, 223–48. Oxford: Oxford University Press.

Schulting, R.J. 2013. War without warriors? The nature of interpersonal conflict before the emergence of formalised warrior élites. In S. Ralph (ed.), *The archaeology of violence: interdisciplinary approaches*, 19–36. Albany: SUNY Press.

Schulting, R.J. and Fibiger, L. (eds) 2012. *Sticks, stones and broken bones: Neolithic violence in a European perspective*. Oxford: Oxford University Press.

Schulting, R.J. and Wysocki, M. 2005. 'In this chambered tumulus were found cleft skulls . . . ': an assessment of the evidence for cranial trauma in the British Neolithic. *Proceedings of the Prehistoric Society* 71, 107–38.

Schulting, R.J., Murphy, E., Jones, C. and Warren, G. 2012. New dates from the north, and a proposed chronology for Irish court tombs. *Proceedings of the Royal Irish Academy* 112C, 1–60.

Sheridan, A. 2010. The Neolithization of Britain and Ireland: the 'big picture'. In B. Finlayson and G. Warren (eds), *Landscapes in transition*, 89–105. Oxford: Oxbow Books.

Shermis, S. 1984. Domestic violence in two skeletal populations. *OSSA* 9–11, 143–51.

Smith, M.J. and Brickley, M.B. 2007. Boles Barrow: witness to ancient violence? *British Archaeology* 93, 22–7.

Smith, M. and Brickley, M. 2009. *People of the long barrows: life, death and burial in earlier Neolithic Britain*. Stroud: The History Press.

Smith, M.J., Brickley, M.B. and Leach, S.L. 2007. Experimental evidence for lithic projectile injuries: improving recognition of an under-recognised phenomenon. *Journal of Archaeological Science* 34, 540–53.

Sørensen, L. and Karg, S. In press. The expansion of agrarian societies towards the north – new evidence for agriculture during the Mesolithic/Neolithic transition in southern Scandinavia. *Journal of Archaeological Science*. http://dx.doi.org/10.1016/j.jas.2012.08.042.

Standen, V.G. and Arriaza, B.T. 2000. Trauma in the preceramic coastal populations of northern Chile: violence or occupational hazards? *American Journal of Physical Anthropology* 112, 239–49.

Teschler-Nicola, M. 2012. The Early Neolithic site Asparn/Schletz (Lower Austria): anthropological evidence of interpersonal violence. In R.J. Schulting and L. Fibiger (eds), *Sticks, stones and broken bones: Neolithic violence in a European perspective*, 101–20. Oxford: Oxford University Press.

Teschler-Nicola, M., Gerold, F., Kanz, F., Lindenbauer, K. and Spannagl, M. 1996. Anthropologische Spurensicherung: Die Traumatischen und Postmortalen Veränderungen an den Linearbandkeramischen Skelettresten von Asparn/Schletz. *Archäologie Österreich* 7, 4–12.

Torres-Rouff, C. and Junqueira, M.A.C. 2006. Interpersonal violence in prehistoric San Pedro de Atacama, Chile: behavioral implications of environmental stress. *American Journal of Physical Anthropology* 130, 60–70.

Tung, T.A. 2008. Violence after imperial collapse: a study of cranial trauma among Late Intermediate Period burials from the former Huari capital, Ayacucho, Peru. *Ñawpa Pacha: Journal of Andean Archaeology* 29, 101–17.

Wahl, J. and König, H. 1987. Anthropologisch-traumologische Untersuchung der menschlichen Skelettreste aus dem bandkeramischen Massengrab bei Talheim, Kreis Heilbronn. *Fundberichte aus Baden-Württemberg* 12, 65–193.

Wahl, J. and Trautmann, I. 2012. The Neolithic massacre at Talheim: a pivotal find in conflict archaeology. In R.J. Schulting and L. Fibiger (eds), *Sticks, stones and broken bones: Neolithic violence in a European perspective*, 77–100. Oxford: Oxford University Press.

Waldron, T. 1991. Rates for the job. Measures of disease frequency in palaeopathology. *International Journal of Osteoarchaeology* 1, 17–25.

Walker, P.L. 1989. Cranial injuries as evidence of violence in prehistoric southern California. *American Journal of Physical Anthropology* 80, 313–23.

Walker, P.L. 1997. Wife beating, boxing, and broken noses: skeletal evidence for the cultural patterning of violence. In D.L. Martin and D.W. Frayer (eds), *Troubled times: violence and warfare in the past*, 145–80. Amsterdam: Gordon and Breach.

Whittle, A., Bayliss, A. and Wysocki, M. 2007. Once in a lifetime: the date of the Wayland's Smithy long barrow. *Cambridge Archaeological Journal* 17.1, supplement, 103–21.

Whittle, A., Healy, F. and Bayliss, A. 2011. *Gathering time: dating the early Neolithic enclosures of southern Britain and Ireland*. Oxford: Oxbow Books.

16

Mass Graves of the LBK

Patterns and Peculiarities

CHRISTIAN MEYER, CHRISTIAN LOHR, OLAF KÜRBIS,
VEIT DRESELY, WOLFGANG HAAK, CHRISTINA J. ADLER,
DETLEF GRONENBORN AND KURT W. ALT

Introduction

THE *LINEARBANDKERAMIK* OR *LINIENBANDKERAMIK* (LBK) is the well known archaeological entity ascribed to the first fully fledged farmers of central Europe. The LBK agriculturists persisted for several centuries predominantly in association with the loess belt, spanning a large part of the continent, and spread along its riverine systems, including the Danube, Rhine, and Elbe within Germany (Modderman 1988). The LBK is subdivided in current research into several phases, which roughly span from 5500 to 4900 cal BC, and its origins seem to lie in the east along the Danube (Bickle *et al.* 2011). The LBK is finally replaced by several different but probably related and/or derivative cultures of the Middle Neolithic in Germany (such as *Hinkelstein*, *Stichbandkeramik*, *Großgartach* and *Rössen*).

Besides settlement sites, graves are generally among the most important sources of knowledge about the ancient past, which also holds true for the LBK. Many cemeteries of differing sizes are known for this culture (Trautmann 2006), and the 'vocabulary of burial practices' (Bickle *et al.* 2011, 1256) is rather well known in comparison to other Neolithic cultures, as it has almost constantly been in the focus of archaeological research (Peschel 1992, Nieszery 1995, Orschiedt 1998, Trautmann 2006, Fritsch *et al.* 2008). Because of the extensive publication record it will be sufficient to quickly summarise the known range of LBK burial rites below, which will then form the basis for a comparison with selected burial features of a rather deviant nature. The focus is put here on the three mass graves of the LBK from Germany (Figure 16.1), including ten or more bodies each, found in Talheim (Baden-Württemberg, in 1983), Wiederstedt (Sachsen-Anhalt, in 1998), and Kilianstädten (Hessen, in 2006), but relevant sites from other regions will be included as well to flesh out the overall patterns.

Proceedings of the British Academy **198**, 307–25.

Figure 16.1 Map of Germany with location of the three mass graves of Talheim (Baden-Württemberg), Wiederstedt (Sachsen-Anhalt) and Kilianstädten (Hessen).

As the current status of (bio)archaeological research is different for each of the mass grave sites, and many analyses are still in progress, the discussion here will mainly target aspects which can be compared with some confidence, based on the published data (Wahl and König 1987, Alt *et al.* 1995, Meyer *et al.* 2004, Price *et al.* 2006, Bentley *et al.* 2008, Wahl and Strien 2007, Wahl and Trautmann 2012, Meyer *et al.* 2012, Meyer *et al.* 2013, Lohr *et al.* in press). The subsequent discussion will focus on a more general interpretative level, not on the precise analytical data themselves, although these are drawn upon when necessary. A comprehensive comparison, including the successfully established triad of osteology, aDNA, and isotope studies for prehistoric mass, multiple, and collective burial features (e.g. Haak *et al.* 2008, Meyer *et al.* 2008, Knipper *et al.* 2014) would be beyond the scope of the present chapter. In fact, the current datasets do not yet allow such a detailed comparison, which will have to be addressed in the future. The reasons and possible cultural mechanics that lay behind the mass fatality events leading up to the utilisation of mass graves in the LBK are touched upon individually per site, but farther-reaching reconstructions of a probable state of 'crisis' put forward by many studies to date (Farruggia 2002, Golitko and Keeley 2007, Gronenborn 2007, Boulestin *et al.* 2009, Zeeb-Lanz 2009; cf. Orschiedt and Haidle 2012), and central to many treatments of later phases of the LBK, shall not be pursued here. Instead, we will try to deduce what the mass graves themselves tell us about the handling of a large number of dead people accumulated within a very short time and LBK burial customs in general.

LBK burial customs

In LBK burial features, the central finds are usually the skeletal remains of the deceased, which are often, but not always, accompanied by grave goods, and personal ornaments made of different materials (Veit 1992). The general picture we have of grave goods today is dominated by the more durable artefacts made of stone, (burnt) clay, shell or bone, for example during the early Neolithic, while everything fashioned out of organic materials like leather, hair or plant materials is usually completely decomposed and therefore largely invisible. Whereas the contents of seemingly empty pots may only have been guessed at for a long time, analytical procedures can nowadays determine some preserved components which leached into the material of the pots themselves (Craig *et al.* 2005, Salque *et al.* 2013). The presence of other organic grave goods besides food matter may sometimes be assumed by the arrangement of partly preserved artefacts. While arrowheads made from flint or bone may be completely preserved, the supposed shafts and quivers are usually not. Nevertheless, it is highly likely that such items were originally present in the graves if the arrowheads appear to be arranged as they would be in a bundle of complete arrows (Gronenborn 2003). The same is presumed for stone tools, which usually possessed a wooden shaft, bindings and maybe other features like ornaments completely invisible in the archaeological record in dry soils. Artefacts from lake-shore dwellings or from unique finds like the Ötztal ice mummy from the Alps (Egg and Spindler 2009) can fill this gap to some degree, by hinting at what could have been present.

Some grave goods, like adzes, seem to indicate a certain social status of the individual buried and are strongly correlated with one sex (Bentley *et al.* 2012), but for our purpose here it is entirely sufficient to state that grave goods made out of durable material are a common occurrence in LBK graves. Numbers vary greatly between sites, but on average about 45–60% of individuals per cemetery are found

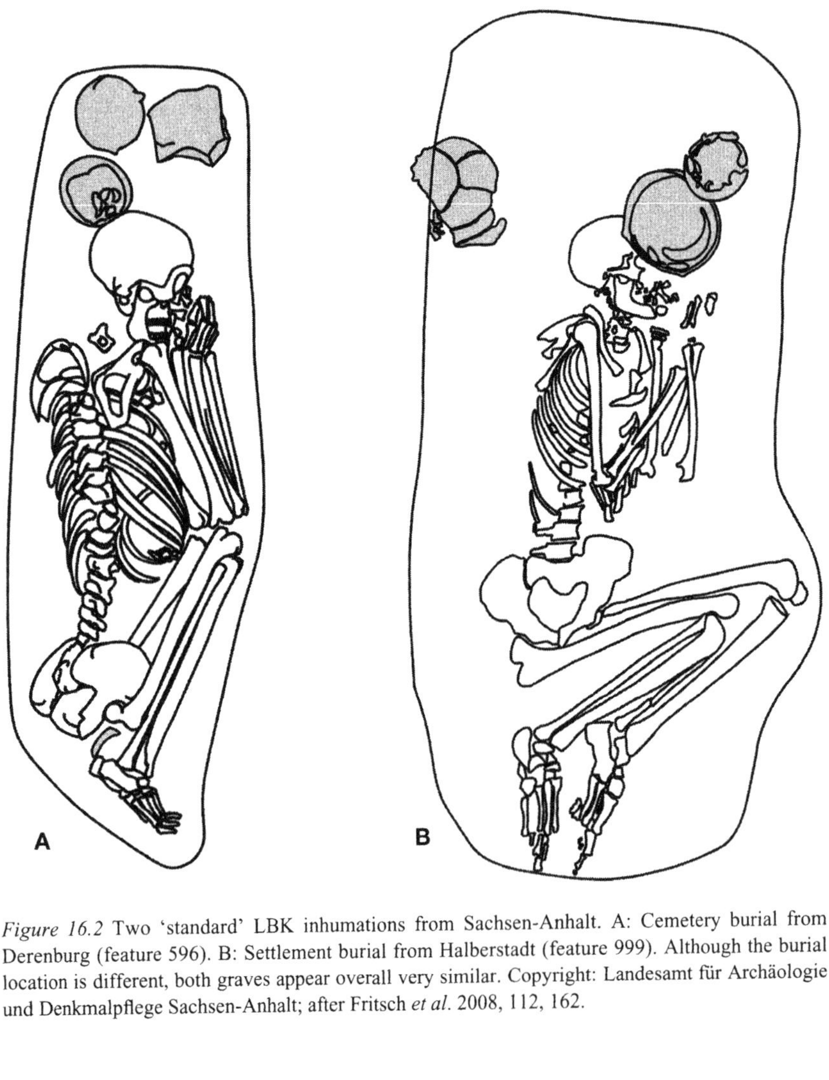

Figure 16.2 Two 'standard' LBK inhumations from Sachsen-Anhalt. A: Cemetery burial from Derenburg (feature 596). B: Settlement burial from Halberstadt (feature 999). Although the burial location is different, both graves appear overall very similar. Copyright: Landesamt für Archäologie und Denkmalpflege Sachsen-Anhalt; after Fritsch *et al.* 2008, 112, 162.

in recognisably furnished graves (Nieszery 1995, 19; Price *et al.* 2003, 28; Bickle *et al.* 2011, 1245). The body position is usually crouched (Figure 16.2A), but alternative variants commonly occur. There is furthermore a clear preference for placing the bodies on their left-hand side. At Aiterhofen, currently the largest Bavarian LBK cemetery known, 85% of graves were those of left-crouched individuals (Bickle *et al.* 2011) and 82% of such were found at Derenburg, Sachsen-Anhalt (Fritsch *et al.* 2008). Lower percentages are known, for example of about 75% from the large cemetery of Schwetzingen in Baden-Württemberg (Gerling 2009) and of 56% at Dillingen, again in Bavaria (Nieszery 1995). Even if there is overall a great variety, including right-crouched inhumations, supine and extended ones, the majority are still left-crouched. Most of these are oriented with their heads towards eastern directions (Peschel 1992, 187), for example 80% at Aiterhofen (Bickle *et al.* 2011, 1246), but again with a considerable diversity in this aspect.

While most LBK inhumations are found in cemeteries, of which several dozen are now known (Trautmann 2006), burials in settlement areas are a common occurrence (Veit 1992). These generally appear quite similar to the cemetery burials, with a grave goods frequency of about 46%, but seem to include more children and females (Veit 1996, Orschiedt 1998). They are also a bit more diverse in their individual characteristics (Hofmann 2009, 231), but when 'extracted' from their locational context, settlement and cemetery burials are largely indistinguishable from one another (Figure 16.2B; Meyer *et al.* 2013).

The actual location of a burial therefore is no good indicator in itself to determine if a grave is to be regarded as 'regular' or 'deviant', since these categories are dependent on the current state of research and therefore should reserve some flexibility in their definitions. Due to their overall similarity to cemetery burials, LBK settlement burials should not be termed *Sonderbestattungen* in every case, as done in the past (cf. Peschel 1992), but should rather be regarded as regular graves (Meyer *et al.* 2013). Although cemeteries and settlement areas were spatially separated in many cases, one could encroach on the other, resulting in an overlap between the spheres of the dead and the living, which apparently were not truly separated in Neolithic times (Veit 1992, Trautmann 2006).

This is also shown by the obvious and intense manipulation of human remains by the living, for example at the enigmatic ritual site of Herxheim, Rheinland-Pfalz, which was clearly part of the LBK behavioural repertoire (Boulestin *et al.* 2009, Zeeb-Lanz 2009, Zeeb-Lanz *et al.* 2009), which at present is not completely understood. At the site of Vaihingen/Enz, Baden-Württemberg, the partly artificial separation of cemetery and settlement was also disregarded, as here the standard place of burial apparently was the ditch surrounding the settlement at one time (Krause 1998), thereby creating an interlocking zone of life and death. Of course the internal chronology of a site has to be taken into account when discussing the contemporaneity of each settlement and cemetery, especially when there is evidence of changing customs between the defined cultural phases (e.g. Price *et al.* 2003).

Most inhumations, whether placed in a settlement or cemetery, are single burials, containing only one individual each. Multiple burials do occur, but are rather rare for the LBK (Peschel 1992, 186). These show a similar care for the deceased on the part of the surviving community as most LBK inhumations do. Only in rare cases do individual corpses appear to have been dumped into pits without much ceremony (Krause 1998), resulting in erratic body positions, in clear contrast to the variety of positions seen in regular LBK burial customs. Cremation burials occur as well, and for these all characteristics based on the placement and orientation of the body are of course unavailable. But as the cremations are found within the regular cemeteries (Trautmann 2006, Gerling 2009) and do contain grave goods (Nieszery 1995), they appear to have been another regular way of respectfully treating the dead within the LBK. Unfortunately, most bioarchaeological methodology is greatly limited when the bodies have been fully cremated, so the information which can be gleaned from them will always lag behind the inhumation burials.

This short summary underlines the left-crouched single inhumation placed with the head to the east in a furnished grave as the classic LBK way of treating the dead (cf. Veit 1992, Hofmann 2009). Although substantial variation existed, most LBK graves are distinctly recognisable, as they tend to occur in clusters, mostly in cemeteries but also in settlements (Meyer *et al.* 2013). As shown, settlement burials should not be labelled by default as deviant (Peschel 1992, Orschiedt 1998). However, clearly deviant graves do exist, which contrast with the previously reconstructed funerary portfolio of the LBK. These are the mass graves, which have been found in Germany in the last 30 years.

LBK mass graves

Talheim

The mass grave of Talheim, Baden-Württemberg, was excavated in 1983–84 after its accidental discovery in a garden plot (Wahl and König 1987). The mostly articulated and well preserved bones were lying in a pit of about 2.9 by 1.5 m, the individuals erratically commingled (Figure 16.3A). There was no evidence of careful arrangement of the deceased, whose number was determined by the osteological analysis as comprising at least 34 individuals. It appeared that all had been dumped into the pit at the same time, but without any of the usual trappings of the standard LBK burial rites. Grave goods were absent, but some pottery sherds were found associated with the skeletons, which are believed to represent chance settlement refuse that was part of the infill. No other graves were found in the vicinity, but settlement features are known to have been nearby (Wahl and Trautmann 2012). This indicates that the mass grave was originally part of a previously settled area and not of a cemetery. The reconstructed demography was very close to the ideal

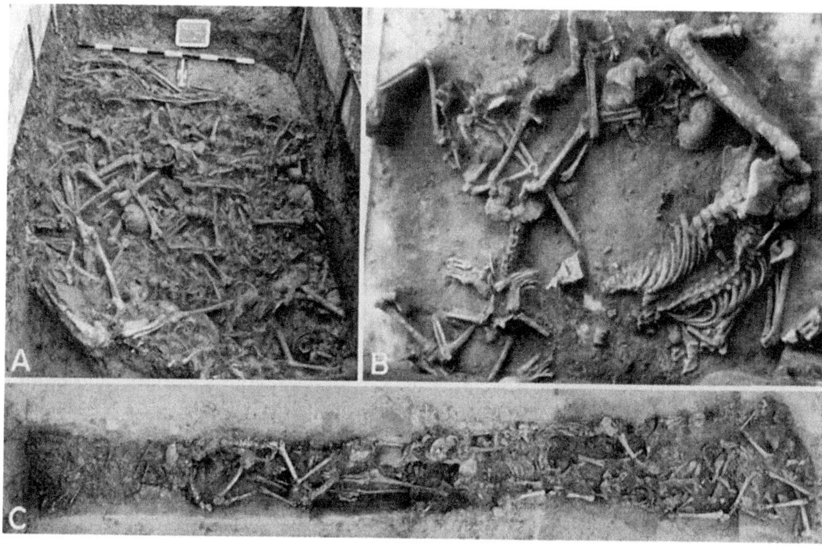

Figure 16.3 Mass graves of the LBK. A: Talheim (Baden-Württemberg), during excavation in 1983. Photo: E. Stauß; Landesamt für Denkmalpflege Baden-Württemberg; after Price *et al.* 2006, 263. B: Wiederstedt (Sachsen-Anhalt), during excavation in 1998. Photo: O. Kürbis; Landesamt für Archäologie und Denkmalpflege Sachsen-Anhalt; after Meyer *et al.* 2004, 32. C: Kilianstädten, (Hessen) during excavation in 2006. Photos: Landesamt für Denkmalpflege Hessen; Arrangement: C. Lohr.

model for prehistoric societies, with 47% sub-adults and 53% adults. The 18 adults were determined as nine males and seven females; two skeletons could not be reliably sexed. The sex and age ratios are therefore quite balanced, and it is widely believed that the individuals from the Talheim mass grave represent a complete population subsample, maybe the inhabitants of a small village, who had been killed at the same time (Wahl and König 1987, Alt *et al.* 1995).

The simultaneous death of all individuals very probably was a violent one, as up to 59% of all skeletons bear the traces of peri-mortem and lethal injuries. These wounds could be traced to the known stone tools and weapons of the LBK, and they were predominantly located on the right or posterior side of the head, suggesting that most people were struck from behind, in some cases repeatedly. The scarcity of defence injuries furthermore supports the notion of a surprise attack, or one by a far superior force, which could have encouraged the victims to flee rather than fight back (Wahl and König 1987, Wahl and Trautmann 2012). Matching the dimensions of the injuries to the known tool sizes of some LBK groups strongly suggests that the attackers also belonged to the LBK, possibly originating not far away along the Neckar river (Wahl and Strien 2007).

As the Talheim mass grave was the first to be discovered, numerous analyses have been carried out on the skeletal remains so far, including an odontological kinship study (Alt *et al.* 1995) and isotope analyses (Price *et al.* 2006, Bentley *et al.* 2008). Studies of ancient DNA have been initiated (Wahl and Trautmann 2012). The results of the dental trait and isotope analyses indicate that at least some of the individuals were likely biologically related, and that the $^{87}Sr/^{86}Sr$ range determined for them is much tighter than at LBK cemetery sites. This is taken as further support for a close-knit community, which died at the same time. The absence of younger adult females which grew up locally is interpreted as a possible indicator for their abduction by the attackers by some researchers (Bentley *et al.* 2008), although these actually should be absent when a viri- or patrilocal residence pattern is assumed for the LBK, as suggested by various authors (Bentley *et al.* 2008, Meyer and Alt 2010, Bickle *et al.* 2011).

Wiederstedt

The mass grave of Wiederstedt, Sachsen-Anhalt, was accidentally discovered in January 1998 by construction work (Meyer *et al.* 2004). Upon notification to the archaeological authorities, a rescue excavation was carried out under adverse winter conditions. Ten erratically commingled skeletons were identified with anatomical connections largely preserved (Figure 16.3B). In addition to recent damage caused by the construction work, the top layer of the feature showed some evidence of older disturbance which could be attributed to an eighteenth-century cemetery located right on top of the Neolithic mass grave. One Neolithic skull, accidentally removed by grave diggers in the eighteenth century, was later found placed next to a coffin of the same period. It clearly matched the facial bones and lower jaw, which had remained *in situ* in the mass grave (Meyer *et al.* 2012).

The osteological analysis revealed a peculiar demography of this population sample, as eight of ten individuals were sub-adults ranging in age from 3 to 15 years. Five of them fell into the range of about 7 to 12 years, which complicated the correct individual assignment and recovery of the commingled bones. Due to this peculiar age structure, only the two oldest individuals could be reliably sexed. The skeleton whose head was temporarily removed in the eighteenth century was a male of about 18–25 years; the remains directly underneath were those of a woman of about 35–45 years. This oldest individual is the only one whose body position can be described as being roughly compatible with the crouched LBK standard, whereas all others appear to have been haphazardly thrown into the pit. There was no evidence of grave goods or other signs of a careful and considerate burial. The pottery sherds and animal bones found in the lower strata appear to be settlement refuse and the pit of about 1.5m in diameter apparently was not originally intended as a mass grave. Rather, it seems, a pit already partly filled with refuse and sediment has been used to deposit the ten corpses (Meyer *et al.* 2004). The causes of death for all individuals remain unknown. In contrast to the frequent injuries encountered in the Talheim skeletons (Wahl and König 1987), or those from the LBK enclosure at Schletz in Austria (Teschler-Nicola *et al.* 1999), the skeletons at Wiederstedt did not bear unequivocal signs of violence. The peculiar age structure of the Wiederstedt mass grave may be a further hint that an epidemic or some sort of accident caused the deaths of mostly children and only some adults. Preliminary genetic data furthermore suggest that some of the individuals might be biologically related to one another and some carry the N1a mtDNA haplotype often found in LBK individuals (Haak *et al.* 2005, Haak *et al.* 2010). Further genetic as well as multiple isotope analyses are underway.

Kilianstädten

The most recent LBK mass grave that was discovered in 2006 is that of Schöneck-Kilianstädten, Hessen (Schwitalla and Schmitt 2006). Here, the rather badly preserved remains of at least 21 erratically commingled individuals were discovered during the excavation of an extensive LBK settlement area, in an oblong pit about 7.5m long and 0.3–1.0m wide (Lohr *et al.* in press). The ongoing analysis of the remains is complicated by their pronounced fragility and the after-effects of some adverse conditions, which negatively affected the bones and the whole feature during excavation. At least half of the skeletons were still articulated (Figure 16.3C), indicating a treatment of the dead similar to Talheim and Wiederstedt, such as the dumping of the collected corpses into a nearby and possibly pre-existing pit very shortly after their demise. Grave goods are absent, but settlement refuse was found with the skeletons, as at the other two sites. This mainly consists of scattered LBK sherds and broken stone artefacts, but at least two triangular bone arrowheads were also part of the mass grave. These were found during the cleaning and preparation of the skeletal remains, and are presumed to stem from the sediment previously adhering to the human bones. The arrowheads possibly represent projectile parts remaining inside some of the bodies, thereby indicating at least two arrow injuries and supporting a scenario involving interpersonal violence for the Kilianstädten mass grave. Further suggestive signs of peri-mortem violence can be found on some of the skulls, but due to their sub-standard state of preservation the extent of cranial injuries in the whole sample cannot reliably be stated at present (Meyer *et al.* 2013, Lohr *et al.* in press). In addition to this, a number of long bones appear to have suffered peri-mortem fractures as well, especially those of the lower extremities. Many of the comminuted fracture fragments were still present, further supporting the peri-mortem nature of these traumas and the notion of a fast burial, as smaller skeletal elements tend to disappear the longer they are unburied (cf. Teschler-Nicola *et al.* 1999, 2006).

The preliminary demography currently reveals a balanced ratio of sub-adults to adults, which is quite similar to Talheim but very different from the skewed demographic profile of Wiederstedt. Among the adults younger persons prevail, and male and female individuals can both be identified from suitable skeletal elements found within the mass grave. As not all bones were retrieved as anatomical connections in excavation, each trait has so far been assessed separately. The analyses have not been fully completed, all results are still preliminary and the numbers may still change. Nevertheless, it can be securely stated that the overall pattern presented by the Kilianstädten mass grave closely resembles that of Talheim, with a balanced age ratio and signs of peri-mortem interpersonal violence, while the number of postcranial fractures seems peculiar so far (Meyer *et al.* 2013, Lohr *et al.* in press). Although the bad preservation of the human remains may hinder some further analytical approaches, these have been initiated to characterise this new population sample as far as possible.

Patterns and peculiarities

The three LBK mass graves of Talheim (Figure 16.3A), Wiederstedt (Figure 16.3B), and Kilianstädten (Figure 16.3C) share a number of common characteristics evident from these short descriptions, but can also be distinguished from each other by some peculiarities. The first shared observation is that all three mass graves have been discovered unexpectedly, either by non-archaeological building activity (Talheim and Wiederstedt) or by excavations connected to road-building (Kilianstädten). In contrast to modern or even historical cases, where the location of mass burials is usually (roughly) known, suspected, or can even be pointed out by living witnesses (Haglund *et al.* 2001), the discovery of prehistoric mass graves obviously cannot be planned in advance. This is a basic but relevant factor for the success of extensive bioarchaeometric analyses, which should always be carried out on such complex features, as the collection, adequate storage, and subsequent handling of samples ideally have to be managed right on the excavation site. This can cause problems if the necessary preparations are not already in place (cf. Brandt *et al.* 2010).

All three mass graves were located within settlement contexts, as evidenced by other features nearby, the refuse found in the infill, and the absence of other graves clearly associated with them. A settlement burial is known from Kilianstädten (Schwitalla and Schmitt 2006), but as shown before, this is a standard occurrence if large areas are excavated and most likely is entirely unrelated to the mass grave itself. Although settlement burials have previously been discussed as possibly containing people of lower social status than those placed in a well defined cemetery (Veit 1996, 185), most of them have obviously been arranged with care and usually contain only single individuals, multiple inhumations being rare during the LBK. While the mass graves could therefore be termed *Siedlungsbestattungen* based on their location alone, they clearly belong to an entirely different category. Currently they represent best the obvious deviance from the standard LBK burial rites in several aspects, with the absence of grave goods, the erratic body positions and the high number of bodies per pit among them. Other sites with large quantities of human remains, like the enclosures at Schletz (Teschler-Nicola *et al.* 1999, 2006) and Herxheim (Zeeb-Lanz 2009, Zeeb-Lanz *et al.* 2009) have to be mentioned here as well, but there, intentions were supposedly different. The highly complex features at Herxheim, containing smashed and intensively manipulated human bones, mixed with pottery and animal remains, most probably are the leftovers of enigmatic rituals, possibly involving cannibalism (Boulestin *et al.* 2009, Orschiedt and Haidle 2012). The purpose at Herxheim apparently was not the primary and supposedly final burial of intact persons, which in fact is represented at the site in the form of 'classic' settlement burials (Zeeb-Lanz *et al.* 2009). In the ditches at Schletz, the scattered remains of dozens of mostly incomplete individuals were found, bearing the traces of peri-mortem violence and subsequent carnivore activity. Burial probably never took place for these individuals, as the dead were left where they had fallen in the aftermath of a violent conflict, finally being covered by sediment by natural processes (Teschler-Nicola *et al.* 2006). Again, settlement burials probably unrelated to the massacre were present in Schletz as well, as to be expected in a village site of this size.

Peri-mortem violence is also one of the key features of the Talheim mass grave, clearly connecting it to the site of Schletz, which appears to be largely contemporaneous (Wild *et al.* 2004). A massacre with complete extermination of a local community is the commonly accepted reconstruction for this site, although it remains unclear who carried out the collection and disposal of the corpses (Wahl and König 1987). Judging from the erratic commingling, any emotional attachment to the dead can be largely excluded, but this remains conjecture. The same chaotic dumping of bodies into a (nearby) pit is encountered at Wiederstedt and Kilianstädten, raising the same question. While Kilianstädten appears quite similar to Talheim at the moment, with evidence of peri-mortem violence, the skeletons from Wiederstedt show no violent injuries. Although these might have been present only in the non-preserved soft tissues, experience tells us that Neolithic violence will often be visible in the skeletal record to some extent, especially if higher numbers of dead people are the result (Wahl and König 1987, Teschler-Nicola *et al.* 1999, Meyer *et al.* 2009, Fibiger *et al.* 2013). Therefore, cause of death was probably different and probably non-violent at Wiederstedt, and the skewed and 'incomplete' demography strongly suggests that there were a number of survivors, which is unlikely for Talheim and possibly Kilianstädten. Nevertheless, mass graves were chosen to receive the dead at all three sites, the bodies were not arranged with care or obvious 'affection' (Peschel 1992, 265), and grave goods and personal ornaments were absent as well as dedicated individual grave pits. The denial of grave goods and of one of the 'standard' LBK burial arrangements clearly shows that the individuality and personhood of the dead as members of LBK communities were less important than their disposal below ground as quickly and effortlessly as possible. The example of Schletz demonstrates that carnivore activity is usually preserved in the skeletal record, if it was present in higher numbers (Teschler-Nicola *et al.* 1999, 2006). As no clear evidence of it was found in the three mass graves, and anatomical articulations were largely preserved, a quick collection of corpses and fast disposal are the most parsimonious explanation.

All of the sites mentioned – the three mass graves as well as the enclosures of Herxheim and Schletz – share the central trait that the individuality of the dead is suppressed by inclusion in a chaotic mass deposit. This can take the form of erratic commingling of complete, semi-complete or entirely disintegrated bodies and fragmented parts thereof. Within a study of deviant burial practices of the LBK, a working typology of mass fatality sites in the widest sense has recently been created (Meyer *et al.* 2013), which aimed at collating the relevant characteristics of truly deviant LBK features containing the remains of numerous individuals (Figure 16.4). Other burial sites exist, which could have been included as well, such as the cave site of Jungfernhöhle, but recent research shows that the older studies are highly unreliable and were therefore omitted here (cf. Seregély 2012). The published mass graves are contrasted with the enclosure sites, and their basic characteristics are shown in Figure 16.4. While the Talheim site can be seen as a mass grave with erratic disposal and signs of violence (type A), the Wiederstedt mass grave is rather similar minus the violent injuries (type B). The enclosure at Schletz shows evidence of violence, but lacks the collection of bodies and their deposition in a single pit (type C). The bones found at Herxheim also stem from enclosure features, but are highly manipulated within a ritualistic context, which surely involved physical

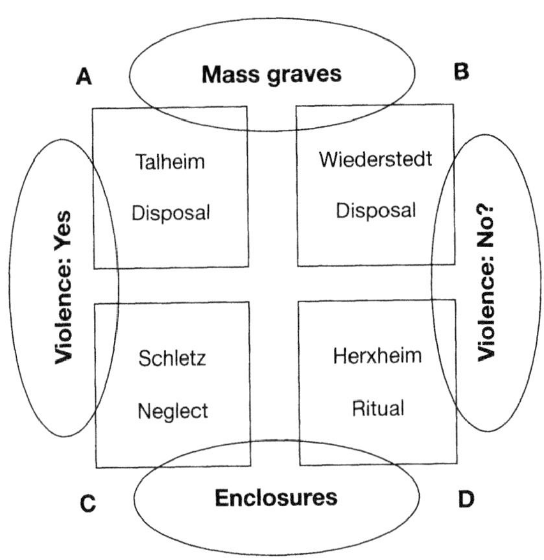

Figure 16.4 Working typology of deviant treatment of deceased in the LBK including the four sites of A: Talheim, B: Wiederstedt, C: Schletz, D: Herxheim. Depicted are the interrelations of shared patterns and individual peculiarities.

force, but it is unknown if violence was also directed at living people there (type D). In any case it has to be remembered that violence in prehistory is not always visible in the skeletal record, and that further research might once again alter our perceptions of these and other sites (cf. Fibiger *et al.* 2013).

Transversal versus longitudinal population transects

When dealing with mass graves or similar sites of contemporary mass fatality, one central difference from cemetery sites has to be singled out specifically. While cemeteries usually receive the deceased members of local communities over a larger span of time, possibly even centuries, mass graves represent a snapshot of a contemporaneously living (and dying) population sample (cf. Bentley *et al.* 2008). This has great implications for analysing population structure from both. Whereas long-term patterns and possible social change are better represented in larger cemetery samples, where the individual is buried with a culture-specific statement expressed in burial position, orientation, and grave goods, the mass grave sample might reveal the number of people, of families, of biological and geographical lineages present at the same place at one single point in time. The cemetery as a longitudinal population transect usually has the advantage of sample size, but averaging might blur smaller changes. The mass fatality sample, although usually smaller in numbers, is a much more precise mirror of contemporary social structure, a transversal transect of a population frozen in time. The absence of grave goods and of a careful burial rite is a drawback, however, as it is unclear how these individuals were originally viewed within their community, what their social status was, and how they would otherwise have been buried. This gap can partly be filled by the application of bioarchaeological methods of investigation, including the assessment of stress markers, trauma, degenerative and infectious disease, dietary and kinship reconstruction, geographical fingerprinting and population genetics which will supply important and highly relevant information regarding the life-histories of individuals and populations alike (Alt *et al.* 1995, Price *et al.* 2006, Bentley *et al.* 2008, Haak *et al.* 2008). For a detailed study of the interrelations of artefacts and people within the burial rite one has to refer to cemeteries again, where each individual is supposedly buried with information regarding their social identity and perception by others (Bickle *et al.* 2011, Bentley *et al.* 2012).

Conclusions

Multiple and simultaneous deaths sometimes led to the utilisation of mass graves in the LBK, whether violence was involved or not. The care usually received by deceased individuals, which was probably geared towards a fitting place in the 'spiritual' afterlife, was apparently overridden by the socio-cultural impact of the mass fatality events in these cases. Judging from the currently available numbers of skeletons from the sites of Talheim, Kilianstädten, Wiederstedt and Schletz, we currently have a lower boundary of about ten individuals to conservatively define such a 'mass' fatality event. Carefully arranged multiple burials with two or three bodies are known from LBK cemetery and settlement contexts (Peschel 1992, Veit 1996), albeit in much smaller numbers than individual inhumations, so it might be assumed that there possibly occurred a conceptual shift from careful burial to careless disposal in the range of five to ten individuals. Whether the usual capacities for carrying out the supposed death rituals were overtaxed by such numbers, or if peculiar conditions were largely responsible for the picture we currently have for deviant burial at each site, cannot be securely stated. One important factor will surely have been the social identity of the persons carrying out the burial, or the non-burial in the case of the Schletz corpses, in regard to the dead. Matching wound diameters to tool sizes may be one way to approach this problem (Wahl and Strien 2007), but such wounds do not always exist or are not sufficiently preserved and in fact do not indicate if attackers and buriers were actually the same. Therefore, the identity of the latter will remain mostly unknown, and attempts to reveal their identity have to come from the graves themselves. In modern forensic practice and cases of genocide, it may be feasible to attribute the more orderly arranged mass graves to the group to which the deceased belonged or to a culturally perceived 'self'. In contrast, erratic commingling and evidently careless disposal indicate 'others' with no positive emotional attachment to the deceased or in some cases a negative one (Komar 2008). But even if it is supposed that there must have been survivors of the same community, like in the case of Wiederstedt with its incomplete and mostly sub-adult demography, we have no knowledge of what exactly happened afterwards, or even before the mass fatality event. Even if kin survived, they might have been unable to arrange a formal burial (cf. Haak *et al.* 2008, Meyer *et al.* 2009). And as no orderly, neatly arranged LBK mass grave is currently known, we cannot decide whether they existed as a further variant of the burial rite portfolio, or if mass fatalities, regardless of their reason, always ended up in order-less piles of dead bodies in settlement or rubbish pits. That human remains sometimes were intensively modified, manipulated, and deposited in ways clearly different from a recognisable individual or multiple burial, and, equally important, clearly different from modern western perceptions of a considerate burial, is obvious from sites like Herxheim (Zeeb-Lanz *et al.* 2009). Future excavations and analyses might reveal if the possible conceptual shift from careful burial to careless disposal is just an artefact of the current state of research, or if it was an actual part of LBK burial customs. One has to bear in mind, however, that mass fatality events might have been masked by individual burial in a regular cemetery, a possibility which would be almost impossible to detect. Although cemeteries are rightly regarded as a more longitudinal population transect, in contrast to the transversal nature of mass graves, mass fatalities might also affect cemetery samples to an unknown degree. Nevertheless, mass graves of the LBK, once thought to be singular, are seemingly more common and probably formed at least one aspect of LBK funerary traditions in the widest sense. Put more specifically, they were already a standard way of dealing with high numbers of dead in the Early Neolithic of central Europe, and the well known characterisation of 'diversity in uniformity' (Modderman 1988) now seems applicable even to the LBK mass graves.

Acknowledgements

We thank C. Knipper, J. Ziegler as well as G. Schwitalla and S. Schade-Lindig for their support in the ongoing study. J. Wahl and the Landesamt für Denkmalpflege Baden-Württemberg kindly provided a photo of the Talheim mass grave.

References

Alt, K.W., Vach, W. and Wahl, J. 1995. Verwandtschaftsanalyse der Skelettreste aus dem bandkeramischen Massengrab von Talheim, Kreis Heilbronn. *Fundberichte aus Baden-Württemberg* 20, 195–217.

Bentley, R.A., Wahl, J., Price, T.D. and Atkinson, T.C. 2008. Isotopic signatures and hereditary traits: snapshot of a Neolithic community in Germany. *Antiquity* 82, 290–304.

Bentley, R.A., Bickle, P., Fibiger, L., Nowell, G.M., Dale, C.D., Hedges, R.E.M., Hamilton, J., Wahl, J., Francken, M., Grupe, G., Lenneis, E., Teschler-Nicola, M., Arbogast, R.-M., Hofmann, D. and Whittle, A. 2012. Community differentiation and kinship among Europe's first farmers. *Proceedings of the National Academy of Sciences of the United States of America* 109, 9326–30.

Bickle, P., Hofmann, D., Bentley, R.A., Hedges, R., Hamilton, J., Laiginhas, F., Nowell, G., Pearson, D.G., Grupe, G. and Whittle, A. 2011. Roots of diversity in a *Linearbandkeramik* community: isotope evidence at Aiterhofen (Bavaria, Germany). *Antiquity* 85, 1243–58.

Boulestin, B., Zeeb-Lanz, A., Jeunesse, C., Haack, F., Arbogast, R.-M. and Denaire, A. 2009. Mass cannibalism in the Linear Pottery culture at Herxheim (Palatinate, Germany). *Antiquity* 83, 968–82.

Brandt, G., Knipper, C., Roth, C., Siebert, A. and Alt, K.W. 2010. Beprobungsstrategien für aDNA und Isotopenanalysen an historischem und prähistorischem Skelettmaterial. In H. Meller and K.W. Alt (eds), *Anthropologie, Isotopie und DNA – biografische Annäherung an namenlose vorgeschichtliche Skelette?*, 17–32. Halle: Landesmuseum für Vorgeschichte.

Craig, O.E., Chapman, J., Heron, C., Willis, L.H., Bartosiewicz, L., Taylor, G., Whittle, A. and Collins, M. 2005. Did the first farmers of central and eastern Europe produce dairy foods? *Antiquity* 79, 882–894.

Egg, M. and Spindler, K. 2009. *Kleidung und Ausrüstung der kupferzeitlichen Gletschermumie aus den Ötztaler Alpen*. Mainz: Römisch-Germanisches Zentralmuseum.

Farruggia, J.-P. 2002. Une crise majeure de la civilisation du Néolithique Danubien des années 5100 avant notre ère. *Archeologické rozhledy* 54, 44–98.

Fibiger, L., Ahlström, T., Bennike, P. and Schulting, R.J. 2013. Patterns of violence-related skull trauma in Neolithic southern Scandinavia. *American Journal of Physical Anthropology* 150, 190–202.

Fritsch, B., Classen, E., Müller, U. and Dresely, V. 2008. Die linienbandkeramischen Gräberfelder von Derenburg 'Meerenstieg II' und Halberstadt 'Sonntagsfeld', Lkr. Harz. *Jahresschrift für Mitteldeutsche Vorgeschichte* 92, 25–229.

Gerling, C. 2009. Schwetzingen, ein 'reguläres' Gräberfeld der jüngeren Linearbandkeramik. In A. Zeeb-Lanz (ed.), *Krisen – Kulturwandel – Kontinuitäten. Zum Ende der Bandkeramik in Mitteleuropa*, 103–10. Rahden: Marie Leidorf.

Golitko, M. and Keeley, L.H. 2007. Beating ploughshares back into swords: warfare in the *Linearbandkeramik*. *Antiquity* 81, 332–42.

Gronenborn, D. 2003. Der 'Jäger/Krieger' aus Schwanfeld. Einige Aspekte der politisch-sozialen Geschichte des mitteleuropäischen Altneolithikums. In J. Eckert, U. Eisenhauer and A. Zimmermann (eds), *Archäologische Perspektiven: Analysen und Interpretationen im Wandel. Festschrift für Jens Lüning zum 65. Geburtstag*, 35–48. Rahden: Marie Leidorf.

Gronenborn, D. 2007. Climate change and socio-political crises: some cases from Neolithic central Europe. In T. Pollard and I. Banks (eds), *War and sacrifice: studies in the archaeology of conflict*, 13–32. Leiden: Brill.

Haak, W., Forster, P., Bramanti, B., Matsumura, S., Brandt, G., Tänzer, M., Villems, R., Renfrew, C., Gronenborn, D., Alt, K.W. and Burger, J. 2005. Ancient DNA from the first European farmers in 7500-year-old Neolithic sites. *Science* 310, 1016–18.

Haak, W., Brandt, G., de Jong, H.N., Meyer, C., Ganslmeier, R., Heyd, V., Hawkesworth, C., Pike, A.W.G., Meller, H. and Alt, K.W. 2008. Ancient DNA, strontium isotopes and osteological analyses shed light on social and kinship organization of the Later Stone Age. *Proceedings of the National Academy of Sciences of the United States of America* 105, 18226–31.

Haak, W., Balanovsky, O., Sanchez, J.J., Koshel, S., Zaporozhchenko, V., Adler, C.J., Der Sarkissian, C.S.I., Brandt, G., Schwarz, C., Nicklisch, N., Dresely, V., Fritsch, B., Balanovska, E., Villems, R., Meller, H., Alt, K.W., Cooper, A. and The Genographic Consortium. 2010. Ancient DNA from European early Neolithic farmers reveals their Near Eastern affinities. *PLoS Biology* 8, e1000536.

Haglund, W.D., Connor, M. and Scott, D.D. 2001. The archaeology of contemporary mass graves. *Historical Archaeology* 35, 57–69.

Hofmann, D. 2009. Cemetery and settlement burials in the Lower Bavarian LBK. In D. Hofmann and P. Bickle (eds), *Creating communities: new advances in central European Neolithic research*, 220–34. Oxford: Oxbow Books.

Knipper, C., Meyer, C., Jacobi, F., Roth, C., Fecher, M., Stephan, E., Schatz, K., Hansen, L., Posluschny, A., Höppner, B., Maus, M., Pare, C.F.E. and Alt, K.W. 2014. Social differentiation and land use at an Early Iron Age 'princely seat': bioarchaeological investigations at the Glauberg (Germany). *Journal of Archaeological Science* 41, 818–25.

Komar, D. 2008. Patterns of mortuary practice associated with genocide. Implications for archaeological research. *Current Anthropology* 49, 123–33.

Krause, R. 1998. Die bandkeramischen Siedlungsgrabungen bei Vaihingen an der Enz, Kreis Ludwigsburg (BW). *Berichte der Römisch-Germanischen Kommission* 79, 5–106.

Lohr, C., Meyer, C., Alt, K.W. and Gronenborn, D. In press. Ein zweites Talheim?! Erste Ergebnisse aus dem Massengrab von Schöneck-Kilianstädten. In H. Meller and S. Friederich (eds), *Salzmünde – Regel oder Ausnahme*. Halle: Landesmuseum für Vorgeschichte.

Meyer, C. and Alt, K.W. 2010. An anthropological perspective of the Early and Middle Neolithic of the Upper Rhine Valley: results of an osteometric study of postcranial skeletal elements. In D. Gronenborn and J. Petrasch (eds), *The spread of the Neolithic to central Europe*, 487–96. Mainz: Römisch-Germanisches Zentralmuseum.

Meyer, C., Kürbis, O. and Alt, K.W. 2004. Das Massengrab von Wiederstedt, Ldkr. Mansfelder Land. Auswertung und Gedanken zur Interpretation im Kontext der Linienbandkeramik. *Jahresschrift für Mitteldeutsche Vorgeschichte* 88, 31–66.

Meyer, C., Kranzbühler, J., Drings, S., Bramanti, B., Nehlich, O., Richards, M.P. and Alt, K.W. 2008. Die menschlichen Skelettfunde aus der neolithischen Totenhütte von Benzingerode. Anthropologische Untersuchungen an den Bestattungen eines Kollektivgrabs der Bernburger Kultur. *Archäologie in Sachsen-Anhalt* Sonderband 7, 107–51.

Meyer, C., Brandt, G., Haak, W., Ganslmeier, R.A., Meller, H. and Alt, K.W. 2009. The Eulau eulogy: bioarchaeological interpretation of lethal violence in Corded Ware multiple burials from Saxony-Anhalt, Germany. *Journal of Anthropological Archaeology* 28, 412–23.

Meyer, C., Ganslmeier, R., Dresely, V. and Alt, K.W. 2012. Die Bedeutung taphonomischer Einflüsse bei der bioarchäologischen Interpretation neolithischer Grabfunde. In T. Link and D. Schimmelpfennig (eds), *Taphonomische Forschungen (nicht nur) zum Neolithikum*, 143–56. Kerpen-Loogh: Welt und Erde.

Meyer, C., Lohr, C., Strien, H.-C., Gronenborn, D. and Alt, K. W. 2013. Interpretationsansätze zu 'irregulären' Bestattungen während der linearbandkeramischen Kultur: Gräber *en masse* und Massengräber. In N. Müller-Scheeßel (ed.), *'Irreguläre' Bestattungen in der Urgeschichte: Norm, Ritual, Strafe . . .?*, 111–22. Bonn: Habelt.

Moddermann, P.J.R. 1988. The Linear Pottery culture: diversity in uniformity. *Berichten van de Rijksdienst voor het Oudheidkundig Bodemonderzoek* 38, 63–139.

Nieszery, N. 1995. *Linearbandkeramische Gräberfelder in Bayern*. Espelkamp: Marie Leidorf.

Orschiedt, J. 1998. *Bandkeramische Siedlungsbestattungen in Südwestdeutschland. Archäologische und anthropologische Untersuchungen*. Rahden: Marie Leidorf.

Orschiedt, J. and Haidle, M.N. 2012. Violence against the living, violence against the dead on the human remains from Herxheim, Germany. Evidence of a crisis and mass cannibalism? In R. Schulting and L. Fibiger (eds), *Sticks, stones, and broken bones: Neolithic violence in a European perspective*, 121–37. Oxford: Oxford University Press.

Peschel, C. 1992. *Regel und Ausnahme. Linearbandkeramische Bestattungssitten in Deutschland und angrenzenden Gebieten, unter besonderer Berücksichtigung der Sonderbestattungen*. Buch am Erlbach: Marie Leidorf.

Price, T.D., Wahl, J. and Bentley R.A. 2006. Isotopic evidence for mobility and group organization among Neolithic farmers at Talheim, Germany, 5000 BC. *European Journal of Archaeology* 9, 259–84.

Price, T.D., Wahl, J., Knipper, C., Burger-Heinrich, E., Kurz, G. and Bentley, R.A. 2003. Das bandkeramische Gräberfeld vom 'Viesenhäuser Hof' bei Stuttgart-Mühlhausen: Untersuchungsergebnisse zum Migrationsverhalten im frühen Neolithikum. *Fundberichte aus Baden-Württemberg* 27, 23–58.

Salque, M., Bogucki, P.I., Pyzel, J., Sobkowiak-Tabaka, I., Grygiel, R., Szmyt, M. and Evershed, R.P. 2013. Earliest evidence for cheese making in the sixth millennium BC in northern Europe. *Nature* 493, 522–5.

Schwitalla, G. and Schmitt, K.M. 2006. Ein Massengrab aus der Umgehungsstraße oder 'Aus der Linearbandkeramik ist alles bekannt'. *hessenArchäologie* 2006, 31–2.

Seregély, T. 2012. Neolithische Siedlungen und vergessene Funde. Neues zur Jungfernhöhle bei Tiefenellern. In F. Falkenstein (ed.), *Hohler Stein, Rothensteine und Jungfernhöhle. Archäologische Forschungen zur prähistorischen Nutzung naturheiliger Plätze auf der nördlichen Frankenalb*, 64–73. Würzburg: Institut für Altertumswissenschaften.

Teschler-Nicola, M., Prohaska, T. and Wild, E.M. 2006. Der Fundkomplex von Asparn/Schletz (Niederösterreich) und seine Bedeutung für den aktuellen Diskurs endlinearbandkeramischer Phänomene in Zentraleuropa. In J. Piek and T. Terberger (eds), *Frühe Spuren der Gewalt – Schädelverletzungen und Wundversorgung an prähistorischen Menschenresten aus interdisziplinärer Sicht*, 61–76. Schwerin: Landesamt für Kultur und Denkmalpflege.

Teschler-Nicola, M., Gerold, F., Bujatti-Narbeshuber, M., Prohaska, T., Latkoczy, C., Stingeder, G. and Watkins, M. 1999. Evidence of genocide 7000 BP – Neolithic paradigm and geo-climatic reality. *Collegium Antropologicum* 23, 437–50.

Trautmann, I. 2006. *The significance of cremations in Early Neolithic communities in central Europe*. University of Tübingen, http://nbn-resolving.de/urn:nbn:de:bsz:21-opus-30056.

Veit, U. 1992. Burials within settlements of the Linienbandkeramik and Stichbandkeramik cultures of central Europe. On the social construction of death in early-Neolithic society. *Journal of European Archaeology* 1, 107–40.

Veit, U. 1996. *Studien zum Problem der Siedlungsbestattung im europäischen Neolithikum*. Münster: Waxmann.

Wahl, J. and König, H.G. 1987. Anthropologisch-traumatologische Untersuchung der menschlichen Skelettreste aus dem bandkeramischen Massengrab bei Talheim, Kreis Heilbronn. *Fundberichte aus Baden-Württemberg* 12, 65–186.

Wahl, J. and Strien, H.-C. 2007. *Tatort Talheim. 7000 Jahre später: Archäologen und Gerichtsmediziner ermitteln*. Heilbronn: Städtische Museen Heilbronn.

Wahl, J. and Trautmann, I. 2012. The Neolithic massacre at Talheim: a pivotal find in conflict archaeology. In R. Schulting and L. Fibiger (eds), *Sticks, stones, and broken bones: Neolithic violence in a European perspective*, 77–100. Oxford: Oxford University Press.

Wild, E.M., Stadler, P., Häußer, A., Kutschera, W., Steier, P., Teschler-Nicola, M., Wahl, J. and Windl, H.J. 2004. Neolithic massacres: local skirmishes or general warfare in Europe? *Radiocarbon* 46, 377–85.

Zeeb-Lanz, A. 2009. Gewaltszenarien oder Sinnkrise? Die Grubenanlage von Herxheim und das Ende der Bandkeramik. In A. Zeeb-Lanz (ed.), *Krisen – Kulturwandel –*

Kontinuitäten. Zum Ende der Bandkeramik in Mitteleuropa, 87–101. Rahden: Marie Leidorf.

Zeeb-Lanz, A., Arbogast, R.-M., Haack, F., Haidle, M.N., Jeunesse, C., Orschiedt, J., Schimmelpfennig, D. and van Willigen, S. 2009. The LBK Settlement with pit enclosure at Herxheim near Landau (Palatinate). First results. In D. Hofmann and P. Bickle (eds), *Creating communities: new advances in central European Neolithic research*, 202–19. Oxford: Oxbow Books.

17

Revealing Our Vibrant Past

Science, Materiality and the Neolithic

OLIVER J.T. HARRIS

Introduction

IT IS NOW MORE THAN TEN YEARS since Andrew Jones (2002) published his book *Archaeological Theory and Scientific Practice*. It has stood the test of time, and indeed now looks rather prophetic; among other things it foreshadows symmetrical archaeology (Witmore 2007, Olsen *et al.* 2012). The central tenet of the book is that archaeology at that time was dominated by two cultures, much as wider academia had been famously described by C.P. Snow in 1959 (1959). On the one hand stood archaeological science and on the other archaeological theory (cf. Parikh and Hall 2012). The former saw itself as dealing with static, inanimate objects bequeathed from the past that could, given the correct application of technique, yield up their secrets (Jones 2002, 169). The latter saw itself as dealing with issues of meaning, human agency and interpretation, and as engaging with fluid, animate subjects (Jones 2002, 169). The book not only diagnosed this condition, but correctly recognised it as hugely disabling. The oppositions on which this divide was predicated – science versus theory, nature versus culture, object versus subject and so on – were modernist constructs, and thus the arguments that flew back and forth across it did not engage with a debate about the actual nature of the past.

Over ten years later it might feel as if these oppositions are less central to archaeological discourse (see Robb this volume, Chapter 2). The theory wars of the 1980s and 1990s appear to have ceased, and most participants on both sides are content to concede that both science and theory have much to offer archaeology. Furthermore, there has been an increased willingness by a new generation of scholars to draw on both complex social theory and detailed scientific analysis (see Robb this volume). However, much like the divisions between North and South Korea, simmering tensions remain, and no real peace has been agreed. The boundaries between the two sides have not been done away with; instead a

Proceedings of the British Academy **198**, 327–45. © The British Academy 2014.

negotiated compromise keeps both happy. In this chapter I want to suggest that in the end it is neither enough simply to keep the peace, nor are efforts at bridging the divide going to be successful (Webmoor and Witmore 2008). What we need is to start from a different, non-dichotomous, perspective, and I will outline what one such version of this might look like, drawing on Jane Bennett's (2010) notion of vibrant matter, as well as the increasingly important concept of assemblage.

Such a move is particularly apposite in the context of the subject of this volume, early farming. The Neolithic perhaps more than any period remains beset by oppositional thought: farmers versus foragers, incomers versus indigenous people, domestic versus wild, and so on. It is also a period which has seen numerous outstanding applications of scientific techniques that open up the possibility for a range of new connections to be drawn, and understandings to be developed. So in this chapter I aim to do three things. First, I will reiterate the evidence that archaeological thought has tended to be dichotomous and oppositional, particularly in relation to the Neolithic, and why this is so problematic. Secondly, I will set out how we can undercut these oppositions by taking up a different position in relation to archaeological material and by embracing a different ontology. Thirdly, I will look at the Late Mesolithic and Early Neolithic of Britain to outline some of the ways in which this new ontology can open up possibilities for different understandings of the past.

The tenacity of oppositional thinking

Dichotomies have become the *bête noire* of archaeological thought in recent years. Emerging from archaeology's relationship with modernity, and drawing on traditions of thought going back via Descartes to Plato, they have been diagnosed as the fundamental stumbling block to understanding the past (Jones 2002, 2012; Olsen *et al.* 2012; Thomas 1996, 2004). The analysis of these dichotomies, and the problems they bring, has been widely debated, and so only a brief summary is offered here. Beginning with the fundamental opposition between person and world, they structure our understandings of nature and culture, body and mind, outside and inside, and so on. Archaeologists have lurched from privileging one side of these dichotomies to the other, arguing that what matters are the ideas in people's heads and how they express them, or that it is people's engagement with the natural world that really drives history, or that it is the way people understand symbols and meanings that really matters (see Robb this volume). Throughout the history of archaeological thought the modernist dichotomies that structure the understandings have remained in place, and archaeologists have simply argued about which side of the divide matters. The separation between archaeological theory and archaeological science matches itself neatly on to these oppositions.

Nowhere has this been more true than in the study of the Neolithic because it is here, above all periods, that archaeologists have tended to identify the past in terms of these dichotomies (Pollard 2004, 56). Think of the major debates that structure our arguments around farming and one finds that they map neatly on to different sides of a dichotomous debate (Borić 2005). As Borić (2005) has pointed out, the very opposition between farmer and forager starts this off, but it runs through our oppositions of wild and domestic foodstuffs, incomers versus locals, and so on. Is the Neolithic primarily a way of getting food (an engagement with the natural world) or a set of belief systems (a cultural package)? Should we pin the blame for its spread on active farmers (subjects) dominating passive hunter gatherers (objects), or is it the other way around (with the identification of subjects and objects reversed)? The reason that debate over the start of agriculture has been so fierce is precisely because it frames itself around the very dichotomies that structure our thought.

Issues with oppositions

As many authors have pointed out, however, these dichotomies are not neutral descriptors but hugely problematic modernist assumptions (Latour 1993, Thomas 2004). The oppositions they describe emerge not from scientific investigation but from a series of compromises reached in the seventeenth century to keep science and politics – or if you like nature and culture – apart from one another (Latour 1993). In terms of the Neolithic this means that the oppositions between domestic and wild, farmer and forager, nature and culture, are our *impositions* on to the material; they are acts of what Latour (1993) calls purification, which disguises the way in which these things are never clearly separate. Imposing these assumptions on to the past has a series of significant issues for us as archaeologists. First, it essentialises and universalises a single way of thinking about the world, imposing it on all times and places. Secondly, it actively prevents us from understanding the past, because of the way in which these essentialised assumptions impose barriers on to our engagements. The emphasis placed upon humanity within modernity, that we (humans) are uniquely special, given culture and language in a way that raises us above all other creatures, means that the contributions of other beings – both 'animate' and 'inanimate' in traditional terms – are lost to us (Ingold 2007, Olsen *et al.* 2012). Thirdly, we waste lots of effort on pointless intellectual debates. We wrestle with how to resolve the dichotomy between forager and farmer, with whether the prime movers were indigenous or outsider, with whether domestic animals or wild ones provided more of an economic resource. The terms of these debates mean they are irresolvable, simply because they spring from our preconceptions, not from our engagement with the material. This does not mean that we should ignore the differences between animals which we currently categorise as domesticated and those which we currently categorise as wild, for example. It does mean that we should attend to the specific differences that exist between different animal species, rather than lumping them together into two purified, and artificial groups (Pollard 2006, 136–7).

Fundamentally, an uncritical reliance on dichotomies also keeps archaeology itself polarised into two camps, with on the one side the scientists providing data, and on the other theorists offering interpretation, even when the same archaeologist plays both roles. This opposition too is of our making. New data alone can never resolve the questions we have about the past and nor can new theories. We need to rethink the very basis on which we think about our discipline and about the past.

New ways forward?

Luckily, archaeology is currently equipped to do exactly that thanks to a range of new ways of engaging with the past that have emerged over the last decade and more. If previous archaeological debates have been fundamentally about epistemology – how we know the world – these new discourses focus on ontology – what the world actually is. Although manifestly variable and different from one another, these approaches share a rejection of the dominant modernist tropes touched on above, and seek to engage with the past in a way that attends to its material and non-representational qualities. A detailed review of the various approaches would be an article – or more likely a book – in itself, but some that might be mentioned within this include: symmetrical archaeology (e.g. Olsen *et al.* 2012, Witmore 2007), drawing on the work of Latour, and to a lesser extent thinkers like Harman and Whitehead; ideas of local ontologies (Alberti and Marshall 2009, Conneller 2011), drawing on Viveiros de Castro, Ingold and Barad; assemblage theory (Harris 2013, 2014; Harrison 2011; Jones 2012; Normark 2008, 2009, 2010; Lucas 2012), drawing on Deleuze, DeLanda and Bennett; phenomenology (Cobb 2008, Thomas 1996, Olsen 2010), drawing on Heidegger; relational realism (Fowler 2013), drawing on Deleuze, Latour and Barad; and entanglement (Hodder 2012). The fact that several authors occur in different sections should indicate that these different approaches have a lot in common with one another. They are all broadly focused on relations and materials, reject idealism, essentialism and human exceptionalism, and attempt to understand the world outside standard ideas of naturalism or social constructivism. There are significant differences between these approaches as well – it would be disingenuous to pretend otherwise – but for the sake of the argument here they can be linked to one another as part of an emergent 'ontological turn' (Alberti *et al.* 2011; see Harris and Robb 2012 for a broader discussion of this phenomenon in anthropology and archaeology). Rather than attempt any further analysis of these approaches, however, the swiftest way forward is to outline those that I find most useful, and which are of particular relevance to the question here. These are approaches drawing on the notion of assemblage, and with it ideas of 'vibrant matter' (Bennett 2010). As Jones (2012, 200) has noted, this offers us the best chance of reconciling our understandings of both science and theory, not as competing enterprises, but two mutually linked ways of attending to the complexities of the past.

Vibrant matter and assemblages

Let us start with matter itself. In the traditional understanding of matter, it is something inanimate, a non-living substance that awaits the enlivening touch of humanity to transform it into something meaningful. Most of our conceptions of archaeological objects rest on this distinction; raw materials (clay, flint, bone) are passive and inert until human beings arrive to shape them in particular ways, producing them as that bizarre mixture of culture and nature which we refer to as material culture (Thomas 2007). This is the 'hylomorphic' model of matter, which goes back to Aristotle (Ingold 2013). This position firmly associates matter with nature and opposes it to culture, the stuff of human beings. Even positions advocating object agency (e.g. Gell 1998) maintain this distinction, requiring as they do the addition of human beings to the mix to add the 'magical mind-dust' of agency to the equation (Ingold 2007, 11). Both scientists and theorists have tended to treat things in this manner. For scientists the properties of materials, the contents of pots, and the isotopic signature of bones are all information encoded within the fixed nature of matter. For theorists, although matter can take on meaningful properties, this is only in conjunction with human beings; it is what they add to the equation, a cultural layer seemingly floating on top of the natural materials underneath (Ingold 2013). So a pot is understood as the imposition of human mind on to malleable clay, whether this is the shape of the pot itself, the contents added to it or the meanings attached to the decoration on the outside (Jones 2012, 100).

What if we took a different view, however? What if we emphasised not the hylomorphic model which separates matter and form, nature and culture, but instead emphasised the morphogenetic capabilities of materials? Both the anthropologist Tim Ingold and the political ecologist Jane Bennett have recently urged us to do precisely this: to think of the way in which materials themselves are animate, always in flow. Materials, in this perspective, are not static, nor are they fixed. Their properties are not somehow innately contained within them – a form of essence – but rather emerge through their relationships with the world. They are, as Ingold (2007, 15) puts it, 'histories'. Focusing on these vibrant materials pulls us away from thinking about the world in terms of people and objects, culture and nature, and asks us to attend to the flow of life itself, to the way in which things are always in the process of becoming something else. From this perspective the pot is no longer a fixed object produced as finished article by the imposition of form on to matter by a human agent. Instead it is a particular moment – an event – in the flow of materials from clay extraction site, temper source, the cow that provided the milk, the human that helped to bring these things together, and so on. This history continues through fragmentation, deposition, excavation, cataloguing, drawing and analysis. Instead of a bounded object, therefore, it is in Heidegger's term a gathering, or as I prefer to see it, an assemblage.

Assemblages, taken broadly from the work of Deleuze and Guattari (2004), offer a particular way of thinking about the world, and in this case about the materials which archaeologists uncover (Lucas 2012). Assemblages are the temporary gathering together of all manner of different kinds of relations, different kinds of materials, signs, utterances, meanings and ideas (Deleuze and Parnet 2007). Assemblages have a number of important characteristics. They are relational, but exceed the terms of their relations – they are more than the sum of their parts (DeLanda 2006). This means that they are also real (cf. Fowler 2013). That is, they are not just the product of their analysis but carry real ontological weight. Assemblages are also four-fold in form, or, put another way, they are tetravalent (Deleuze and Guittari 2004, 556; Dewsbury 2011). They have physical (or machinic) form (dimension one), but also expressive qualities (dimension two) and are also both simultaneously both coming together and moving apart (territorialising and deterritorialising: two further dimensions; see Harris 2013; Lucas 2012, 199). Thus our pot brings together physical bodies of clay, temper and milk; expressive decorations that can evoke meanings and emotions as well as symbolising ideas of connectedness and community; and it is the coming together of the substances listed, and the slow process of their breaking apart. Thinking about assemblages in this way has a series of critical advantages. First, it forces us to concentrate on relations over essences; that is, we are not concerned with some essential form of our pot – some kind of 'pothood' – but with the way in which it gathers together these different elements. Secondly, because this process is in flow, this is not a static measuring of a fixed thing, but the identification of something emergent in the world; in this it differs from the more static network analysis of Bruno Latour (1999; cf. Harman 2009; Harris 2013, 2014). Thirdly, assemblages work at multiple scales (DeLanda 2006; see Robb this volume on the importance of a multi-scalar approach), so we can think not only of the pot but also pottery traditions and archaeological sites as an assemblage; even concepts like 'the European Neolithic' are assemblages. This allows us additionally to engage with the reality of these concepts rather than dismissing them as imposed conceptions of the archaeologist. Fourthly, and perhaps most crucially here, assemblages 'bridge work between philosophy, science and social theory' (Dewsbury 2011, 149). Rather than treating these as different elements, assemblages place science and theory, nature and culture, within a single flat ontology amenable to analysis and engagement in the same way (DeLanda 2002, 2006, 2010).

Changing the focus of enquiry

Thus switching our focus to assemblages allows us to begin from a different ontological position. There are four key consequences of this for my argument here.

First, we can begin from a non-dichotomous standpoint. As emphasised above, this is not about reducing all elements of the world to being the same, but rather about understanding difference as emergent from relationships rather than prefiguring them as terms of analysis. Thus cows and aurochs are different because of the relations in which they are enmeshed (which include relations with human beings, but also landscapes, plants, DNA and other factors). Equally pots and clay are different not because one is 'cultural' and the other 'natural', but because the former are caught up in relations (with humans, substances used as temper, traditions of potting, or animals through the liquids they hold) with which the latter is not.

Secondly, this allows us to escape from some of the most debilitating conundrums of research on the Neolithic. Whether wild or domestic plants were of particular importance, for example, becomes replaced by questions about whether barley, wheat, crab apples, hazelnuts or any other plant one cares to mention were part of more or less intensive relationships during this period, and how this changed through time. As Ingold (2000, chapter 5) has so cogently argued, the key difference between what we think of as wild and domestic is the degree to which humans are involved in the conditions of growth – and the degree therefore to which other organisms help grow humans. Different forms of care are crucial here, and rather than examining generic categories, we have a new relationship between specific plants and specific groups of people to explore.

Thirdly, more broadly, this approach places science and theory on a single ontological footing, and does not regard them as two separate parts of an enterprise. That this can have significant impact is demonstrated by Joanna Sofaer's recent book (2006). Here Sofaer demonstrates that only by understanding the human body both through theory and through osteological information (both understood as similar forms of engagement with the material) can we hope to understand bodies in the past (and see Robb and Harris 2013). This means that archaeology is primarily about attending to the relations which we can trace in the present that survive from the past, the relations that endure (Fowler 2014). Tracing these relations happens in the field, when we excavate a feature and record the items that have been placed in a particular context together. It happens in the library when we link different sites together through reading reports and comparing other ideas and interpretations. It also – crucially – happens in the laboratory as we trace the vibrancy of matter. Different isotope signatures do not 'represent' different dietary practices in the past, but are parts of the relationships produced through those past practices which endure in the present. The decay of ^{14}C atoms measured does not solely 'represent' evidence for the date of the death of a piece of organic matter, but also is one way in which we as archaeologists can attend to the histories of vibrant matter. All these practices are thus ontologically equivalent.

Finally, authorship of the past is now more evenly distributed. The past is neither a passive object out there awaiting the enlivening touch of archaeologists, nor is it something purely constructed in the present, a relativist nightmare in which anything goes. Instead the past itself is a relational assemblage produced by all manner of materials, humans, animals, plants, scientific machinery and so on, past and present, coming together (cf. Olsen *et al.* 2012, Witmore 2012). The residues of the past that stretch into the present (Lucas 2012) mean that we as archaeologists are not free to build assemblages as we see fit. Instead we take our part in the active making of the past through the movement of our trowels, the testing of our machinery, our explorations of social and philosophical theory (cf. Latour 1999). These things bring us into correspondence (Ingold 2013) with the pot sherds, isotope signatures, animal bone assemblages, monuments and people of the past and it is through the collective negotiation and endeavour of *all* these elements that potential pasts emerge – never fixed or final, but never utterly fluid either. This democratic vision of past-making places data and interpretation on the same ontological level, both emergent in the present, and neither are simply free to be bent to the archaeologist's will (Witmore 2012).

Enlivening the past

How might these ideas work in practice? As in other areas of Europe, the debate in Britain around the transition to agriculture has been locked into opposed, dualistic camps, arguing back and forth about whether it represent a radical economic disjuncture with a new fully farming, settled economy coming from the outside, or new world view taken up with new material culture (for a review see Cummings and Harris 2011). Whilst we now know much more about the temporality of this transition (after Whittle *et al.* 2011), and have seen a number of attempts at mediating the dispute (Cummings and Harris 2011, Garrow and Sturt 2011, Whittle *et al.* 2011), we may be no closer to a resolution. Rather than focus on the transition itself here, however, I want to concern myself with how the line of thinking developed in this chapter can add to (note not replace) our current understandings of the differences and similarities between the Mesolithic and the Neolithic in Britain. To do this I want to compare two sites from either side of the divide, the shell midden of Cnoc Coig on Oronsay (Mellars 1987), and the midden underlying the chambered tomb of Ascott-under-Wychwood (Benson and Whittle 2007). In both cases I will examine these sites as tetravalent assemblages, in order to consider both similarities and differences in how materials and companion species (*sensu* Haraway 2008) emerged in these periods and how an understanding of vibrant matter allows us to situate both science and theory on an equal footing. Rethinking the Mesolithic and the Neolithic in these terms will allow us to conceptualise the differences between them that do not rest on artificial modernist dichotomies.

Assembling Cnoc Coig

The shell midden at Cnoc Coig most probably dates to between 4300 and 3800 cal BC,[1] and was in use for perhaps 200 years (Richards and Sheridan 2000, Meiklejohn *et al.* 2005). The site is made up of a large deposit of shells and other materials as well as features including over 50 hearths and two possible structures (Mellars 1987). Deposits of cowrie shell beads, human bones, remnants of butchery, antler tools and so on were also included within the midden. The dominant resources were fish and shellfish (supported by the stable isotope analysis (Richards and Mellars 1998)) but large numbers of seals were also consumed on site. Other animals present included otters, which appear largely to have been hunted for their fur rather than for meat (Grigson and Mellars 1987, 277), and parts of red deer and wild boar. The deer were not native to Oronsay and are likely to have come from further afield, perhaps from the island of Jura (Mithen and Finlayson 1991). Notably two different populations of this animal seem to have been drawn upon, and the parts of their bodies incorporated into the site suggest that they were mainly present for use as tools rather than as foodstuffs (Grigson and Mellars 1987, 252, 260). As an archaeological assemblage, the Cnoc Coig midden can be seen as incorporating multiple materials, from wood and stone through to shell and bone (Cobb 2008, 262).

What about the more theoretical take on assemblage which I have advocated here? Assemblages, as I noted above, have four dimensions (Dewsbury 2011), and each of these can be detected in the process of the midden's becoming. The first of these dimensions are the physical or machinic relationships. Here we can think about the physical connections formed by interweaving materials, the acts of placing substances together, the gathering of multiple elements of the Mesolithic world into a single place. The processes and practices of fishing, hunting and consumption are revealed in this element of the assemblage. The second dimension of an assemblage is the way in which it allows 'the collective assemblage of enunciation of acts and statements' (Dewsbury 2011, 150), what I called above the expressive dimension. This can be seen most clearly in Cnoc Coig in the famous juxtaposition of a human hand and seal flipper (Meiklejohn *et al.* 2005). As authors such as Warren (2005, 147) have noted, this seems to express an equivalence

[1] Two sets of radiocarbon dates are available, those from charcoal obtained as part of the original publication and more recent AMS dates from human bone (Richards and Sheridan 2000). The former suggest a date of between 4700 and 4000 cal BC, whilst the latter suggest the dating above. Although the discrepancy in these dates is worrying, the lack of offsets of the latter dates, and the spatial placing of human bones deep within the midden, suggest that for the moment the more recent dates, 4300–3800 cal BC, are more reliable (Meiklejohn *et al.* 2005, 98).

between human beings and seals. Specifically, the way in which the assemblage is gathered together here reveals the manner in which the seal flipper is treated like a human hand (and judging by the normal treatment of seal bones, not vice versa (Meiklejohn *et al.* 2005, 99)). Here we have seal-becoming-human expressed through the physical assembling of these items, and these represent just some of the meaningful juxtapositions created throughout the midden's history (Cobb 2008, 262–3). It should be clear here that the physical juxtaposition and the expressive content of the assemblage are not opposed to one another but form a dimension along which we can analyse different assemblages.

The other two dimensions are the way in which assemblages are stabilised (territorialised) and also simultaneously break apart and are destabilised (deterritorialised). In the midden at Cnoc Coig we can detect the ways in which multiple other places are tied together and linked into the assemblage: the origins of at least two different deer populations, the places where seals basked and were hunted, the rocks on which shellfish were gathered, or the beaches where flint pebbles could be found. These all indicate the way in which different temporalities were tied into these different materials yet bound into a single place. The similarity of diet of the people whose partial remains were interred in the midden (Richards and Mellars 1998) reveals the way in which a community territorialised itself through shared consumption. Simultaneously the decay of material within the midden helped it to territorialise, to reach out from itself and to leave the assemblage. Other parts of the people who were deposited here were taken away and disposed of elsewhere, separating out lines of relations that escaped away from the boundaries of this assemblage (cf. Cobb 2008, 281).

The assemblage of animals and humans, the materials they produce together, and the materials brought in from elsewhere produced the midden at Cnoc Coig. This assemblage reveals these animals, most explicitly – or expressively – seals, as companion species with human beings in this period, a becoming-with that led to these different bodies helping to produce the Mesolithic site as we understand it today. This approach does not divorce cultural and natural perspectives, social and economic practices, or humans and animals. Rather it recognises the way in which shared practices involving and authored by a host of 'animate' and 'inanimate' materials allowed this kind of becoming to emerge, and break apart, through time.

Assembling Ascott-under-Wychwood

At Ascott-under-Wychwood, a Neolithic midden was formed prior to the construction of a megalithic tomb. This latter event will not be part of the analysis here.[2] The midden was formed from the gathering of multiple materials including

[2] Though of course it is similarly amenable to an examination rooted in assemblage theory.

contemporary flint tools, older microliths, animal bone, pottery, fire, timber and earth (McFadyen *et al.* 2007; Whittle 2007a, 345). It was probably assembled between 3950 and 3800 cal BC and was probably in use for less than a century (Bayliss *et al.* 2007, 33; McFadyen *et al.* 2007, 35). It formed above a series of earlier pits and above and around two linear post structures (McFadyen *et al.* 2007). Specific connections were drawn here between different materials and different parts of the site, and care was taking in the assembly of these different materials (McFadyen 2007, 351). As McFadyen (2007, 351) has powerfully argued, the practices at the timber structure and in the formation of the midden were connected to multiple elements of people's lives including:

> grassland, the cutting of turf, woodland, the cutting and working of wood, the setting of posts through cut turf, flint, the working of flint, animals, the butchery of animals, charcoal, the burning of wood on hearths and the roasting of red deer and pig, clay, the making of pottery, the keeping of milk in pots, stone, the making of quern stones, and perhaps the growing and managing of plants.

Can we add to this through our analysis of tetravalent assemblages? The physical connections are well drawn out in the analysis by McFadyen (2007, 351–2), showing the way in which the material structure of timber, pits and midden gathered together the disparate elements of Neolithic life (cf. Pollard 2008, 56). The expressive elements are revealed in some of these connections, in particular the association of flint axe flakes with deer bone, and in the comparable burning of flint and pig bones. Alongside deer and pig, parts of other animals became entwined in the midden including cows, which were the most dominant animal in terms of numbers. In recent analyses authors have often promoted the social understandings of animals, particularly cattle, and their roles as metaphors for kinship (e.g. Ray and Thomas 2003), as ancestors (e.g. Jones and Richards 2003) or as means for thinking through the Neolithic world (e.g. Whittle 2003). It is not my intention to deny the importance of these interpretations. The association of deer and axes in particular speaks of deliberate associations being expressed between these materials (McFadyen 2007).

The other dimensions of the assemblage can be traced as well. Territorialisation worked at multiple scales. On the one hand specific connections were bound together in the pits, notably between burnt pig bones and burnt flints. At a higher scale we see the manner in which multiple clay sources were drawn upon (Barclay and Case 2007), and these vessels in turn brought together not only milk, but also pig fats (notably in vessel 3, which was found in pits 9 and 13; Barclay and Case 2007, 269). In smaller numbers other animals were of importance too, including fox, boar and wild cat, each tied into the assemblage through single canines (Mulville and Grigson 2007, 239). One can dwell further on the potential expressive qualities of these elements of the assemblage (Pollard 2004). The territorialisation of multiple animals is also indicated not only between species (whether wild or domestic in our terms) but in variation *between* animals of the same species. One pig sampled for stable isotopes revealed that it had an unusually low nitrogen isotopic value, suggesting a markedly different diet from others, whilst another had significantly higher than normal values for the same isotope (Hedges *et al.* 2007, 259–60). Thus different animals here – defined not by species but by relationships to food and people – were gathered into the assemblage.

Further acts of territorialisation bound not only multiple materials but multiple temporalities together; as Whittle (2007b) has rightly recognised, the inclusion of older microliths forms perhaps the best example of this. Here we see the importance of the differential axes of the assemblage. Whether people meant to include the microliths or not (and this can be debated) will tell us whether this forms part of the *expressive* dimension of the assemblage, but their physical, machinic, connection demonstrates their territorialisation regardless. Other temporalities were bound in through the decay of the wood in the post lines, the rotting organic materials and so on. These acted simultaneously to allow elements to leave the assemblage, to be deterritorialised, as did the way fire consumed the flesh around animal bones and the removal of parts of the axes not deposited.

The midden at Ascott-under-Wychwood was an assemblage of multiple materials, once again produced through the actions of those materials themselves, and the animals and humans they bound together. Multiple animals were in the process of becoming-with people, of being companion species (Haraway 2008). Cattle and pigs in particular, but also red deer and other animals, reveal their roles in the making of Neolithic life, and the isotopic analysis of their material remains also suggests that different animals of the same species played differing roles. It is the vibrancy of these materials that allows us to detect these enduring relations. Thus I would disagree with Whittle when he argues that 'it is not the materials themselves but their contexts and transformations which matter' (2007a, 345). The materials do matter – as do their contexts and transformations – because they gather together and assemble the Neolithic, and we do not require people at the time to have been aware of all of their relationships and connections to trace these in the present.

Changing assemblages

What are the differences and similarities between these assemblages, and how does this figure in the context of the larger scale assemblages that we refer to as the Mesolithic and Neolithic? Both Whittle (2007b) and Pollard (1999) have commented on the similarities between Mesolithic and Neolithic middens. More broadly, Vicki Cummings and I have also argued for continuities in practices across the division (Cummings and Harris 2011). Viewing these sites as tetravalent assemblages means that we can push this further; both act to territorialise the diverse elements of the worlds in similar ways. At both sites deer and boar play minor roles, and at both people expressed themselves through the juxtaposition of alternative material elements, but in both cases differing kinds of association were in play.[3] This is not, however, a switch between nature and culture, but concerns the different kinds of specific animals and plants brought together, and particularly the different intensities of engagement with them. As Pollard (2006, 139) has noted, this is captured in the less intense ways that 'wild' animals were 'entwined in the fabric of social life'.

Yet dwelling on the similarities between these middens is not to deny that simultaneously there are crucial differences. The practices at both are clearly very different. One involves large-scale fishing, hunting seals, gathering shellfish and skinning otters. The other involves herding cattle, milking cattle, flaking polished flint axes, and making and using pottery. There are huge differences in landscape setting and the kinds of environments that can be drawn into the assemblages at each site. Yet the differences are greater even than this. As we move up in scale we can detect how these two sites capture elements of the broader differing character of Mesolithic and Neolithic worlds, and of the way in which different relationships of varying intensities produced differing kinds of assemblages that came together and moved apart in the past. One is Mesolithic and one is Neolithic, and these differences do matter; they are real (cf. Lucas 2012, 177). How can we avoid denying the drama of this change, its historical importance, whilst still seeing both sites as relational assemblages with similarities? How can we do this without lurching back to seeing one as nature and the other as culture?

The answer is that the change between the Mesolithic and the Neolithic represents a 'phase transition' (DeLanda 2002, 2006) in the kinds of assemblages which people, plants, places, animals and materials were in the process of becoming. A phase transition is the term used to describe how relatively minor changes can amount to significant differences in material form (DeLanda 2002). For example, heating water two degrees from 19 to 21°C produces no significant change in the material. However, heating it from 99 to 101°C has a much more significant effect, although the increase in energy itself is no greater in the second case. Throughout the Mesolithic there would have been changes in the nature of the assemblages and relationships that existed. The heavy marine-dominated lifestyles evidenced at Cnoc Coig, for example, may reveal a specifically Late Mesolithic way of living, untypical of earlier forms (Mithen 2000). This was not, however, a phase transition; the practices at Cnoc Coig remain Mesolithic. The transition to the Neolithic, however, had longer lasting, more substantial, and, in the end, irreversible

[3] One other element is worth commenting on. Neolithic assemblages are often seen as being far more wide-ranging, or heterogeneous than their Mesolithic forebears (e.g. Hodder 2006). Of course, in terms of weight of artefacts this is true, but in overall terms we do not see a simple progression from less to more complex relationships, but rather the replacement of one set of things with another. There is no teleological sense of progress here (cf. Fowles 2010).

consequences (cf. Robb 2013). Just like the differences and similarities between water and steam, we can appreciate how the Mesolithic and the Neolithic are very different from one another, without imposing an ontological gap between the two, and thus also embrace their manifest similarities. Although I have not discussed actual transition itself here, I suggest that starting from this non-dichotomous position, from a place able to appreciate the processes that produced both the similarities and differences between the periods in a way that does not rest on tired old dichotomies, is an essential first step in building towards a more complex understanding of the mechanism of change itself. That, however, is for another time.

Conclusion

Turning to assemblages and vibrant matter allows us to approach our understandings of the past in new ways. Taking a fundamentally relational approach, which embraces both science and theory on an equal footing, allows us to rethink our imposed categories and to develop different ways of discussing key moments of change. Nowhere is this more crucial than in the transition to farming. No other period in archaeology, with the possible exception of the emergence of modern humans at the start of the Palaeolithic, maps itself quite so neatly on to our assumptions about how the world works: that nature is opposed to culture, that data require interpretation, that change emerges from a single cause with an active agent (a subject) acting on a passive recipient (an object), and that the world of materials is mute and lifeless requiring the animating touch of the human to bring it to life. We have spent the best part of 60 years debating this transition and yet still seem stuck in the same terms. To get out of this, to explore other ways of thinking, we do not need more data; we need a different starting point.

One possibility for such a starting point has been outlined here, and many others are now available in the archaeological literature (see discussion above). What such approaches are *not* about is fourfold. First, they are not about suppressing difference, or denying the very real changes that took place in the past. These changes need to be understood as material outcomes of altered relationships, however, not as transformations of idealised groups. Secondly, they are not about denying the vital role of science to the archaeological endeavour. On the contrary, assemblage theory draws explicitly on complexity theory, chaos theory, non-Euclidian geometry and others; it is precisely suited to undercutting the great divide between science and the arts (DeLanda 2002, 2006, 2010; Webmoor and Witmoor 2008). Thirdly, they are not about denying the progress that archaeological work has undoubtedly made over the last 150 years, or about demanding that we start all over again. Finally, this is not about ignoring the importance of what people thought and felt, since the expressive elements of an assemblage are as important as any other, but it is about placing things, people, animals, places, plants, their association, territorialisation and deterritorialisation on an equal ontological footing, to reveal the different kinds of becoming, and the different kinds of becoming-with that existed in the past (cf. Pollard 2004, 66). Fundamentally, what a turn to assemblages and vibrant matter does offer is the potential for a new engagement that does not write off half our knowledge as theoretical waffling and the other half as empiricist essentialism. In turning to this approach, we will not produce data that subsequently require interpretation, facts that need theory, or nature that needs culture, but rather narratives of the vibrant materials that form our pasts, presents and futures.

Acknowledgements

I would very much like to acknowledge the multiple discussions on this topic which I have had with both Rachel Crellin and Chris Fowler, which have greatly influenced my thoughts. Rachel, Hannah Cobb and the editors also offered very helpful comments on a previous draft. Aspects of this chapter were presented to the European Later Prehistory Group at Cambridge, and discussions there and afterwards with the members of the group were also extremely useful in clarifying my thinking. Nick Overton was kind enough to interrupt writing his PhD to answer questions about the Mesolithic. Finally, I must acknowledge the Leverhulme Trust and the Universities of Newcastle and Leicester who funded my Early Career Fellowship during which these ideas were developed.

References

Alberti, B. and Marshall, Y. 2009. Animating archaeology: local theories and conceptually open-ended methodologies. *Cambridge Archaeological Journal* 19, 344–56.

Alberti, B., Fowles, S., Holbraad, M., Marshall, Y. and Witmore, C.L. 2011. Worlds otherwise: archaeology, anthropology, and ontological difference. *Current Anthropology* 52, 896–912.

Barclay, A. and Case, H. 2007. The Early Neolithic pottery and fired clay. In D. Benson and A. Whittle (eds), *Building memories: the Neolithic Cotswold long barrow at Ascott-under-Wychwood, Oxfordshire*, 263–81. Oxford: Oxbow Books.

Bayliss, A., Benson, D. Gales, D. Humprey, L. McFadyen, L. and Whittle, A. 2007. One thing after another: the date of the Ascott-under-Wychwood long barrow. *Cambridge Archaeological Journal* 17(1), supplement, 29–44.

Bennett, J. 2010. *Vibrant matter: a political ecology of things*. London: Duke University Press.

Benson, D. and Whittle, A. (eds) 2007. *Building memories: the Neolithic Cotswold long barrow at Ascott-under-Wychwood, Oxfordshire*. Oxford: Oxbow Books.

Borić, D. 2005. Deconstructing essentialisms: unsettling frontiers of the Mesolithic–Neolithic Balkans. In D. Bailey, A. Whittle and V. Cummings (eds), *(Un)settling the Neolithic*, 16–31. Oxford: Oxbow Books.

Cobb, H.L. 2008. *Media for movement and making the world: an examination of the Mesolithic experience of the world and the Mesolithic to Neolithic transition in the northern Irish Sea basin*. Unpublished PhD Thesis, Manchester University.

Conneller, C. 2011. *An archaeology of materials: substantial transformations in early prehistoric Europe*. London: Routledge.

Cummings, V. and Harris, O.J.T. 2011. Animals, people and places: the continuity of hunting and gathering practices across the Mesolithic–Neolithic transition in Britain. *European Journal of Archaeology* 14, 361–82.

DeLanda, M. 2002. *Intensive science and virtual philosophy*. London: Continuum.

DeLanda, M. 2006. *A new philosophy of society: assemblage theory and social complexity*. London: Continuum.

DeLanda, M. 2010. *Deleuze: history and science*. New York: Atropos Press.

Deleuze, G. and Guattari, F. 2004. *A thousand plateaus: capitalism and schizophrenia*. London: Continuum.

Deleuze, G. and Parnet, C. 2007. *Dialogues II*. New York: Columbia University Press.

Dewsbury, J.-D. 2011. The Deleuze-Guattarian assemblage: plastic habits. *Area* 43, 148–53.

Fowler, C. 2013. *The emergent past: a relational realist archaeology of Early Bronze Age mortuary practices*. Oxford: Oxford University Press.

Fowler, C. 2014. Dynamic assemblages, or the past is what endures: change and duration of relations. In B. Alberti, A.M. Jones and J. Pollard (eds), *Archaeology after interpretation: returning materials to archaeological theory*, 235–56. Walnut Creek: Left Coast Press.

Fowles, S. 2010. People without things. In M. Bille, F. Hastrop and T.F. Sørensen (eds), *An anthropology of absence: materializations of transcendence and loss*, 23–41. New York: Springer.

Garrow, D. and Sturt, F. 2011. Grey waters bright with Neolithic argonauts? Maritime connections and the Mesolithic–Neolithic transition within the 'western seaways' of Britain, c. 5000–3500 BC. *Antiquity* 85, 59–72.

Gell, A. 1998. *Art and agency: an anthropological theory*. London: Clarendon Press.

Grigson, C. and Mellars, P.A. 1987. The mammalian remains. In P.A. Mellars, *Excavations on Oronsay: prehistoric human ecology on a small island*, 243–89. Edinburgh: Edinburgh University Press.

Haraway, D. 2008. *When species meet*. London: University of Minnesota Press.

Harman, G. 2009. *Prince of networks: Bruno Latour and metaphysics*. Melbourne: Re.Press.

Harris, O.J.T. 2013. Relational communities in prehistoric Britain. In C. Watts (ed.), *Relational archaeologies: humans, animals, things*, 173–89. London: Routledge.

Harris, O.J. T. 2014. (Re)assembling communities. *Journal of Archaeological Method and Theory* 21, 76–97.

Harris, O.J.T. and Robb, J.E. 2012. Multiple ontologies and the problem of the body in history. *American Anthropologist* 114, 668–79.

Harrison, R. 2011. Surface assemblages: towards an archaeology *in* and *of* the present. *Archaeological Dialogues* 18, 141–61.

Hedges, R.E.M., Stevens, R.E. and Pearson, J.A. 2007. Carbon and nitrogen stable isotope compositions of human and animal bone. In D. Benson and A. Whittle (eds), *Building memories: the Neolithic Cotswold long barrow at Ascott-under-Wychwood, Oxfordshire*, 255–62. Oxford: Oxbow Books.

Hodder, I. 2006. *Çatalhöyük: the leopard's tale. Revealing the mysteries of Turkey's ancient 'town'*. London: Thames and Hudson.

Hodder, I. 2012. *Entanglement: an archaeology of the relationships between humans and things*. Oxford: Wylie-Blackwell.

Ingold, T. 2000. *Perceptions of the environment: essays in livelihood, dwelling and skill*. London: Routledge.

Ingold, T. 2007. Materials against materiality. *Archaeological Dialogues* 14, 1–16.

Ingold, T. 2013. *Making: anthropology, archaeology, art and architecture*. London: Routledge.

Jones, A.M. 2002. *Archaeological theory and scientific practice*. Cambridge: Cambridge University Press.

Jones, A.M. 2012. *Prehistoric materialities: becoming material in prehistoric Britain and Ireland*. Oxford: Oxford University Press.

Jones, A. and Richards, C. 2003. Animals into ancestors: domestication, food and identity in Late Neolithic Orkney. In M. Parker Pearson (ed.), *Food, culture and identity in the Neolithic and Early Bronze Age*, 45–51. Oxford: British Archaeological Reports.

Latour, B. 1993. *We have never been modern*. Cambridge, MA: Harvard University Press.

Latour, B. 1999. *Pandora's hope: essays on the reality of science studies*. Cambridge, MA: Harvard University Press.

Lucas, G. 2012. *Understanding the archaeological record*. Cambridge: Cambridge University Press.

McFadyen, L. 2007. Making architecture. In D. Benson and A. Whittle (eds), *Building memories: the Neolithic Cotswold long barrow at Ascott-under-Wychwood, Oxfordshire*, 348–54. Oxford: Oxbow Books.

McFadyen, L. Benson, D. and Whittle, A. 2007. The pre-barrow contexts. In D. Benson and A. Whittle (eds), *Building memories: the Neolithic Cotswold long barrow at Ascott-under-Wychwood, Oxfordshire*, 23–77. Oxford: Oxbow Books.

Mellars, P.A. 1987. *Excavations on Oronsay: prehistoric human ecology on a small island*. Edinburgh: Edinburgh University Press.

Meiklejohn, C., Merrett, D.C., Nolan, R.W., Richards, M.P. and Mellars, P.A. 2005. Spatial relationships, dating and taphonomy of the human bone from the Mesolithic site of Cnoc Coig, Oronsay, Argyll, Scotland. *Proceedings of the Prehistoric Society* 71, 85–105.

Mithen, S.J. 2000. Mesolithic sedentism on Oronsay: chronological evidence from adjacent islands in the southern Hebrides. *Antiquity* 74, 298–304.

Mithen, S.J. and Finlayson, B. 1991. Red deer hunters on Colonsay? The implications of Staosnaig for the interpretation of the Oronsay middens. *Proceedings of the Prehistoric Society* 52(2), 1–8.

Mulville, J. and Grigson, C. 2007. The animal bones. In D. Benson and A. Whittle (eds), *Building memories: the Neolithic Cotswold long barrow at Ascott-under-Wychwood, Oxfordshire*, 237–53. Oxford: Oxbow Books.

Normark, J. 2008. The triadic causeways of Ichmul: virtual highways becoming actual roads. *Cambridge Archaeological Journal* 18, 215–38.

Normark, J. 2009. The making of a home: assembling houses at Nohcacab, Mexico. *World Archaeology* 41, 430–44.

Normark, J. 2010. Involutions of materiality: operationalising a neo-materialist perspective through the causeways at Ichmul and Yo'okop. *Journal of Archaeological Method and Theory* 17, 132–73.

Olsen, B. 2010. *In defense of things: archaeology and the ontology of objects*. Plymouth: Altamira Press.

Olsen, B., Shanks, M., Webmoor, T. and Witmore, C.L. 2012. *Archaeology: the discipline of things*. Berkeley: University of California Press.

Parikh, D. and Hall, K. 2012. The two archaeologies: crossing the great divide. *Archaeological Review from Cambridge* 27, 1–8.

Pollard, J. 1999. 'These places have their moments': thoughts on settlement practice in the British Neolithic. In J. Brück and M. Goodman (eds), *Making places in the prehistoric world: themes in settlement archaeology*, 76–93: London: UCL Press.

Pollard, J. 2004. A 'movement of becoming': realms of existence in the Early Neolithic of southern Britain. In A. Chadwick (ed.), *Stories from the landscape: archaeologies of inhabitation*, 55–69. Oxford: British Archaeological Reports.

Pollard, J. 2006. A community of beings: animals and people in the Neolithic of southern Britain. In D. Serjeantson and D. Field (eds), *Animals in the Neolithic of Britain and Europe*, 135–48. Oxford: Oxbow Books.

Pollard, J. 2008. Deposition and material agency in the Early Neolithic of Southern Britain. In B.J. Mills and W.H. Walker (eds), *Memory work: archaeologies of material practices*, 41–59. Sante Fe: SAR Press.

Ray, K. and Thomas, J. 2003. In the kinship of cows: the social centrality of cattle in the earlier Neolithic of southern Britain. In M. Parker Pearson (ed.), *Food, culture and identity in the Neolithic and Early Bronze Age*, 37–44. Oxford: British Archaeological Reports.

Richards, M.P. and Mellars, P.A. 1998. Stable isotopes and the seasonality of the Oronsay middens. *Antiquity* 72, 178–84.

Richards, M.P. and Sheridan, A. 2000. New AMS dates on human bone from Mesolithic Oronsay. *Antiquity* 74, 313–15.

Robb, J. 2013. Material culture, landscapes of action and emergent causation: a new model for the origins of the European Neolithic. *Current Anthropology* 54, 657–83.

Robb, J. and Harris, O.J.T. 2013. *The body in history: Europe from the Palaeolithic to the future*. Cambridge: Cambridge University Press.

Snow, C.P. 1959. *The two cultures and the scientific revolution*. Cambridge: Cambridge University Press.

Sofaer, J. 2006. *The body as material culture: a theoretical osteoarchaeology*. Cambridge: Cambridge University Press.

Thomas, J. 1996. *Time, culture and identity: an interpretive archaeology*. London: Routledge.

Thomas, J. 2004. *Archaeology and modernity*. London: Routledge.

Thomas, J. 2007. The trouble with material culture. *Journal of Iberian Archaeology* 9/10, 11–23.

Warren, G. 2005. *Mesolithic lives in Scotland*. Stroud: Tempus.

Webmoor, T. and Witmore, C.L. 2008. Things Are Us! A commentary on human/things relations under the banner of a 'social archaeology'. *Norwegian Archaeological Review* 41, 1–18.

Whittle, A. 2003. *The archaeology of people: dimensions of Neolithic life*. London: Routledge.

Whittle, A. 2007a. Uses and continuities of place: Mesolithic into Neolithic. In D. Benson and A. Whittle (eds), *Building memories: the Neolithic Cotswold long barrow at Ascott-under-Wychwood, Oxfordshire*, 344–8. Oxford: Oxbow Books.

Whittle, A. 2007b. Building memories. In D. Benson and A. Whittle (eds), *Building memories: the Neolithic Cotswold long barrow at Ascott-under-Wychwood, Oxfordshire*, 361–4. Oxford: Oxbow Books.

Whittle, A., Healy F. and Bayliss, A. 2011. *Gathering time: dating the early Neolithic enclosures of southern Britain and Ireland*. Oxford: Oxbow Books.

Witmore, C.L. 2007. Symmetrical archaeology: excerpts of a manifesto. *World Archaeology* 39, 546–62.

Witmore, C.L. 2012. The realities of the past: archaeology, object-orientations, pragmatology. In B. Fortenberry and L. McAtackney (eds), *Modern materials: proceedings from the Contemporary and Historical Archaeology in Theory Conference 2009*, 25–36. Oxford: British Archaeological Reports.

18

Pottery, Archaeology and Chemistry

Contents and Context

JESSICA SMYTH AND RICHARD P. EVERSHED

Introduction

LIKE OTHER CONTRIBUTIONS IN THIS VOLUME (such as by Harris, Chapter 17), the starting point for this chapter has been a revisiting of Andrew Jones's *Archaeological Theory and Scientific Practice* (2002). In that work Jones details the seeming paradox of archaeological science. In order to carry out science[1] we must believe in the constancy of the natural world and take up a detached position to observe it; yet, if we are to study society, we cannot appeal to an objectified and constant nature. It is, we feel, and as Jones himself admits, a thesis founded more on science philosophy than anything resembling actual scientific practice; few (if any) scientists today would recognise or adhere to the very positivist approach outlined in Jones's book. Likewise, many of the problems said to spring from this 'misfit' (for example, the poor connections between the interpretation of site and artefact; scientific analyses unstructured by contextual considerations; or the lack of contact between specialists and excavators through all stages of analysis) took little account of the dynamism that twenty-first century science was already bringing to archaeological research. In this sense, the archaeology/science opposition was, and perhaps continues to be, a bit of a straw man. For some time, both archaeologists and scientists have been initiating increasing numbers of interdisciplinary collaborations, accompanied by a growing number of conference sessions (indeed, entire conferences), books and special journal volumes devoted to archaeological science (for example, Bogaard and Outram 2013). How are we tackling Jones's apparent paradox, if at all? We aim to answer this question below, focusing on a particular example of how interdisciplinary undertakings are routinely and successfully brought to fruition.

[1] In the Aristotelian, rationalist, mainstream sense of science.

Proceedings of the British Academy **198**, 347–67.

Getting started

The SCHERD project started with a cold call, as research collaborations frequently do. The archaeologist in this pairing had a research background in the Irish Neolithic, specifically settlement remains and the archaeology of what she likes to refer to as 'daily life'. Having previously examined the domestic architecture of the period, she was looking to explore settlement activity from a different angle and to build on earlier observations about social practice among early farming communities (Smyth 2011, 2014). A colleague suggested lipid analysis of residues in pottery – material frequently encountered on Irish Neolithic sites – and this seemed ideal. What object is more central to domestic life than pottery, or better able to provide insight into the wide range of social behaviour structured around food preparation, consumption and storage? After a quick literature review and some internet searching, a cautious email was sent. This was answered almost immediately with a telephone call from a scientist keen to talk further. Members of his Organic Geochemistry Unit were at that point involved in two similar projects: a PhD investigating the emergence of dairying practices in early Neolithic mainland Europe, with a focus on the northern Mediterranean and the regions of the LBK, by analysing lipid residues preserved in potsherds – part of a Marie Curie Initial Training Network examining lactase persistence and the early cultural history of Europe (LeCHE; Feulner *et al.* 2011); and a three-year NERC-funded project to characterise marine resource exploitation from the Neolithic to the Norse period in the North Atlantic, combining pottery lipid residue analysis, zooarchaeological data and human and animal isotope data (Cramp *et al.* 2014). The proposal to undertake a systematic investigation of a large body of material from the western edge of Europe, in Ireland, was considered an important complement to these projects. There followed a few months of dialogue via email, framing the research questions facing the Irish Neolithic within the experience of more than 30 years of archaeological lipid analysis, and scaling some ideas up and others down. These ideas eventually became SCHERD, a Study of Cuisine and animal Husbandry among Early farmers via Residue analysis and radiocarbon Dating, a research proposal successfully funded under the Marie Curie Intra-European Fellowship scheme in 2010.

Archaeo-chemical dialogues

The terms of the funding stipulated provision of scientific training, that is, for the archaeologist[2] to undertake all aspects of the lipid analysis herself. However, the

[2] Although, it should be noted, not the primary, excavating archaeologist.

process of interlocution started long before any Neolithic pottery reached the laboratory. While the likelihood of Irish pottery yielding relatively high concentrations of lipid was suspected (Cramp and Evershed in Cleary and Kelleher 2011), careful site and sherd sampling criteria still needed to be applied to maximise the chances of obtaining meaningful, statistically robust results. From a chemistry perspective, a combination of amassed experience and targeted experimental work has been crucial in identifying the optimal conditions for lipid preservation in pottery sherds (for review see Evershed 2008). The degradation of lipids derived from a variety of sources has been extensively studied in both field and laboratory experiments (Charters 1996, Bland 1999, Dudd 1999, Aillaud 2001), while additional work has estimated the lipid capacity of a pottery sherd to be around 10 mg/g (Evershed 2008).[3]

The analysis of several thousand archaeological potsherds over many years has provided a mean lipid yield of *c.* 100 μg/g, demonstrating that only *c.* 1% of lipid originally absorbed survives burial. Experience has also shown that total lipid extract (TLE) concentrations below 5 μg/g (based upon an initial extraction of a fragment of potsherd) cannot be reliably interpreted, other than in special cases, due to problems of contamination from a range of sources (Evershed 2008). Further experiments have been designed to examine where in pottery vessels lipids are most abundant, and this appears to be in the rim and upper body of vessels, with concentrations in cooking vessels decreasing down the vessel profile from the top to bottom (Charters *et al.* 1993, 1997). Finally, analysis of sites across prehistoric Europe and beyond has revealed quite variable rates of lipid recovery, based on the percentage of sherds in a sampled assemblage with lipid concentrations greater than 5 μg/g. This has ranged from 70–80% on some sites to as low as 10% on others. Background chemistry knowledge and experience thus had an impact on the project design in a number of ways; wherever possible, sites with assemblages of more than 30 vessels were targeted for sampling,[4] as if recovery rates were to be 30%, results from 10 sherds would provide a statistically acceptable minimum. Adopting this criterion did mean that several archaeologically interesting sites were passed over, in particular megalithic monuments, which generally have small pottery assemblages. Rim sherds and upper body sherds were also preferred, ideally weighing 5g or more, which would allow for re-sampling if necessary and easier manipulation when recording and cleaning. All in all, some 448 sherds from 15 sites were analysed for SCHERD, yielding a recovery rate of approximately 90% and a mean lipid concentration of 300 μg/g.

[3] The maximum concentration of absorbed lipid ever observed in an archaeological potsherd at the Organic Geochemistry Unit at Bristol is 17.8 mg/g (Copley *et al.* 2005).

[4] Lambay Island, Lowpark, Balregan and Ballynahatty are exceptions; sherds representing just 18–23 vessels were available from these sites.

The SCHERD sampling strategy was also shaped by archaeological considerations; it was not just a question of selecting material from sites where pottery assemblages were large enough and permissions were forthcoming. The Irish Neolithic can be divided into three broad phases – early, middle and late – that are defined by changes in domestic architecture, monuments and artefact types (including pottery), and analysis thus sought to examine whether there were discernible differences in vessel use and contents through the period. Assemblages from different types of sites were also selected, such as causewayed enclosures, houses and pit complexes, and from different parts of the island, to explore whether site function and geography had an impact on what pottery was being used for (Figure 18.1). We were mindful too of the potential of patterning at an intra-site level. Pottery is a culturally generated and culturally situated material, and meaningful interpretation depends on post-excavation analyses retaining and reconstructing the connections that defined the relationships between site and artefact. For SCHERD, this involved putting in place procedures that might not be routinely thought necessary in chemical analyses. As much information as possible about each pottery sherd was documented, from spatio-temporal information such as context, feature, phasing and other artefactual associations, to information on vessel shape and decoration, on paste, temper and construction methods and vessel completeness, gleaned from specialist reports where they existed and from personal inspection. These measures were about considering the archaeological narrative alongside the scientific findings: how the compositions and concentrations of lipid residues in vessels might relate to how they were constructed and decorated, how they were used, or where they were ultimately deposited.

A science-based archaeological narrative where the science lacks an acceptable level of rigour is a meaningless narrative. Sampling is the most crucial part of any analytical protocol and there are some basic rules that have to be followed, but sensitivity to wider archaeological and ceramic analytical practices is also essential. In sub-sampling sherds, for example, care was taken to leave sufficient *diagnostic* material (rims, shoulders, decoration) to aid in future vessel characterisation. Critical to this was the scientific and archaeological imperative to preserve the link between object (sample) and archive, grounded in experience of how excavated materials are stored long-term in museums and how susceptible to human interference this system can be. As much as possible, sample preparation procedures were worked around museum and/or excavators' archiving systems. The surface cleaning of sherds avoided pottery finds numbers, often written on top of a daub of Tippex or covered by a layer of acrylic varnish – a nuisance and possible contaminant in the extraction of lipids but a crucial link to an object's wider relationships. Maintaining this balance between scientific sample and archivable object was not always easy, as finds numbers were frequently located in the very place one would hope to sample, for example, on rim sherds at the furthest point from vessel rims, and on decorated sherds at the furthest point from decoration (Figure 18.2).

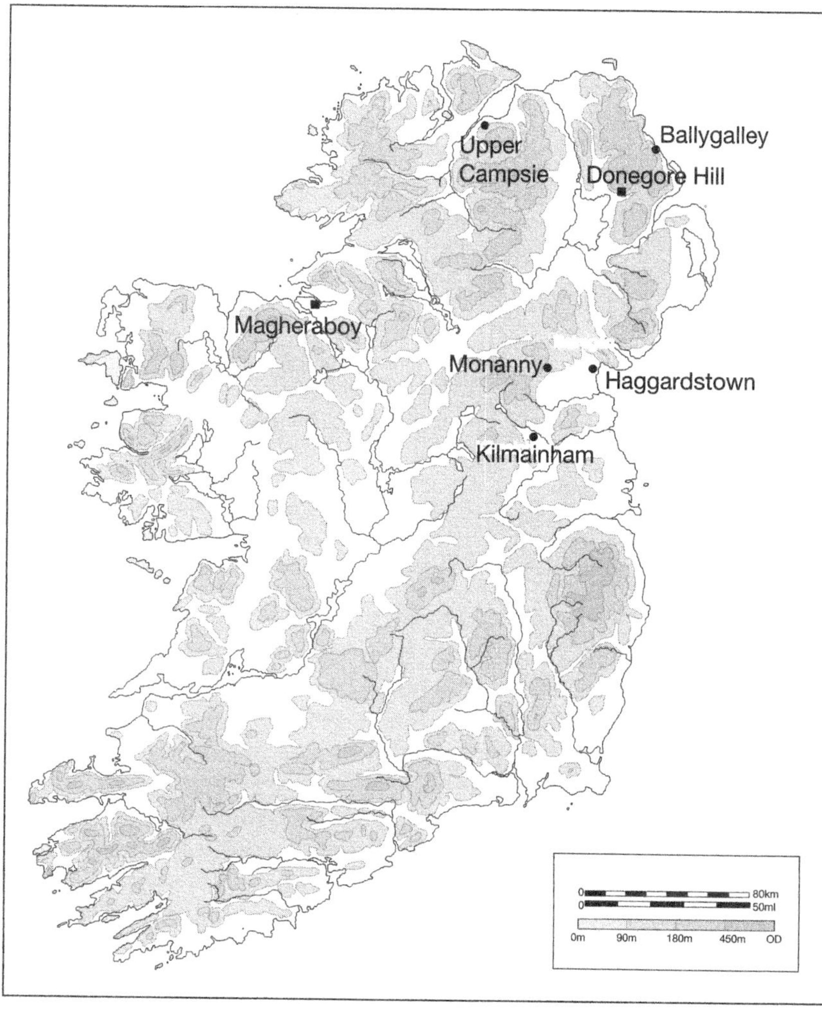

Figure 18.1 Map showing Early Neolithic sites sampled as part of SCHERD (squares: causewayed enclosures; circles: house sites/settlements).

Making it look easy

The first investigations into absorbed lipids in archaeological ceramics from northern Europe were undertaken in the early 1990s (Evershed *et al.* 1990, 1991, 1992), and such analyses are now routinely carried out on pottery assemblages across the globe (Craig *et al.* 2013, Dunne *et al.* 2012, Mukherjee *et al.* 2008, Reber and Evershed 2004). Degraded animal fats are by far the most common class of residue observed in archaeological pottery, and in some regions of the world, such

Figure 18.2 Shoulder sherd from an Early Neolithic Carinated Bowl from Monanny; (*left*) photographed in profile; (*right*) the finds number applied with ink and varnish and located away from the diagnostic shoulder section towards the upper edge of the vessel, an area ordinarily targeted for sampling.

as north-west temperate Europe, dairy fats seem especially predominant. The apparent ease with which lipid analysis can be carried out has, we have observed, occasionally led archaeological audiences to become blasé about results obtained. In reality, the biomarker approach – attempting to correlate molecular components, particularly lipids, preserved within pottery residues with biomolecules present in modern-day plant and animal tissues – is far from straightforward and has required decades of 'behind-the-scenes' research for scientists to be confident about commodity and species attributions.

One of the first issues to be tackled was whether absorbed lipids observed in pottery vessels and other archaeological materials were in fact anthropogenic (Evershed and Connolly 1988). Evaluated and subsequently ruled out was the possibility that the organic matter in potsherds accumulates post-deposition, through migration from the surrounding burial soil (Heron *et al.* 1991). This is explained by the hydrophobic properties of lipids that limit their transfer from the soil to the clay wall of the pottery and *vice versa*. The clay matrix of the pottery vessel also protects the lipid residues from microbial degradation and minimises decay over time. The anthropogenic origin of absorbed lipids has been further confirmed through the identification of certain lipid markers in pottery vessels, the result of chemical transformations caused by different human activities. Prime among these marker molecules are ketones, specifically those containing 31, 33 and 35 carbon atoms, which have been shown to form when $C_{16:0}$ and $C_{18:0}$ fatty acids in pottery are heated in excess of 300°C in the presence of the metal salts present in some pottery clays (Evershed *et al.* 1995a, Raven *et al.* 1997). As all archaeological lipids are degraded lipids, research has also been needed on how fresh animal and plant fats, oils and waxes break down over time, if the commodities originally processed in pots are to be successfully identified (Dudd *et al.* 1998, Evershed *et al.* 1995b, Evershed 2008).

With the advent of compound-specific isotope analysis, and the possibility of distinguishing ruminant from non-ruminant meat fats, a bank of compound-specific carbon stable isotope values has been carefully generated from modern reference fats, making sure that the diet and environment of reference animals match prehistoric conditions as closely as possible (Mottram *et al.* 1999). The additional development of detecting dairy fats in archaeological potsherds has emerged through exploiting biosynthetic and metabolic phenomena originally discovered by researchers in the field of animal science (McDonald *et al.* 1988, Christie 1981); the $\delta^{13}C$ values of the $C_{18:0}$ fatty acid in ruminant dairy fats are *c.* 2.3‰ lower than in ruminant meat fats. This has required not only a further set of reference values on modern ruminant milks, but extensive investigations of $\delta^{13}C$ values of the major fatty acids and carbohydrates of a range of plants representative of the diets of grazing and browsing animals (Docherty 2002; Dungait *et al.* 2008, 2010, 2011), as it is the lower $\delta^{13}C$ values of dietary plant fatty acids that contribute to the lower $\delta^{13}C$ values of the $C_{18:0}$ fatty acids in dairy fats. Further experimental work has then been required to confirm that these $\delta^{13}C$ values do not shift as the fats degrade, as would happen archaeologically (Aillaud 2001; Evershed 2008, 38–9). To ensure that the relative $\delta^{13}C$ values of ruminant to non-ruminant and meat to milk fats are a truly global proxy and can be used to distinguish between fats from animals raised on different diets in different environments, a substantial database of reference values has been built up in recent years and includes ranges of values from animals from Britain, France, Switzerland, Turkey, the Levant, Kazakhstan, Kenya and Libya (Salque 2012, Dunne *et al.* 2012, Salque *et al.* 2013). Finally, recent field and laboratory experiments indicate that the dairy signal from absorbed lipids in archaeological potsherds is more likely reflecting the processing of milk products in vessels during their use-life than any waterproofing or glazing of pottery vessels with milk prior to use (Šoberl 2011).

Scope and scale

The addition of compound-specific stable isotope analysis to the suite of techniques comprising archaeological lipid analysis has been an undeniably important advance, allowing us to say much more about commodity use in the past (Evershed *et al.* 1997, Mottram *et al.* 1999, Dudd and Evershed 1998). However, the subtleties of isotopic variation and the fact that the distinguishing characteristics of plant and animal tissues lie within a *range* of values rather than a single, absolute reading make it necessary to obtain a cluster of data points before allowing identification of the most important domesticates, such as ruminant versus non-ruminant, or ruminant carcass versus dairy fat. A single data point derived from the $\delta^{13}C$ values of $C_{16:0}$ and $C_{18:0}$ fatty acids from a single potsherd is rarely, if at all, statistically significant. This is an important point to stress, because requests to analyse a specific pot are often made by archaeologists, who may have uncovered a particularly finely decorated vessel, for example, or one deposited in an unusual context. The significance of the organic residue within a single vessel can only be adequately judged against a background of knowledge of general use within a vessel type and between types. Hence, there is no escaping the need to analyse statistically valid numbers of vessels in order to draw significant conclusions aimed at building a robust narrative.

The need to interrogate large datasets and to plot relatively large numbers of data points makes compound-specific stable isotope analysis particularly suited to a certain type and scale of archaeological narrative. Within research projects, providing direct evidence for commodity use in ever more distant temporal and geographical locations, or charting the relative increase or decrease of commodities in an area over centuries or millennia, generates almost ready-made, headline-grabbing stories. Certainly, these types of studies remain valuable. In the case of temperate Europe, for example, compound-specific isotope analysis has helped to establish that Early Neolithic farmers were producing milk at a time when they probably lacked the biochemical machinery to effectively digest it (Burger *et al.* 2007, Itan *et al.* 2009). This is an enormously significant finding, which overturns the long-held assumption that milk is a 'secondary product' of the agricultural revolution (Sherratt 1981). Yet can archaeological lipid analysis – specifically compound-specific stable isotope analysis – do more than describe broad trends in the past? Within commercial archaeology, the interpretation of organic residue data can be even more coarse-grained. Frequently, little beyond the presence or absence of certain commodities is noted in the main site publication, with plots, tables and other 'technical' data relegated to the appendix, and the appropriateness of this practice needs review.

The relationship between scales of analysis and scales of resolution requires careful consideration at the research design stage. Large-scale regional studies using organic residues in pottery, such as that described in Evershed *et al.* (2008), can certainly address key questions related to diet and animal management over time and space. Shifting the narrative to individual sites, contexts and vessels greatly increases the analytical demands, especially if statistical rigour is to be maintained. Nevertheless, this is an area where much untapped potential lies. Each archaeological site has its own trajectory, marked by an array of potentially heterogeneous archaeological contexts, and exploring the differences and similarities between different sites will provide us with a greater understanding of past consumption and technological practices involving pottery vessels. As we begin to interpret the results from the SCHERD project, such an approach, and scale of resolution, are looking increasingly achievable. This is partly because the project has analysed a large number of vessels from a range of sites of different function and geographic location dating to a relatively short time period of *c.* 1500 years (cf. Copley *et al.* 2005), but also because of the way in which archaeological and scientific knowledge combine to define the limits of interpretation. Below we provide preliminary sketches of early farming activity in Neolithic Ireland by way of illustration.

As mentioned above, just under 450 sherds from 15 Neolithic sites were examined as part of SCHERD. The Early Neolithic was represented by 239 sherds from seven sites: two causewayed enclosures – Magheraboy and Donegore Hill – and five house sites or settlements – Ballygalley, Upper Campsie, Haggardstown, Monanny and Kilmainham 1C (Figure 18.3). A very narrow date range for the construction and use of houses centring on the thirty-seventh century cal BC has recently been provided through Bayesian analysis[5] and suggests that this particular form of architecture (and its architects) spread across the island within decades, perhaps even years. Additional Bayesian modelling indicates that both the Magheraboy and Donegore Hill causewayed enclosures (the only two excavated to date in Ireland) had longer histories, of several centuries' duration, built before the start of the 'house horizon' and continuing in use afterwards (Cooney *et al.* 2011).[6]

Analysing these sites together, *c.* 90% of pottery sherds (n = 215) yielded appreciable amounts of lipid residues, displaying distributions of fatty acids and other acyl lipids consistent with degraded animal fats (Figure 18.4). Of the samples carried forward for compound-specific stable isotope analysis (n = 111), 89% were found to contain lipid mixtures of predominantly dairy origin – unequivocal evidence that dairying was being practised by the earliest farming communities across at least half of the island.[7]

When results from the lipid analyses are resolved by site type, vessels from both causewayed enclosures and houses remain dominated by milk fats, the former at 96% (n = 50) and the latter at ~84% (n = 61). The remainder of residues exhibit values predominantly plotting within ranges expected for ruminant meat fats, or a mixture of milk and meat, although on house sites this is more prominent (Figure 18.5). While further investigation of taphonomic factors is needed,

[5] Whitehouse *et al.* 2013: starting *3720–3680 cal BC*, ending *3640–3620 cal BC*; Cooney *et al.* 2011: starting *3730–3660 cal BC*, ending *3640–3605 cal BC*; McSparron 2008: starting *3715–3650 cal BC*, ending *3690–3625 cal BC*; all estimates at 95% probability.

[6] Current posterior density estimates (preferred models) are Magheraboy: starting *4115–3850 cal BC*, ending *3615–3355 cal BC* and Donegore Hill: starting *3855–3665 cal BC*, ending *3590–3430 cal BC* (95% probability; Cooney *et al.* 2011). The start date for Magheraboy is acknowledged by Cooney *et al.* as being anomalously early, but is based on the secure dating of oak sapwood from the enclosure ditch. Further investigation and dating of the primary levels of the enclosure ditches at Magheraboy are required to confirm or discount this curious, but statistically robust, estimate.

[7] $\Delta^{13}C$ values ($\delta^{13}C_{18:0} - \delta^{13}C_{16:0}$) of <–3.4‰, which is the 2 sigma range of UK reference milk fats (Copley *et al.* 2003).

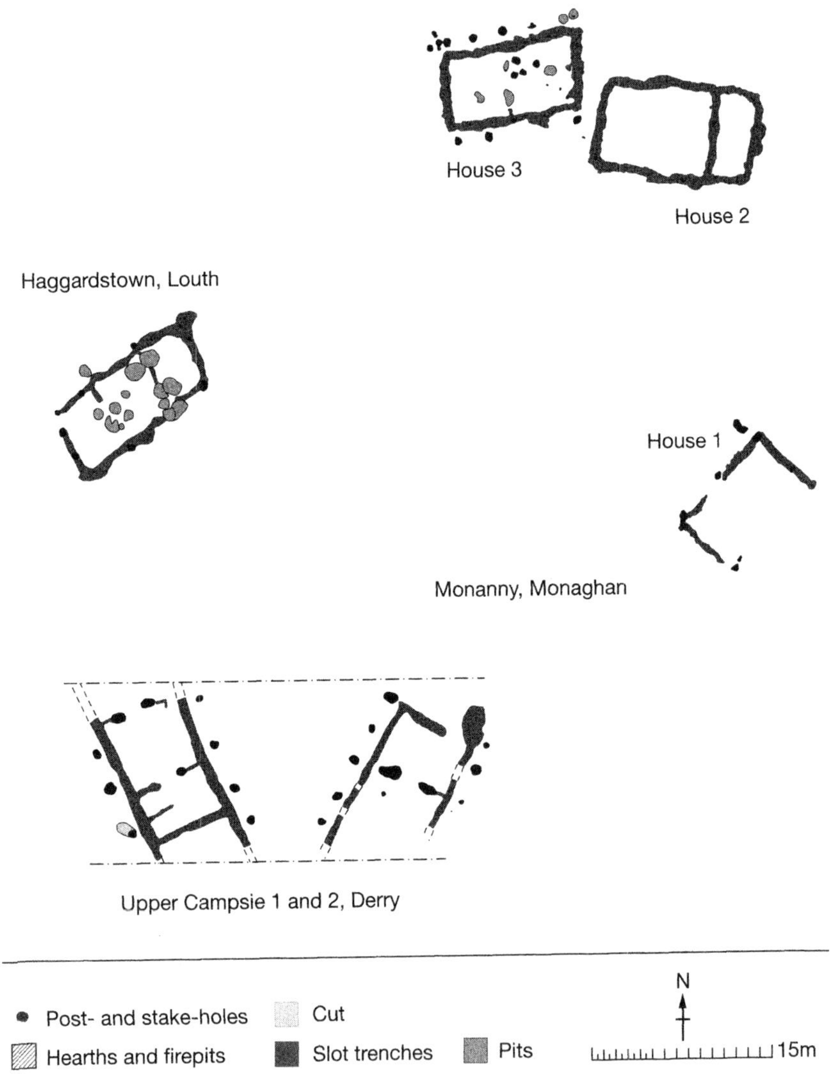

Figure 18.3 Plans of the Early Neolithic houses from Haggardstown, county Louth, Upper Campsie, County Derry and Monanny, County Monaghan.

variability might also be reflected in lipid concentrations across individual sites. The mean lipid concentration of the sampled Magheraboy assemblage was 480 μg/g, for example, while at the house sites of Monanny and Haggardstown, mean lipid concentration was around 80μg/g. Moreover, just 25–30% of samples from Monanny and Haggardstown (n = 8 and n = 9, respectively) contained sufficient concentrations of $C_{16:0}$ and $C_{18:0}$ acids for stable carbon isotope analysis. Across

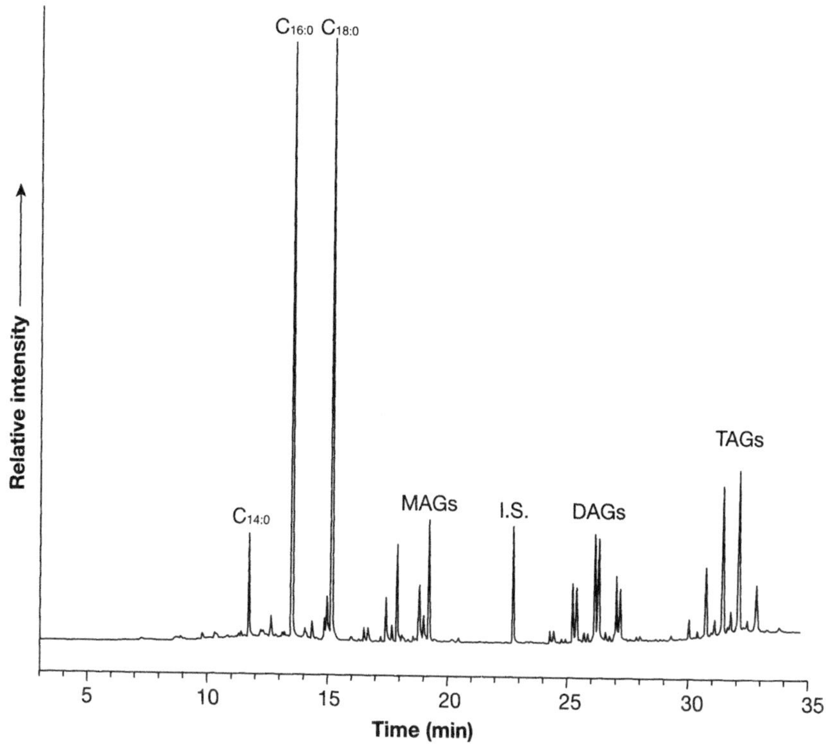

Figure 18.4 High Temperature Gas Chromatogram showing the trimethylsilylated total lipid extract containing lipid components characteristic of a degraded animal fat, from sherd UPC-8 (112:46a) from the Early Neolithic settlement at Upper Campsie, Derry. Key: Cx:y are free fatty acids of carbon length x, and degree of unsaturation y. MAGs are monoacylglycerols, DAGs are diacylglycerols and TAGs are triacylglycerols. I.S. is the internal standard (C_{34} n-alkane).

the Magheraboy assemblage, this figure was over 70%. There is much work still to be done on integrating scientific data with the archaeological remains of the Irish Early Neolithic but such variance in lipid concentrations and apparent differences in commodity use across site types raises intriguing possibilities. Was meat consumed at the level of the household, for example, but not in the context of activity conducted at causewayed enclosures, where it is either not consumed or consumed in different ways, after being spit roasted, for example? What are the social implications of such practices? While the potential for taphonomic bias in lipid preservation remains poorly understood and must be considered in final interpretations, low lipid concentrations may mean that pottery vessels from house sites were not used as intensively as those from causewayed enclosures, or that a greater percentage were not used for cooking or processing, but for storage instead.

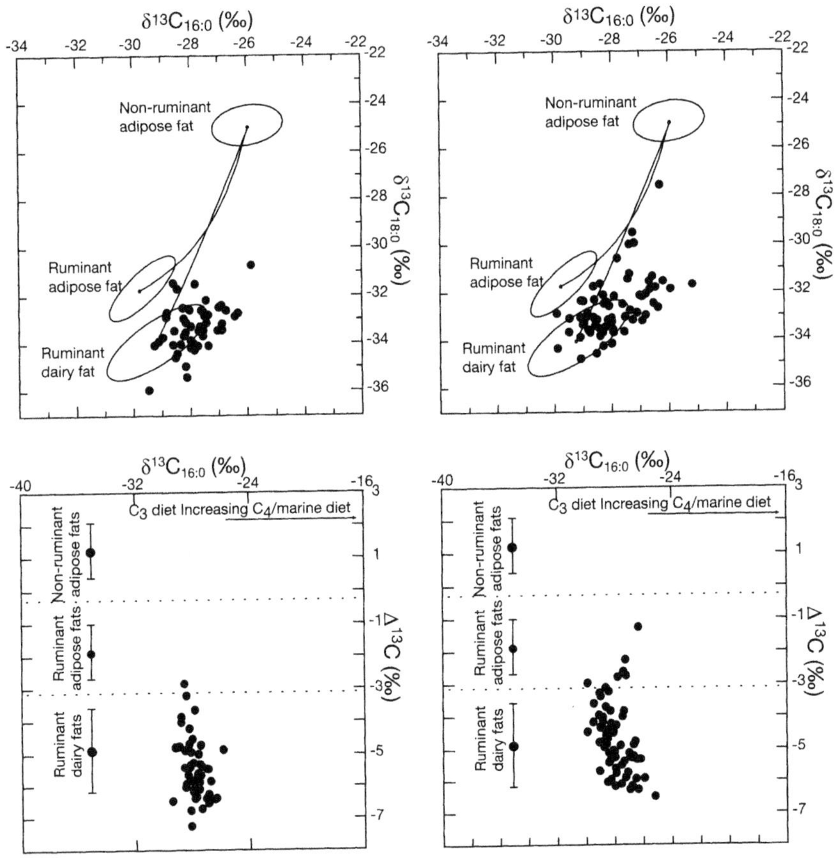

Figure 18.5 Scatter plots showing $\delta^{13}C$ values determined from $C_{16:0}$ and $C_{18:0}$ fatty acids preserved in pottery from Irish Early Neolithic causewayed enclosures (left) and house sites (right). Ellipses show 1 standard deviation confidence ellipses from modern reference terrestrial species from the UK (Copley *et al.* 2003) (*top*). Δδ13C values (= δ13C18:0 – δ13C16:0) plotted against δ13C16:0 values. Ranges show the mean ± 1 s.d. of the Δδ13C values for a global database comprising modern reference animal fats from United Kingdom (animals raised on a pure C_3 diet), Africa, Kazakhstan, Switzerland and the Near East (*bottom*). Archaeological and modern data are corrected for the addition of a methyl carbon during derivatisation using a mass balance equation (Rieley 1994) and the reference fats are corrected for the contribution of post-industrial carbon (Friedli *et al.* 1986).

Filter and proxy

Returning to the larger scale, the lipid residues analysed as part of SCHERD point to milk being a very important element of the Irish Neolithic, a commodity found in a range of vessels from Early Neolithic Carinated Bowl pottery to Late Neolithic Grooved Ware, over a span of approximately 1500 years (Smyth and Evershed forthcoming). In the context of a Europe-wide research tradition that has in the main viewed milk as a secondary or peripheral product of the agricultural 'revolution', it is a striking finding and an important one to be able to defend. Communicating the preliminary results of SCHERD to colleagues has elicited varying responses. Anecdotally, milk fats in pots are either now so commonplace that they can be seen by some as uninteresting, or feature so prominently in assemblages that they arouse suspicion with others – the dairy signal deriving from an as yet unidentified source that is somehow skewing results. How to combat these attitudes? One of the benefits of directly involving the archaeologist in the preparation and running of samples has been the first-hand experience of how scientific data are generated and a subsequent appreciation of the measures in place to flag error and ensure accuracy, for example, running samples in duplicate or triplicate, comparing values with standards, and working within the accepted error margins of instruments and the probability ranges of data sets. Coming thus from a position of analytical confidence, we have been better able to focus on the significance of the results and to reflect on the wider archaeological implications of a strong dairy signal at the start of the Irish Neolithic. There is a growing body of data from northern Atlantic Europe, based on contemporary pottery assemblages from southern and northern England, mainland Scotland and the western and northern isles, that echoes the 'milkiness' of lipid residues from the Irish Neolithic (Copley *et al.* 2005, Mukherjee *et al.* 2008, Šoberl and Evershed in Murray *et al.* 2009, Cramp *et al.* 2014). Given this emerging scenario, we feel the more – or at least equally – appropriate question is 'why *not* milk?' Why might we question a strong dairy signal? What does not fit with our established narratives of the Neolithic in Ireland, Britain and beyond and *why* does it not fit? Have we perhaps under-appreciated dairy consumption in the Early Neolithic, uncritically assuming that domesticated animals were farmed, and valued, more for their meat than for their milk? In Ireland at least, there is very compelling documentary evidence from the medieval period onwards that milk and milk products could constitute a major, if not the main, part of people's diet (Lucas 1960). Indeed, the high abundance of lactase persistence amongst northern European populations is predicated on an evolutionary history underpinned by dairy farming, and it would be surprising if milk fats were not commonplace in archaeological food vessels in Ireland and other parts of northern Europe (Gerbault *et al.* 2013).

At the same time, we recognise that pottery is a cultural object. It emerges out of a web of social relations, spatially and temporally contingent, and in turn acts to reinforce and solidify these social relations (e.g., Latour 2005, 1991). As such, it provides its own perspective on past social practices, filtering out certain types of information whilst letting through others. In terms of the processing, consumption and storage of commodities, how much of the Early Neolithic 'store cupboard' passed through pottery vessels? The relative impermeability of fired clay makes pottery vessels an obvious choice for collecting and storing milk and other liquids, whilst its heat resistance makes it very useful for boiling or stewing. However, wooden or bark containers could have fulfilled at least some of these functions, with woven items providing additional ways to store, steep and strain foodstuffs and other items. Woven baskets or bags are occasionally recovered from wetland locations in Ireland – the alder basket found preserved in clays at Carrigdirty Rock along the Shannon estuary, for example (O'Sullivan 2001, 78–82; Figure 18.6) – and offer glimpses of the range of organic containers that must also have been utilised in the Neolithic (see also Raftery 1970). In Ireland at least there is a close cultural association between pottery and other novelties such as cereals, domesticates and timber houses, which increasingly seem to have been introduced as part of the same 'package' in the early centuries of the fourth millennium cal BC (Cooney *et al.* 2011, Whitehouse *et al.* 2013). Apart from practical considerations, pottery and milk products may have been especially closely linked, perhaps even sharing similar symbolic categories, as two totally and very obviously new substances. An alternative argument, though perhaps less convincing, is that the value attached to pottery upon its introduction, as an item of prestige or a marker of identity (or both), was such that pottery quickly became the means by which most food was processed and the lipid residues surviving in these pots are thus more or less representative of the Neolithic diet. Both of these scenarios, and others, remain to be more fully worked through and, crucially, to be incorporated into our reconstructions of early farming life.

Notwithstanding the above, SCHERD has shown particularly well that while pottery may not be a *traitless* proxy, it remains a proxy nonetheless. On many

Figure 18.6 The alder basket from Carrigdirty Rock, County Limerick, dated to 3710–3510 cal BC (4820±50 BP, both GrA-6520 and Beta-102087, at 94% probability; from O'Sullivan 2001, reproduced with permission of the author).

Irish Neolithic sites – indeed, prehistoric sites in general – soil acidity corrodes animal bone to the extent that species identification is often extremely challenging. The construction of kill-off profiles, from which the management of dairy herds is inferred, is near-impossible. At Donegore Hill causewayed enclosure, for example, only very small fragments of burnt bone were recovered from excavation, none attributable to species, human or animal (Mallory *et al.* 2011, 227). As we have seen, $\delta^{13}C$ values of C_{16} and C_{18} fatty acids in the absorbed lipid residues from the Donegore Hill pottery are consistent with those from reference milk fats. In the absence of diagnostic faunal remains the animal fat residues in pottery give a strong indication that ruminant animals, either cattle or sheep/goat or both, were present within the enclosure, as well as providing conclusive evidence for dairying, if not actually on-site then by groups using the site, in the early fourth millennium cal BC, at or near the very start of the Irish Neolithic.

In considering pottery as a proxy, one can also usefully further explore the materiality of pottery, down to the molecular level. As we have mentioned above, the lipid recovery rate from Irish Neolithic potsherds has been exceptionally high, at *c.* 90%, with a mean lipid concentration of 300 μg/g and from individual potsherds sometimes as high as 4 mg/g. Our working hypothesis is that such good lipid preservation is due to the combination of acidic soils and a cool, temperate climate retarding microbial action and lipid degradation. Preservation is also greatly helped by the micro-structure of potsherds, as lipid molecules become trapped in the clay matrix of the pottery vessel 'out of reach' of microbes. What SCHERD has consistently shown is that the carbon in lipids, absorbed into the matrix of clay vessels, frequently survives where other sources of carbon do not (such as charred plant macrofossils or bone collagen). Thus, given the right soil and climatic conditions, and pottery with a porosity conducive to trapping lipids, archaeological potsherds can in theory represent a potentially important source of ^{14}C information, of enormous use in dating archaeological sites with otherwise poor preservation of organic remains. The dating of absorbed lipids from pottery has been attempted previously, with promising results (Stott *et al.* 2003, Berstan *et al.* 2008), but is still in a relatively early stage of development. A complementary objective of the SCHERD project has thus been to further investigate whether absorbed lipids can be used as an accurate dating tool. Twenty samples from Irish Neolithic vessels comprising both absorbed lipid residues and charred surface residues for comparative purposes have been submitted via the NERC Radiocarbon Facility to the Oxford radiocarbon laboratory. Results will be discussed, along with the project methodology, in detail elsewhere.

The perfect marriage?

Above are outlined what we hope are some honest and useful perspectives on the challenges and the potential of interdisciplinary collaboration. With regard to archaeological lipid analysis, such collaborations have been taking place since at least the early 1990s, often bringing research in new, unexpected directions. What began, for example, as a straightforward request from the commercial sector for post-excavation analysis of medieval ceramic lamps and 'dripping dishes' from Causeway Lane, Leicester, led to the first comparison of molecular, stable isotopic and distributional data obtained for archaeological lipids with those of modern reference fats, and to the development of the fatty acid-based stable isotope proxy (Evershed *et al.* 1997, Dudd and Evershed 1998, Mottram *et al.* 1999). A similar request for analysis of two different Neolithic ceramic types – Grooved Ware and Peterborough Ware – recovered from the excavation of a barrow at Upper Ninepence in the Welsh Borderlands, part of the Walton Basin Project (Gibson 1999), led to another fruitful collaboration, with the post-excavation report turning into a peer-reviewed paper (Dudd *et al.* 1999) and generating a PhD on the extent of pig exploitation in the Late British Neolithic Grooved Ware tradition (Mukherjee 2004, Mukherjee *et al.* 2008). In the ten or so years since the publication of *Archaeological Theory and Scientific Practice*, several similar initiatives have come to fruition, SCHERD being the most recent. SCHERD might seem to be an unusual collaboration, in that it provided time for an archaeologist, based firmly in a humanities tradition, to become familiar with scientific knowledge and procedures. However, it was in many ways the logical next step in marrying the social and natural sciences.

We recognise that this level of interdisciplinary cross-over is not always feasible, either within commercial archaeology or academia, where time is an increasingly precious resource. By the same token, in tackling questions concerning the nature, spread and local character of early farming, neither archaeology nor science can afford to continue without the other and continued meaningful and inclusive dialogue is critical. From the scientific side, there are still perhaps too few groups properly engaging with archaeologists in exploring the potential of excavated materials – developing new methods, for example. From the archaeology side, lipid analysis can certainly still be too readily relegated to the appendices in publications, falling foul of an attitude that sees results from organic residue and other scientific analyses as providing interesting but isolated morsels of information. In other instances, attempts at integrating scientific data with archaeological research questions can be hampered by financial constraints or other limits on the collection of statistically significant datasets (e.g., Cleary and Kelleher 2011, Fenton-Thomas 2009). As a final point, we might also add that as multi-authored interdisciplinary projects become more common, with papers presented at an increasing number of conference sessions geared towards, for example, biomolecular archaeology, it is striking how often results are presented by team members lacking a basic understanding of the methods by which their data has been generated and the issues surrounding such data. As the science applied to archaeological problems becomes increasingly complex, this is hardly surprising. Nevertheless, when new interpretations based on scientific results are presented to peers, within the scientific or the archaeological community, both scientists *and* archaeologists should bear responsibility for articulating and defending the investigative process. Complex issues such as those that surround the transition to and development of farming will require the perspectives of both traditions of learning. To paraphrase Marshall Sahlins (2013), we perhaps should *all* now aspire to become human scientists.

Acknowledgements

The research leading to these results has received funding from the People Programme (Marie Curie Actions) of the European Union's Seventh Framework Programme FP7/2007-2013/ under REA grant agreement n°273462. The NERC Radiocarbon Facility awarded project support for 20 radiocarbon dates on pottery lipids and surface residues and the help and advice of ORAU staff is gratefully acknowledged. Jessica Smyth wishes to thank the staff of museums, university departments and commercial archaeology companies across Ireland who facilitated access to and sampling of pottery assemblages. Colleagues at the Organic Geochemistry Unit, particularly Alison Kuhl and James Williams, are also thanked for their patience and kindness in the laboratory throughout the Marie Curie fellowship.

References

Aillaud, S. 2001. *Field and laboratory studies of diagenetic reactions affecting lipid residues absorbed in unglazed archaeological pottery vessels*. Unpublished PhD thesis, University of Bristol.

Berstan, R., Stott, A.W., Minnitt, S., Bronk Ramsey, C., Hedges, R.E.M. and Evershed, R.P. 2008. Linking ^{14}C dating and use of the Sweet Track pottery via compound-specific radiocarbon and stable isotope analysis. *Antiquity* 82, 702–71.

Bland, H.A. 1999. *Chemical investigations of ancient biomolecules in artefacts and ecofacts from Qasr Ibrim, Egyptian Nubia*. Unpublished PhD thesis, University of Bristol.

Bogaard, A. and Outram, A. (eds) 2013. Special issue: stable isotopes. *World Archaeology* 45, 333–518.

Burger, J., Kirchner, M., Bramanti, B., Haak, W. and Thomas, M.G. 2007. Absence of the lactase-persistence-associated allele in early Neolithic Europeans. *Proceedings of the National Academy of Sciences of the USA* 104, 3736–41.

Charters, S. 1996. *Chemical analysis of absorbed lipids and laboratory simulation experiments to interpret archaeological pottery vessel contents and use*. Unpublished PhD thesis, University of Bristol.

Charters, S., Evershed, R.P., Goad, L.J., Leyde, N.A., Blinkhorn, P.W. and Denham, V. 1993. Quantification and distribution of lipid in archaeological ceramics: implications for sampling potsherds for organic residue analysis and the classification of vessel use. *Archaeometry* 35, 211–23.

Charters, S., Evershed, R.P., Quye, A., Blinkhorn, P.W. and Reeves, V. 1997. Simulation experiments for determining the use of ancient pottery vessels: the behaviour of epicuticular leaf wax during boiling of a leafy vegetable. *Journal of Archaeological Science* 24, 1–7.

Christie, W.W. 1981. Effects of diet and other factors on the lipid composition of ruminant tissues and milk. In W.W. Christie (ed.), *Lipid metabolism in ruminant animals*, 193–226. Oxford: Pergamon.

Cleary, R. and Kelleher, H. 2011. *Archaeological excavations at Tullahedy, county Tipperary. Neolithic settlement in North Munster*. Cork: The Collins Press.

Cooney, G., Bayliss, A., Healy, F., Whittle, A., Danaher, E., Cagney, L., Mallory, J., Smyth, J., Kador, T. and O'Sullivan, M. 2011. Chapter 12: Ireland. In A. Whittle, F. Healy and A. Bayliss, *Gathering time: dating the early Neolithic enclosures of southern Britain and Ireland*, 562–669. Oxford: Oxbow Books.

Copley, M.S., Berstan, R., Dudd, S.N., Docherty, G., Mukherjee, A.J., Straker, V., Payne, S. and Evershed, R.P. 2003. Direct chemical evidence for widespread dairying in prehistoric Britain. *Proceedings of the National Academy of Sciences of the USA* 100, 1524–9.

Copley, M.S., Berstan, R., Dudd, S.N., Straker, V., Payne, S. and Evershed, R.P. 2005. Dairying in antiquity. III. Evidence from absorbed lipid residues dating to the British Neolithic. *Journal of Archaeological Science* 32, 523–46.

Craig, O.E., Saul, H., Lucquin, A., Nishida, Y., Taché, K., Clarke, L., Thompson, A., Altoft, D.T., Uchiyama, J., Ajimoto, M., Gibbs, K., Isaksson, S., Heron, C.P. and Jordan, P. 2013. Earliest evidence for the use of pottery. *Nature* 496, 351–4.

Cramp, L.J.E., Jones, J., Sheridan, A., Smyth, J., Whelton, H., Mulville, J., Sharples, N. and Evershed, R.P. 2014. Immediate replacement of fishing with dairying by the earliest farmers of the NE Atlantic archipelagos. *Proceedings of the Royal Society B* 281, doi: 10.1098/rspb.2013.2372.

Docherty, G. 2002. *Natural variations in plant tissue individual fatty acid and monosaccharide $\delta^{13}C$ values: implications for defining the origins of animal fats in archaeological pottery*. Unpublished PhD thesis, University of Bristol.

Dudd, S.N. 1999. *Molecular and isotopic characterisation of animal fats in archaeological pottery*. Unpublished PhD thesis, University of Bristol.

Dudd, S.N. and Evershed, R.P. 1998. Direct demonstration of milk as an element of archaeological economies. *Science* 282, 1478–81.

Dudd, S.N., Evershed, R.P. and Gibson, A. 1999. Evidence for varying patterns of exploitation of animal products in different prehistoric pottery traditions based on lipids preserved in surface and absorbed residues. *Journal of Archaeological Science* 26, 1473–82.

Dudd, S.N., Regert, M. and Evershed, R.P. 1998. Assessing microbial lipid contributions during laboratory degradations of fats and oils and pure triacylglycerols absorbed in ceramic potsherds. *Organic Geochemistry* 29, 1345–54.

Dungait, J.A.J., Docherty, G., Straker, V. and Evershed, R.P. 2008. Interspecific variation in bulk tissue, fatty acid and monosaccharide $\delta^{13}C$ values of leaves from a mesotrophic grassland plant community. *Phytochemistry* 69, 2041–51.

Dungait, J.A.J., Docherty, G., Straker, V. and Evershed, R.P. 2010. Seasonal variations in bulk tissue, fatty acid and monosaccharide $\delta^{13}C$ values of leaves from mesotrophic grassland plant communities under different grazing managements. *Phytochemistry* 71, 415–28.

Dungait, J.A.J., Docherty, G., Straker, V. and Evershed, R.P. 2011. Variation in bulk tissue, fatty acid and monosaccharide $\delta^{13}C$ values between autotrophic and heterotrophic plant organs. *Phytochemistry* 72, 2130–8.

Dunne, J., Evershed, R.P., Salque, M., Cramp, L., Bruni, S., Ryan, K., Biagetti, S. and di Lernia, S. 2012. First dairying in green Saharan Africa in the fifth millennium BC. *Nature* 486, 390–4.

Evershed, R.P. 2008. Experimental approaches to the interpretation of absorbed organic residues in archaeological ceramics. *World Archaeology* 40, 26–47.

Evershed, R.P. and Connolly, R.C. 1988. Lipid preservation in Lindow Man. *Naturwissenschaften* 75, 143–5.

Evershed, R.P., Heron, C. and Goad, L.J. 1990. Analysis of organic residues of archaeological origin by high temperature gas chromatography/mass spectrometry. *Analyst* 115, 1339–42.

Evershed, R.P., Heron, C. and Goad, L.J. 1991. Epicuticular wax components preserved in potsherds as chemical indicators of leafy vegetables in ancient diets. *Antiquity* 65, 540–4.

Evershed, R.P., Heron, C., Charters, S. and Goad, L.J. 1992. The survival of food residues: new methods of analysis, interpretation and application. In A.M. Pollard (ed.), *New developments in archaeological science*, Proceedings of the British Academy 77, 187–208. Oxford: Oxford University Press/The British Academy.

Evershed, R.P., Stott, A.W., Raven, A., Dudd, S.N., Charters, S. and Leyden, A. 1995a. Formation of long-chain ketones in ancient pottery vessels by pyrolysis of acyl lipids. *Tetrahedron Letters* 36, 8875–8.

Evershed, R.P., Charters, S. and Quye, A. 1995b. Interpreting lipid residues in archaeological ceramics: preliminary results from laboratory simulations of vessel use and burial. In P.B. Vandiver, J. Druzik, G.S. Wheeler and I.C. Freestone (eds), *Materials Research Society Symposium Proceedings*, 85–95. Pittsburgh, PA: Materials Research Society 352.

Evershed, R.P., Mottram, H.R., Dudd, S.N., Charters, S., Stott, A.W. and Lawrence G.J. 1997. New criteria for the identification of animal fats preserved in archaeological pottery. *Naturwissenschaften* 84, 402–6.

Evershed, R.P., Payne, S., Sherratt, A.G., Copley, M.S., Coolidge, J., Urem-Kotsu, D., Kotsakis, K., Ozdogan, M., Ozdogan, A., Nieuwenhuyse, O., Akermans, P.M.M.G., Bailey, D.K., Andeescu, R-R., Campbell, S., Farid, S., Hodder, I., Yalman, N., Ozbarasan, M., Bicakci, E., Garfinkel, Y., Levy, T. and Burton, M.M. 2008. Earliest date for milk use in the Near East and southeastern Europe linked to cattle herding. *Nature* 455, 528–31.

Fenton-Thomas, C. 2009. *A place by the sea: excavations at Sewerby Cottage Farm, Bridlington, Yorkshire*. York: On-Site Archaeology.

Feulner, F., Doorn, N.L. and Leonardi, M. (eds) 2011. *May contain traces of milk: investigating the role of dairy farming and milk consumption in the European Neolithic*. York: University of York.

Friedli, H., Loetscher, H., Oeschger, H., Siegenthaler, U. and Stauffer, B. 1986. Ice core record of the $^{13}C/^{12}C$ ratio of atmospheric CO_2 in the past two centuries. *Nature* 324, 237–8.

Gerbault, P., Roffet-Salque, M., Evershed, R.P. and Thomas, M.G. 2013. How long have adult humans been consuming milk? *International Union of Biochemistry and Molecular Biology* 65, 983–90.

Gibson, A. 1999. *The Walton Basin project. Excavation and survey in a prehistoric landscape 1993–7.* York: Council for British Archaeology.

Heron, C., Evershed, R.P. and Goad, L.J. 1991. Effects of migration of soil lipids on organic residues associated with buried potsherds. *Journal of Archaeological Science* 18, 641–59.

Itan, Y., Powell, A., Beaumont, M.A., Burger, J. and Thomas, M.G. 2009. The origins of lactase persistence in Europe. *PLoS Computational Biology* 5, 1–13.

Jones, A. 2002. *Archaeological theory and scientific practice.* Cambridge: Cambridge University Press.

Latour, B. 1991. Technology is society made durable. In J. Law (ed.), *A sociology of monsters: essays on power, technology and domination*, 103–31. London: Routledge.

Latour, B. 2005. *Reassembling the social: an introduction to actor-network theory.* Oxford: Oxford University Press.

Lucas, A.T. 1960. Irish food before the potato. *Gwerin* 3, 8–43.

Mallory, J.P., Nelis, E. and Hartwell, B. 2011. *Excavations on Donegore Hill, Co. Antrim.* Dublin: Wordwell.

McDonald, P., Edwards, R.A. and Greenhalgh, J.F.D. 1988. *Animal nutrition.* Essex: Longman.

McSparron, C. 2008. Have you no homes to go to? Neolithic housing. *Archaeology Ireland* 22, 18–21.

Mottram, H.R., Dudd, S.N., Lawrence, G.J., Stott, A.W. and Evershed, R.P. 1999. New chromatographic, mass spectrometric and stable isotope approaches to the classification of degraded animal fats preserved in archaeological pottery. *Journal of Chromatography* A833, 209–21.

Mukherjee, A. 2004. *The importance of pigs in the later British Neolithic: integrating stable isotope evidence from lipid residues in archaeological potsherds, animal bone, and modern animal tissues.* Unpublished PhD thesis, University of Bristol.

Mukherjee, A.J., Gibson, A.M. and Evershed, R.P. 2008. Trends in pig product processing at British Neolithic Grooved Ware sites traced through organic residues in potsherds. *Journal of Archaeological Science* 35, 2059–73.

Murray, H.K, Murray, J.C. and Fraser, S.M. 2009. *A tale of unknown unknowns: a Mesolithic pit alignment and a Neolithic timber hall at Warren Field, Crathes, Aberdeenshire.* Oxford: Oxbow Books.

O'Sullivan, A. 2001. *Foragers, farmers and fishers in a coastal landscape: an intertidal archaeological survey of the Shannon estuary.* Dublin: Royal Irish Academy.

Raftery, J. 1970. Prehistoric coiled basketry bags. *Journal of the Royal Society of Antiquaries of Ireland* 100, 167–8.

Raven, A.M., van Bergen, P.F., Stott, A.W., Dudd, S.N. and Evershed, R.P. 1997. Formation of long-chain ketones in archaeological pottery vessels by pyrolysis of acyl lipids. *Journal of Analytical and Applied Pyrolysis* 40–1, 267–85.

Reber, E.A. and Evershed, R.P. 2004. How did Mississippians prepare maize? The application of compound specific carbon isotopic analysis to absorbed pottery residues from several Mississippi Valley sites. *Archaeometry* 46, 19–33.

Rieley, G. 1994. Derivatization of organic-compounds prior to gas-chromatographic combustion-isotope ratio mass-spectrometric analysis – identification of isotope fractionation processes. *Analyst* 119, 915–19.

Sahlins, M. 2013. Human science. *London Review of Books* 35, 29.

Salque, M. 2012. *Regional and chronological trends in milk use in Prehistoric Europe traced through organic residues preserved in pottery vessels*. Unpublished PhD thesis, University of Bristol.

Salque, M., Bogucki, P.I., Pyzel, J., Sobkowiak-Tabaka, I., Grygiel, R., Szmyt, M. and Evershed, R.P. 2013. Earliest evidence for cheese making in the sixth millennium BC in northern Europe. *Nature* 493, 522–5.

Sherratt, A. 1981. Plough and pastoralism: aspects of the secondary products revolution. In I. Hodder, G. Isaac and N. Hammond (eds), *Pattern of the past: studies in honour of David Clarke*, 261–305. Cambridge: Cambridge University Press.

Smyth, J. 2011. The house and group identity in the Irish Neolithic. *Proceedings of the Royal Irish Academy* 111C, 1–31.

Smyth, J. 2014. *Settlement in Neolithic Ireland: new discoveries at the edge of Europe*. Oxford: Prehistoric Society/Oxbow Books.

Smyth, J. and Evershed. R.P. Forthcoming. The molecules of meals: new insight into Neolithic foodways. 'Food and Drink in Ireland'. *Proceedings of the Royal Irish Academy* 115C.

Šoberl, L. 2011. *Pots for the afterlife: organic residue analysis of British Bronze Age burial vessels*. Unpublished PhD thesis, University of Bristol.

Stott, A.W., Berstan, R., Evershed, R.P., Bronk-Ramsey, C., Hedges, R.E. and Humm, M.J. 2003. Direct dating of archaeological pottery by compound-specific ^{14}C analysis of preserved lipids. *Analytical Chemistry* 75, 5037–45.

Whitehouse, N.J., Schulting, R., McClatchie, M., Barratt, P., McLaughlin, T.R, Bogaard, A., Colledge, S., Marchant, R., Gaffrey, J. and Bunting, M.J. 2013. Neolithic agriculture on the European western frontier: the boom and bust of early farming in Ireland. *Journal of Archaeological Science*, online 22 August 2013 (http://dx.doi.org/10.1016/j.jas.2013.08.009).

19

Constructing a Narrative for the Neolithic of Britain and Ireland

The Use of 'Hard Science' and Archaeological Reasoning

ALISON SHERIDAN AND PIERRE PÉTREQUIN

Introduction

THERE CAN BE NO DOUBT that the application of techniques from physics, chemistry and other branches of so-called 'hard' science plays a vital, and growing, role in the construction of archaeological narratives for the British and Irish Neolithic, as it does for other periods in the past. Radiocarbon dating, along with other forms of absolute dating, has provided the basic building blocks of our chronology for well over half a century, and the recent judicious application of Bayesian modelling to radiocarbon dates has allowed the construction of fine-grained narratives for several parts of Neolithic Britain and Ireland (Whittle *et al.* 2011). Similarly, advances in palaeoenvironmental science have allowed increasingly detailed landscape reconstruction to be achieved. We can now assess landscape use and cultivation practices from the level of the individual settlement (as at Warren Field, Crathes, Aberdeenshire: Lancaster *et al.* 2009) to the level of the entire island of Ireland, over several centuries, as the recent Heritage Council-funded Cultivating Societies project has demonstrated (Whitehouse *et al.* 2013). The burgeoning field of stable isotope analysis, as applied to Neolithic human and animal bone and tooth enamel, has yielded invaluable information on diet (for example, Schulting and Richards 2002) and mobility (Montgomery *et al.* 2007), while advances in the analysis of absorbed lipids in pottery have allowed the determination of pots' former contents, and in particular have clarified whether the pots had been used to cook marine foodstuffs (for example, Evershed 2007; Smyth and Evershed, this volume Chapter 18). Meanwhile, advances in zoological studies have provided new information on the proximate origin of some of the domestic animals found in Neolithic Britain (Tresset and Vigne 2007) and have permitted the detailed characterisation of animal lifestyles (as in the case of the

Proceedings of the British Academy **198**, 369–90. © The British Academy 2014.

seaweed-eating sheep of Papa Westray, for instance: Balasse *et al.* 2006). And recent human osteological studies have highlighted not only the incidence of disease among the individuals deposited in chamber tombs in Neolithic Orkney (Lawrence 2006, 2012; Schulting *et al.* 2010) but have also shed light on inter-personal conflict in Neolithic Britain (Schulting 2012; Schulting and Fibiger, this volume Chapter 15). These, and other applications of scientific methods, allow us to create narratives that are significantly more detailed, nuanced and reliable than the culture-historical accounts of just a few decades ago.

It would be wrong, however, to regard science as offering a 'magic bullet' to solve all the mysteries of archaeology; and its application to archaeological questions is not as uncomplicated, or as objective, as some of its proponents would like to believe. Because of the highly specialised nature of the disciplines involved, it can be difficult for archaeologists to assess the reliability of a specific analytical technique or to follow the finer points of a methodological debate; similarly, it cannot be assumed that scientists are fully *au fait* with the current state of knowledge about Neolithic Britain and Ireland. Furthermore, because the various disciplines tend to publish the results of their work in specialist publications with relatively little cross-disciplinary readership, there can be a basic lack of awareness of current and past research. It is nevertheless necessary for archaeologists and scientists alike to familiarise themselves with the full range of information that is currently available, to maintain a critical awareness of the processes used to build and test hypotheses about the past, and to be aware of the debates surrounding issues of methodology and the interpretation of analytical results. That is why interdisciplinary initiatives such as the aforementioned Cultivating Societies project, and the conference that gave rise to the current publication, are so important in fostering information exchange and debate.

This contribution offers a few case studies from current and recent research on the Neolithic of Britain and Ireland to highlight some of the issues involved in using scientific data to build archaeological narratives.

Using radiocarbon dates to characterise the Neolithisation process

The question of whether the principal agents responsible for the introduction of a Neolithic way of life to Britain and Ireland had been immigrant farming communities from the Continent or indigenous groups of hunter-gatherer-fishers has been debated intensively for the past two decades, and it is not intended to rehearse all the arguments here, since they are well documented elsewhere (by, among others: Thomas 2008; Sheridan 2010, 2011a, 2011b; Sheridan *et al.* 2008; Whittle *et al.* 2011; each with an extensive bibliography). Diverse approaches have been used to address this question of change, with some (for example, Richards and Schulting 2006) focusing on evidence for dietary change as revealed through isotopic analysis of bone collagen, while others have concentrated more on the evidence from material culture and monuments (Sheridan 2010), and yet others have considered the role of climate change, either as a contributory factor (Tipping 2010) or even as a determining factor (Bonsall *et al.* 2002). The resulting characterisations of the Mesolithic–Neolithic transition in Britain and Ireland remain resolutely diverse. Thus, for example, Julian Thomas has characterised the process as a sudden and widespread adoption, around 4000 cal BC, of a 'new kind of identity . . . a new means of self-identification' by indigenous communities who had allegedly long been in contact with Continental farmers (Thomas 2008, 80). By contrast, one of the current authors (AS) has posited a complex, multi-strand process of Neolithisation, featuring various small-scale movements of farmers from different parts of the French coast to different parts of Britain and Ireland at various times between *c.* 4300 cal BC and *c.* 3800 cal BC, coming for different reasons and resulting in differing outcomes (Sheridan 2010, 2011a; Sheridan and Pailler 2011).

A key element in this debate has been the question of the timing and tempo of the change, and here radiocarbon dating has a vital role to play. Two recent contributions have foregrounded the existing radiocarbon evidence in their narratives of the Mesolithic–Neolithic transition, and these are the focus of critical scrutiny here.

The first, by Mark Collard and colleagues (Collard *et al.* 2010), arises from a broad-brush, demographic approach to the spread of agriculture and stock-keeping across Europe and explicitly uses radiocarbon date frequency as a proxy for population density – a technique that has been, and continues to be, used by several researchers (Shennan and Edinborough 2007). The underlying rationale for using date densities to track the spread of farming is the claim, based on ethnographic observation, that 'with the exception of certain coastal and riparian areas where rich aquatic resources are available, farming usually supports much higher population densities in temperate regions than hunting and gathering' (Collard *et al.* 2010, 867). As the population rises, so the number of sites with evidence for farmers in a given time period should also increase. As Collard *et al.* express it, 'the number of site phases in a given time period can be expected to relate monotonically to population size, [and therefore] changes in summed probability distributions of calibrated 14C dates derived from different site phases serve as a proxy for changes in population' (Collard *et al.* 2010, 867). Having applied this approach to radiocarbon dates relating to 'Neolithic' activity in Britain – running the model for all the dates, then screening out those relating to monuments and finally modelling just the 76 cereal grain dates – Collard *et al.* concluded that there had been a sharp population rise between 6000 and 5610 cal BP (4000–3610 cal BC), and that its start had coincided with the arrival of domesticated plants. They argued that the rate of this increase offered a better match for the hypothetical rate of population rise within a 'migrating farmers' model than for a 'rapid indigenous adoption' model. Next, looking at the geographical distribution of dates for

'Neolithic' activity within 100-year time slices between 6100 and 5300 cal BP (4100–3300 cal BC), they concluded that the immigrant farmers had first arrived in south-west England (that is, Wessex and surrounding counties: Collard *et al.* 2010, 867, fig. 2), around 4000 cal BC, and that there had been a second wave of immigrants, arriving in Scotland south of the Great Glen, around a century later.

The second model that foregrounds radiocarbon dates, formulated by Alasdair Whittle, Frances Healy and Alex Bayliss in their magisterial *Gathering Time* publication (2011), arrived at a significantly different conclusion, even though some of the same dates will have featured in both studies. Here, by applying Bayesian modelling, Whittle *et al.* concluded that 'Neolithic things and practices' probably first appeared in south-east England from across the Channel in the forty-first century cal BC, and then spread westwards and northwards, by a process of 'chain migration' and indigenous acculturation, picking up speed around 3800 cal BC (Whittle *et al.* 2011, fig. 14.16). Elsewhere in the study, a more complex and speculative model is offered, featuring cross-Channel movements of resources, practices and ideas from several points along the north French coast (Whittle *et al.* 2011, fig. 15.8); but neither version corresponds with the model proposed by Collard *et al.*, and indeed all of these models differ significantly from the multi-strand model proposed by one of the current authors (AS), and from that propounded by Julian Thomas. How did this come about, and which is closer to the truth?

The discrepancy between Collard *et al.*'s version of events and that postulated by Whittle *et al.* is partly due to the fact that, among the radiocarbon dates that were modelled in the Gathering Time project was a set of 27 recently obtained dates for individuals buried at the chamber tomb at Coldrum in Kent (Whittle *et al.* 2011, 381–3), whose calibrated range extended back as far as the forty-first century cal BC. Together with a handful of other early dates from Kent and the Thames Estuary, these offer the earliest *radiocarbon-dated* evidence (our emphasis) for the presence of 'Neolithic things and practices' in Britain. While the nine early dates for the large house at White Horse Stone, Kent (Whittle *et al.* 2011, 379–81) had been available to Collard *et al.* when they wrote their article, it may be that these authors' focus on tracking clusters of dates as a proxy for population size led them to overlook this particular piece of evidence. The key factor responsible for the discrepancy between the two teams' models will have been their approach to the set of available radiocarbon dates. Collard *et al.*'s inclusive, and relatively uncritical, approach to using the dates contrasts with Whittle *et al.*'s rigorous source-critical approach and their use of Bayesian modelling to constrain the chronological scatter that is inherent in any set of radiocarbon dates. Added to this, the basic assumptions underpinning the construction of their models of Neolithisation, and informing the inferences drawn from the dates, differ.

In the current authors' view, however, both teams' models are flawed by their over-reliance on radiocarbon dating at the expense of a more holistic approach (although see below regarding Whittle *et al.*'s modified version of their model).

Collard *et al.* make no real attempt to engage with the artefactual and structural evidence for the earliest Neolithic in Britain (and indeed show scant understanding of it), and their basic assumptions – that radiocarbon date densities can be used as a proxy for population size, and that areas with high densities of early dates will be the points of origin for incoming farming groups – are highly contentious. Using dates as a demographic proxy can only work if there is an even coverage of dates – that is, if every site or community (or at least a representative sample thereof) has been radiocarbon dated, and to the same proportional extent, over time. This is patently not the case for prehistoric Britain. There are indeed relatively few visible late Mesolithic sites, but even fewer are associated with any radiocarbon date. By contrast, a relatively small number of Neolithic monuments in southern England have been intensively radiocarbon dated. Thus, the distribution of dates will reflect not so much prehistoric population size, but instead the degree of investment in radiocarbon dating by public bodies and other funders of excavation and research. By the same token, it seems simplistic to equate areas with high densities of early dates with points of arrival by immigrant farmers.

As for Whittle *et al.*'s model, the points of contention in their reasoning, and in their use of the archaeological evidence, have already been discussed elsewhere so will not be repeated in detail here (Sheridan 2012a). A positive aspect is that a greater attempt has been made to engage with the archaeological evidence than was the case with Collard *et al.*'s model, even if that has led to a curious, and in our view insufficiently explained, disjunction between the two versions of their model as presented in their figures 14.16 and 15.8 (with the latter coming slightly closer to the multi-strand model of Neolithisation as proposed by one of us (AS)). One can argue the toss with Whittle *et al.* as to whether the 'Carinated Bowl Neolithic' was introduced to Kent and the Thames Estuary and spread northwards and westwards from there, or else was introduced as part of a wider diasporic episode (or episodes) of small-scale immigration from northern France extending far up the east coast of Britain. Similarly, the evidence regarding the nature, dating and significance of cross-Channel links between Normandy (and perhaps also north Brittany) and southern and south-west England will no doubt continue to be debated (Sheridan 2011a); all agree, however, that more information is required about what happened between 4000 cal BC and 3800 cal BC in those parts of England, since the existing crumbs of evidence are tantalising (Sheridan 2011a, 24–9). However, what both Whittle *et al.* and Collard *et al.* fail to do is to deal with the one strand of Neolithisation that is not yet linked with any radiocarbon dates in Britain – that is, the Atlantic façade strand, featuring the arrival of small numbers of farmers from the Morbihan region of south-east Brittany in coastal areas in south-west and north-west Wales, western Scotland and around the northern half of Ireland (Sheridan 2010). Whittle *et al.* do indeed discuss this strand, if only to reject Sheridan's hypothesis; but, as argued elsewhere (Sheridan 2012a), the arguments used to dismiss this hypothesis are themselves contentious. In the opinion of one of us (AS), these hypothetical incoming farmers brought the practice of making Breton-style pottery (including Late Castellic ware), and of building Breton-style, closed, polygonal megalithic chambers and simple passage tombs, to those parts of Britain and Ireland. In the Morbihan the Late Castellic period is well dated, at a complex of sites in Locmariaquer, to between 4400/4300 cal BC and *c*. 3900 cal BC (Cassen and François 2009, 761, fig. 13). In Ireland, the similarly early radiocarbon dates obtained by Göran Burenhult for the construction of the closed chamber tombs and simple passage tombs at the Carrowmore cemetery, Co. Sligo, have rightly been challenged on the grounds of inadequate taphonomic and stratigraphic documentation (Hensey and Bergh 2013), but that does not actually prove that the monuments were *not* built during that date bracket. The Bayesian-modelled dates that Hensey and Bergh (2013) have recently obtained from 25 bone and/or antler pins from tombs at Carrowmore, ranging from *3775–3520 cal BC* (start) to *3305–2950 cal BC* (end) *at 95.4% probability*, do not negate the '4400/4300–3900 BC construction date' model since, as this author has repeatedly pointed out, the cemetery was clearly in use over several centuries, and these pins need not represent the earliest use of the monuments.

In basing their models principally or exclusively on evidence from radiocarbon dates, both Whittle *et al.* and Collard *et al.* privilege those sites which have been radiocarbon-dated over others that have not. Dealing with a site such as Achnacreebeag, where no radiocarbon-datable material survived and yet which potentially informs us about an entire strand of Neolithisation, requires an approach that takes a broader view of all the evidence available to us – something that Whittle *et al.*'s more speculative version of their model (that is, their fig. 15.8) starts to do.

The Mesolithic–Neolithic diet change debate

Once again, it is not intended to describe this debate in detail, since it is well documented (Schulting and Richards 2002; Richards *et al.* 2003; Milner 2010; Milner *et al.* 2004, 2006; Schulting 2013; Schulting *et al.* 2013). The reason for featuring it is to show how other kinds of scientific evidence have been deployed in constructing narratives of the Mesolithic–Neolithic transition in Britain, and how the 'diet change' debate has progressed in the decade since the publication of Schulting and Richards' initial study in 2002.

Essentially, Schulting and Richards sought to use carbon and nitrogen isotope data from human bone collagen to characterise the diet of late Mesolithic and early Neolithic individuals whose remains have been found in – and who had probably lived in – coastal areas, in particular on the west coast of Scotland. They wished to determine whether any shift from a marine-dominated diet to a terrestrially dominated diet could be identified. In the absence of human remains relating to the very earliest Neolithic, they used skeletons from Clyde cairns – a particular type of megalithic chamber tomb – in west mainland Scotland to represent the earliest available 'Neolithic' skeletal material. Several generations will have passed between the initial appearance of Neolithic 'things and practices' and the deposition of those individuals in the Clyde cairns. Their findings appeared to show a definitive contrast, with no obvious sign of a marine element in the diet of the Neolithic individuals. In fact, all subsequent isotopic analysis of British and Irish Neolithic individuals – with the exception of some of those found in a communal cist at Sumburgh in Shetland, dating mostly to the second half of the fourth millennium cal BC (Melton and Montgomery 2009, Montgomery *et al.* 2013) – have produced a similarly marked 'terrestrial diet' signature, which contrasts with that of many Mesolithic individuals whose remains have been found on or near the coast (Schulting 2013, Schulting *et al.* 2013). Schulting and Richards concluded, from their initial study and their subsequent research, that there had indeed been a rapid and more or less complete dietary shift away from the consumption of marine resources upon the appearance of domesticated plants and animals. Without wishing to commit themselves to too detailed a model, they acknowledged that some small-scale immigration of farming groups is the most plausible explanation, with rapid subsequent acculturation of indigenous Mesolithic communities being responsible for the apparent speed of the dietary switch.

The critiques of Schulting and Richards' work have focused on issues such as the fact that it is possible to consume marine resources for up to as much as 10% of one's diet before this will show in one's isotopic signature (a fact that Schulting and Richards had acknowledged); on the fact that not all Mesolithic communities or individuals relied on a marine diet (again, recognised by Schulting and Richards); and on the pace of the diet shift, with Schulting and Richards' model of sudden change being challenged by recalibrations of the Mesolithic bone dates from Oronsay (Milner 2010). Such critiques are mostly informed by assumptions about the Mesolithic–Neolithic transition being a gradual affair, with hunter-fisher-gatherer groups being the principal agents of change. Richards and Schulting's initial rebuttal of the critiques (2006), which acknowledges the likely complexity of the prehistoric situation, was robust; but there was a perceived need for corroborative evidence (cf. Schulting 2013). This is where the application of another scientific technique – the absorbed lipid analysis of Neolithic sherds using organic chemistry techniques – was able to move the debate on.

In a NERC-funded study focusing on the identification and use of marine biomarkers in pottery (*Changing patterns of marine product exploitation in human prehistory via biomarker proxies in archaeological pottery)*, Lucy Cramp analysed around 340 Neolithic sherds, including numerous examples belonging to the Early Neolithic, Carinated Bowl tradition and to the Late Neolithic Grooved Ware tradition between 2009 and 2012 (Cramp and Evershed 2014, Cramp *et al.* 2014). Only two Neolithic pots produced any marine signature: one Carinated Bowl sherd from an otherwise 'terrestrial' assemblage from Elgin, Moray, and a decorated

Hebridean Neolithic bowl from Bharpa Langais, North Uist. In a complementary study of Irish Neolithic pottery undertaken by Jessica Smyth (Smyth and Evershed, this volume Chapter 18), the results were similarly conclusive; no evidence was found for the presence of a marine signature. This evidence could be taken to support the idea that the Neolithic diet was indeed overwhelmingly terrestrially based.

It could be argued, however, that the ceramic data relate only to one way of cooking food; we know that others, including dry-roasting in pits and boiling in water-filled troughs, were also used during the Neolithic (for example, at Maybole, Ayrshire, and possibly at Clowanstown, Co. Meath: Becket and MacGregor 2009, 118; Schulting 2013, 325). The 'marine biomarkers in pottery' project therefore sought to obtain a rounder picture of resource use by considering faunal assemblages from the sites represented among the analysed sherds. While many of the sites in Cramp's and Smyth's studies, most notably those associated with Carinated Bowl tradition pottery, have produced disappointingly sparse faunal assemblages, a broader review of Early Neolithic faunal assemblages in Britain and Ireland (Schulting 2013) has confirmed the overwhelmingly terrestrial-species composition of these assemblages (including the use of some wild terrestrial species). A similar conclusion has been reached in Jennifer Jones' examination of slightly later faunal assemblages from the Northern and Western Isles of Scotland (Jones 2013).

This is not to say that *no* marine resources were consumed during the Neolithic; in addition to the isotopic and lipid data referred to above, there is the evidence from several shell middens, including those on the River Forth, which have produced radiocarbon dates in the fourth millennium cal BC, or even later (MacKie 1972). Nevertheless, even in Late Neolithic Orkney, where there is abundant evidence for the exploitation of marine resources in one way or another (such as whalebone for tools and roofing, ivory for jewellery, and oyster shells for ochre-grinding palettes: Clarke 2012), the evidence for their *consumption* remains notably elusive. Cramp's analysis of numerous sherds of Grooved Ware pottery from Orkney (especially from the settlements at Skara Brae and the Links of Noltland) produced no evidence that they had been used to cook fish, shellfish or marine mammals. Furthermore, the isotopic analysis of human remains from the Quanterness passage tomb (Schulting *et al.* 2010, Schulting 2013) and the Isbister and Banks chamber tombs (Lawrence 2012) – that is, the analysed remains that are geographically and chronologically closest to these sites – do not show a noticeable marine signature, which suggests that marine resources did not constitute more than a small or episodic part of these people's diet for the last decade or so of their lives. The presence of fish bones, not only from these Late Neolithic sites, but also from the Knap of Howar in Orkney, suggests that people were indeed probably eating some fish. Furthermore, they may have eaten the meat of the whales and other sea creatures that were occasionally stranded. However, any consumption of such marine resources does not seem to have been sufficiently substantial or regular to have produced a noticeable marine isotope signature in their bones.

Clearly, there is much more that needs to be found out about the diet of the inhabitants of Britain and Ireland during the fourth and third millennia cal BC, especially as regards defining regional and chronological patterning. It could well be, for example, that in Shetland the descendants of the earliest farming communities were forced by the relatively challenging environmental conditions there to turn to marine resources as a way of maintaining a sustainable subsistence strategy (Sheridan 2012b, Montgomery *et al.* 2013). Whether the Bharpa Langais pot is symptomatic of a similar process in the Western Isles remains to be seen; other analysed Hebridean Neolithic pottery has yielded 'terrestrial' rather than 'marine' lipid data. And as for whether the fourth millennium cal BC and later shell middens indicate a decision by the descendants of the first farmers to start eating marine resources, or else the presence of hunter-fisher-gatherer groups who had stubbornly refused to acculturate fully, is a matter for debate. What is clear is that all the currently available data – isotopic, lipid and faunal – support Richards' and Schulting's hypothesis that a significant drop-off in the consumption of marine resources seems to have occurred once domesticated plants and animals had appeared in Britain and Ireland. Whether this occurred as rapidly as they suggest (that is, within a century or so) is hard to gauge since, irrespective of whether the recalibration of the Oronsay dates is correct or not, we do not know when, between 4400 cal BC and 3900 cal BC, the Atlantic façade strand of Neolithisation appeared.

The right kinds of question seem to have been posed in exploring what people ate, and the interdisciplinary nature of the research (particularly that of the lipid analysis project) has ensured that archaeological considerations are given due weight in evaluating the scientific evidence. What is needed now is a substantial amount of new information, of all kinds – isotopic, organic residue, faunal, and other – relating to the first few generations after the initial introduction of cereal growing and stock keeping to different parts of Britain and Ireland.

Orkney voles and red herrings

One example of where the application of scientific techniques to an issue has resulted in conclusions being drawn that are widely discrepant with archaeologists' interpretations is offered by recently published research on the common vole, *Microtus arvalis,* in Orkney (Martínovká *et al.* 2013). This work had originated in a major AHRC-funded project, *The prehistoric origins of Orcadian cultural networks: biomolecular and morphometric studies of Orkney voles*, led by Keith Dobney, Jeremy Searle and Paul O'Higgins. It involved analysing mitochondrial DNA and mapping variability in dental morphology in order to examine the process of colonisation of Orkney by, and the subsequent development of, the Orkney vole. This mammal is of particular interest to biologists and archaeologists alike because it is today found only in Orkney and on the European continent (and on the Channel Islands), but not on the main landmass of Britain or on its other satellite islands. Since its presence in Orkney cannot be accounted for in terms of long-term survival from an ancient, pre-Last Glacial Maximum population or through 'sweepstake colonization' from Britain, the most likely agency for its arrival is long-distance sea travel by humans, from somewhere on the continent (Martínovká *et al.* 2013, 5215). Radiocarbon dating of archaeological specimens from Orkney (Martínovká *et al.* 2013, table 2) indicates a date within the last quarter of the fourth millennium cal BC for its arrival, and the observed genetic variability within archaeological and modern populations in Orkney suggests that 'substantial numbers of female voles must have been introduced' (Martínovká *et al.* 2013, 5215).

While a previous genetic study (Haynes *et al.* 2003) had concluded that the most likely area whence the Orkney vole had travelled to Orkney was France or Spain, Martínovká *et al.* contend that the Belgian coast (and the adjacent north French coast) was the candidate point of departure, since this is 'the closest coastal part of the continental European range of the same mitochondrial lineage as found on Orkney' (Martínovká *et al.* 2013, 5206). They argue that it had been 'introduced by Neolithic settlers [from Belgium] about 5000 BP', either accidentally, among 'plentiful grass livestock bedding/fodder in which the vole stowaways could have survived', or deliberately, 'as food items, pets or for cultural/religious purposes' (Martínovká *et al.* 2013, 5215).

Not surprisingly, this publication has attracted a great deal of interest, with BBC News for north-east Scotland, Orkney and Shetland announcing that 'Orkney vole "came from Belgium with farmers 5,000 years ago"' and quoting Keith Dobney as saying 'Where in Europe they came from and exactly when they were introduced has been a mystery for decades, but now genetic techniques and direct dating of their bones have finally allowed us to answer these questions' (BBC News 2013).

Given that not a shred of independent archaeological evidence exists to corroborate this remarkable claim, and that the arrival of farmers from Belgium around 3100 cal BC conflicts starkly with the currently understood archaeology of Middle to Late Neolithic Orkney (for which, see Schulting *et al.* 2010), it would be prudent to examine the basis for Martínovká *et al.*'s conclusions.

Most of the samples examined were of modern *Microtus arvalis*, with geometric dental morphometrics applied to 553 modern specimens of first lower molar teeth from 27 localities extending from northern Spain to Hungary (shown on Martínovká *et al.* 2013, fig. 1, and listed in their supplementary table S1) and modern DNA being analysed from 651 specimens from 70 localities (Martínovká *et al.* 2013, fig. 2). The area that was assumed to be the point of origin was particularly heavily sampled for modern DNA. Ancient DNA was obtained from only 37 specimens, taken from Neolithic to Viking Age Orkney (30 specimens) and from Late Medieval Belgium (seven specimens). No samples, either modern or ancient, were taken from the west coast of Iberia, a region where, archaeologically, there is archaeological evidence for late fourth-millennium cal BC northward connections along the Atlantic façade (see below).

The morphometric results confirmed that the Orkney vole teeth, like those on Guernsey, differ significantly from modern *Microtus arvalis* teeth on the continent, and it appears that no meaningful information on the point of origin of the Orcadian specimens could be obtained through this approach.

The genetic results, when closely interrogated, do not in fact prove that the Orkney vole arrived from the coast of Belgium/northern France. They 'confirm that *M. arvalis* of the same (Western-North) *cytb* lineage found today in France, Belgium, the Netherlands, Germany and Switzerland also occurs on Orkney' *but* 'despite detailed sampling of the Western-North lineage in general and along the coast of France and Belgium in particular (fig. 2), we found no *cytb* sequences there that are clearly ancestral to those of the Orkney voles' (Martínovká *et al.* 2013, 5212). Their figure 3, which presents a median-joining network showing all modern and all ancient analysed DNA haplotypes in the Western-North phylogroup, actually emphasises the distinctiveness of the Orcadian results. The basis of Martínovká *et al.*'s claim that the Orkney vole arrived from the coast of Belgium or northern France is a goodness-of-fit statistical modelling of the data, based on the assumption that 'one of the sampled continental populations is closest to the source of the Orkney voles'. What they discovered was that, for northern France, Belgium and nearby areas of Germany, the modern sampled populations *that are least genetically dissimilar to the ancient Orkney vole population* are to be found on the Belgian coast and on adjacent parts of the north French coast. This is most emphatically not the same as demonstrating that the Orkney vole arrived in Orkney from that region.

A particular weakness in the work of Martínovká *et al.* is that there are so few results for ancient non-Orcadian *Microtus arvalis*. Practically all of the analysis of the continental genetic data has been based on modern DNA and has consequently had to take into account the possibility of variation over time (including that relating to a dramatic population fluctuation on the continent: Martínovká *et al.* 2013, 5216–17). The reason for this paucity of ancient data was the difficulty of finding the relevant archaeological specimens on the continent (Keith Dobney, pers. comm.). Similarly, the paucity of specimens taken from Iberia, and the absence of specimens from Brittany west of the Morbihan, is to be regretted. Until and unless more ancient specimens can be located and the geographical gaps in the ancient and modern coverage can be filled, it cannot be concluded that the question of the origin of the Orkney vole has been resolved.

Archaeologically, there is indeed evidence for long-distance seaborne travel during the late fourth millennium cal BC, linking Orkney with the wider world. The idea that the ambitious elite of Orkney visited the Boyne Valley on the east coast of Ireland, and brought back ideas about monument design and cosmology which they used in their strategy of competitive conspicuous consumption, is well established (Schulting *et al.* 2010, 39–41). Similarly, evidence from passage tomb design, from megalithic art and from specific artefacts found at the major Newgrange and Knowth passage tombs in the Boyne Valley, strongly indicates that the builders and users of these massive monuments also engaged in 'cosmological acquisition' through long-distance journeys to the Morbihan region of Brittany and to the west coast of Iberia (Eogan 1979; Eogan 1986, 166–72; Stout and Stout 2008, 67–83). This evidence includes the remodelling of the Knowth 1 West tomb into the shape of an *allée coudée* (angled passage tomb); the inclusion in that tomb of a carving strikingly reminiscent of the 'guardian' figure from the *allée coudée* at Luffang in the Morbihan (Herity 1974, fig. 136); the presence of a carved stone 'phallus', comparable to examples around the Tagus estuary, near the entrance to the Knowth 1 West tomb (Eogan 1979); and the discovery, in the cairn of the main tomb at Newgrange, of a carved pebble comparable to 'idols' found in north-west Iberia (Stout and Stout 2008, fig. 36).

Within this scenario, in which long-distance sea travel was undertaken as a deliberate strategy by elites to maintain and enhance their position in society, it is not beyond the bounds of possibility that a sea journey between Brittany or Iberia and Orkney had occurred. The deliberate transportation of voles as a source of food would, in the circumstances, be a sensible strategy, comparable to that used by Roman seafarers using edible dormice, and perhaps also Polynesians in the case of Pacific rats (Martínovká *et al.* 2013, 5215–16). By contrast, there is no archaeological evidence suggesting any connection whatsoever between the late fourth millennium cal BC inhabitants of the Belgian/Nord-Pas de Calais coast and Orkney; the archaeological record in both areas is sufficiently clear for it to be unnecessary to follow Martínovká *et al.*'s advice to make 'cultural linkages between Belgium and Orkney . . . [a] worthwhile focus for future archaeological investigation' (Martínovká *et al.* 2013, 5215). From an archaeological perspective, therefore, it seems justified to take the conclusions of Martínovká *et al.*'s study with more than just a pinch of salt.

Projet JADE: the benefits of a holistic approach

In contrast with the Orkney vole study, where a scientific programme of analysis appears to have been undertaken without sufficient consideration of current archaeological knowledge, Projet JADE (Pétrequin *et al.* 2012a) set out to construct a meaningful research programme that integrated 'hard science' with archaeological and anthropological perspectives. Funded by the Agence Nationale de la Recherche between 2006 and 2009 but building on over 15 years' previous fieldwork, and involving an international team of researchers, Projet JADE investigated the production, circulation, date, use, deposition and meaning of artefacts (principally axeheads) made from jadeitite and other Alpine rocks. In formulating their research strategy, Pierre and Anne-Marie Pétrequin drew on their many years of ethno-archaeological research among the last makers and users of stone axeheads in New Guinea, paying particular attention to the special significance held by certain axeheads within their belief system (Pétrequin and Pétrequin 2000; 2012). Axeheads made from stone extracted from high in the mountains were believed to be imbued with the power of the ancestral beings who resided there. It was for this reason that the search for the Alpine source areas was not restricted to the torrent beds well below the summits of Mont Viso and Mont Beigua in north Italy – the area where geologists (for example, Ricq-de Bouard 1996) had previously argued that raw material had been collected, assuming a 'least-effort' approach to resource procurement by the axehead makers – but instead extended up to the highest slopes. It was there that the first traces of exploitation in the Mont Viso and Mont Beigua massifs were discovered in 2003, at altitudes between 1500 and 2400 metres above sea level; further high-altitude (up to 2500m) extraction and working sites have been discovered ever since, not only around Mont Viso and Mont Beigua but also, as regards nephrite use, in the Swiss Valais (Pétrequin *et al.* 2012b, 2012c). Similarly, knowing that relatively few Alpine axeheads had been found in conventional archaeological contexts such as settlements and graves, they analysed the landscape setting of so-called 'stray' finds and inferred that the deposition pattern was not random, but instead tended to show an association with watery areas and other locations that may have been attributed a special significance in prehistory. Using this insight, they correctly predicted where further deposits of finished objects and roughouts were likely to be found.

The overall approach of the project was systematic, taking the entirety of the extensive European distribution of large axeheads (over 135mm long) and of other Alpine artefacts as its study area and involving intensive, valley-by-valley prospection within the Alpine source areas, allied with intensive recording of all artefacts made of Alpine rock. For Britain, Ireland, the Isle of Man and the Channel Islands, all Alpine axeheads, not just long examples, were included. All existing dating and contextual information relating to the use and deposition of Alpine stone artefacts was collated, and new radiocarbon dates were obtained for the extraction and working of the rock, thereby allowing an overall picture to be built up showing chronological trends in the manufacture of specific types of Alpine stone artefact and in the use of jadeitite, eclogite, omphacitite and (in Valais and Grisons) nephrite. The *chaîne opératoire* of the manufacture of axeheads and other Alpine stone objects was not only investigated but also replicated through experimental work. And the significance of the Alpine artefacts to their users was investigated through a fine-grained study of their cultural contexts across Europe, and by examining their impact on local axehead manufacture.

'Hard science' was deployed in Projet JADE not only in the use of radiocarbon dating but also in the non-destructive identification of the raw materials. This made it possible to determine how far individual objects had travelled and, in combination with information on shape and dating, to compile distribution maps showing the changing patterns of resource use, as noted above. The technique used was scattered reflectance spectroradiometry (Errera *et al.* 2012), a method adopted from the field of remote sensing, in which the amount of light reflected by the specimen is compared with that reflected by a pure white reference material. An extensive database was compiled, featuring around 2500 readings taken from well provenanced raw material samples, from specimens of working debris and from over 1800 axeheads from across Europe. The results were analysed and synthesised into a set of 'spectrofacies' that were characteristic of specific source areas and locations. An individual artefact's composition can be compared with this database, using curve-fitting statistical software to arrive at best-fit matches and to assess the most likely source area. The validity of this approach was confirmed by cross-checking the results with those obtained in a number of cases by other methods – principally specific gravity measurement, compositional analysis and petrological thin-sectioning. Many axeheads had previously been thin-sectioned, especially during the 1950s and 1960s (see, for example, Smith 1963) and Projet JADE was able to provide the all-important thin-section slides of raw material samples. While it is clear that not every deposit of jadeitite and other Alpine rock has been sampled, the spectroradiometric database is sufficiently extensive to permit the reliable provenancing of raw materials, often to the level of an individual valley, in most cases.

The results obtained by Projet JADE have revolutionised our understanding of Alpine stone artefacts and their users. Since these results have been published in detail (Pétrequin *et al.* 2012a and see Pétrequin *et al.* 2013 for the most recent statement published in English), only the highlights will be offered here. An initial period of relatively local, utilitarian use from around 5300 cal BC (mostly using dark-coloured eclogites and omphacitites) was followed by an expansion in the distribution of Alpine stone artefacts around 4700 cal BC when long, probably non-utilitarian adzeheads (of Bégude type), again mostly of eclogites and omphacitites, circulated between Italy and Brittany, probably as a response to the similarly shaped adzeheads of amphibolite that were circulating between the Czech Republic and the Netherlands (Pétrequin *et al.* 2012d, 216). A further expansion, and an increase in the use of jadeitite, can be traced during the peak period of Alpine stone exploitation, between 4600 cal BC and 3700 cal BC. During this time, some axeheads travelled up to 1700 km from their raw material source areas, moving both to the north-west and to the east of the Alps. They crossed cultural (and probably also linguistic) boundaries and were used by diverse societies, some highly inegalitarian – as in the Morbihan region of south-east Brittany (Cassen *et al.* 2012) and, at the other end of Europe, at Varna in Bulgaria (Pétrequin *et al.* 2012e). Exploitation of the finest, most sought after, light-green jadeitite was probably controlled and, during the second half of the fifth millennium cal BC, jadeitite seems to have played a major role in the value system that operated in western Europe, comparable to that played by copper in eastern and south-eastern Europe (Pétrequin *et al.* 2013). As far as the appearance of Alpine axeheads in Britain and Ireland is concerned, there is no unequivocal evidence to suggest that this occurred much before 3807/3806 BC, when one was deposited beside the Sweet Track in Somerset (Figure 19.1); the most plausible scenario, at least in the opinion of one of us (AS), is that most arrived, during the first two centuries of the fourth millennium cal BC, as the treasured, and in many cases ancient, possessions of small immigrant farming communities from various points along the north French coast (Sheridan *et al.* 2010, 2011; Sheridan and Pailler 2012). While these people probably did not know where the Alps were, their Alpine axeheads may well have been accompanied by legends about their origin; and the decision to seek out rock, preferably greenish, from mountainous sources including Great Langdale in Cumbria may well have resulted from the desire to perpetuate the belief in the divine power of stone from such liminal areas, between the world of the living and that of the ancestors and gods.

In addition to delineating the 'big picture' of Alpine stone use, Projet JADE has also been able to trace detailed biographies of individual axeheads, thanks partly

Figure 19.1 Left: Axehead of jadeitite (omphacitic, quartzo-feldspathic, micaceous and slightly retromorphosed) from Mont Viso, found beside the Sweet Track, and deposited around 3807/3806 BC (according to dendro-dates for the trackway's construction). The axehead had probably been reshaped in the Morbihan and had travelled around northern France before being brought to England. Photo: Pierre Pétrequin/Projet JADE. Right: The Sweet Track axehead *in situ* beside the trackway. Photo: Bryony and John Coles.

to the identification of their raw materials and partly to the systematic recording of shape, size and degree of polish across Europe. It is clear that a significant number were modified during their use-life, usually through re-shaping and often through polishing to a glassy finish (which was the case with many of the examples found in Britain and Ireland: Figure 19.2). This is particularly obvious with the examples that were acquired by the elite in the Morbihan from around 4500 cal BC, who had many of them thinned until they became partly translucent, and also had some perforated near their butt, for suspension. It has been suggested that they played a particular role as sacred objects in the rituals associated with leadership by a small number of men. While many of these axeheads ended in the tombs of the elite or as hoard-offerings, some travelled away from the Morbihan, ending up as far away as north-west and north-east Spain, northern Germany, and even southern Italy (Pétrequin *et al.* 2012d, fig. 2). Similarly, copies of Morbihannais axeheads, made in local stone, are known from north-west Iberia (Type Cangas) and in Switzerland (the Zug family of axeheads: Pétrequin *et al.* 2012d, fig. 3).

It is clear, therefore, that Projet JADE offers an outstanding example of where 'hard science' has been integrated successfully and seamlessly within an approach that is informed and guided by archaeological and anthropological considerations. The concept of axeheads as 'object-signs', with specific social and ideological meanings that could change over time and space, is as important to our understanding of these objects as is the information on their raw material and provenance. The reason why the approach taken by Projet JADE has been so fruitful is its genuinely interdisciplinary nature. The right questions have been posed, and the right methods have been applied to addressing those questions, with due concern for maintaining rigour and testing working hypotheses at all stages of the project.

Conclusions

It is hoped that this contribution has shown the benefits, and some of the pitfalls, of using 'hard science' techniques to construct archaeological narratives. The contribution that can be made by scientific analysis is immense; what is needed is a critical awareness of what it (and indeed archaeology) can, and cannot, deliver. At the heart of the matter is the fundamental need to understand both the archaeological issue and the scientific technique or techniques that can be brought to bear on any given problem. The appropriate questions need to be formulated, the appropriate methods need to be used to address those questions, and multiple working hypotheses need to be constructed and tested rigorously.

Figure 19.2 Examples of Alpine axeheads with glasssy polish, found in Britain. The polishing is likely to have been undertaken before the axeheads arrived in Britain but a considerable time after they had originally been made; perhaps it related to beliefs about enhancing the supernatural power attributed to the axeheads. Top: Allegedly found in a logboat on the banks of the Clyde near St. Enoch's Church, Glasgow. Of eclogite from the Bulè valley, Mont Viso, north Italian Alps, this axehead was unusually (and uniquely) kept in its haft when parts of the surface were polished to a high sheen. Bottom: Of jadeitite from Mont Viso, allegedly found in Canterbury, Kent. Photos: Pierre Pétrequin/Projet JADE.

Acknowledgements

John and Bryony Coles are thanked for allowing their photograph of the Sweet Track axehead *in situ* to be reproduced, and Rick Schulting is thanked for his comments on an earlier draft of this chapter.

References

Balasse, M., Tresset, A. and Ambrose, S.H. 2006. Stable isotope evidence ($\delta^{13}C$, $\delta^{18}O$) for winter feeding on seaweed by Neolithic sheep of Scotland. *Journal of Zoology* 270, 170–6.

BBC News for north-east Scotland, Orkney and Shetland. 2013. http://www.bbc.co.uk/news/uk-scotland-north-east-orkney-shetland-23942934. Accessed November 2013.

Becket, A. and MacGregor, G. 2009. Forest grazing and seaweed foddering: early Neolithic occupation at Maybole, South Ayrshire. *Proceedings of the Society of Antiquaries of Scotland* 139, 105–22.

Bonsall, C., Macklin, M.G., Anderson, D.E. and Payton, R.W. 2002. Climate change and the adoption of agriculture in north-west Europe. *European Journal of Archaeology* 5, 9–23.

Cassen, S. and François, P. 2009. Classements et diagnoses de la production céramique à la Table des Marchands. In S. Cassen (ed.), *Autour de la Table: explorations archéologiques et discours savants sur des architectures néolithiques à Locmariaquer, Morbihan (Table des Marchands et Grand Menhir)*, 491–567. Nantes: Laboratoire de recherches archéologiques, CNRS et Université de Nantes.

Cassen, S., Boujot, C., Dominguez Bella, S., Guiavarc'h, M., Le Pennec, C., Prieto Martinez, M.P., Querré, G., Santrot, M.-H. and Vigier, E. 2012. Dépôts bretons, tumulus carnacéens et circulations à longue distance. In P. Pétrequin, S. Cassen, M. Errera, L. Klassen, J.A. Sheridan and A.-M. Pétrequin (eds), *JADE. Grandes haches alpines du Néolithique européen. Ve et IVe millénaires av. J.-C.*, 918–95. Besançon: Presses Universitaires de Franche-Comté (Les cahiers de la MSHE Ledoux nº 17).

Clarke, D.V. 2012. *Skara Brae: the official souvenir guide.* Edinburgh: Historic Scotland.

Collard, M., Edinborough, K., Shennan, S. and Thomas, M. G. 2010. Radiocarbon evidence indicates that migrants introduced farming to Britain. *Journal of Archaeological Science* 37, 866–70.

Cramp, L.J.E. and Evershed, R.P. 2014. Reconstructing aquatic resource exploitation in human prehistory using lipid biomarkers and stable isotopes. In T.E. Cerling (ed.), *Treatise on geochemistry. Volume 12: Archaeology and anthropology* 14, 319–39. Oxford: Elsevier.

Cramp, L.J.E., Jones, J., Sheridan, J.A., Smyth, J., Whelton, H., Mulville, J., Sharples, N.M. and Evershed, R.P. 2014. Immediate replacement of fishing with dairying by the earliest farmers of the NE Atlantic archipelagos. *Proceedings of the Royal Society B* 281, no. 1780, http://rspb.royalsocietypublishing.org/content/281/1780/20132372.full.

Eogan, G. 1979. Objects with Iberian affinities from Knowth. *Revista de Guimarães* 89, 275–9.

Eogan, G. 1986. *Knowth and the passage tombs of Ireland.* London: Thames and Hudson.

Errera, M., Pétrequin, P. and Pétrequin, A.-M. 2012. Spectroradiométrie, réferentiel naturel et étude de la diffusion des haches alpines. In P. Pétrequin, S. Cassen, M. Errera, L. Klassen, J.A. Sheridan and A.-M. Pétrequin (eds), *JADE. Grandes haches alpines du Néolithique européen. Ve et IVe millénaires av. J.-C.*, 440–533. Besançon: Presses Universitaires de Franche-Comté (Les cahiers de la MSHE Ledoux n° 17).

Evershed, R.P. 2007. Exploiting molecular and isotopic signals at the Mesolithic–Neolithic transition. In A. Whittle and V. Cumming (eds), *Going over: the Mesolithic–Neolithic Transition in north-west Europe*, 141–64. London: Oxford University Press for the British Academy.

Haynes S., Jaarola, M. and Searle, J.B. 2003. Phylogeography of the common vole (Microtus arvalis) with particular emphasis on the colonization of the Orkney archipelago. *Molecular Ecology* 12, 951–6.

Hensey, R. and Bergh, S. 2013. 'The inns at Sligo are better than those at Auray . . . and the scenery far more beautiful': Carrowmore re-visited. In M.A. Timoney (ed.), *Dedicated to Sligo: thirty-four essays on Sligo's past*, 41–3. Keash: Publishing Sligo's Past.

Herity, M. 1974. *Irish passage graves: Neolithic tomb-builders in Ireland and Britain 2500 BC*. Dublin: Irish University Press.

Jones, J.R. 2013. *Land and sea: understanding diet and economies through time in the North Atlantic Islands*. Unpublished PhD thesis, Cardiff University.

Lancaster, S., Davis, A., Hastie, M., McCulloch, R., Smith, C. Timpany, S. and Tipping, R. 2009. Palaeoenvironmental synthesis. In H. K. Murray, J. C. Murray and S. M. Fraser, *A tale of the unknown unknowns: a Mesolithic pit alignment and a Neolithic timber hall at Warren Field, Crathes, Aberdeenshire*, 42–50. Oxford: Oxbow Books.

Lawrence, D.M. 2006. Neolithic mortuary practice in Orkney. *Proceedings of the Society of Antiquaries of Scotland* 136, 47–59.

Lawrence, D.M. 2012. *Orkney's first farmers. Reconstructing biographies from osteological analysis to gain insights into life and society in a Neolithic community on the edge of Atlantic Europe*. Unpublished PhD thesis, University of Bradford.

MacKie, E. 1972. Radiocarbon dates for two Mesolithic shell heaps and a neolithic axe-factory in Scotland. *Proceedings of the Prehistoric Society* 38, 413–15.

Martínovká, N., Barnett, R., Cucchi, T., Struchen, R., Pascal, M., Pascal, M., Fischer, M.C., Higham, T., Brace, S., Ho, S.Y.W., Quéré, J.-P., O'Higgins, P., Excoffier, L., Heckel, G., Hoelzel, A.R., Dobney, K.M. and Searle, J.B. 2013. Divergent evolutionary processes associated with colonization of offshore islands. *Molecular Ecology* 22, 5205–20.

Melton, N.D. and Montgomery, J. 2009. Sumburgh Early Neolithic burial cist: combined isotope analysis. *Discovery and Excavation in Scotland* 10, 166.

Milner, N., 2010. Subsistence at 4000–3700 cal BC: landscapes of change or continuity? In B. Finlayson and G. Warren (eds), *Landscapes in transition*, 46–54. Oxford: Oxbow Books.

Milner, N., Craig, O.E., Bailey, G.N., Pedersen, K. and Anderson, S.H. 2004. Something fishy in the Neolithic? A re-evaluation of stable isotope analysis of Mesolithic and Neolithic coastal populations. *Antiquity* 78, 9–22.

Milner, N., Craig, O.E., Bailey, G.N. and Anderson, S.H. 2006. A response to Richards and Schulting. *Antiquity* 80, 456–8.

Montgomery, J., Cooper, R.E. and Evans, J.E. 2007. Foragers, farmers or foreigners? An assessment of dietary strontium isotope variation in Middle Neolithic and Early Bronze Age East Yorkshire. In M. Larsson and M. Parker Pearson (eds), *From Stonehenge to the Baltic: living with cultural diversity in the third millennium BC*, 65–75. Oxford: Archaeopress.

Montgomery, J., Beaumont, J., Jay, M., Keefe, K., Gledhill, A.R., Cook, G.T., Dockrill, S.J. and Melton, N.D. 2013. Strategic and sporadic marine consumption at the onset of the Neolithic: increasing temporal resolution in the isotope evidence. *Antiquity* 87, 1060–72.

Pétrequin, P. and Pétrequin, A.-M. 2000. *Ecologie d'un outil : la hache de pierre polie en Nouvelle-Guinée*. Paris: CNRS Editions, first edition 1993 (Monographies du CRA n° 12).

Pétrequin, P. and Pétrequin, A.-M. 2012. Les modèles ethnoarchéologiques de Nouvelle-Guinée. In P. Pétrequin, S. Cassen, M. Errera, L. Klassen, J.A. Sheridan and A.-M. Pétrequin (eds), *JADE. Grandes haches alpines du Néolithique européen. Ve et IVe millénaires av. J.-C.*, 27–45. Besançon: Presses Universitaires de Franche-Comté (Les Cahiers de la MSHE Ledoux n° 17).

Pétrequin, P., Cassen, S., Errera, M., Klassen, L., Sheridan, J.A. and Pétrequin, A.-M. (eds) 2012a. *JADE. Grandes haches alpines du Néolithique européen. Ve et IVe millénaires av. J.-C.* Besançon: Presses Universitaires de Franche-Comté (Les Cahiers de la MSHE Ledoux n° 17).

Pétrequin, P., Pétrequin, A.-M., Errera, M. and Prodéo, F. 2012b. Prospections alpines et sources de matières premières. Historique et résultats. In P. Pétrequin, S. Cassen, M. Errera, L. Klassen, J.A. Sheridan and A.-M. Pétrequin (eds), *JADE. Grandes haches alpines du Néolithique européen. Ve et IVe millénaires av. J.-C.*, 46–183. Besançon: Presses Universitaires de Franche-Comté (Les Cahiers de la MSHE Ledoux n° 17).

Pétrequin, P., Croutsch, C., Errera, M., Honegger, M., Jaccottey, L., Mariétoz, F. and Rey, P.-J. 2012c. Approche des productions valaisannes en amphibolite calcique (néphrite). In P. Pétrequin, S. Cassen, M. Errera, L. Klassen, J.A. Sheridan and A.-M. Pétrequin (eds), *JADE. Grandes haches alpines du Néolithique européen. Ve et IVe millénaires av. J.-C.*, 184–213. Besançon: Presses Universitaires de Franche-Comté (Les Cahiers de la MSHE Ledoux n° 17).

Pétrequin, P., Errera, M., Martin, A., Fábregas Valcarce, R. and Vaquer, J. 2012d. Les haches en jades alpins pendant les Ve et IVe millénaires. L'exemple de l'Espagne et du Portugal dans une perspective européenne. In M. Borrell, F. Borrell, J. Bosch, X. Clop and M. Molist (eds), *Actes Congrés Internacional, 'Xarxes al neolític. Circulació i intercanvi de matèries, productes i idees a la Mediterrània occidental (VII–III millenni aC), Gavà/Bellaterra, 2–4/2/11*, 213–22. Gavà: Institut Municipal de Gestió del Patrimoni Cultural i Natural (*Rubricatum. Revista del Museu de Gavà*).

Pétrequin, P., Cassen, S., Errera, M., Tsonev, T., Dimitrov, K., Klassen, L. and Mitkova, R. 2012e. Les haches en « jades alpins » en Bulgarie. In P. Pétrequin, S. Cassen, M. Errera, L. Klassen, J.A. Sheridan and A.-M. Pétrequin (eds), *JADE. Grandes haches alpines du Néolithique européen. Ve et IVe millénaires av. J.-C.*, 1231–79. Besançon: Presses Universitaires de Franche-Comté (Les Cahiers de la MSHE Ledoux n° 17).

Pétrequin, P., Cassen, S., Errera, M., Klassen, L., Pétrequin, A.-M. and Sheridan, J.A. 2013. The value of things : the production and circulation of Alpine jade axes during the 5th–4th millennia in a European perspective. In T. Kerig and A. Zimmermann (eds), *Economic archaeology: from structure to performance in European archaeology*, 65–82. Bonn: Habelt.

Richards, M.P. and Schulting, R.J. 2006. Touch not the fish: the Mesolithic–Neolithic change of diet and its significance. *Antiquity* 80, 444–56.

Richards, M.P., Schulting, R.J. and Hedges, R.E.M. 2003. Sharp shift in diet at onset of Neolithic. *Nature* 425, 366.

Ricq-de Bouard, M. 1996. *Pétrographie et sociétés néolithiques en France méditerranéenne. L'outillage en pierre polie*. Paris: CNRS editions (Monographie du CRA 16).

Schulting, R.J. 2012. Skeletal evidence for interpersonal violence: beyond mortuary monuments in southern Britain. In R.J. Schulting and L. Fibiger (eds), *Sticks, stones and broken bones: Neolithic violence in a European perspective*, 223–48. Oxford: Oxford University Press.

Schulting, R.J. 2013. On the northwestern fringes: earlier Neolithic subsistence in Britain and Ireland as seen through faunal remains and stable isotopes. In S. Colledge, J. Conolly, K. Dobney, K. Manning and S.J. Shennan (eds), *The origins and spread of domestic animals in southwest Asia and Europe*, 313–38. Walnut Creek: Left Coast Press.

Schulting, R.J. and Richards, M.P. 2002. The wet, the wild and the domesticated: the Mesolithic–Neolithic transition on the west coast of Scotland. *European Journal of Archaeology* 5, 147–89.

Schulting, R.J., Sheridan, J.A., Crozier, R. and Murphy, E. 2010. Revisiting Quanterness: new AMS dates and stable isotope data from an Orcadian chamber tomb. *Proceedings of the Society of Antiquaries of Scotland* 140, 1–50.

Schulting, R.J., Fibiger, L., Macphail, R., McLaughlin, R., Murray, E., Price, C. and Walker, E.A. 2013. Mesolithic and Neolithic human remains from Foxhole Cave, Gower, South Wales. *The Antiquaries Journal* 93, 1–23.

Shennan, S.J. and Edinborough, K. 2007. Prehistoric population history: from the Late Glacial to the Late Neolithic in central and northern Europe. *Journal of Archaeological Science* 34, 1339–45.

Sheridan, J.A. 2010. The Neolithisation of Britain and Ireland: the big picture. In B. Finlayson and G. Warren (eds), *Landscapes in transition*, 89–105. Oxford: Oxbow Books.

Sheridan, J.A. 2011a. The Early Neolithic of south-west England: new insights and new questions. In S. Pearce (ed.), *Recent archaeological work in south western Britain: papers in honour of Henrietta Quinnell*, 21–40. Oxford: Archaeopress.

Sheridan, J.A. 2011b. Comments on V. Cummings and O. Harris, Animals, people and places: the continuity of hunting and gathering practices across the Mesolithic–Neolithic transition in Britain. *European Journal of Archaeology* 14, 389–91.

Sheridan, J.A. 2012a. Review of A. Whittle, F. Healy and A. Bayliss, *Gathering time: dating the early Neolithic enclosures of southern Britain and Ireland*. *Antiquity* 86, 262–4.

Sheridan, J.A. 2012b. Neolithic Shetland: a view from the 'mainland'. In D.L. Mahler (ed.), *The border of farming and the cultural markers*, 6–36. Copenhagen: Nationalmuseet.

Sheridan, J.A. and Pailler, Y. 2011. La Néolithisation de la Grande-Bretagne et de l'Irlande: plusieurs processus, plusieurs modèles et des questions à l'attention de nos collègues français. In F. Bostyn, E. Martial and I. Praud (eds), *Le Néolithique du nord de la France dans son contexte européen. Habitat et économie aux 4e et 3^e^millénaires avant notre ère. (Actes du 29e colloque interrégional sur le Néolithique, Villeneuve-d'Ascq 2–3 octobre 2009)*, 13–30. Senlis: Société Archéologique de Picardie (*Revue archéologique de Picardie, numéro spécial* 28).

Sheridan, J.A. and Pailler, Y. 2012. Les haches alpines et leurs imitations en Grande-Bretagne, dans l'île de Man, en Irlande et dans les îles Anglo-Normandes. In P. Pétrequin,

S. Cassen, M. Errera, L. Klassen, J.A. Sheridan and A.-M. Pétrequin (eds), *JADE. Grandes haches alpines du Néolithique européen. V^e^ et IVe millénaires av. J.-C.*, 1046–87. Besançon: Presses Universitaires de Franche-Comté (Les Cahiers de la MSHE Ledoux n° 17).

Sheridan, J.A., Schulting, R.J., Quinnell, H. and Taylor, R., 2008. Revisiting a small passage tomb at Broadsands, Devon. *Proceedings of the Devon Archaeological Society* 66, 1–26.

Sheridan, J.A., Field, D, Pailler, Y., Pétrequin, P., Errera, M. and Cassen, S., 2010. The Breamore jadeitite axehead and other Neolithic axeheads of Alpine rock from central southern England. *Wiltshire Archaeological and Natural History Magazine* 103, 16–34.

Sheridan, J.A., Pailler, Y., Pétrequin, P. and Errera, M. 2011. Old friends, new friends, a long-lost friend and false friends: tales from *Projet JADE*. In R.V. Davis and M. Edmonds (eds), *Stone Axe Studies III*, 411–26. Oxford: Oxbow Books.

Smith, W.C. 1963. Jade axes from sites in the British Isles. *Proceedings of the Prehistoric Society* 29, 133–72.

Stout, G. and Stout, M. 2008. *Newgrange*. Cork: Cork University Press.

Thomas, J.S. 2008. The Mesolithic–Neolithic transition in Britain. In J. Pollard (ed.), *Prehistoric Britain*, 58–89. Oxford: Blackwell.

Tipping, R. 2010. The case for climatic stress forcing change in the adoption of agriculture in the British Isles. In B. Finlayson and G. Warren (eds), *Landscapes in transition*, 66–76. Oxford: Oxbow Books.

Tresset, A. and Vigne, J.-D. 2007. Substitution of species, techniques and symbols at the Mesolithic–Neolithic transition in Western Europe. In A. Whittle and V. Cumming (eds), *Going over: the Mesolithic–Neolithic transition in north-west Europe*, 189–210. London: Oxford University Press for the British Academy.

Whitehouse, N.J., Schulting, R.J., McClatchie, M., Barratt, P., McLaughlin, T.R., Bogaard, A., Colledge, S., Marchant, R., Gaffrey, J. and Bunting, M.J. 2013. Neolithic agriculture on the European western frontier: the boom and bust of early farming in Ireland. *Journal of Archaeological Science*, online 22 August 2013, http://dx.doi.org/10.1016/j/jas/2013.08.009.

Whittle, A., Healy, F. and Bayliss, A. 2011. *Gathering time: dating the early Neolithic enclosures of southern Britain and Ireland*. Oxford: Oxbow Books.

20

Doing Science in the Mesolithic, Neolithic and Copper Age

An Insider's Perspective

JOHN CHAPMAN

Introduction

FOR ARCHAEOLOGISTS, it was V. Gordon Childe (1936) who first identified the collective traditions of science, which, for him, were often synonymous with technology. For Childe (1936, 27), humans were born with relatively few inherited instincts – hence the importance of learning. Fresh experiences led to additions and/or modifications to the community's store of knowledge, with positive tests leading to their incorporation into a collective tradition and novel techniques added to the expanding rules of social tradition. More specifically, Childe (1936, 88) recognised the outstanding common features of a Neolithic economy – woodworking, pottery manufacture and a textile industry. For Childe (1936, 96), legitimate deductions from experience were inextricably mixed up with magic, with each operation of every craft accompanied by prescribed ritual acts.

In an outstanding paper, Childe (1949a, 305) proposed that 'a science then is not just a set of rules and formulae that describe things and their relations, but rather formulae that yield rules for action. The truth of the rules is guaranteed by the success of action'. He added: 'The structure of knowledge is social. The general principles – the categories of space, time, class, order, cause – on which the data of sensory experience are systematised and built up into sciences . . . have been elaborated by society and transmitted, ready-made, to all its members' (Childe 1949a, 305). In this way, Childe (1949a, 308) believed, science developed 'by a dual process of addition and subtraction', meaning that those actions that worked were retained while those ideas that failed to produce a result were abandoned. Despite the formulation of these fundamental concepts, Childe did not, however, discuss the importance of scientific or mathematical principles for Neolithic societies.

Lemonnier (1993) took a rather different direction in his discussion of the development of relations between technology and social action. For Lemonnier

Proceedings of the British Academy **198**, 391–417.

(1993, 5), meanings were embedded in the reproduction of an artefact. His example of the meanings added by one group to an exotic artefact traded from another group could be matched by the scientific principles embedded in the making of an object. Lemonnier (1993, 15) identifies the translation occurring between the socially defined technical aim of the artefact and the physical actions used by a culture to perform that function. He goes on (1993, 25) to elaborate two kinds of translation: (a) the translation of social strategies into the particular characteristics of the material culture to produce techniques and a technical system; and (b) the translation of physical laws (relations between material actions and material effects) into their own categories. It is this second translation of the physical characteristics of materials such as stone and clay into scientific principles summarising these characteristics that is of interest here. The relationship between material actions and material effects underlies notions of scientific causation that were recognised, in their own culture-bound ways, by prehistoric persons.

In this chapter, it is my aim to examine the practice of what I take to have been 'prehistoric science' through examples of a wide range of material culture, for the most part from the Balkan Mesolithic, Neolithic and Chalcolithic (for chronological parameters, see Table 20.1). After a review of dimensional order in hunter-gatherer material culture, which includes a discussion of the transition from the 'age of dwelling' to the 'age of building', I turn to David Keightley's (1987) outstanding article on 'The making of China', which enshrines an approach linking material culture to the cognitive. I examine examples of the four key design principles that

Table 20.1 Simplified chronological framework for the Balkan Neolithic and Chalcolithic

Period	*Dates (cal BC)*	*Principal social networks*
Late Mesolithic	8000–5500	Lepenski Vir, Schela Cladovei, Pobiti Kamuni, Soroki, Jászság
Early farmers	6300–5300	Karanovo I & II, First Macedonian Neolithic, West Bulgarian Painted wares, Starčevo, Körös, Criş
Mature farmers	5300–4600	Karanovo III & IV, Usoe, North Bulgarian MN, Vinča, Butmir, Hungarian LBK, Dudeşti, Vădastră, Boian, Hamangia, Pre-Cucuteni, Trypillia A
Climax Copper Age	4600–4000	Karanovo V & VI, Varna, Sava, Gumelniţa, Tisza-Herpály- Csőszhalom, Tiszapolgár, Lengyel, Cucuteni A, AB and B; Trypillia B1 & B2
Post-Climax Copper Age	4000–3000	Karanovo VII, Cernavodă I, Bodrogkeresztúr, latest Cucuteni, Trypillia C1 & C2, Coţofeni, Salcuţa IV, Baden

Keightley enunciates: standardisation, compartmentalisation, symmetry and precision. I then consider the domains of weaving, numerology and astronomy before seeking to integrate some of the implications of these social practices.

Dimensional order in hunter-gatherer societies

In an important contribution to the debate on personhood, Hernando *et al.* (2011, 193) have argued that

> societies where the only functional divisions are those determined by gender also tend to be characterized by other important features: namely, the lack of scientific models capable of explaining the world on mechanistic principles, and a level of technology that does not afford mastery over natural phenomena. Lack of control over nonhuman nature, combined with behavioural homogeneity within each gender, results in a relational or interdependent construction of personhood.

There is a sense in which Hernando and her co-authors link these notions to hunter-gatherers in their evolutionary scheme of the origins of gender inequality. But can this be supported with recourse to data from the Palaeolithic and Mesolithic?

A consideration of geometric precision in the Palaeolithic takes us to the three-dimensionally symmetrical designs well beyond functional requirements of Late Acheulean hand-axe making (700K–250K BP; Gowlett 1984). These hand-axes exemplify the use of the two kinds of body symmetry (left–right, back–front) as a template for forms of symmetrical material culture (Gaydarska *et al.* 2007). Even more stunning bifacially symmetrical tools ('laurel-leaves') were produced in the Solutrean (21K–17K BP; Gamble 1999). Two aspects of geometric order are found in the Eastern Gravettian of Moldavia and the Ukraine, dated to 30K–20K BP (Soffer 2003) – both containers as well as instruments (*pace* Gamble 2007). The mammoth-bone structures are some of the world's earliest outdoor, free-standing structures, with symmetrical, rectangular ground-plans. The collection of basket and cord impressions from Pavlov and Dolní Vestonice (Soffer *et al.* 2000) represents another form of precision design based upon cordage and plaited techniques (Soffer *et al.* 2000).

In the European Mesolithic, in addition to the symmetry and precision of Azilian bone and antler harpoons, the widespread use of composite tools with individual microlithic elements relied upon precision and compartmentalisation in the bilateral symmetry of the design of the slots for the inserts and the geometric regularity of the inserts themselves. It does seem curious that past emphasis on the typology of geometric microliths (Rozoy 1989, Fischer 1989) has diverted attention away from the cognitive implications of materialisation itself.

This brief selection of Pleistocene and Holocene examples indicates that geometric order and symmetrical design principles were widespread in hunter-gatherer lifeways well before early farming times and that a narrow range of particularly fine objects relied on geometric order for their efficacy and visual impact.

It is axiomatic that the change in use from instruments to containers (Gamble 2007, 193–204) led to a major increase in the importance of geometric design principles, themselves embedded in the form of three-dimensional containers. The development of containers created in their users a range of appropriate practices which consistent repetition would convert into learned behaviour. As long as the practices continued, there was a likelihood of continuity in the making of such containers. There can be little doubt that such *habitus* was present in at least some communities of Balkan Mesolithic hunter-gatherers before the advent of farming, as we see at Lepenski Vir.

From the 'age of dwelling' to the 'age of building'

The development of building traditions was one of the defining characteristics of Neolithic lifeways, with Borić (2008) viewing the Neolithic as a change from dwelling to building. Yet most archaeologists take the basic structure of Neolithic houses for granted in descriptive accounts that emphasise shared cultural traits rather than geometric ways of constructing the world (e.g. Benac 1973). Yet, for phenomenologists such as Husserl, carpentry is a geometrical praxis (cited in Vastokas 1994), while, for Ihde (1990), the processes and products of carpentry gave rise to geometric thought without the intervention of language. Vastokas (1994) contrasts the world of the North-West Coast Indians, in which carpenters made plank-built rectangular houses, leading to the configuration of the cosmos as rectangular, with the circular and open-ended worlds of the interior North American tribes, with their sewing and lashing of flexible hides, barks and poles, leading to a circular cosmology. Such a similarly significant cosmological principle is surely materialised in the circularity of houses and sites in Neolithic to medieval north-west Europe (Bradley 1998). Crump (1990, 139) recognises that the underlying principle of mediaeval architecture was divine order, seen from the time of Augustine of Hippo in terms of numbers. This principle understood modulation as a concrete form of abstract numeration, generating its own aesthetic, with inherent beauty resulting from measure and number.

Harmonious ratios

The discovery of the Iron Gates Mesolithic in the 1960s (Srejović 1969, Boroneanţ 1970) was a breakthrough of enormous significance for our understanding of the Balkan Mesolithic, which hitherto had been known from a handful of cave sites and surface collections from sandy areas (Chapman 1989). The most important site, Lepenski Vir, on the right bank of the Danube, was also the site with the largest and best preserved set of trapezoidal houses (Figure 20.1). Unusual in form for

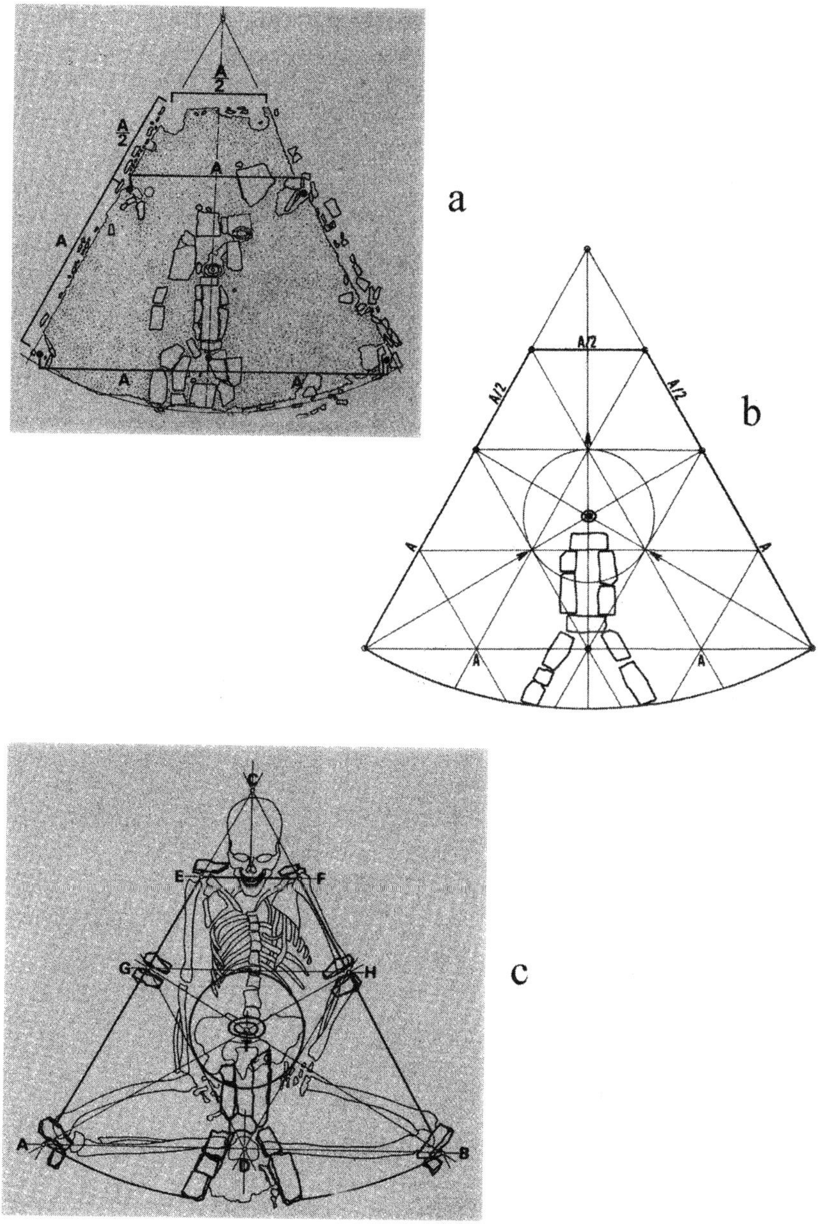

Figure 20.1 Trapezoidal houses at Lepenski Vir: (a) plan of House 37, as excavated; (b) schematic rendering of symmetries of House 37; (c) 'alla turca' burial superimposed upon the plan of House 37, showing links between inhumation and house plan (source: Srejović and Babović 1983, Crteža 17).

either Mesolithic or Neolithic structures, these houses were symmetrically planned with a central long axis marked by a stone-built rectangular hearth, a short rear wall, two straight side walls of the same dimensions and a curved front wall (Srejović 1972, fig. 8). The famous monumental stone sculptures were often placed on the long axis of the houses (Srejović and Babović 1983).

It was the excavator, Dragoslav Srejović, who first noticed that the dimensions of the houses at Lepenski Vir possessed harmonious proportions akin to that of the Golden Section. All of the structures at Lepenski Vir were made with the same trapezoidal form, based upon the sector of a circle with a subtended angle of 60° (Srejović 1972, 50–2). Srejović suggests that the inhabitants probably used a system of triangulation to produce a regular isosceles triangle, with the back wall in a ratio of 1:3 to the length of the side walls and 1:4 in relation to the front wall (Srejović 1972, fig. 8). A length:width plot of the dimensions of a sample of 47 well preserved houses from Lepenski Vir (Figure 20.2) indicates the high degree of dimensional order maintained over a period of over 700 years, during which the trapezoidal form presenced both the gorge's landscape and its mortuary zone.

David Keightley and 'the making of China'

In his essay on 'The making of China', David Keightley made the striking claim that 'people who make their pottery differently live differently' (Keightley 1987, 110). Keightley attempted to infer from the contrasting ceramic assemblages a sense of the differing structure and content of the cognitive activities that underlay the behaviour of China's Neolithic inhabitants on the East Coast and in the inland North West regions (1987, 93). He identifies a correlation between enhanced differentiation and increased order between material and cognitive realms in the East Coast region and an absence of such change in the North West. Specifically, two notions

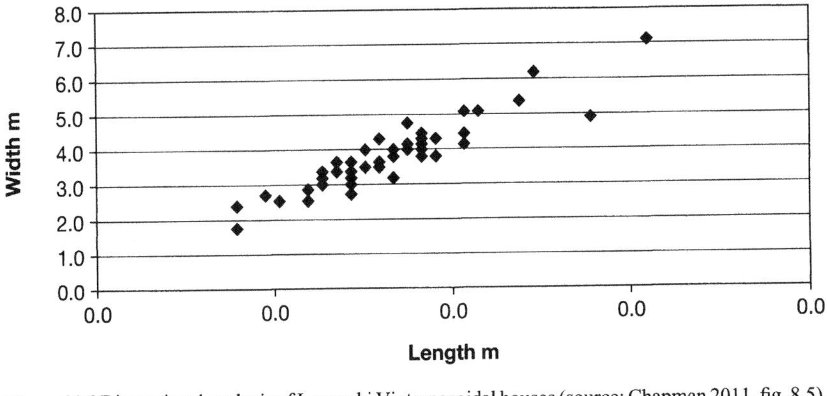

Figure 20.2 Dimensional analysis of Lepenski Vir trapezoidal houses (source: Chapman 2011, fig. 8.5).

are introduced which would imply the emergence of new concepts of cultural order. First, there is the notion that the addition of lips, spouts, handles, lids and lugs to basic ceramic forms introduces constraints on pottery use, leading to choices between appropriate and incorrect behaviour. Secondly, there is the idea that the production of carefully fitting lids and legs of exactly the same dimensions led to a more mathematical view of the world, in which specifications are followed more precisely than before (cf. Wright 1993 for Harappan India; Chapman and Gaydarska 2006, Chapter 2 for the Balkan Neolithic).

Standardisation

An excellent example of the design principle of standardisation concerns the Boldogkőváralja lithic hoard, Hungary, dated *c.* 5000 cal BC – a set of 567 limno-quartzite blade blanks found in a large Bükk storage vessel (Vértes 1965, Mester and Tixier 2013). Vértes' length vs. frequency plot of the blades (1965, fig. 3) shows a normal (unimodal) distribution, indicating not only a desired end-product to the knapping sequence but also a high level of skill in attaining this end-product in such a high proportion of cases.

Compartmentalisation

After the First Temperate Neolithic period of pottery production, there were dramatic changes in the mature farming period (Karanovo II–III and III), with the making of multi-part vessels, often using inversions to attach the legs and featuring varied handles and lugs and tight-fitting lids. These vessels betokened a major increase in compartmentalisation and precision, as exemplified by two 'type-fossils' of the Karanovo III period – high polypod stands and handled mugs (Nikolov 2006). The high stands, sometimes made as black-topped ware, are typically made in several parts – the shallow dish and the three or four long, tapering legs, which are moulded onto the inverted dish. It is obviously essential to fashion the legs to identical lengths to avoid an unstable placement on a feasting table. Design precision is thus part and parcel of the compartmentalisation of ceramic design. The same is true of the handled mugs ('tankards'), whose cylindrical body is made first, after which a separately made handle is added to the mug; occasionally, the mugs have added feet as well (Figure 20.3). It is conceivable that the handle itself was made in two parts – the basic handle and the long cylindrical upper part, with the upper part to extend no higher than the rim of the mug. These prototypical drinking-vessels were made in various sizes, including two-litre tankards measuring 0.80m in height, as seen in the pit deposits of Chavdarova Chesma (unpublished, Yambol Museum). Thus precision is built into the design of the Karanovo III mugs in three ways: the appropriate fitting of the lower part of the handle, the length of the cylindrical part of the upper handle and the size of the mug itself.

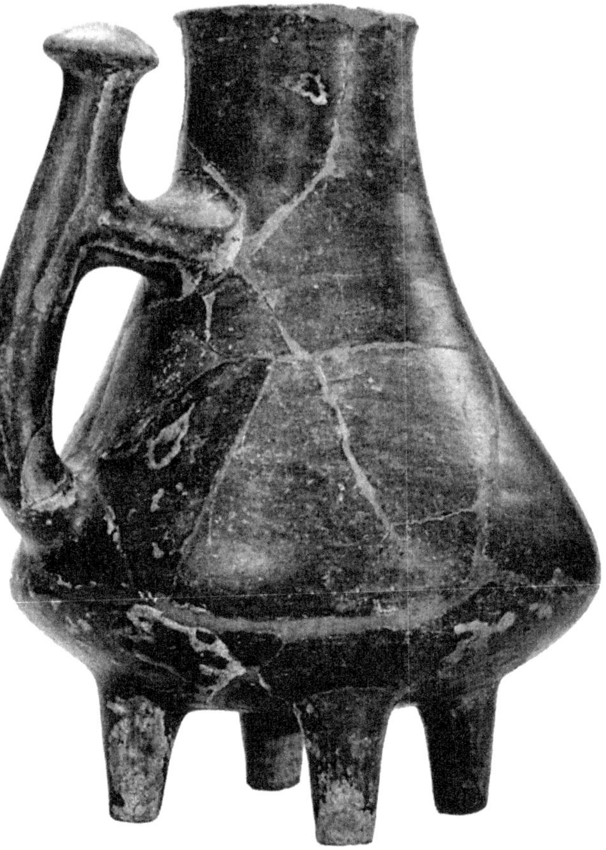

Figure 20.3 Footed handled mug, Yasa Tepe (Nikolov 2006, 24).

Symmetrical design

An intriguing example of polished stone geometry concerns the ornament from the Early Neolithic settlement of Kărdzhali, in the Eastern Rhodopes, South East Bulgaria (Peïkov 1972). This ornament forms one of an Early Neolithic series specific to the Macedo-Bulgarian transition zone (Nandris 1970) and described alternately as 'frogs' or 'swastikas'. All examples were made of similar fine-grained volcanic rocks, using similar techniques of grinding, polishing and facetting (Todorova and Vajsov 1993, 228–9, Ris. 154; Kostov 2007, 41–7; Figure 20.4). The Kărdzhali swastika represents a complex example of four-fold rotational symmetry, with a high degree of polish and an attractive light green finish. The maker would surely have gained enormous prestige in the transformation of nephrite into an object replete with such cultural symbolism. The combination of complex geometric shape, exoticity, colour and brilliance makes the Kărdzhali swastika one of the most striking objects in the East Balkan Early Neolithic.

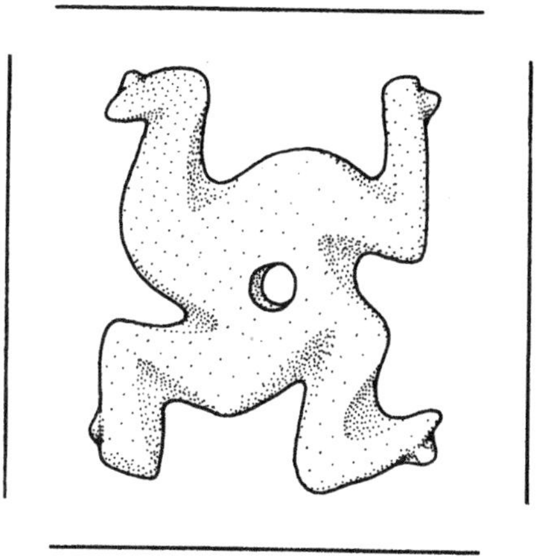

Figure 20.4 Nephrite polished stone 'swastika'. Early Neolithic Kărdzhali (Todorova and Vajsov 1993, 181, fig. 154).

It is almost impossible to select a single example of symmetrical design from the huge mass of pottery from any one of a dozen Balkan Neolithic or Chalcolithic 'cultures'. However, the complexity and the spectacular aesthetic results of Cucuteni-Tripolye pottery are probably without parallel in the Climax Chalcolithic period (for colour illustrations, see Mantu *et al.* 1997). Ellis (1984, chapter 3) has made the case for specialist potters in Cucuteni settlements. The finest large Cucuteni vessels were made using turntables of the kind found in the Varvareuca pottery workshop (Ellis 1984, 115–16), with firing temperatures of 1000°C or higher achieved in updraught kilns to produce a fine red-slipped and burnished surface on which motifs were painted using white and black paints (Ellis 1984, 120, 157).

The vessel specially chosen for symmetry analysis is the stemmed globular vessel with a lid from the Cucuteni A3 phase of the Scânteia settlement, dated 4200–4050 cal BC (Mantu and Dumitroaia 1997, 189, fig. 46). The vessel was made in three parts: a hollow flaring cylindrical pedestral base, a globular body with a short vertical neck and a bell-shaped flaring-rimmed lid with a button handle (Mantu and Dumitroaia 1997, fig. 46; Figure 20.5). The pattern flows across the three-dimensional vessel surface, despite the continuous changes in angle and two sharp breaks. There can be few better examples in a single artefact of Keightley's four principles of compartmentalisation, standardisation, precision and symmetry.

According to Washburn and Crowe's (2004) scheme, three types of symmetry are represented on this pot: (a) mirror reflection (the big spirals), (b) glide reflection (the small spiral), and (c) two-dimensional translation of motifs as one moves up and along the vessel surface. This is in itself unusual, for most symmetrical decorative patterns show one or two forms of symmetry. However, Cucuteni trichrome vessels stand out for having a three-colour pattern – one of only 23 variants known in art history, most created by M.C. Escher (Washburn and Crowe 2004).

Figure 20.5 Trichrome painted ware vessel, Scânteia (Mantu and Dumitroaia 1997, 189, fig. 46).

The implications of this complex decorative design are not only that specialist pot-painters were almost certainly involved in its production but that an appreciation of its symmetrical design effects was widespread among Cucuteni communities. Marinescu-Bîlcu (2000, 110) indicates that painted wares formed well over 50% of potters' total output at the Cucuteni A4 site of Draguşeni, implying that these richly painted, brightly coloured and shining, geometrically designed pots were a central part of the domestic visual domain.

High-precision execution

The earliest painted ware in the Balkan peninsula comprises the so-called 'White-on-Red' painted style, made locally in all of the regional stylistic zones of the Balkan Early Neolithic in the period 6200–5700 cal BC (Oross 2007, 571–85; Spataro 2007). Many examples exist in the Macedo-Bulgarian transitional zone – the same area as the Kărdzhali swastika – at sites such as Anza (Mock 1976) and Rakitovo

(Raduntcheva *et al.* 2002). At the latter site, the white-painted decoration on many sherds comprised a series of fine oblique painted lines, each of which was no more than 1mm in width and with a gap of 1mm between lines. The technical question was: How were such thin lines painted on a red slipped and burnished background?

Raduntcheva (pers. comm., 1998) used an experimental programme to demonstrate that the only material that was at the same time hard enough to hold the paint and slender enough to produce 1mm lines were the bristles of a wild boar. The fascinating interpretative aspect of this example concerns the juxtaposition of one of the most ferocious animals in the Balkan forest with some of the finest painted, slipped and burnished ware vessel series in the Balkan Early Neolithic, used in the home for hospitality and as display ceramics. A comparable case concerns the finely polished bone spoons of the Early Neolithic, made from the metapodia of mostly aurochs (*Bos primigenius* Boj.; Nandris 1972) or, sometimes, domestic cattle (*Bos taurus* L.; Choyke 2007). The associations of the wild, the hunters and the forest added an extra symbolic dimension to white-on-red painted vessels and bone spoons.

A second example of high-precision execution dates from the two best known cemeteries of the East Balkan Climax Chalcolithic – the Varna and Durankulak cemeteries. The Varna cemetery is best known for the earliest accumulation of gold ornaments in the world (Ivanov 1988, 1991; Chapman *et al.* 2006; Higham *et al.* 2007). Carnelian and agate beads were buried in a restricted number of graves, both at Varna and Durankulak (Todorova 2002; Kostov 2007, 68–71), where there is a high correlation between the presence of carnelian beads and outstandingly 'rich' graves.

Kostov used a gemmologist's lens to detect up to 16 fine facets on each half of the barrel-shaped beads (Kostov 2007, 66–77, fig. 26; Figure 20.6). There is no evidence for a Balkan source of carnelian, nor for any kind of carnelian bead-making in this region. There are therefore at least two possibilities. First, the facetted beads may have been exchanged into the western Black Sea region from the South Pontic zone. Secondly, small lumps of carnelian may have been exchanged by the same network for production of beads by local specialists. Kostov (2007, 73) maintains that the wear traces on the edges of the facets could only have been made by rotary motion – that is, some form of fast or lap wheel. The current evidence does not allow us to differentiate between the two ideas of local versus exotic making of facetted carnelian beads. Whichever is the case, the precision necessary for this ultimate transformation of nature into cultural order is matched only by the eyesight and hand-eye co-ordination necessary to produce the facets. The making of facetted carnelian beads surely indicates a craft specialist whose powers of creation may well have given him or her a supernatural dimension linked to brilliance, for the *only* reason for the creation of fine facets on a bead is to make the bead shine even more brightly than it does after stage 3 polishing.

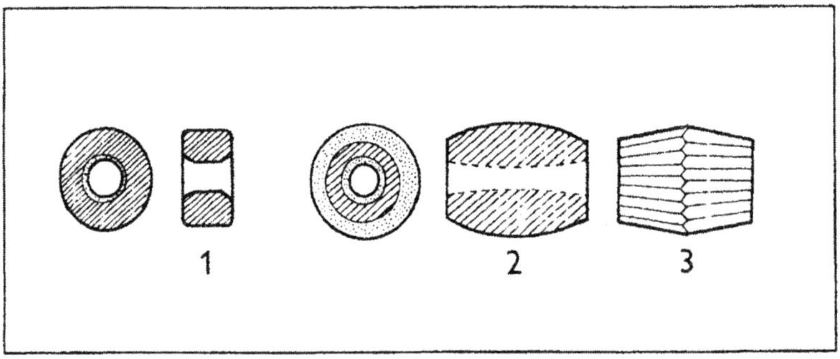

Figure 20.6 Facetted carnelian beads, Varna Cemetery I (Kostov 2007, 66, fig. 26).

Neolithic textiles

Childe (1936, 95) remarked on the invention of the loom as one of the great triumphs of human ingenuity. The complex operations of the loom allow cloth to represent the key dilemmas of social and political life (Weiner and Schneider 1989). In an ethnographic example of the values of cloth, Vastokas (1994, 344, 356) has demonstrated the key function of textiles in Lithuanian folk culture, where spinning and weaving constitute the praxis for cognition, since weaving, with its movements of warps and wefts, mirrors a contrasting movement across social networks, territories and time (that is, from birth to death). In Lithuanian culture, the female self was constructed through spinning and weaving, with the 'self' forming a precipitate of dialogue with other 'selves', including objects.

In many Neolithic and Copper Age cultures in the Balkans and central Europe, the sole evidence for weaving consists of the remains of loom-weights and spindle-whorls. Barber (1991, 92–9) takes this evidence to mean an extensive distribution of warp-weighted looms in these regions. Three recent investigations of textile impressions have produced additional knowledge of the history of weaving technology.

A recent compilation of textile impressions from the Hungarian Early Neolithic Körös culture (Makkay 2001) indicates that plain weave cloth was made with S-spun and I-spun techniques on a narrow or a warp-weighted vertical loom, probably using flax or hemp. Loom-weights and fired clay spindle-whorls were found in each domestic assemblage on these Körös sites.

Two Late Neolithic collections have produced an impressive range of textile and basket/mat impressions: the Late Vinča site of Divostin, Serbia (Adovasio and Maslowski 1988) and the impressions from Late Neolithic sites in Hungary (Kalicz 1985, 85; Füzes 1990; Bíró 1992; Richter 2003, 2005).

Adovasio and Maslowski (1988) identified three main and nine structural variant types of Late Neolithic textile and mat/basket impressions from a sample of 110 Late Vinča impressions, dated *c.* 4000 cal BC:

1 *Twining textiles:* Among the twinings, the most common was a Close Diagonal Twining with a Z-twist weft (n = 56), with fewer examples of Close Diagonal Twinings with an S-twist weft (n = 5), Open Diagonal Twinings with a Z-twist weft (n = 4), Open Diagonal Twinings with an S-twist weft (n = 2) and Close Simple Twinings with an S-twist Weft (n = 15).
2 *Plain woven textiles:* Unbalanced Plain Weave (n = 20). Balanced Plain Weave (n = 4).
3 *Plaited mats/baskets:* Twill Plating (n = 3).

According to the authors, many of these textiles examples from pot bases were so coarse-textured that they represented impressions from mats, although the open-twined examples were probably from bags. Adovasio and Maslowski (1988, 353) argue that all of these impressions were made by textiles that could have been produced without a heddle loom! However, if a loom was indeed used, it was probably a belt loom or backstrap loom suitable principally for plain weave and its derivative types. Rasson (1988, 337–8), studied a collection of 71 loom-weights from Divostin, comprising house groups of 3–12 pieces (1988, Table 11.14), but found no post-holes of any loom-frame in houses.

Impressions on sherds from various Late Neolithic groups show that there were two techniques consistently used in Hungary – plaiting and plain weaving (Richter 2003, 98–100; Richter 2005, 123–44). It is clear from their impressions that the principles of symmetry are inherent in these two similar techniques. It is possible to identify two main structure types: the basic plain/tabby and its derivative twill binding (diagonal, chevron, lozenge and complex) variations (Figure 20.7). While only the tabby/plain binding structure was found among the textile impressions, all these motif types were found as plaited mat/basket fragment impressions in different Neolithic cultures as well as forming pottery decoration in the Late Neolithic Tisza culture (Richter in press).

We have no evidence that geometrical woven patterns were used on cloth or other textiles but we can see such motifs on the surfaces of anthropomorphic figurines of the Tisza group (Trogmayer 1987). The motifs of the Tisza group could also appear as painting on clothes; a pattern-making clay roller for use on textiles or leather was found at Kökénydomb, with its surface covered with repeated lozenge motifs (Csalog 1955, fig. 5.1).

Textiles were made by using plain/tabby weaving with the warp (vertical) and the weft (horizontal) yarns interlacing in a one-under and a one-over structure without twisting or looping. The use of warp-weighted loom throughout the Hungarian Neolithic is indicated by the heaps of clay loom-weights found on many house floors (e.g., Selmeczi 1969, Szénászki 1979).

To summarise, we can say that textiles from Divostin exhibited more technical types and variants (twinings and twills) but were less finely textured than those of Late Neolithic Hungary. Some of the structure is less dense and the threads are thicker, while the spun thread is not visible on the surface of the threads (Adovasio

Figure 20.7
Textile impressions, Hódmezővásárhely-Kökénydomb, Late Neolithic Tisza culture (Richter 2005, fig. 8).

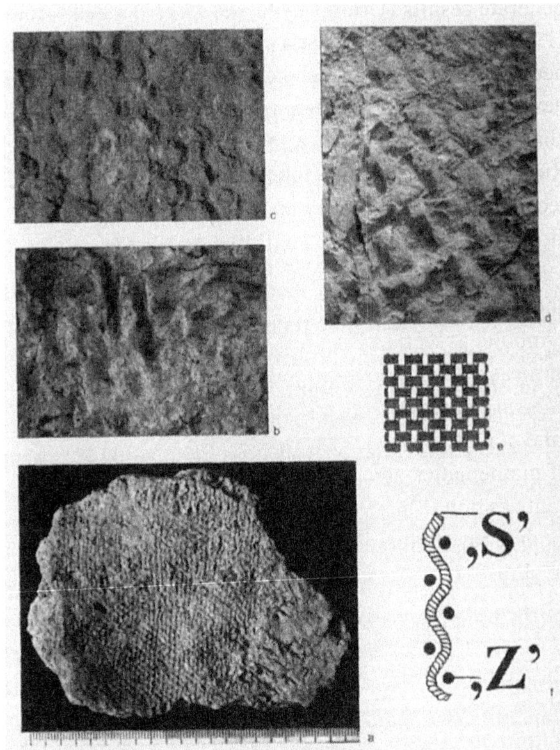

and Maslowski 1988, 352–3). The lack of twining and twill binding on the textile impressions is an important feature during the entire Hungarian Neolithic. It is interesting that such great diversity in cloth production techniques and quality of execution occurred in two regions scarcely 500km apart.

Numerology

In a short discussion of mathematics in the ancient world, Childe (1949b) quotes the important finding of Leslie White (1947, 16) that mathematics is as much part of a society's culture as tools and machines. This observation, supported by Menninger (1969, 10–12), implies not only the presence of mathematics at a level appropriate to a society's cultural development but the differentiation of mathematical skills and practices between various cultures. Crump (1990, 3) defines numeracy as 'the ability to understand and work with the structure of mathematics, at such a level of development as is feasible given the linguistic resources of the local culture'. Crump identifies the only experimental techniques in mathematics as counting and ordering; these can be based upon simple algorithms: ones that generate results (Crump 1990, 4). To these two we can add proportion. Crump (1990, 2) recognised that the cognitive basis of early numbers was defined by the need to record quantities of concrete objects, not by that of facilitating the development of arithmetic in the abstract. However, the use of proportions to characterise the form of a building takes numeracy beyond the pragmatic. For Russell (1920), order is a property of the members of any set in which there is a recognised relation of precedence and succession (see above for the Golden Section at Lepenski Vir).

We can thus expand Crump's list of two basic mathematical operations, counting and ordering, to a third – the use of proportions – while asserting that there are symbolic as well as pragmatic uses for such notions. This and many other aspects of measurement are discussed in the recent edited volume on *The Archaeology of Measurement* (Morley and Renfrew 2010). For the editors, measurement is a new kind of material and cognitive engagement with the world; it encapsulates the seeds of mathematics and of science (Renfrew and Morley 2010, 1). In his thoughtful contribution, Malafouris (2010) seems to have problems with the alternative to the hypothesis that language enables humans to move beyond approximation; as he phrases it, 'if not language, then what?' The obvious alternative which this study would prioritise is material culture.

Over the past 30 years, there has been a steady stream of specialists, sometimes with home bases in disciplines other than archaeology, who have sought to prove that a common system of weights and measures or distances was current at a particular juncture of prehistory (for a consideration of the 'megalithic yard' in north-west Europe, see Ruggles 1999). The usually implicit assumption guiding this research is that the existence of such a common measure indicates a higher level of culture or sophistication among the user communities. Peroni (1998) has published the most inclusive claim for a common system of weights for bronze and gold in the Early Bronze Age to the Early Iron Age for an area spanning the Aegean to Scandinavia. Despite critical comments on the details of the argument, even Turk (2001, 249) admits that 'since the Early Bronze Age, inhabitants throughout Europe may be ascribed with the knowledge of designating precise values for bronze and gold according to weight systems'. Is there any such evidence in the Balkan Neolithic or Copper Age?

The only site where there has been an investigation of large samples of metal – in this case gold – is the Varna I cemetery, on the Bulgarian Black Sea coast, dated to *c.* 4750–4450 cal BC (Higham *et al.* 2007). Two minor comparative studies were also added for the Varna II and Devnja cemeteries, dated to the Middle and Late Copper Age respectively but without radiocarbon dates. Here the gemmologist Ruslan Kostov (2007, 79–82) has weighed many hundreds of the gold ornaments in the main cemetery, identifying three groups of objects according to weight: (a) golden beads and appliqués, up to 10g; (b) golden pectorals, 10–20g; and (c) golden diadems and bracelets, over 20g. Kostov claims two minimal weight units of 0.145g

Table 20.2 Multiples of 'basic Copper Age unit' of weight in gold objects, Varna Cemetery I (source: Kostov 2007,78–82)

Multiple of 'basic CA unit'	*Category of object*	*Varna I grave No.*
× 2	6 trapezoidal appliqués	Grave 1
× 2	6 circular appliqués	Grave 3
× 3	Needle (? Nail)	Grave 15
× 4	30 zoomorphic appliqués	Grave 36
× 5	7 anthropomorphic amulets	Grave 36
× 6	2 plaques used to repair shell bracelet	Grave 43
× 10	Anthropomorphic amulet	Grave 15
× 13	3 rings	Grave 5
× 13	Sickle	Grave 36
× 20	Bow fitting	Grave 1
× 21	Bow fitting	Grave 1
× 22	Bow fitting	Grave 1

and 0.415g, proposing that the 0.415g unit is the basic Chalcolithic unit for the Varna cemetery and, by extension, a wider area of the east Balkans. Kostov's (2007) measurements show that multiples of the basic unit are found for ×2, ×3, ×4, ×20, ×28, ×30, ×40 and ×60 (Table 20.2).

These calculations share the same problem as Eiwanger's (1989) Eurobalcanic Gold Standard of 0.93571g – a multiplicity of decimal points that no Copper or Bronze Age goldsmith could possibly have measured on the crude balances of these periods. The notion of excessive precision has no place in prehistoric numerology, especially when variations in the weights of individual examples of types such as the Devnja gold rings amount to weights hundreds of times greater than the claimed 'Basic Unit'. We still await convincing evidence for a numerological system in the Balkan Chalcolithic.

Astronomy in the Balkan Chalcolithic

In his general account of prehistoric astronomy, Clive Ruggles emphasises that, in ancient societies, astronomy was an integral part of cosmology (Ruggles 1999, 163). An important characteristic of prehistoric cosmology was its finality, as part of sacrosanct applications of mathematical principles. People did not just move through the landscape but through the whole cosmos as they perceived it (Ruggles 1999, 153). Every culture made an appraisal of its own environment by selecting certain aspects of its physical surroundings and assigning meaning to them and to their relations with them (Ruggles 1999, 145). The resulting worldview may have included other people, animals, mountains, weather and astronomical phenomena. Celestial objects could be used only metaphorically, since the sky is immutable; probably astronomical elements were of great importance in worldviews.

The forms of evidence pertaining to prehistoric astronomy show marked regional variation across Europe, with the monumental structures of north-west Europe giving rise to a multitude of solar and lunar calendrical observations, in contrast to the paucity of evidence from the Balkan Chalcolithic.

Each of the three Balkan objects with potential calendrical notation dates to the Copper Age, as is the case with the more ambiguous scenes from two Bulgarian caves (Stoev and Stoychev 1991). The first object comes from the Early Copper Age settlement of Slatino, in the Struma valley, West Bulgaria, and dates to *c.* 4800 cal BC (Chokadzhiev 2006). An oven model with complex designs incised on the base (Figure 20.8) was found in a house in Horizon 6. Chokadzhiev (2006) interprets

Figure 20.8 Calendrical objects: the Slatino oven model (Chokadzhiev 2006, fig. 193b).

this as a mixed calendar, with elements from the lunar calendar (moon phases) and from the solar calendar (the seasons). He sees it as possibly not an 'eternal' calendar but rather as presenting ancestral astronomical knowledge.

The second object is the well known 'scene' from Ovcharovo, comprising 26 miniature objects found under the fragments of a house model made of unbaked clay, itself found in a burnt house in Horizon IX, which is dated to *c.* 4750–4550 cal BC (Todorova 1983). The scene comprises: three altars (Figure 20.9); three short tables, red-painted on all sides; three miniature pots, each with a lid; three chairs, also red-painted, with each table; and four miniature anthropomorphic figurines, each with a miniature drum or a loaf of bread. Of the three interpretations of the Ovcharovo scene, only Todorova's narrative takes account of all of the objects.

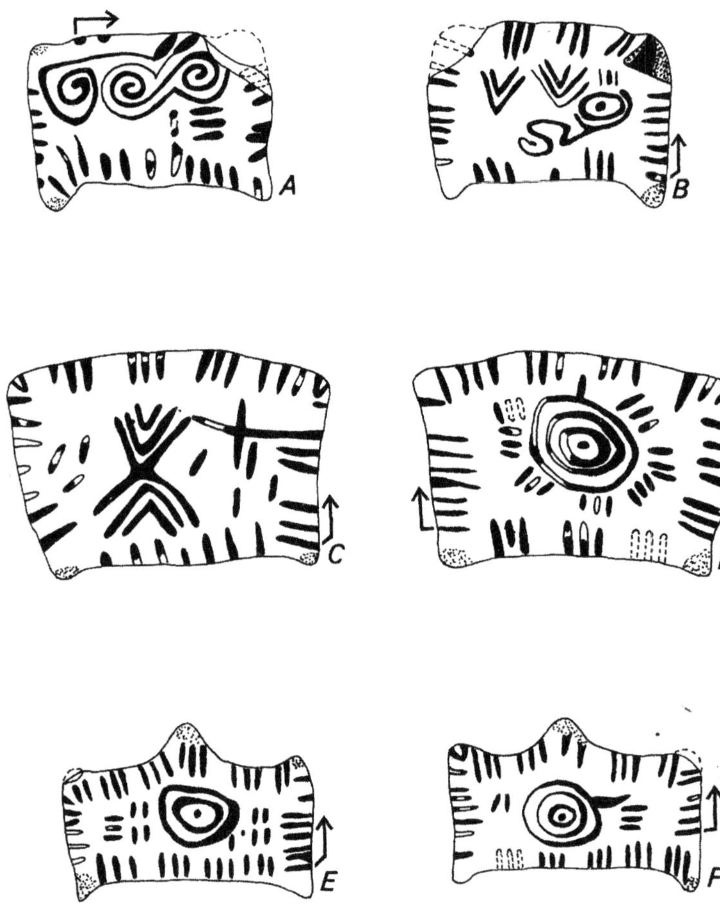

Figure 20.9 Calendrical objects: ends of three altars, from a set of ritual objects, Ovcharovo (Koleva 1986, 138).

She sees the scene as the representation of an ancient temple to the sun, the moon and nature, of which the solar cult was central. The figures were mediators rather than mother goddesses, with the older woman as the cult leader. By contrast, Koleva's (1986) interpretation of the Ovcharovo scene was based entirely on the three main altars which were held to represent three lunar months. Thirdly, Nikolov (1991) posits the main function of three altars as safeguarding crops in the hazardous last month before the harvest, in rituals related to spring and early summer.

The third object concerns an elaborately decorated clay vessel found as part of an unusual collection of objects deposited in a pit in a Vučedol house at the tell of Vinkovci, with dendro-dates of 2600–2400 BC for the horizon (Durman 2000, 93). This set comprised five sets of two-part clay moulds for copper axes and a crucible for metalworking, with several almost complete pots. The calendrical vessel (Durman 2000, fig. 56) has extensive decoration in four parallel bands each comprising 12 or 13 panels (Figure 20.10). Each band is interpreted as one of the

Figure 20.10 Calendrical objects: the Vučedol calendrical vessel (Durman 2000, 103).

four seasons. Durman (2000, 119) interprets this calendar as more astral than solar, using this extraordinary vessel to interpret the well known and plentiful signs on other vessels in terms of celestial and solar symbols.

The common features of the three calendrical objects concern their domestic settings, their transformations from everyday to calendrical objects, and the mixed basis of the calendars, whether lunar, solar or astral.

Discussion

Reflecting on, for example, the regularities of limnoquartzite blades, the fine detail of facetted carnelian beads, the calendrical motifs on Vučedol pottery and the geometric motifs on Tisza basketry, what characteristics could possibly have been shared by these disparate objects from different times and places in Balkan and Carpathian prehistory? And even if there is a positive answer to this question, are these objects typical of a wider range of objects, as material citations of more general cognitive practices? Finally, how did these objects work their Childean-Gellian magic in their everyday and sacred contexts of use?

There are two obvious shared characteristics of these objects: their material ontologies of a high-visibility aesthetic of colour, brilliance and exoticity, and their embodiment of the principles of Neolithic science underpinning key design concepts.

Combining the insights of Childe and Lemonnier, we see that science constitutes the formulae that yield the rules for action and it is the relationship between material actions and material effect that gave prehistoric people insights into scientific causation. These formulae could be directly mathematical, as in the case of the ratios of the trapezoidal house dimensions at Lepenski Vir, analogical, as in the case of the four-fold symmetry of the Cucuteni vessel from Scânteia and the Tisza textiles, or indeed representational, as when the Vučedol tell-dwellers replicated a vision of the heavens on a fineware vessel. All of these formulae are ultimately based upon a sense of harmonious geometric order enshrined in the four design principles of standardisation, compartmentalisation, symmetry and precision. For example, mathematical precision was expressed at different scales in the facets on Varna carnelian beads, the nephrite 'swastika' from Kărdzhali, and the complex textiles on a Divostin impression.

Childe also realised that it was the systematisation of sensory experiences that led to scientific formulae. These sensory experiences lay at the heart of the aesthetic of colour, brilliance and exoticity so frequently materialised in prehistoric objects from central and eastern Europe. The high visibility of these things is constant, whether in the centrality of the Lepenski Vir trapezoidal houses to long-term dwelling at the site, the Varna carnelian beads worn round the neck of a prominent person in the land of the living, the multitude of Cucuteni-Tripolye painted vessels that were central to household ritual and entertaining, or the elaborately decorated Tisza costumes for ceremonial use. The use of any such object in its appropriate context created a performative power that gave the object its own agency, framing the actions of the persons around it as much as *vice versa*.

If there was a consistent set of shared characteristics linking these everyday and special objects and structures, were they indeed typical of a much wider range of objects found throughout the Balkan Mesolithic, Neolithic and Chalcolithic? It is clear that the lives of many of the communities of these periods were framed, indeed formed, by a high density of material objects, which display, as well as regional variations, a consistent inner homogeneity and a marked external differentiation from other communities in, for example, Anatolia or the Alpine zone (see John Nandris' (1970, 1972) discussion of the 'First Temperate Neolithic'). In a study of the external analogies for the surface artefact collection from Orlovo, south-east Bulgaria, the strongest links with its figurines, polished stone axes and personal ornaments were documented with north-east Bulgaria, with weaker ties extending to Romania and Moldavia but with moderately strong links to northern Greece and even western Hungary (Chapman 2010, chapter 5, fig. 5.3). While the material evidence for local differentiation is indisputable, there is a sense of communities linked over wide areas in a social network of common materiality, in which individuals' senses of themselves were woven into a wider-than-local web. Although the individual objects and structures are widely separated in time and place, there is a common background in artefact-based networks providing a shared background of both technological expertise and aesthetic symbolism.

Despite these shared inter-regional networks, it is apparent that there is some strong time–space patterning in the occurrence of such objects and structures. It is striking that most of the examples discussed above (the Varna beads, the calendrical objects from Slatino and Ovcharovo, the Divostin and Tisza textiles) dated to the fifth millennium cal BC – the period of the so-called 'Climax Copper Age', denoting the apogee of material diversity and object quantity in many regions of central and south-east Europe (Chapman 2000, 1–3). The two chronological exceptions also marked local peaks in artefact diversity, with the trapezoidal houses at the Lepenski Vir site expressing the greatest artefact variability of any regional foraging group in central and south-east Europe (Radovanović 1996), and the material culture of the Vučedol group in Croatia, including the calendrical vessels, being richer and more diverse than in most other Late Copper Age groups, especially in terms of metallurgy and decorated ceramics (Tasić 1995). Thus, all these examples of objects denoting a more complex form of geometric order than was usual in these *milieux* were made, used and deposited (or abandoned in the case of the Lepenski Vir houses) in a period of peak local material culture diversity.

Andrew Jones (2007, 61–2) has emphasised that objects indexed past experience and acted as a form of material citation, whether on a daily basis of habitual practice or on a longer timescale. This was possible because of the sensual qualities of objects, which link the present not only to the past but also to future experiences. This means that material culture not only brought absent people, places and things into presence but also established a framework for continuity of action. Material citation led to a diversification of the metaphorical web linking everyday and 'special' objects to each other and to makers, users and other related persons. This is part of the process which Gell (1992) termed 'the enchantment of technology', by which artistic and scientific objects are constituted by the way they transcend the technical schemas of the spectator, who posited a divine and/or ancestral spirit filling these things. These practices were found not only in so-called ritual contexts (for example, the Varna cemetery) but also in the domestic arena, as in the calendrical objects, the Karanovo III mugs and standfoots, the Bükk lithic hoard and the textile remains.

A final point should be made concerning these harmonious material relationships. Such relationships are idealisations designed to represent the notion of cultural order in its full materiality. The widest range of harmonious objects discussed above were made in the Late Chalcolithic, precisely in the fifth millennium cal BC which attests to the greatest prevalence of defensive structures, weapons and (presumably) warfare and internal tensions (Chapman 1999). The concept of a peace-loving matriarchal 'Old Europe' in the Late Chalcolithic, subsequently destroyed by heavily-armed North Pontic horse-riding warriors in the post-Climax Chalcolithic (Gimbutas 1977), could hardly have been more opposed by the material evidence!

Conclusions

After this exemplification of practical science and aesthetics conjoined in the material record of later Balkan prehistory, I offer four short concluding thoughts for archaeological scientists and others. First, this chapter has exemplified the conclusions of V. Gordon Childe that each industry was made possible through the accumulation of experience and the application of deductions, that is, practical science with a base of reliable knowledge. Secondly, these examples support Mannheim's argument that the cognitive act is an instrument for dealing with life situations. Thirdly, the materials used in practical Neolithic science are co-constitutive of that science with the persons using them. And, finally, my hope is that the preceding discussion has shown that the materialisation of Neolithic science is widespread and has great potential for further study.

Acknowledgements

I am indebted to Alice Choyke for introducing me to the Neolithic textile research of Éva Richter. I am grateful to my Durham colleagues Tom McLeish and Janet

Montgomery for sharing their views of what constitutes science, as well as to Robert Hedges for his thoughts on science in the past. My most grateful thanks go to Bisserka, as ever, for her critical comments on the first draft and for countless discussions on Balkan prehistory.

References

Adovasio, J.M. and Maslowski, R.F. 1988. Textile impressions on ceramic vessels at Divostin. In A. McPherron and D. Srejović (eds), *Divostin: a Neolithic village in Central Serbia*, 345–57. Pittsburgh: University of Pittsburgh Press.

Barber, E.J.W. 1991. *Prehistoric textiles: the development of cloth in the Neolithic and Bronze Ages*. Princeton: Princeton University Press.

Benac, A. 1973. Obre II. *Wissenschaftliche Mitteilungen des Bosnisch-Herzegowinisch Landesmuseum (Sarajevo), Band III, Heft A. Archäologie*, 5–191.

Bíró, K.T. 1992. Adatok a korai baltakészítés technológiájához. – Data on the technology of early axe production. *Pápai Múzeumi Értesítő* 3–4, 33–79.

Borić, D. 2008. First households and 'house societies' in European prehistory. In A. Jones (ed.), *Prehistoric Europe: theory and practice*, 109–42. Oxford: Blackwell.

Boroneanţ, V. 1970. La période épipaléolithique sur la rive roumaine des Portes de Fer du Danube. *Prähistorisches Zeitschrift* 45, 1–25.

Bradley, R. 1998. *The significance of monuments: on the shaping of human experience in Neolithic and Bronze Age Europe*. London and New York: Routledge.

Chapman, J. 1989. The early Balkan village. *Varia Archaeologica Hungarica* 2, 33–53.

Chapman, J. 1999. The origins of warfare in the prehistory of Central and Eastern Europe. In J. Carman and A. Harding (eds), *Ancient warfare: archaeological perspectives*, 101–42. Stroud: Alan Sutton.

Chapman, J. 2000. *Fragmentation in archaeology: people, places and broken objects in the prehistory of South Eastern Europe*. London: Routledge.

Chapman, J., with É. Richter, 2009. Geometric order and scientific principles: a view from the Mesolithic, Neolithic and Chalcolithic of Central and South East Europe. In G. Bodi (ed.), *In medias res praehistoria. Miscellanea in honorem annos LXV peragentis Professoris Dan Monah oblate*, 21–58. Iaşi: Editura Universităţii 'Alexandru Ioan Cuza'.

Chapman, J. (ed.) 2010. *From surface finds to past lifeways: the prehistoric site of Orlovo, South East Bulgaria*. Oxford: Oxbow Books.

Chapman, J. 2011. Enchantment and enchainment in later Balkan prehistory: towards an aesthetic of precision and geometric order. In A. Hadjikoumis, E. Robinson and S. Viner (eds), *The dynamics of Neolithisation in Europe: studies in honour of Andrew Sherratt*, 153–76. Oxford: Oxbow Books.

Chapman, J. and Gaydarska, B. 2006. *Parts and wholes: fragmentation in prehistoric context*. Oxford: Oxbow Books.

Chapman, J., Higham, T., Slavchev, V., Gaydarska, B. and Honch, N. 2006. The social context of the emergence, development and abandonment of the Varna cemetery, Bulgaria. *European Journal of Archaeology* 9, 159–83.

Childe, V.G. 1936. *Man makes himself* (fourth edition). London: Watts & Co.

Childe, V.G. 1949a. The sociology of knowledge. *The Modern Quarterly* 4, 302–9.
Childe, V.G. 1949b. *Social worlds of knowledge*. The L.T. Hobhouse Lecture (delivered King's College, London, 12 May 1949).
Chokadzhiev, S. 2006. *Slatino: praistoricheski selishta*. Veliko Trnovo: Faber.
Choyke, A.M. 2007. Objects for a lifetime – tools for a season: the bone tools from Ecsegfalva 23. In A. Whittle (ed.), *The Early Neolithic on the Great Hungarian Plain: investigations of the Körös culture site of Ecsegfalva 23, County Békés*, 641–66. Budapest: Institute of Archaeology, Hungarian Academy of Sciences.
Crump, T. 1990. *The anthropology of numbers*. Cambridge: Cambridge University Press.
Csalog, J. 1955. A tiszai műveltség viszonya a szomszédos újkőkori műveltségekhez/ Die Beziehungen der Theiss-Kultur zu den neolitischen Nachbarkulturen. *Folia Archaeologica* 7, 23–44, 227–30.
Durman, A. 2000. *The Vučedol Orion and the oldest European calendar*. Zagreb: Arheološki Muzej Zagreb/Gradski Muzej Vinkovci.
Eiwanger, J. 1989. Talanton. Ein bronzezeitlicher Goldstandard zwischen Ägäis und Mitteleuropa. *Germania* 67, 443–462.
Ellis, L. 1984. *The Cucuteni-Tripolye culture: a study in technology and the origins of complex society*. Oxford: British Archaeological Reports, International Series 217.
Fischer, A. 1989. Hunting with flint-tipped arrows: results and experiences from practical experiments. In C. Bonsall (ed.), *The Mesolithic in Europe*, 29–39. Edinburgh: J. Donald.
Füzes, M. 1990. A földmívelés kezdeti sza kaszának (neolitikum és rézkor) növényleletei Magyarországo/Die Pflanzenfunden in Ungarn der anfänglichen Entwicklungsphase des Ackerbaues (Neolithikum und Kupferzeit). *Tapolcai Városi Múzeum Évkönyve* 1, 1989/1990, 139–298.
Gamble, C. 1999. *The Palaeolithic settlement of Europe*. Cambridge: Cambridge University Press.
Gamble, C. 2007. *Origins and revolutions: human identity in earliest prehistory*. Cambridge: Cambridge University Press.
Gaydarska, B., Chapman, J., Raduntcheva, A. and Koleva, B. 2007.The châine opératoire approach to prehistoric figurines: an example from Dolnoslav, Bulgaria. In C. Renfrew and I. Morley (eds), *Image and imagination: a global prehistory of figurative representation*, 171–84. Cambridge: McDonald Institute for Archaeological Research.
Gell, A. 1992. The technology of enchantment and the enchantment of technology. In J. Coote and A. Shelton (eds), *Anthropology and aesthetics*, 40–63. Oxford: Clarendon Press.
Gimbutas, M. 1977. The first wave of Eurasian steppe pastoralists into Copper Age Europe. *Journal of Indo-European Studies* 5, 277–338.
Gowlett, J. 1984. Mental abilities of early man: a look at some hard evidence. In R. Foley (ed.), *Hominid evolution and community ecology*, 167–92. London: Academic Press.
Hernando, A., Politis, G., González Ruibal, A. and Beserra Coelho, E. 2011. Gender, power and mobility among the Awá-Guajá (Maranhão, Brazil). *Journal of Anthropological Research* 67, 189–211.
Higham, T., Chapman, J., Slavchev, V., Gaydarska, B., Honch, N., Yordanov, Y. and Dimitrova, B. 2007. New perspectives on the Varna cemetery (Bulgaria): AMS dates and social implications. *Antiquity* 81, 640–54.

Ihde, D. 1990. *Technology and the lifeworld: from garden to earth*. Bloomington: University of Indiana Press.

Ivanov, I. 1988. Die Ausgrabungen des Gräberfeldes von Varna. In A. Fol and J. Lichardus (eds), *Macht, Herrschaft und Gold*, 49–66. Saarbrücken: Moderne Galerie des Saarland-Museums.

Ivanov, I. 1991. Der Bestattungsritus in der chalkolitischen Nekropole von Varna (mit einem Katalog des wichstigsten Gräber). In J. Lichardus (ed.), *Die Kupferzeit als historische Epoche*, 125–50. Saarbrücken: Saarland Museum.

Jones, A. 2007. *Memory and material culture*. Cambridge: Cambridge University Press.

Kalicz, N. 1985. *Kőkori falu Aszódon*. Aszód: Múzeumi Füzetek 32.

Keightley, D. 1987. Archaeology and mentality: the making of China. *Representations* 18, 91–128.

Koleva, V. 1986. Dania za kalendar v kultovata tsena ot Ovcharovo. *Interdistsiplinarni Issledvania* XIV, 131–52.

Kostov, R.I. 2007. *Arheomineralogiya na neolitni i halkolitni artefakti ot Bulgaria i tyahnoto znachenie v gemologyata*. Sofia: Publising House 'Sv. Ivan Rilski'.

Lemonnier, P. (ed.) 1993. *Technological choices: transformations in material cultures since the Neolithic*. London: Routledge.

Makkay, J. 2001. *Textile impressions and related finds of the Early Neolithic Körös culture in Hungary. With an Appendix on ritual spinning*. Tractata Minuscula 27. Budapest: privately printed.

Malafouris, L. 2010. Grasping the concept of number: how did the sapient mind move beyond approximation? In I. Morley and A.C. Renfrew (eds), *The archaeology of measurement*, 35–42. Cambridge: Cambridge University Press.

Mantu, C.-M. and Dumitroaia, G. 1997. Catalogue. In C.-M. Mantu, G. Dumitroaia and A. Tsaravopoulos (eds), *Cucuteni – the last great Chalcolithic civilization of Europe*, 101–239. Thessaloniki: Archaeological Museum of Thessaloniki.

Mantu, C.-M., Dumitroaia, G. and Tsaravopoulos, A. (eds) 1997. *Cucuteni – the last great Chalcolithic civilization of Europe*. Thessaloniki: Archaeological Museum of Thessaloniki.

Marinescu-Bîlcu, S. 2000. The pottery, tradition and innovation. In S. Marinescu-Bîlcu and A. Bolomey (eds), *Draguşeni: a Cucuteni community*, 91–110. Bucureşti: Editura Enciclopedică.

Menninger, K. 1969. *Number words and number symbols: a cultural history of numbers* (transl. P. Broneer). Cambridge, MA: MIT Press.

Mester, Z. and Tixier, J. 2013. Pot à lames: the Neolithic blade depot from Boldogkőváralja (northeast Hungary). In A. Anders, and G.Kulscar (eds), *Moments in time*, 173–85. Budapest: L'Harmattan.

Mock, R. 1976. Anza I–III (pre-Vinča) pottery. In M. Gimbutas (ed.), *Neolithic Macedonia*, 78–116. Los Angeles: University of California Press.

Morley, I. and Renfrew, A.C. (eds) 2010. *The archaeology of measurement*. Cambridge: Cambridge University Press.

Nandris, J. 1970. The development and relationships of the earlier Greek Neolithic. *Man* 5, 192–213.

Nandris, J. 1972. *Bos primigenius* and the bone spoon. *Bulletin of the Institute of Archaeology, University of London* 10, 63–83.

Nikolov, V. 1991. Kalendarna interpretatsiya na kultovata tsena ot Ovcharovo. *Interdistsiplinarni Issledvania* 18, 97–103.

Nikolov, V. 2006. *Kultura i izkustvo na praistoricheska Trakia*. Sofia: Letera.

Oross, K. 2007. The pottery from Ecsegfalva 23. In A. Whittle (ed.), *The Early Neolithic on the Great Hungarian Plain: investigations of the Körös culture site of Ecsegfalva 23, County Békés*, 491–620. Budapest: Institute of Archaeology, Hungarian Academy of Sciences.

Peĭkov, A. 1972. Sondazhni razkopki na neolitnoto selishte v Kărdzhali prez 1972 g. *Ahrid* 1, 7–44.

Peroni, R. 1998. Bronzezeitliche Gewichtssysteme im Metallhandel zwischen Mittelmeer und Ostsee. In B. Hänsel (ed.), *Mensch und Umwelt in der Bronzezeit Europas*, 217–24. Kiel: Oetker-Voges Verlag.

Radovanović, I. 1996. *The Iron Gates Mesolithic*. Ann Arbor: International Monographs in Prehistory.

Raduntcheva, A., Matsanova, V., Gatsov, I., Kovachev, G., Georgiev, G., Chakalova, E. and Bozhilova, E. 2002. *Neolitno selishte do grad Rakitovo*. Sofia: Gal-iko.

Rasson, J. 1988. Loom weights. In A. McPherron and D. Srejović (eds), *Divostin: a Neolithic village in Central Serbia*, 337–8. Pittsburgh: University of Pittsburgh Press.

Renfrew, A.C. and Morley, I. 2010. Introduction. Measure: the construction of our world. In I. Morley and A.C. Renfrew (eds), *The archaeology of measurement*, 1–4. Cambridge: Cambridge University Press.

Richter, É. 2003. A tiszai kultúra szőttes jellegű díszítésvilágának kapcsolata a szövés és gyékényfonás motívumaival/The connection between the ornamental repertoire of the Tisza culture with the motifs of plaiting and weaving. *Prehistoric Newsletter (Budapest)* 5, 98–106.

Richter, É. 2005. Textil és négyzetrendszeres fonatlenyomatok az Alföld Neolitikumából/ Woven and Plaited Fabrics in the Neolithic of Great Hungarian Plain. In G. Lőrinczy and L. Bende (eds), *Hétköznapokvénuszai*, 123–44. Hódmezővásárhely: János Tornyai Múzeum.

Richter, É. In press. Reconstructions of the plaiting and weaving techniques of the Neolithic Carpathian Basin. Submitted for M. Vicze (ed.), *Conference volume*. Százhalombatta: Matrica Museum.

Rozoy, J. 1989. The revolution of the bowmen. In C. Bonsall (ed.), *The Mesolithic in Europe*, 13–28. Edinburgh: John Donald.

Ruggles, C. 1999. *Astronomy in prehistoric Britain and Ireland*. New Haven: Yale University Press.

Russell, B. 1920. *Introduction to mathematical philosophy*, London: Allen and Unwin.

Selmeczi, L. 1969. Das Wohnhaus der Körös Gruppe von Tiszajenő/Neuere Ausgaben zu den Haustypen des Frühneolitikums. *Móra Ferenc Múzeum Évkönyve* 1969, 19–22.

Soffer, O. 2003. Mammoth bone accumulations: death sites? kill sites? dwellings? In S. A. Vasil'ev, O.Soffer and J.K. Kozłowski (eds), *Perceived landscapes and built environments: the cultural geography of Late Paleolithic Eurasia*, 39–46. Oxford: Archaeopress.

Soffer, O., Adovasio, J.M. and Hyland, D.C. 2000. The 'Venus' figurines: textiles, basketry, gender, and status in the Upper Palaeolithic. *Current Anthropology* 41, 511–37.

Spataro, M. 2007. Everyday ceramics and cult objects: a millennium of cultural transmission. In M. Spataro and P. Biagi (eds), *A short walk through the Balkans: the first farmers of the Carpathian Basin and adjacent regions. Società per la Preistoria e protostoria della regione Friuli-Venezia Giulia. Quaderno* 12, 149–60.

Srejović, D. 1969. *Lepenski Vir: nova praistorijska kultura u Podunavlju*, Beograd: Srpska Kniževna Zadruga.

Srejović, D. 1972. *Europe's first monumental sculpture. New discoveries at Lepenski Vir.* London: Thames and Hudson.

Srejović, D. and Babović, L. 1983. *Umetnost Lepenskog Vira.* Beograd: Jugoslavija.

Stoev, A. and Stoychev, T. 1991. Moon observatories in the Bulgarian lands. *Interdistsiplinarni Izsledvaniya* 18, 137–44.

Szénászki, J. 1979. A korai szakálháti csoport települése Battonyán/The settlement of the early Szakálhát group at Battonya. *Archaeologiai Értesítő* 106, 67–77.

Tasić, N. 1995. *Eneolithic cultures of central and west Balkans.* Belgrade: Institute for Balkan Studies.

Todorova, H. 1983. Duhoven zhivot na naselenieto. In H. Todorova, V. Vasilev, Z. Yanushevich, M. Kovacheva and P. Valev (eds), *Ovcharovo. Razkopki i Prouchvania IX*, 91–6. Sofia: Bulgarska Akademiya na Naukite.

Todorova, H. (ed.) 2002. *Durankulak Band II. Die prähistorischen Gräberfelder.* Berlin-Sofia: Anubis.

Todorova, H. and Vajsov, I. 1993. *Novo-kamennata epoha v Bulgaria.* Sofia: Nauka i Izkustvo.

Trogmayer, O. 1987. The god with the axe. In P. Raczky (ed.), *The Late Neolithic in the Tisza region.* Budapest-Szolnok: Szolnok County Museums.

Turk, P. 2001. The weight of objects in Late Bronze Age hoards in Slovenia and possibilities for determining weight standards. *Arheološki Vestnik* 52, 249–79.

Vastokas J. M. 1994. Are artefacts texts? Lithuanian woven sashes as social and cosmic transactions. In S.H. Riggins (ed.), *The socialness of things: essays on the socio-semiotics of objects*, 337–62. Berlin-New York: Mouton de Gruyter.

Vértes, L. 1965. The depot of silex blades from Boldogkőváralja. *Acta Archaeologica Hungarica* 17, 128–36.

Washburn, D. and Crowe, D. (eds) 2004. *Symmetry comes of age.* Seattle: University of Washington Press.

Weiner, A.B. and Schneider, J. (eds) 1989. *Cloth and human experience.* Washington, DC: Smithsonian Institution Press.

White, L. 1947. The locus of mathematical reality. *Philosophy of Science* 14, 289–303.

Wright, R.P. 1993. Technological styles: transforming a natural world into a cultural object. In S. Lubar and W.D. Kingery (eds), *History from things: essays on material culture*, 242–69.Washington, DC: Smithsonian Institution Press.

21

Archaeological Science and the Neolithic

The Power and Perils of Proxy Measures

PAUL HALSTEAD

Introduction

JUST HOW REMARKABLE IS THE WEALTH OF INSIGHTS from the application of scientific techniques, both new and not-so-new, is highlighted by the limitations of the data available to Neolithic researchers a generation or so ago. A well known example is Andrew Sherratt's model of a fourth–third millennium cal BC 'secondary products revolution' that enabled exploitation of domestic animals for milk, wool and traction, in addition to their presumed original use for meat (Sherratt 1981, 1983). Iconographic representations, ard-marks under burial mounds, and waterlogged textiles, wheels, yokes and ards, which provided Sherratt's principal empirical underpinning, are preserved very patchily in time and space. Absence of evidence is thus insecure evidence of absence, and its presence largely indicates *knowledge* rather than *routine use* of secondary products. Sherratt sought supplementary support in novel ceramic shapes, that he tentatively attributed to the handling and consumption of milk (1981, 280), and in changing site distribution relative to terrain and soils, that he linked to a shift from hoe- to ard-cultivation (1981, 292–3). Ceramics and site distributions are widely available evidence and ostensible indicators of routine use, but represent very indirect 'proxies' for secondary products. In addition, he drew attention to the sparse zooarchaeological evidence for potentially traction-related stress in cattle limb bones (Sherratt 1983, 91–2) and for differential culling ages of male and female domestic ruminants that indicated husbandry prioritising meat, milk, traction or wool (Sherratt 1981, 283–5; 1983, 94, 98).

Today, by contrast, there is relatively abundant macroscopic zooarchaeological evidence for Neolithic 'traction pathologies' (Bălăşescu *et al.* 2006, Helmer and Gourichon 2008, Isaakidou 2006) and especially culling patterns (Arbuckle *et al.* 2009, Greenfield 2005, Vigne and Helmer 2007). In addition, a range of biomolecular applications have explored age of weaning (affecting availability of milk for human consumption) in livestock (Balasse and Tresset 2002), documented

Proceedings of the British Academy **198**, 419–33. © The British Academy 2014.

heating of ruminant milk in ceramics (Copley *et al.* 2005, Craig *et al.* 2005, Evershed *et al.* 2008), and shed light on the evolution of lactase persistence (enabling digestion of fresh milk) in humans (Burger *et al.* 2007, Itan *et al.* 2009). Whether the increase in the volume and range of evidence supports (Greenfield 2010) or undermines (Halstead and Isaakidou 2011a, Vigne and Helmer 2007) Sherratt's model, it provides a dramatically more detailed and reliable picture of prehistoric use of secondary products, arguably for two reasons. First, macroscopic and biomolecular data alike provide more direct proxies for *aspects of* secondary products use than did most of Sherratt's original evidence. Secondly, multiple lines of evidence now shed complementary light on different aspects of this problem. Organic residues in ceramics document heat-processing of ruminant milk, but not the scale or nutritional significance of milk consumption; culling patterns and weaning ages reveal the extent to which livestock management made milk available to humans and thus measure *potential* intensity of exploitation, but do not confirm use; and the genetic history of lactase persistence suggests that ability to consume milk conferred strong selective advantage on some early farmers, but leaves open whether milk was a routine dietary staple or a source of sustenance in scarcity (Gerbault *et al.* 2011). Likewise, although ceramic typology cannot demonstrate consumption of dairy produce, coupled with organic residue evidence for heating of milk in vessels of particular categories (Salque *et al.* 2012) or deposited in particular archaeological contexts (Kotsakis *et al.* 2008), it may illuminate the cultural and social significance of consumption. The current methodological armoury is thus extraordinarily powerful, provided we remember the limitations of each method in isolation and the rich potential of combining several complementary methods.

A recurring feature of the expanding analytical repertoire is periodic recognition of unforeseen sources of ambiguity or equifinality. For example, human consumption of heavily manured staple cereals may replicate the raised $\delta^{15}N$ levels previously thought indicative of a diet rich in animal products (Bogaard *et al.* 2007); and consumption of cervid adipose fat may *partly* replicate $\delta^{13}C$ values thought indicative of consumption of domestic ruminant dairy fat (Craig *et al.* 2012). These sources of equifinality may of course lead to further analytical tools, as with use of $\delta^{15}N$ levels in cereal remains to identify manuring of crops (Fraser *et al.* 2011), and are not confined to biomolecular methods, as illustrated by the need to distinguish between selective culling (Legge 1981, Payne 1973, cf. Balasse 2003), seasonal occupation (Helmer *et al.* 2005, Mlekuz 2005, 35) and epidemic disease (Halstead 1998) as cause(s) of apparently high infant ('milk') mortality in assemblages of domestic ruminant bones. Even without non-cultural sources of equifinality, the relationship between scientific data and past human activity is often filtered by archaeological formation processes or the complexities of human – or indeed livestock – behaviour (cf. Pollard 2011). The remainder of this commentary explores these latter issues.

Subsistence proxies

In their recent demonstration that heat-processing of milk, and hence milking, dates back to at least the seventh millennium cal BC in Anatolia, Evershed and collaborators interpret variation in the proportion of ceramic samples yielding traces of milk fat as evidence of 'regional variation in the importance of milk use' and further argue that a positive correlation between milk traces and the proportion of cattle bones in faunal assemblages identifies cattle as the main source of dairy produce (Evershed *et al.* 2008, 530). Leaving aside the difficulty of sampling pottery assemblages consistently across regions with different traditions of ancient ceramic production and modern ceramic research, selective use of particular vessel types for preparing different foods has been demonstrated for the Neolithic (for example, food with and without animal fat in Greece: Urem-Kotsou and Kotsakis 2007; animal dairy and adipose fats in Poland: Salque *et al.* 2012). In addition, milk may have escaped detection by being processed without fire (cf. Mukherjee *et al.* 2005), heated in organic containers or even consumed fresh, and, of course, the nature of any heating regime and the fabric of any pottery containers used will have affected not only the preservation of dairy fats but also the use-life of the vessels and thus their relative frequency in ceramic assemblages. The proportion of a ceramic sample yielding milk traces, therefore, is categorically not a measure of the importance of milk use or production. The suggested correlation between frequency of milk residues and proportion of cattle bones may also be misleading. The *potential* availability of milk from a given species is related to the number of adult females among *live*stock and the age of weaning/slaughter of their male offspring, but neither of these variables necessarily correlates with that species' (including all ages and both sexes) percentage contribution to faunal *dead*stock – not least because this percentage is affected by changes in the culling profile or 'turnover' of other species represented in the assemblage (Halstead 2001). Nor can it be assumed that the most *numerous* domestic ruminant species was the main milk provider. On the contrary, because milking and processing of milk for storage are labour-intensive, one might equally expect a mixed farmer keeping, say, a few ewes and many cows to exploit the former intensively for milk and the latter extensively for meat, as Rowley-Conwy (2000) suggests (from mortality data) for sheep and goats at Arene Candide in north-west Italy. In practice, the choice of which domestic ruminant(s) to milk was presumably influenced also, as in the recent past, by other considerations: for example, the relative value placed on dairy produce and lambs, kids or calves (for which loss of milk is detrimental); the suitability of local pasture for species with divergent feeding requirements; the use of cows for draught (difficult to combine with milking unless these animals were well fed); and the intended use of milk (that of particular species being widely favoured for different products: butter, cream, cheese, yoghurt, and so on).

Animal husbandry proxies

Staying with livestock but moving beyond milking, many early applications of isotopic analysis in bioarchaeology treated a few animal specimens as local 'controls' for interpreting larger samples of human populations, making the tacit assumption that livestock are necessarily less mobile and less variable in diet than humans. The analysis of large livestock samples, for their own sake, is thus very welcome and has highlighted, for example, the variable histories of (probably local) mobility among young Neolithic cattle in southern Germany (Knipper 2009) and differences of grazing or foddering regime not only between species, but also between domesticates and wild relatives (for example, cattle and aurochs in north-west Europe: Noe-Nygaard *et al.* 2005) and, within a single domestic species, between age/sex groups (e.g., for cattle in Rumania: Gillis *et al.* 2013), individuals (for example, for cattle and sheep in Rumania: Balasse *et al.* 2013), and seasons (for example, for cattle and sheep in France: Balasse *et al.* 2012). Dental microwear in sheep and goats (Mainland and Halstead 2005) and *speed* of macroscopic tooth wear in sheep and pigs (Tzevelekidi 2012, 96–100, 108–11) from northern Greece further indicate dietary differences between contexts of deposition and so, probably, contexts of consumption. Failure to explore such variability risks making misleadingly generalised statements about livestock management and also confounding the target variables of diet and mobility. For example, at Çatalhöyük, a broad range of $\delta^{13}C$ values in sheep and goats, implying variable consumption of C_4 plants, has been interpreted in terms of extensive herding over distances large enough to encompass diverse pasture types (Pearson *et al.* 2007, 2178). C_4 plants are characteristic of warm rather than cool climates and abound in saline environments, but also flourish in summer row-crops and, especially after summer rainfall around the northern margins of the Mediterranean, in stubble of harvested winter crops (Oberdorfer 1954). In northern Greece in moist summers, pigs and sheep enthusiastically select the green C_4 plants in stubble and shepherds report immediate improvements in milk yields. Variable $\delta^{13}C$ values in livestock may thus represent small herds grazing locally in years of variable summer rainfall rather than large herds ranging extensively over different types of terrain. On the other hand, by combining complementary macroscopic, microscopic and biomolecular analyses, there is now a real prospect of exploring in detail how Neolithic husbandry and consumption of livestock responded to ecological constraints on various timescales and to cultural demands at various social scales.

Proxies for the Neolithic transition

Neolithisation also exemplifies the problems posed by incautious use of proxy measures, new and old. Probably the most striking recent innovation is the harnessing of aDNA and historical modelling of modern DNA data to explore the genetic, and by extension geographical, origins of early farmers and their domesticates. This has confirmed the exotic origins of the earliest crop and livestock species in southeast Europe, including those (einkorn, cattle, pigs) for which potential wild progenitors existed in the region (Edwards *et al.* 2007, Heun *et al.* 1997, Larson *et al.* 2007). The proliferation of such (palaeo)genetic evidence has coincided with a resurgence in support for colonisation as the principal vehicle for European Neolithisation, but does not strictly shed light on whether the first crops and livestock were carried by migrating farmers or adopted through exchange by indigenous foragers. The issue *may* be resolved once much more human aDNA is available for terminal foragers and early farmers in both Europe and potential source areas, but common sense (see Robb and Miracle 2007) and initial results alike suggest a mixed picture as the most likely outcome, with regionally variable genetic contributions by both locals and immigrants. If so, it will be difficult to disentangle the relative contributions of different populations to *initial* Neolithisation from subsequent inequalities in the social and biological success of different lineages. Much the same problem arises with palaeogenetic evidence that, while most early domestic pigs in southeast and perhaps central Europe were of Anatolian origin, their later counterparts were descended from European wild boar. In interpreting this in terms of local domestication, rather than interbreeding between 'oriental' domesticates and European boar, Larson *et al.* (2007) make assumptions about both the demography of domestic pigs and wild boar and the management of the former (Burger and Thomas 2011, Halstead and Isaakidou 2011b), testing of which will require more traditional zooarchaeological and palaeoecological evidence. Jones *et al.* (2012) are more circumspect in using modern DNA to chart two routes (south and north of the Alps) for the westwards spread of barley across Europe and to document apparent adaptations to regionally variable climatic conditions, without making unsubstantiated claims about human agency.

Human demography has long been central to debate on the Neolithic transition, with claims both of absent or sparse terminal foragers (Clarke 1978 for central Europe; Biagi and Spataro 2002 for the west-central Mediterranean) and rapidly growing early farming populations (Bocquet-Appel 2011) mobilised in support of colonisation. The density of foraging sites is difficult to assess, however, given diachronic variation in human use of archaeologically visible caves (Brochier and Livache 2003) and the vulnerability of open-air encampments to rising sea-level, alluviation, erosion, and non-detection (Runnels *et al.* 2009). Even occupation levels in caves exhibit uneven survival as a result of diachronic changes in natural and anthropogenic sedimentation and post-depositional disturbance (Brochier 1999, Mlekuz *et al.* 2008). Bocquet-Appel (2011) uses an ostensibly more direct proxy measure of demographic trends in treating the proportion of young individuals in cemetery assemblages as an index of population growth. Even back-of the-envelope calculations, however, make clear that most inhabitants of Mesolithic and Neolithic

Europe never received archaeologically recognisable burial, while skeletal analysis shows that the influence of age at death on mortuary practices varied both temporally and geographically (Hofmann 2009, Triantaphyllou 2008). Conceivably the mortuary treatment of young individuals reflected, *inter alia*, their value as future marriage partners or labourers and this in turn was related to the rate of population growth and/or speed of agricultural expansion (in which case, Bocquet-Appel *might* be right), but there are obvious perils in treating the proportion of young individuals among those receiving archaeologically visible burial as a measure of population growth.

Perhaps because of the difficulties that formation processes pose for inferring population trends from the Neolithic settlement or burial record, an alternative approach has used the temporal and spatial distribution of radiocarbon dates as a measure of relative population density (Collard *et al.* 2010, Shennan and Edinborough 2007). This approach, in which problems of differential survival and detection of sites are compounded by uneven investment in dating programmes and variable survival and recovery of dateable material, was subject to robust criticism at the Cardiff conference. Even if radiocarbon dates are accepted as a proxy for relative population density, however, does the apparently sharp demographic upturn at the start of the Neolithic support migration as the mechanism of Neolithisation, as claimed (Collard *et al.* 2010)? For reasons long appreciated, the transition from foraging to farming in most regions of Europe arguably entailed a significant increase in female fertility (Bocquet-Appel 2011), while the heavy labour demands of farming presumably provided an incentive to rear numerous offspring. A sharp increase in population might be expected at the inception of farming, therefore, regardless of the relative contribution thereto of external migration and indigenous adoption.

For similar reasons, impressively precise modelling of the radiocarbon evidence for the Neolithisation of Britain beginning in the south-east and spreading thence north and west is certainly compatible with initial colonisation (Whittle *et al.* 2011, 853), but adoption (with or without some migration) might plausibly have resulted in the same pattern (see also Whittle *et al.* 2011, 864–6). In principle, this refinement of Early Neolithic chronology offers the tantalising prospect of being able to identify the material traces of the first *generation* of Neolithic Britons and, if some of them received archaeologically recognisable burial, determining isotopically whether some or all were migrants from continental Europe. In practice, however, if the members of this first generation – whether migrants or acculturated locals – were very few (above; cf. Whittle *et al.* 2011, 861), they may elude archaeological prospection or be found in such small numbers that their isotopic signatures could not be treated as representative of their cohort (cf. Bickle and Hofmann 2007), especially if both migrants and local foragers were involved but treated differently in death (cf. Bentley *et al.* 2003). A rival interpretation of the related ceramic and mortuary evidence, as indicating that colonising farmers reached the British Isles from Brittany via the Irish Sea (Sheridan 2010, 2012), in turn begs the question why piecemeal similarities of material culture need represent movement of colonists, rather than appeals to 'continental' identity by acculturated foragers.

It is sometimes suggested that the need for relevant technical knowhow to accompany the dispersal of domesticates (Burger and Thomas 2011, 375; Vigne 2008, 186) favours Neolithisation by colonisation, although a mixture of immigration and adoption seems an equally viable mechanism for transmitting such knowledge. Anyway, terminal foragers probably had sufficiently detailed understanding of plant ecology and animal behaviour to make sense of guidance received with any exchanged domesticates or, if the latter were acquired by theft, to work out how to manage them within a timespan too short for archaeological detection. Arguably a more fundamental obstacle to indigenous adoption of farming was the radical ideological shift from foraging of communal resources to husbandry of domestic animals, cultivated crops and cultivated land that were in some sense private property (Ingold 1986) and, to explore this issue, contextualised archaeological investigation of material culture remains our best hope (Bogaard *et al.* 2009, 2011; Chapman 1997; Halstead 2011; Kotsakis 1999).

Conclusions

The applications of archaeological science presented and discussed in this volume are revolutionising our capacity to understand the Neolithic, but we must not lose sight of their limitations. They provide proxy measures of past human behaviour that are often quite indirect and thus open to significant ambiguity of interpretation. In criticising several recent claims that genetic, isotopic, chronometric, demographic or material culture evidence supports migration as the principal vehicle of European Neolithisation, this chapter is not arguing for the primacy of indigenous adoption, nor even for a regionally variable blend of migration and adoption (although this seems most likely). It suggests instead that much of the research cited sheds no light on the relative contributions of migration and adoption to Neolithisation, but merely repeats *a priori* assumptions. A recent study of the distribution of early crops and weeds in the east Mediterranean was unusually candid in admitting that 'we rely on population movement as a primary factor in the initial spread of agriculture [into south-east Europe] only because the evidence [prior interpretations of other classes of data] strongly favours demic diffusion for this phase of the early Neolithization' (Colledge *et al.* 2004, S55); their own results were equally compatible with adoption.

With due attention to the limitations of new and old proxy measures, however, and by focusing on questions that attract fewer headlines but are answerable, we can significantly advance understanding of Neolithic societies in Europe. Divergent dispersal routes and adaptation to regional climate are emerging in distribution patterns of early crop species and varieties (Colledge *et al.* 2005, Jacomet 2007, Jones *et al.* 2012, Zapata *et al.* 2004), although arguably not in percentages of livestock species (*pace* Vigne 2008, Conolly *et al.* 2012), given their adaptability and apparent integration with crop husbandry through manuring (Bogaard *et al.* 2013, Bogaard, this volume Chapter 10), ploughing (Isaakidou 2006) and grazing cultivated land (Mainland 2007). More locally, herding of livestock was clearly flexible (Balasse *et al.* 2013, Knipper 2009), while detailed contextualised analyses reveal the complexity of early farming society (Zvelebil and Pettitt 2008). Local communities included 'insiders' and 'outsiders'; women were more mobile, probably through both peaceful intermarriage (Bentley 2007) and violent abduction (Schulting and Fibiger 2012); and men with distinctive right-arm pathologies at Stuttgart-Mühlhausen possibly trained for combat (Villotte and Knüsel 2014). Material culture signalled these and other dimensions of identity in both life (Bogaard *et al.* 2011, Mina 2008) and death (Hofmann 2009, Bickle *et al.* 2011). Household control of agricultural production (Maier 1999, Marinova 2007) was balanced by collective labour in some agricultural tasks (Maier and Harwath 2011), in 'public' works such as digging enclosure ditches (Bernabeu Aubán *et al.* 2006) and probably in house building (Marciniak 2006). Different social groups within local communities maintained differential access to land (Bogaard *et al.* 2011), livestock (Hachem 2000; Hachem and Hamon, this volume Chapter 9) and perhaps animal produce (Urem-Kotsou and Kotsakis 2007), while commensality (Pappa *et al.* 2004), shared rites of passage (Robb 2002) and collective mortuary customs (Triantaphyllou 2008) variously countered centrifugal tensions within local communities.

In exploring how regional populations and local communities of early European farmers, whatever their descent, created new ecological niches and social worlds, such studies are remaining faithful to the scientific methods and their limitations without losing sight of the archaeological end-game.

Acknowledgements

I thank Amy Bogaard, Valasia Isaakidou and Duska Urem-Kotsou for comments on a draft; and Sébastien Villotte and Chris Knüsel for permission to cite unpublished work.

References

Arbuckle, B.S., Öztan, A. and Gülçür, S. 2009. The evolution of sheep and goat husbandry in central Anatolia. *Anthropozoologica* 44, 129–57.

Bălăşescu, A., Moise, D. and Radu, V. 2006. Une utilisation des bovins pour la traction pendant le Chalcolithique en Roumanie? In P. Pétrequin, R.-M. Arbogast, A.-M.

Pétrequin, S. van Willigen and M. Bailly (eds), *Premiers chariots, premiers araires. La diffusion de la traction animale en Europe pendant les IVe et IIIe millénaires avant notre ère*, 269–73. Paris: CNRS.

Balasse, M. 2003. Keeping the young alive to stimulate milk production? Differences between cattle and small stock. *Anthropozoologica* 37, 3–10.

Balasse, M. and Tresset, A. 2002. Early weaning of Neolithic domestic cattle (Bercy, France) revealed by intra-tooth variation in Nitrogen isotope ratios. *Journal of Archaeological Science* 29, 853–9.

Balasse, M., Boury, L., Ughetto-Monfrin, J. and Tresset, A. 2012. Stable isotope insights ($\delta^{18}O$, $\delta^{13}C$) into cattle and sheep husbandry at Bercy (Paris, France, 4th millennium BC): birth seasonality and winter leaf foddering. *Environmental Archaeology* 17, 29–44.

Balasse, M., Bălăşescu, A., Janzen, A., Ughetto-Monfrin, J., Mirea, P. and Andreescu, R. 2013. Early herding at Măgura-Boldul lui Moş Ivănuş (early sixth millennium BC, Romania): environments and seasonality from stable isotope analysis. *European Journal of Archaeology* 16, 221–46.

Bentley, A. 2007. Mobility, specialisation and community diversity in the Linearbandkeramik: isotopic evidence from the skeletons. In A. Whittle and V. Cummings (eds), *Going over: the Mesolithic–Neolithic transition in north-west Europe*, 117–40. London: Oxford University Press for The British Academy.

Bentley, R.A., Krause, R., Price, T.D. and Kaufmann, B. 2003. Human mobility at the Early Neolithic settlement of Vaihingen, Germany: evidence from strontium isotope analysis. *Archaeometry* 45, 471–86.

Bernabeu Aubán, J., Molina Balaguer, L., Diez Castillo, A. and Orozco Köhler, T. 2006. Inequalities and power. Three millennia of prehistory in Mediterranean Spain (5600–2000 cal BC). In P. Díaz-del-Río and L. García Sanjuán (eds), *Social inequality in Iberian late prehistory*, 97–116. Oxford: Archaeopress.

Biagi, P. and Spataro, M. 2002. The Mesolithic/Neolithic transition in north eastern Italy and in the Adriatic Basin. In E. Badal, J. Bernabeu and B. Martí (eds), *El paisaje en el Neolitico mediterráneo*, 167–78. Valencia: Universitat de València.

Bickle, P. and Hofmann, D. 2007. Moving on: the contribution of isotope studies to the early Neolithic of Central Europe. *Antiquity* 81, 1029–41.

Bickle, P., Hofmann, D., Bentley, R. A., Hedges, R., Hamilton, J., Laiginhas, F., Nowell, G., Pearson, D.G., Grupe, G. and Whittle, A. 2011. Roots of diversity in a Linearbandkeramik community: isotope evidence at Aiterhofen (Bavaria, Germany). *Antiquity* 85, 1243–58.

Bocquet-Appel, J.-P. 2011. The agricultural demographic transition during and after the agriculture inventions. *Current Anthropology* 52, S497–510.

Bogaard, A., Heaton, T.H.E., Poulton, P. and Merbach, I. 2007. The impact of manuring on nitrogen isotope ratios in cereals: archaeological implications for reconstruction of diet and crop management practices. *Journal of Archaeological Science* 34, 335–43.

Bogaard, A., Charles, M., Twiss, K. C., Fairbairn, A., Yalman, N., Filipović, D., Arzu Demirergi, G., Ertuğ, F., Russell, N. and Henecke, J. 2009. Private pantries and celebrated surplus: storing and sharing food at Neolithic Çatalhöyük, Central Anatolia. *Antiquity* 83, 649–68.

Bogaard, A., Krause, R. and Strien, H.-C. 2011. Towards a social geography of cultivation and plant use in an early farming community: Vaihingen an der Enz, south-west Germany. *Antiquity* 85, 395–416.

Bogaard, A., Fraser, R.A., Heaton, T.H.E., Wallace, M., Vaiglova, P., Charles, M., Jones, G., Evershed, R.P., Styring, A.K., Andersen, N.H., Arbogast, R.-M., Bartosiewicz, L., Gardeisen, A., Kanstrup, M., Maier, U., Marinova, E., Ninov, L., Schäfer, M. and Stephan, E. 2013. Crop manuring and intensive land management by Europe's first farmers. *Proceedings of the National Academy of Sciences of the United States of America* 110, 12589–94.

Brochier, J.É. 1999. Couche archéologique, sol archéologique et distributions spatiales: quelques réflexions (géo)archéologiques sur un vieux problème. In Universitat de València (eds), *Geoarqueologia i Quaternari litoral: memorial Maria Pilar Fumanal*, 91–5. Valencia: Universitat de València.

Brochier, J.E. and Livache, M. 2003. Un faciès sédimentaire anthropique original du mésolithique vauclusien: les terres noires à petits galets calcaires. *L'Anthropologie* 107, 153–65.

Burger, J. and Thomas, M.G. 2011. The palaeopopulationgenetics of humans, cattle and dairying in Neolithic Europe. In R. Pinhasi and J. T. Stock (eds), *Human bioarchaeology of the transition to agriculture*, 371–84. Chichester: Wiley.

Burger, J., Kirchner, M., Bramanti, B., Haak, W. and Thomas, M.G. 2007. Absence of the lactase-persistence-associated allele in early Neolithic Europeans. *Proceedings of the National Academy of Sciences of the United States of America* 104, 3736–41.

Chapman, J. 1997. The origin of tells in eastern Hungary. In P. Topping (ed.), *Neolithic landscapes*, 139–64. Oxford: Oxbow Books.

Clarke, D.L. 1978. *Mesolithic Europe: the economic basis*. London: Duckworth.

Collard, M., Edinborough, K., Shennan, S. and Thomas, M.G. 2010. Radiocarbon evidence indicates that migrants introduced farming to Britain. *Journal of Archaeological Science* 37, 866–70.

Colledge, S., Conolly, J. and Shennan, S. 2004. Archaeobotanical evidence for the spread of farming in the Eastern Mediterranean. *Current Anthropology* 45, S35–47.

Colledge, S., Conolly, J. and Shennan, S. 2005. The evolution of Neolithic farming from SW Asian origins to NW European limits. *European Journal of Archaeology* 8, 137–56.

Conolly, J., Manning, K., Colledge, S., Dobney, K. and Shennan, S. 2012. Species distribution modelling of ancient cattle from early Neolithic sites in SW Asia and Europe. *The Holocene* 22, 997–1010.

Copley, M.S., Berstan, R., Dudd, S.N., Aillaud, S., Mukherjee, A.J., Straker, V., Payne, S. and Evershed, R.P. 2005. Processing of milk products in pottery vessels through British prehistory. *Antiquity* 79, 895–908.

Craig, O.E., Chapman, J., Heron, C., Willis, L.H., Bartosiewicz, L., Taylor, G., Whittle, A. and Collins, M. 2005. Did the first farmers of central and eastern Europe produce dairy foods? *Antiquity* 79, 882–94.

Craig, O.E., Allen, R.B., Thompson, A., Stevens, R.E., Steele, V.J. and Heron, C. 2012. Distinguishing wild ruminant lipids by gas chromatography/combustion/isotope ratio mass spectrometry. *Rapid Communications in Mass Spectrometry* 26, 2359–64.

Edwards, C.J., Bollongino, R., Scheu, A., Chamberlain, A., Tresset, A., Vigne, J.-D., Baird, J.F., Larson, G., Ho, S.Y.W., Heupink, T. H., Shapiro, B., Freeman, A.R., Thomas, M.G., Arbogast, R.-M., Arndt, B., Bartosiewicz, L., Benecke, N., Budja, M., Chaix, L., Choyke, A.M., Coqueugniot, E., Döhle, H.-J., Göldner, H., Hartz, S., Helmer, D., Herzig, B.,

Hongo, H., Mashkour, M., Özdogan, M., Pucher, E., Roth, G., Schade-Lindig, S., Schmölcke, U., Schulting, R.J., Stephan, E., Uerpmann, H.-P., Vörös, I., Voytek, B., Bradley, D.G. and Burger, J. 2007. Mitochondrial DNA analysis shows a Near Eastern Neolithic origin for domestic cattle and no indication of domestication of European aurochs. *Proceedings of the Royal Society B* 274, 1377–85.

Evershed, R.P., Payne, S., Sherratt, A.G., Copley, M.S., Coolidge, J., Urem-Kotsu, D., Kotsakis, K., Özdogan, M., Özdogan, A.E., Nieuwenhuyse, O., Akkermans, P.M.M.G., Bailey, D., Andeescu, R.-R., Campbell, S., Farid, S., Hodder, I., Yalman, N., Özbasaran, M., Bıçakcı, E., Garfinkel, Y., Levy, T. and Burton, M.M. 2008. Earliest date for milk use in the Near East and southeastern Europe linked to cattle herding. *Nature* 455, 528–31.

Fraser, R., Bogaard, A., Heaton, T., Charles, M., Jones, G., Christensen, B.T., Halstead, P., Merbach, I., Poulton, P.R., Sparkes, D. and Styring, A.K. 2011. Manuring and stable nitrogen isotope ratios in cereals and pulses: towards a new archaeobotanical approach to the inference of land use and dietary practices. *Journal of Archaeological Science* 38, 2790–804.

Gerbault, P., Liebert, A., Itan, Y., Powell, A., Currat, M., Burger, J., Swallow, D.M. and Thomas, M.G. 2011. Evolution of lactase persistence: an example of human niche construction. *Philosophical Transactions of the Royal Society B* 366, 863–77.

Gillis, R., Bréhard, S., Bălăşescu, A., Ughetto-Monfrin, J., Popovici, D., Vigne, J.-D. and Balasse, M. 2013. Linking cattle diet history and mortality profile to milk production through stable isotope analysis ($\delta^{13}C$, $\delta^{15}N$ and $\delta^{18}O$) of dental rows: a case study at Borduşani-Popină (Romania, 5th millennium BC). *World Archaeology* 45, 447–72.

Greenfield, H.J. 2005. A reconsideration of the secondary products revolution: 20 years of research in the central Balkans. In J. Mulville and A. Outram (eds), *The zooarchaeology of milk and fats*, 14–31. Oxford: Oxbow Books.

Greenfield, H.J. 2010. The Secondary Products Revolution: the past, the present and the future. *World Archaeology* 42, 29–54.

Hachem, L. 2000. New observations on the Bandkeramik house and social organization. *Antiquity* 74, 308–12.

Halstead, P. 1998. Mortality models and milking: problems of uniformitarianism, optimality and equifinality reconsidered. *Anthropozoologica* 27, 3–20.

Halstead, P. 2001. Texts, bones and herders: approaches to animal husbandry in Late Bronze Age Greece. *Minos* 33–34, 149–89.

Halstead, P. 2011. Farming, material culture and ideology: repackaging the Neolithic of Greece (and Europe). In A. Hadjikoumis, E. Robinson and S. Viner (eds), *The dynamics of Neolithisation in Europe: studies in honour of Andrew Sherratt*, 131–51. Oxford: Oxbow Books.

Halstead, P. and Isaakidou, V. 2011a. Revolutionary secondary products: the development and significance of milking, animal-traction and wool-gathering in later prehistoric Europe and the Near East. In T. Wilkinson, S. Sherratt and J. Bennet (eds), *Interweaving worlds: systemic interactions in Eurasia, 7th to 1st millennia* BC, 61–76. Oxford: Oxbow Books.

Halstead, P. and Isaakidou, V. 2011b. A pig fed by hand is worth two in the bush: ethnoarchaeology of pig husbandry in Greece and its archaeological implications. In U. Albarella and A. Trentacoste (eds), *Ethnozooarchaeology: the present and past of human-animal relationships*, 160–74. Oxford: Oxbow Books.

Helmer, D. and Gourichon, L. 2008. Premières données sur les modalités de subsistance à Tell Aswad (Syrie, PPNB Moyen et Récent, Néolithique Céramique Ancien) – Fouilles 2001–2005. In E. Vila, L. Gourichon, H. Buitenhuis and A. Choyke (eds), *Archaeozoology of the Near East 8*, 119–51. Lyon: Maison de l'Orient et de la Méditerranée.

Helmer, D., Gourichon, L., Maamar, H.S. and Vigne, J.-D. 2005. L'élevage des caprinés néolithiques dans le sud-est de la France: saisonnalité des abattages, relations entre grottes-bergeries et sites de plein air. *Anthropozoologica* 40, 167–89.

Heun, M., Schäfer-Pregl, R., Klawan, D., Castagna, R., Accerbi, M., Borghi, B. and Salamini, F. 1997. Site of einkorn wheat domestication Identified by DNA fingerprinting. *Science* 278, 1312–14.

Hofmann, D. 2009. Cemetery and settlement burial in the Lower Bavarian LBK. In D. Hofmann and P. Bickle (eds), *Creating communities: new advances in central European Neolithic research*, 220–34. Oxford: Oxbow Books.

Ingold, T. 1986. *The appropriation of nature: essays on human ecology and social relations*. Manchester: Manchester University Press.

Isaakidou, V. 2006. Ploughing with cows: Knossos and the 'secondary products revolution'. In D. Serjeantson and D. Field (eds), *Animals in the Neolithic of Britain and Europe*, 95–112. Oxford: Oxbow Books.

Itan, Y., Powell, A., Beaumont, M.A., Burger, J. and Thomas, M.G. 2009. The origins of lactase persistence in Europe. *PLoS Computational Biology* 5, 1–13.

Jacomet, S. 2007. Neolithic plant economies in the northern Alpine Foreland from 5500–3500 cal BC. In S. Colledge and J. Conolly (eds), *The origins and spread of domestic plants in southwest Asia and Europe*, 221–58. Walnut Creek, CA: Left Coast Press.

Jones, G., Jones, H., Charles, M.P., Jones, M.K., Colledge, S., Leigh, F.J., Lister, D. A., Smith, L.M.J., Powell, W. and Brown, T.A. 2012. Phylogeographic analysis of barley DNA as evidence for the spread of Neolithic agriculture through Europe. *Journal of Archaeological Science* 39, 3230–8.

Knipper, C. 2009. Mobility in a sedentary society: insights from isotope analysis of LBK human and animal teeth. In D. Hofmann and P. Bickle (eds), *Creating communities: new advances in central European Neolithic research*, 142–58. Oxford: Oxbow Books.

Kotsakis, K. 1999. What tells can tell: social space and settlement in the Greek Neolithic. In P. Halstead (ed.), *Neolithic society in Greece*, 66–76. Sheffield: Sheffield Academic Press.

Kotsakis, K., Urem-Kotsou, D., Evershed, R.P., Sherratt, A. and Copley, M. 2008. Epinan oi neolithikoi georgoktinotrofoi gala? I marturia gia tin arkhaioteri katanalosi galaktos stin Ellada. In E. Beneki (ed.), *I istoria tou Ellinikou galaktos kai ton proionton tou*, 83–8. Xanthi: Politistiko Idrima Omilou Peiraios.

Larson, G., Albarella, U., Dobney, K., Rowley-Conwy, P., Schibler, J., Tresset, A., Vigne, J.-D., Edwards, C.J., Schlumbaum, A., Dinui, A., Balaçsescu, A., Dolman, G., Tagliacozzo, A., Manaseryan, N., Miracle, P., Wijngaarden-Bakker, L.V., Masseti, M., Bradley, D.G. and Cooper, A. 2007. Ancient DNA, pig domestication, and the spread of the Neolithic into Europe. *Proceedings of the National Academy of Sciences of the United States of America* 104, 15276–81.

Legge, A.J. 1981. The agricultural economy. In R. J. Mercer (ed.), *Grimes Graves excavations 1971–72*, 79–103. London: Her Majesty's Stationery Office.

Maier, U. 1999. Agricultural activities and land use in a neolithic village around 3900 BC: Hornstaad Hörnle 1A, Lake Constance, Germany. *Vegetation History and Archaeobotany* 8, 87–94.

Maier, U. and Harwath, A. 2011. Detecting intra-site patterns with systematic sampling strategies. Archaeobotanical grid sampling of the lakeshore settlement Bad Buchau-Torwiesen II, southwest Germany. *Vegetation History and Archaeobotany* 20, 349–65.

Mainland, I.L. 2007. A microwear analysis of selected sheep and goat mandibles. In A. Whittle (ed.), *The Early Neolithic on the Great Hungarian Plain: investigations of the Körös culture site of Ecsegfalva 23, Co. Békés*, 343–8. Budapest: Hungarian Academy of Sciences.

Mainland, I.L. and Halstead, P. 2005. The diet and management of domestic sheep and goats at Neolithic Makriyalos. In J. Davies, M. Fabis, I.L. Mainland, M. Richards and R. Thomas (eds), *Diet and health in past animal populations: current research and future directions*, 104–12. Oxford: Oxbow Books.

Marciniak, A. 2006. From animals and food in space to bones in context: social zooarchaeology of the Neolithic farming settlements. In D. Papaconstantinou (ed.), *Deconstructing context: a critical approach to archaeological practice*, 34–49. Oxford: Oxbow Books.

Marinova, E. 2007. Archaeobotanical data from the early Neolithic of Bulgaria. In S. Colledge and J. Conolly (eds), *The origins and spread of domestic plants in southwest Asia and Europe*, 93–109. Walnut Creek, CA: Left Coast Press.

Mina, M. 2008. Figurin' out Cretan Neolithic society: anthropomorphic figurines, symbolism and gender dialectics. In V. Isaakidou and P. Tomkins (eds), *Escaping the labyrinth: the Cretan Neolithic in context*, 115–35. Oxford: Oxbow Books.

Mlekuz, D. 2005. The ethnography of the Cyclops: Neolithic pastoralists in the eastern Adriatic. *Documenta Praehistorica* 32, 15–51.

Mlekuz, D., Budja, M., Payton, R. and Bonsall, C. 2008. 'Mind the gap': caves, radiocarbon sequences, and the Mesolithic–Neolithic transition in Europe – lessons from the Mala Triglavca rockshelter site. *Geoarchaeology* 23, 398–416.

Mukherjee, A. J., Copley, M.S., Berstan, R., Clark, K.A. and Evershed, R.P. 2005. Interpretation of $\delta^{13}C$ values of fatty acids in relation to animal husbandry: food processing and consumption in prehistory. In J. Mulville and A. K. Outram (eds), *The zooarchaeology of fats, oils, milk and dairying*, 77–93. Oxford: Oxbow Books.

Noe-Nygaard, N., Price, T.D. and Hede, S.U. 2005. Diet of aurochs and early cattle in southern Scandinavia: evidence from ^{15}N and ^{13}C stable isotopes. *Journal of Archaeological Science* 32, 855–71.

Oberdorfer, E. 1954. Über Unkrautgesellschaften der Balkanhalbinsel. *Vegetatio* 4, 379–411.

Pappa, M., Halstead, P., Kotsakis, K. and Urem-Kotsou, D. 2004. Evidence for large-scale feasting at Late Neolithic Makriyalos, N Greece. In P. Halstead and J. C. Barrett (eds), *Food, cuisine and society in Prehistoric Greece*, 16–44. Oxford: Oxbow Books.

Payne, S. 1973. Kill-off patterns in sheep and goats: the mandibles from Aşvan Kale. *Anatolian Studies* 23, 281–303.

Pearson, J.A., Buitenhuis, H., Hedges, R.E.M., Martin, L., Russell, N. and Twiss, K. C. 2007. New light on early caprine herding strategies from isotope analysis: a case study from Neolithic Anatolia. *Journal of Archaeological Science* 34, 2170–9.

Pollard, A.M. 2011. Isotopes and impact: a cautionary tale. *Antiquity* 85, 631–8.
Robb, J. 2002. Time and biography: osteobiography of the Italian Neolithic lifespan. In Y. Hamilakis, M. Pluciennik and S. Tarlow (eds), *Thinking through the body: archaeologies of corporeality*, 153–71. New York: Kluwer/Plenum.
Robb, J. and Miracle, P. 2007. Beyond 'migration' versus 'acculturation'; new models for the spread of agriculture. In A. Whittle and V. Cummings (eds), *Going over: the Mesolithic–Neolithic transition in north-west Europe*, 99–115. London: Oxford University Press for The British Academy.
Rowley-Conwy, P. 2000. Milking caprines, hunting pigs: the Neolithic economy of Arene Candide in its West Mediterranean context. In P. Rowley-Conwy (ed.), *Animal bones, human societies*, 124–32. Oxford: Oxbow Books.
Runnels, C., Korkuti, M., Galaty, M.L., Timpson, M.E., Stocker, S.R., Davis, J.L., Bejko, L. and Maçaj, S. 2009. Early prehistoric landscape and landuse in the Fier region of Albania. *Journal of Mediterranean Archaeology* 22, 151–82.
Salque, M., Bogucki, P.I., Pyzel, J., Sobkowiak-Tabaka, I., Grygiel, R., Szmyt, M. and Evershed, R.P. 2012. Earliest evidence for cheese making in the sixth millennium BC in northern Europe. *Nature* 493, 522–5.
Schulting, R. J. and Fibiger, L. (eds) 2012. *Sticks, stones, and broken bones: Neolithic violence in a European perspective*. Oxford: Oxford University Press.
Shennan, S. and Edinborough, K. 2007. Prehistoric population history: from the Late Glacial to the Late Neolithic in central and northern Europe. *Journal of Archaeological Science* 34, 1339–45.
Sheridan, A. 2010. The Neolithization of Britain and Ireland: the 'big picture'. In B. Finlayson and G. Warren (eds), *Landscapes in transition*, 89–105. Oxford: Oxbow Books.
Sheridan, A. 2012. Review of A. Whittle, F. Healy and A. Bayliss (2011), *Gathering time: dating the Early Neolithic enclosures of southern Britain and Ireland*. *Antiquity* 86, 262–4.
Sherratt, A.G. 1981. Plough and pastoralism: aspects of the secondary products revolution. In I. Hodder, G. Isaac and N. Hammond (eds), *Pattern of the past: studies in honour of David Clarke*, 261–305. Cambridge: Cambridge University Press.
Sherratt, A. 1983. The secondary exploitation of animals in the Old World. *World Archaeology* 15, 90–104.
Triantaphyllou, S. 2008. Living with the dead: a re-consideration of mortuary practices in the Greek Neolithic. In V. Isaakidou and P. Tomkins (eds), *Escaping the labyrinth: the Cretan Neolithic in context*, 136–54. Oxford: Oxbow Books.
Tzevelekidi, V. 2012. *Dressing for dinner: butchery and bone deposition at Late Neolithic Toumba Kremastis-Koiladas, northern Greece.* Oxford: Archaeopress.
Urem-Kotsou, D. and Kotsakis, K. 2007. Pottery, cuisine and community in the Neolithic of north Greece. In C. Mee and J. Renard (eds), *Cooking up the past: food and culinary practices in the Neolithic and Bronze Age Aegean*, 225–46. Oxford: Oxbow Books.
Vigne, J.-D. 2008. Zooarchaeological aspects of the Neolithic diet transition in the Near East and Europe, and their putative relationships with the Neolithic demographic transition. In J.-P. Bocquet Appel and O. Bar-Yosef (eds), *The Neolithic demographic transition and its consequences*, 179–205. New York: Springer.
Vigne, J.-D. and Helmer, D. 2007. Was milk a 'secondary product' in the Old World Neolithisation process? Its role in the domestication of cattle, sheep and goats. *Anthropozoologica* 42, 9–40.

Villotte, S. and Knüsel, C. 2014. 'I sing of arms and of a man . . .': medial epicondylitis and the sexual division of labour in prehistoric Europe. *Journal of Archaeological Science* 43, 168–74.

Whittle, A., Healy, F. and Bayliss, A. 2011. *Gathering time: dating the Early Neolithic enclosures of southern Britain and Ireland.* Oxford: Oxbow Books.

Zapata, L., Peña-Chocarro, L., Pérez-Jordá, G. and Stika, H.-P. 2004. Early Neolithic agriculture in the Iberian Peninsula. *Journal of World Prehistory* 18, 283–325.

Zvelebil, M. and Pettitt, P. 2008. Human condition, life, and death at an Early Neolithic settlement: bioarchaeological analyses of the Vedrovice cemetery and their biosocial implications for the spread of agriculture in central Europe. *Anthropologie* 46, 195–218.

Index